APPLIED TIME SERIES ANALYSIS

APPLIED TIME SERIES ANALYSIS

Proceedings of the International Conference
held at Houston, Texas, August 1981

Edited by

O. D. ANDERSON
TSA & F, Nottingham, UK

and

M. R. PERRYMAN
Baylor University, Texas, USA

1982

NORTH-HOLLAND PUBLISHING COMPANY – AMSTERDAM • NEW YORK • OXFORD

ISBN: 0 444 86424 5

Published by:

NORTH-HOLLAND PUBLISHING COMPANY
AMSTERDAM • NEW YORK • OXFORD

Sole distributors for the U.S.A. and Canada:

ELSEVIER SCIENCE PUBLISHING COMPANY, INC.
52 VANDERBILT AVENUE
NEW YORK, N.Y. 10017

Library of Congress Cataloging in Publication Data

Main entry under title:

Applied time series analysis.

Papers from the second American Time Series Meeting (5th ITSM), Houston, Tex., Aug. 6-7, 1981.
1. Time-series analysis--Congresses. 2. Forecasting--Congresses. I. Anderson, O. D. (Oliver Duncan), 1940- . II. Perryman, M. Ray (Marlin Ray), 1952- . III. American Time Series Meeting (2nd : 1981 : Houston, Tex.)
QA280.A64 519.5'5 82-7842
ISBN 0-444-86424-5 AACR2

PRINTED IN THE NETHERLANDS

CONTENTS

APPLIED TIME SERIES ANALYSIS
O.D. Anderson & M.R. Perryman (eds.)

THE SECOND HOUSTON TIME SERIES MEETING

Oliver D. Anderson
TSA&F
9 Ingham Grove, Lenton Gardens, Nottingham NG7 2LQ, England

1. THESE PROCEEDINGS

This book contains many of the papers which were presented at the Second American Time Series Meeting (5th ITSM) held at Houston, Texas, 6-7 August 1981. It includes a most important plenary contribution by D.W. Marquardt and S.K. Acuff on Direct Quadratic Spectrum Estimation, and the very welcome return to Time Series of Professor J.S. White, who treats the Distribution of the Serial Correlation Coefficient. The majority of the remaining articles are workmanlike practical applications of Time Series Analysis and Forecasting (TSA&F), mainly in the Time Domain.

Apart from the papers by the first editor, the contents run by alphabetical order of author. Again this was for practical reasons: the uncertainty of who exactly would make final deadlines made it important to have a simple nontechnical criterion for ordering papers, and alphabetical was thought to be rather more useful (and aesthetically pleasing) than just the order in which the manuscripts reached the publisher.

O.D. Anderson was responsible for all the editing, but M.R. Perryman arranged more of the refereeing.

2. THE AMERICAN MEETINGS

The sequence of International Time Series Meetings (ITSM), initiated by the first editor, has been quite fully discussed in O.D. Anderson (1982), where they are divided into the three types: those in Europe, those in North America and Special Topics ITSM.

This second Houston Meeting is regarded as the final pilot conference before establishing a fixed series of events satelliting the American Statistical Association's (ASA) annual Summer Meetings. ITSM is therefore most grateful to Professor M. Ray Perryman for taking on the bulk of the local organisation for the first pair of North American Meetings. For this he was ably aided by his wife, Nancy Perryman, who has been the major administrative supervisor. Their staff of helpers were Roni Carter, Marianne Clack, Christy and Leigh Humphrey, and Becky Payette, to whom we are most appreciative. Also Dean Young, Fred Hulme and, especially, Kris Moore (all of Baylor) aided substantially with the planning and execution of the Meetings. In Nottingham, the typing skills of Jane Clerbaut have been invaluable.

We are now poised for a program of major meetings in the US and Canada. The next few are:

<u>3rd American General Interest Time Series Meeting (8th ITSM): Cincinnati (Ohio) 19-22 August, 1982</u>

<u>North American Time Series Meetings, Toronto (Canada) 1983</u>

Special Topics ITSM on Hydrological, Geophysical and Spatial Time Series:
11-14 August, 1983
General Interest ITSM: 18-21 August, 1983

Philadelphia (Pennsylvania) ITSM, 1984

Special Topics: 9-12 August, 1984
General Interest: 16-19 August, 1984.

The Special Topics Meetings, evidently, are highly specific in nature; and, for details, one should watch the TSA&F News and TSA&F Flyer. However, like all the North American Events, they are relatively low-cost, and feature both invited and contributed papers. For the other Meetings, a brief outline is now given.

The objects of the General Interest ITSM are to discuss recent developments in the theory and practice of Time Series Analysis and Forecasting, and to bring practitioners together from diverse parent disciplines, work environments and geographical locations.

Each Conference will convene directly after the ASA (American Statistical Association) annual summer Meeting, and use the same headquarters hotel as ASA, at very competitive rates. Thus, interested participants will be able to conveniently and economically attend both events in a single trip without changing accommodation.

Major Time Series themes will be: Statistical Methodology; Applications to Economics and in Econometrics; Government, Business and Industrial Examples; the Hydrosciences, such as Limnology, Hydrology, Water Quality Regulation and Control, and the Modelling of Marine Environments; the Geosciences, especially such areas as Oil Exploration and Seismology; Civil Engineering and allied disciplines; Spatial and Space-Time Processes - their theory and application - especially in Geography and related areas, such as City Planning or Energy Demand Forecasting; Biology and Ecology; Environmental Studies - Air and River Pollution; Medical Applications and Biomedical Engineering; Irregularly Spaced Data (including Outliers and Missing Observations); Robust and Nonparametric Methods; Seasonal Modelling and Adjustment, Calendar Effects; Causality; Bayesian Approaches; Box-Jenkins Univariate ARIMA, Transfer-Function, Intervention and Multivariate Modelling; State Space; Nonlinear Modelling; Estimation; Diagnostic Checking; Signal Extraction; Comparative Studies; Spectral Analysis, especially for the Physical Sciences; Business Cycle and Expectations Data; Data Revisions; Computer Software and Numerical Analysis; Forecasting (including new topics such as Traffic Forecasting and Safety, and Forecasting in Agriculture); and, no doubt, many other areas of the subject.

Abstracts for contributed papers are solicited, as are suggestions for complete sessions (with offers to organise and chair these). Other help will also be most welcome, especially that connected with local arrangements and with refereeing papers and editing the Proceedings. All the ITSM are organised independently of ASA by Oliver Anderson - from whom further details may be obtained.

3. RETURNING TO THIS VOLUME

The Editors would like to thank the Brown Foundation and the Hankamer School of Business (Baylor University), and Oliver Anderson Consultants, for released time and other financial support. Apart from the Editors, manuscripts were reviewed by Jan G. de Gooijer (Amsterdam), Timothy Koch (Texas Tech), Kris Moore (Baylor), and James R. Schmidt and Suan-Boon Tan (both of Nebraska) - all of whom we warmly thank.

Also, we would like to appreciatively acknowledge those Registrants who ran the gauntlet of the US Air Traffic Controllers Strike to reach the Meeting; and, finally, our thanks go to the Authors who presented their work and then patiently endured the rigours of a tight schedule for getting this book into print. To all we are most grateful.

REFERENCES

ANDERSON, O.D. (1982). International Time Series Meetings and this ITSM (Valencia, Spain, 22-26 June 1981). In Time Series Analysis: Theory and Practice 1 (Proceedings of the International Conference held at Valencia, June 1981). Ed: O.D. Anderson, North-Holland, Amsterdam & New York, 1-6.

TSA&F NEWS (Quarterly Newsletter), 1979 onwards. Ed: O.D. Anderson. ISSN 0143-0505.

TSA&F FLYER (Monthly Information Bulletin), 1980 onwards. Ed: O.D. Anderson. ISSN 0260-9053.

APPLIED TIME SERIES ANALYSIS
O.D. Anderson & M.R. Perryman (eds.)

SAMPLED SERIAL CORRELATIONS FROM ARIMA PROCESSES

O.D. ANDERSON
9 Ingham Grove, Lenton Gardens, Nottingham NG7 2LQ, England

The behaviour of the k-th sample serial correlation, $r_k^{(n)}$ $(0<k<n)$, for a series realisation of length n from any autoregressive integrated moving average (ARIMA) model is investigated, and comparisons made with recent results reported by Hasza (1980). Other related finite sample work on ARMA and ARUMA processes is referenced.

1. INTRODUCTION

Consider a time series realisation of length n, denoted by $\{z_1,\ldots,z_n\}$, and define its k-th serial covariance as

$$c_k^{(n)} = \frac{1}{n}\sum_{i=1}^{n-k} (z_i-\bar{z})(z_{i+k}-\bar{z}) \qquad (k=0,\ldots,n-1)$$

where $\bar{z} = (z_1+\ldots+z_n)/n$ is the sample mean of the observed series. Define the k-th serial correlation by $r_k^{(n)} = c_k^{(n)}/c_0^{(n)}$. In both these relations, the superfix (n) serves to emphasise the length of the series.

Box and Jenkins (1976) have greatly popularised the use of ARMA(p,q) and ARIMA(p,d,q) processes for modelling time series. These are defined by $\{Z_i\}$ satisfying

$$(1-\phi_1B-\ldots-\phi_pB^p)(1-B)^dZ_i = (1-\theta_1B-\ldots-\theta_qB^q)A_i \qquad (1)$$

with d = 0 for the ARMA cases and $d \geq 1$ for those ARIMA models which are excluded from the ARMA class. In (1), $(\phi_1,\ldots,\phi_p)$ and $(\theta_1,\ldots,\theta_q)$ are two sets of real parameters, with the first subject to the stationarity condition, namely that the polynomial $\phi_p(\zeta) \equiv 1 - \phi_1\zeta - \ldots - \phi_p\zeta^p$ in the complex variable ζ has no zero within or on the unit circle; and B is the backshift operator, such that B^j operating on any X_i, for instance A_i or Z_i, produces X_{i-j}. $\{A_i\}$ is a white noise sequence of independent but identically distributed normal zero-mean random variables, all with variance σ^2 say. Note that it is unnecessary (for our purposes) to impose any further restrictions on the real θ-parameters.

In this paper we will wish to extend the ARIMA class to the so-called ARUMA models, obtained by replacing the d-times differencing operator $(1-B)^d$, in (1), with $U_d(B) = 1-u_1B-\ldots-u_dB^d$, all of whose zeros lie precisely on the unit circle, but not necessarily all (or any) taking the value 1. We then study certain serial dependence properties for the whole ARUMA class. In particular, we consider the first two moments of $c_k^{(n)}$ and $r_k^{(n)}$.

Hasza (1980) has discussed the asymptotic distribution of the sample serial correlations for ARIMA(p,1,q) models as $n \to \infty$. His theory is rigorous and provides *inter alia* formulae for the population mean and variance of a finite-lagged serial correlation from an infinitely long realisation, namely

$$E\left[r_k^{(\infty)}\right] = 1 \tag{2}$$

$$\text{Var}\left[r_k^{(\infty)}\right] = 0 \tag{3}$$

in agreement with Roy and Lefrancois (1978).

However, when application is made to finite-lengthed series realisations, approximations of unknown validity are introduced and Hasza concentrates on just the mean and variance of $r_k^{(n)}$, for the IMA(1,1) case,

$$(1-B)Z_i = (1-\theta B)A_i \qquad |\theta|<1 \tag{4}$$

to get

$$E\left[r_k^{(n)}\right] \simeq E(n,k,\theta) = 1-n^{-1}\{7.335k+10.652\theta/(1-\theta)^2\} \tag{5}$$

$$\text{Var}\left[r_k^{(n)}\right] \simeq S^2(n,k,\theta) = n^{-2}\{19.501k^2+69.588k\theta/(1-\theta)^2+69.420\theta^2/(1-\theta)^4\}. \tag{6}$$

In this note, we advocate an alternative approach to predicting finite series behaviour, which also gives rise to formulae for $E\left[r_k^{(n)}\right]$ and $\text{Var}\left[r_k^{(n)}\right]$. Again approximations are employed, but we would suggest that this does not make the theory any less rigorous than Hasza's, for the practical purpose of gaining insight as to how real series behave. Moreover, these formulae appear superior to Hasza's for short series, and are certainly in better agreement with the simulations which he gave.

In what follows, γ_k will always denote the k-th theoretical autocovariance for that ARMA part of the model which remains after any I or U factor has been removed by appropriate simplification, from what was originally an ARMA, ARIMA or ARUMA process.

Finally, we define

$$E_{k,1}^{(n)} = E\left[c_k^{(n)}\right]/E\left[c_o^{(n)}\right] \tag{7}$$

which can be considered as a first approximation to $E\left[r_k^{(n)}\right]$ (with an $O(n^{-1})$ error) - compare Wichern (1973) - and its second approximation correct to order n^{-1} (with an error of $O(n^{-3/2})$)

$$E_{k,2}^{(n)} = E_{k,1}^{(n)}\{1 - \frac{\text{Cov}[c_k^{(n)}, c_o^{(n)}]}{E[c_k^{(n)}]E[c_o^{(n)}]} + \frac{\text{Var}[c_o^{(n)}]}{E^2[c_o^{(n)}]}\} \quad (8)$$

and, to order n^{-1}, an approximation for $\text{Var}[r_k^{(n)}]$

$$V_{k,2}^{(n)} = \{S_{k,2}^{(n)}\}^2 = \{E_{k,1}^{(n)}\}^2\{\frac{\text{Var}[c_k^{(n)}]}{E^2[c_k^{(n)}]} - \frac{2\text{Cov}[c_k^{(n)}, c_o^{(n)}]}{E[c_k^{(n)}]E[c_o^{(n)}]} + \frac{\text{Var}[c_o^{(n)}]}{E^2[c_o^{(n)}]}\}.^* \quad (9)$$

2. EXACT COVARIANCE RESULTS

Explicit formulae for $E[c_k^{(n)}]$ or $E_{k,1}^{(n)}$ have been obtained for all ARMA and ARIMA models by Anderson (1979a), who also gave the results for the other ARUMA cases in Anderson (1979b). Similar results for $\text{Var}[c_k^{(n)}]$ and $\text{Cov}[c_k^{(n)}, c_o^{(n)}]$ ($0 \le k < n$), given an ARMA or ARIMA(p,1,q), have been recorded in Anderson and de Gooijer (1979a) and, for other general homogeneously nonstationary ARUMA models, in Anderson (1980a). Finally, Anderson and de Gooijer (1980b) give $\text{Cov}[c_k^{(n)}, c_h^{(n)}]$ ($h \le k < n$), whilst Anderson and de Gooijer (1981) show how similar results can be obtained for space-time systems.

Thus, for ARMA(p,q),

$$E[c_k^{(n)}] = \frac{1}{n^3}\{n(n-k)(n\gamma_k - \gamma_o) + 2n\sum_{j=1}^{k-1}(k-j)\gamma_j - 2n\sum_{j=1}^{n-k-1}(n-k-j)\gamma_j - 2k\sum_{j=1}^{n-1}(n-j)\gamma_j\}. \quad (10)$$

Again, for any ARIMA(p,1,q) process, we have that

$$E[c_k^{(n)}] = \frac{1}{6n^3}[n(n-k)\{(n^2-4kn+2k^2-1)\gamma_o - 6n\sum_{j=1}^{k-1}(k-j)\gamma_j\} - 2n\sum_{j=1}^{k-2}(k-j)\{(k-j)^2-1\}\gamma_j$$

$$+2n\sum_{j=1}^{n-k-2}(n-k-j)\{(n-k-j)^2-1\}\gamma_j + 2k\sum_{j=1}^{n-2}(n-j)\{(n-j)^2-1\}\gamma_j] \quad (11)$$

and, in particular, for model (4) this reduces to

$$E[c_k^{(n)}] = (n-k)\{n\lambda_k - 1 - (n-1)(1-\theta) + (n^2-4kn+2k^2-1)(1-\theta)^2/6\}\sigma^2 n^{-2} \quad (12)$$

where $\lambda_k = (1-\theta)$, whenever $k \geq 1$, but $\lambda_o = 1$. This agrees with formulae given by Roy (1977) - and yields the result for $E_{k,1}^{(n)}$, previously deduced by Wichern (1973), except that we disagree with him as to its precision when used to approximate $E[r_k^{(n)}]$.

Given any ARIMA(p,d,q) with $d > 1$, we find that

$$E_{k,1}^{(n)} = (n-k)(n^2-2nk-2k^2-1)/\{n(n^2-1)\} \quad (13)$$

---

* $V_{k,1}^{(n)} = 0$, rather uninterestingly.

whilst, for any ARUMA model of the form (1) with $(1-B)^d$ replaced by $(1+B)$, we get

$$E_{k,1}^{(n)} = \{n^2(n-k)(-1)^k - n\delta_{n,k} - k\delta_{n,k}^2\}/(n^3 - n\delta_{n,o}) \tag{14}$$

where $\delta_{n,k} = 0$ (when n is even), $= 1$ (n odd, k even) and $= -1$ (otherwise). If, instead, we replace the $(1-B)^d$ by $(1-2B\cos\omega+B^2)$, to order n^{-2} we have that

$$E_{k,1}^{(n)} \simeq (1-k/n)\cos k\omega. \tag{15}$$

The corresponding second moment formulae, for ARMA and ARIMA(p,1,q) models, are much more involved but have the simple form

$$\mathrm{Var}\left[c_k^{(n)}\right] = \sum_{i,j=0}^{n-1}\sum g_{i,j}\gamma_i\gamma_j$$

where the $g_{i,j}$ coefficients are all known, and similarly for $\mathrm{Cov}\left[c_k^{(n)}, c_o^{(n)}\right]$ - see Anderson and de Gooijer (1979a) - or, more generally, $\mathrm{Cov}\left[c_k^{(n)}, c_h^{(n)}\right]$ which subsumes both the previous results and was more recently derived in Anderson and de Gooijer (1980b). Those for other ARUMA models are less complicated. For instance, the (1+B) case yields

$$\frac{\mathrm{Var}\left[c_k^{(n)}\right]}{E^2\left[c_k^{(n)}\right]} = \frac{2\left[n^4(n-k)^2 - 2\{n^2(n^2-k^2) - nk\}\delta_{n,k} + (n^2+k^2)\delta_{n,k}^2\right]}{\{n^2(n-k)(-1)^k - n\delta_{n,k} - k\delta_{n,k}^2\}^2} \tag{16}$$

$$\frac{\mathrm{Cov}\left[c_k^{(n)}, c_o^{(n)}\right]}{E\left[c_k^{(n)}\right]E\left[c_o^{(n)}\right]} = \frac{2\left[n^2(n-k)\{n^2(-1)^k - \delta_{n,k}\} + (1-n^2)(n\delta_{n,k} + k\delta_{n,k}^2)\right]}{\{n^2(n-k)(-1)^k - n\delta_{n,k} - k\delta_{n,k}^2\}(n^2 - \delta_{n,o})}. \tag{17}$$

So, on simplifying for the various special cases, (16) and (17) give

$$\frac{\mathrm{Var}\left[c_k^{(n)}\right]}{E^2\left[c_k^{(n)}\right]} = \begin{cases} 2 + 8n^3/\{(n-k)(n^2-1)^2\} & \text{(when n and k are both odd)} \\ 2 & \text{(otherwise)} \end{cases} \tag{18}$$

$$\mathrm{Cov}\left[c_k^{(n)}, c_o^{(n)}\right]/\{E\left[c_k^{(n)}\right]E\left[c_o^{(n)}\right]\} = 2. \tag{19}$$

Then, using (7), (8) and (9), we can obtain approximations to $E\left[r_k^{(n)}\right]$ and $\mathrm{Var}\left[r_k^{(n)}\right]$. Thus, for instance, with the (1+B) ARUMA we have

$$E_{k,2}^{(n)} = E_{k,1}^{(n)} = \begin{cases} (-1)^k(1-k/n) - 2k\{n(n^2-1)\} & \text{(when n is odd and k even)} \\ (-1)^k(1-k/n) & \text{(otherwise)} \end{cases} \tag{20}$$

$$V_{k,2}^{(n)} = \begin{cases} 8n(n-k)/(n^2-1)^2 & \text{(when n and k are both odd)} \\ 0 & \text{(otherwise)} \end{cases} \tag{21}$$

Since the approximations, $E_{k,2}^{(n)}$ and $V_{k,2}^{(n)}$, are both correct to $O(n^{-1})$, it follows that, for this ARUMA subclass,

$$E\left[r_k^{(n)}\right] = (-1)^k(1-k/n)+O(n^{-3/2}) \tag{22}$$

$$\text{Var}\left[r_k^{(n)}\right] = 0 + O(n^{-3/2}) \tag{23}$$

and, thus, actually observed serial correlations will be given by

$$r_k^{(n)} = (-1)^k(1-k/n)+O_p(n^{-3/4}). \tag{24}$$

Using $\text{Cov}\left[c_k^{(n)},c_h^{(n)}\right]$, we can similarly obtain $\text{Cov}\left[r_k^{(n)},r_h^{(n)}\right]$ approximately.

3. AGREEMENT WITH SIMULATION

The theory of the last section gives rise to approximations which provide good agreement with simulation in all the cases studied so far. For instance, see Anderson and de Gooijer (1979b, 1980a) and Anderson (1980a).

In particular, using our formulae, we get much better agreement with the Monte Carto results, based on 1000 replications and reported by Hasza (1980), than he did with his asymptotic approximations. Thus, when n = 50 and $\theta = \pm.8$, we have comparisons, for k = 3, as shown in table 1. Note that, there, the exact

Table 1. Comparison between Hasza's and our Results for $(1-B)Z_i = (1-\theta B)A_i$

	$\theta = -.8$	$\theta = .8$
Hasza's observed $\bar{r}_3^{(50)}$	.65	.16
Hasza's $E(50,3,\theta)$	.61	-3.70
Our $E_{3,1}^{(50)}$	.74	.20
Our $E_{3,2)}^{(50)}$	.65	.15
Exact $E\left[\bar{r}_3^{(50)}\right]$		.1653
Hasza's observed s.e. $\left[r_3^{(50)}\right]$	.17	.19
Hasza's $S(50,3,\theta)$	.23	3.58
Our $S_{3,2}^{(50)}$	.08	.21
Exact s.d. $\left[\bar{r}_3^{(50)}\right]$		.1839

values quoted are those obtained from numerical integration by a method discussed in Anderson and de Gooijer (1980a) and de Gooijer (1980).

4. DISCUSSION

Also note that although the asymptotic distribution of, say, $r_1^{(n)}$ for the IMA(1,1) model with $\theta = .8$ is nonnormal, as demonstrated by Hasza (1980), the short series distributions are not highly negative skew, as intuition might suggest. For instance, when $n = 50$, the exact p.d.f is virtually symmetric as shown in table 2, its expectation being .2151. Also see Anderson and de Gooijer (1980a).

Table 2. Percentages of First Serial Correlations falling within various Ranges for Length-50 Realisations from $(1-B)Z_i = (1-.8B)A_i$

Range, $r_1^{(50)}$	$\leq -.05$	$(-.05,.00]$	$(.00,.05]$	$(.05,.10]$	$(.10,.15]$	$(.15,.20]$	$(.20,.25]$
Percentage	6.65	4.54	6.27	7.98	9.40	10.31	10.55
Percentage	6.90	4.50	6.09	7.68	9.12	10.12	
Range, $r_1^{(50)}$	$>.50$	$(.45,.50]$	$(.40,.45]$	$(.35,.40]$	$(.30,.35]$	$(.25,.30]$	

Again, the typical shape for the run of serial correlations $\{r_k^{(n)}\}$ from a finite series realisation of an ARIMA(p,d,q), with $d > 0$, is not a slow linear decline - as, for instance, suggested by Box and Jenkins (1976). Rather, it is a smooth curve that starts with positive values that decrease to a negative minimum and then increase again towards zero. Compare (12) or (13). That something like this must happen is also indicated by noting that, for any series realisation whatsoever,

$$\sum_{k=1}^{\infty} r_k^{(n)} = -\tfrac{1}{2} \qquad (25)$$

where $r_k^{(n)}$ is taken as zero whenever $k \geq n$.

Finally, the motivation for our study stems from a general belief that, for observed series, the finite sample behaviour of the serial correlations can be very different from what asymptotic theory has previously led practitioners to expect. Thus we believe that our results provide us with a means of recognising specific ARUMA models, when they occur; and then, by a method analogous to differencing d-times for the ARIMA case, give us a way of reducing the process to just its ARMA part, by application of an appropriate simplifying operator. For instance, see Anderson (1980b).

POSTSCRIPT

Recently, a different approach for obtaining approximations to the moments of the serial correlations for finite series realisations from stationary ARMA(p,q) models has been discussed by Davies and Newbold (1980) and Anderson and de Gooijer (1981). Davies and Newbold obtained (with some errors) $E[r_k^{(n)}]$ and $\mathrm{Var}[r_k^{(n)}]$ approximately, for an MA(q) process, whilst by first extending their method and then using a rather different means, Anderson and de Gooijer deduced the analogous general formulae, given any ARMA(p,q) model, and also quoted the corresponding $\mathrm{Cov}[r_k^{(n)}, r_h^{(n)}]$ result, for the MA(q) case. Here, we outline how this last quantity (which of course specialises to $\mathrm{Var}[r_k^{(n)}]$) can be derived for a general ARMA(p,q).

For simplicity, we drop the superscript (n) and we also work with the non-centred serial correlations for both the observed series $\{z_i\}$ and its driving shocks $\{a_i\}$, namely

$$r_k(z) = \sum_{i=1}^{n-k} z_i z_{i+k} / \sum_{i=1}^{n} z_i^2 \qquad (0<k<n)$$

$$r_k(a) = \sum_{i=1}^{n-k} a_i a_{i+k} / \sum_{i=1}^{n} a_i^2 \qquad (0<k<n).$$

First we note some results for the $r_k(a)$, which tidies up work reported in Davies (1977). For $0<j<n$ and $m=1,2,\ldots,$

$$E[r_j^{2m-1}(a)] = 0$$

$$E[r_j^{2m}(a)] = O(n^{-m})$$

with, in particular,

$$E[r_j^2(a)] = (n-j)/\{n(n+2)\}$$

$$E[r_j^4(a)] = 3\{(n-j)^2+2(n-j)+4\langle n-2j\rangle\}/\{n(n+2)(n+4)(n+6)\}$$

where $\langle x\rangle$ denotes the positive part of x. Finally $E[r_i^{\ell}(a)r_j^{m}(a)] = O(n^{-2})$ or less, whenever $\ell>0$. (For instance, should ℓ or m (or both of them) be odd, then

$$E[r_i^{\ell}(a)r_j^{m}(a)] = 0 \qquad (i\neq j)$$

and

$$E[r_i^2(a)r_j^2(a)] = \{(n-i)(n-j)+4(n-j)+8\langle n-i-j\rangle\}/\{n(n+2)(n+4)(n+6)\} \qquad (i<j).$$

Thus we see that all powers or cross products of the $r_j(a)$'s, other than the $r_j^2(a)$'s, only give rise to terms of $O(n^{-2})$ or less, on taking expectations.

Then, following Davies and Newbold (1980), we can write

$$r_k(z) \simeq \{1+2\sum_{j=1}^{n-1} \rho_j r_j(a)\}^{-1}\{r_k(a)+\sum_{j=1}^{n-1} \rho_j[r_{k+j}(a)+r_{k-j}(a)]\}. \qquad (26)$$

So, ignoring powers or cross products which will only yield $O(n^{-2})$ or less, we have that

$$r_k(z) \simeq \{1-2\sum_{j=1}^{n-1} \rho_j r_j(a)+4\sum_{j=1}^{n-1} \rho_j^2 r_j^2(a)\}\{\rho_k+\sum_{j=1}^{n-1}(\rho_{k+j}+\rho_{k-j})r_j(a)\}$$

$$\simeq \rho_k+\sum_{j=1}^{n-1} F_{k,j}r_j(a)-2\sum_{j=1}^{n-1} \rho_j F_{k,j} r_j^2(a)$$

where $F_{k,j} = \rho_{k-j}+\rho_{k+j}-2\rho_k\rho_j$. So

$$r_k(z)r_h(z) \simeq \rho_k\rho_h + \sum_{j=1}^{n-1}(\rho_k F_{h,j}+\rho_h F_{k,j})r_j(a)$$

$$-2\sum_{j=1}^{n-1}\{\rho_j(\rho_k F_{h,j}+\rho_h F_{k,j})+F_{k,j}F_{h,j}\}r_j^2(a).$$

Thus

$$E[r_k(z)] \simeq \rho_k - \frac{2}{n(n+2)}\sum_{j=1}^{n-1}(n-j)\rho_j(\rho_{k-j}+\rho_{k+j}-2\rho_k\rho_j)$$

and

$$\mathrm{Cov}[r_k(z),r_h(z)] = E[r_k(z)r_h(z)] - E[r_k(z)]E[r_h(z)]$$

$$\simeq \{\sum_{j=1}^{n-1}(n-j)(\rho_{k-j}+\rho_{k+j}-2\rho_k\rho_j)(\rho_{h-j}+\rho_{h+j}-2\rho_h\rho_j)\}/\{n(n+2)\}.$$

This last relation, which represents an improvement on Bartlett's famous 1946 approximation, is in fact implicit in the result stated (without proof) in Anderson and de Gooijer (1980b, eqn. 29) for an MA(q) model, given the other work contained in that paper. (We also note that the seven line equation (2.74) of Davies (1977) for $E[r_k(z)r_h(z)]$, from an MA(q), even if it was correct, is quite unnecessarily complicated.) However, even for stationary ARMA(p,q) processes, as explained in Anderson and de Gooijer (1980b), we still prefer our original type of approximation to that of this postscript, (but perhaps with the expansion, giving (9), taken rather further then).

REFERENCES

ANDERSON, O.D. (1979a). Formulae for the Expected Values of the Sampled Variance and Covariances from Series Generated by General Autoregressive Integrated

Moving Average Processes of Order (p,d,q). *Sankhyā B* 41, 177-195.

ANDERSON, O.D. (1979b). The Autocovariance Structures Associated with General Unit Circle Nonstationarity Factors in the Autoregressive Operators of Otherwise Stationary ARMA Time Series Models. *Cahiers du CERO* 21, 221-237.

ANDERSON, O.D. (1980a). Serial Dependence Properties of ARUMA Models. *Cahiers du CERO* 22, 309-323.

ANDERSON, O.D. (1980b). An Augmented Box-Jenkins Approach with Applications to Economic Time Series Data. Paper presented at Joint Statistical Meetings, Houston (Texas), August 1980. In: Proceedings of Business and Economic Statistics Section, American Statistical Association, 432-437.

ANDERSON, O.D. and DE GOOIJER, J.G. (1979a). Formulae for the Covariance Structure of the Sampled Autocovariances from Series Generated by General Autoregressive Integrated Moving Average Processes or Order (p,d,q), d = 0 or 1. To appear in Sankhyā.

ANDERSON, O.D. and DE GOOIJER, J.G. (1979b). On Discriminating between IMA(1,1) and ARMA(1,1) Processes: some extensions to a paper by Wichern. *Statistician* 28, 119-133.

ANDERSON, O.D. and DE GOOIJER, J.G. (1980a). Distinguishing between IMA(1,1) and ARMA(1,1) Models: a large scale Simulation Study of two particular Box-Jenkins Time Processes. In *Time Series* (ed O.D. Anderson), North-Holland, Amsterdam, (Proceedings of International Time Series Meeting held at Nottingham University, England, March 1979), 15-40.

ANDERSON, O.D. and DE GOOIJER, J.G. (1980b). The Covariances between Sampled Autocovariances and between Serial Correlations for Finite Realisations from ARUMA Time Series Models. To appear in *Time Series Analysis: Theory and Practice 1* (ed O.D. Anderson), North-Holland, Amsterdam, (Proceedings of International Time Series Meeting held at Valencia, Spain, June 1981), 7-22.

ANDERSON, O.D. and DE GOOIJER, J.G. (1981). Approximate Moments of the Sample Space-Time Autocorrelation Function. Submitted for publication.

BARTLETT, M.S. (1946). On the theoretical specification and sampling properties of autocorrelated time series. Supplement *J. Roy. Statist. Soc. B* 8, 27-41. Correction (1948) 10, 1.

BOX, G.E.P. and JENKINS, G.M. (1976). *Time Series Analysis, Forecasting and Control*. San Francisco: Holden Day.

DE GOOIJER, J.G. (1980). Exact moments of the sample autocorrelations from series generated by general ARIMA processes of order (p,d,q), d = 0 or 1. *J. Econometrics* 14, 365-379.

DAVIES, N. (1977). Model misspecification in time series analysis. Ph.D thesis, University of Nottingham.

DAVIES, N. and NEWBOLD, P. (1980). Sample moments of the autocorrelations of moving average processes and a modification to Bartlett's asymptotic variance formula. *Commun. Statist. A* 9, 1473-1481.

HASZA, D.P. (1980). The Asymptotic Distribution of the Sample Autocorrelations for an Integrated ARMA Process. *J. Amer. Statist. Assoc.* 75, 349-352.

ROY, R. (1977). On the Asymptotic Behaviour of the Sample Autocovariance Function for an Integrated Moving Average Process. *Biometrika* 64, 419-421.

ROY, R. and LEFRANCOIS, P. (1978). On the Sample Autocovariance and Autocorrelation Functions for an Autoregressive Integrated Moving Average Process. Paper presented at IMS Special Topics Meetings on Time Series Analysis, Iowa, May 1978.

WICHERN, D.W. (1973). The Behaviour of the Sample Autocorrelation Function for an Integrated Moving Average Process. *Biometrika* 60, 235-239.

APPLIED TIME SERIES ANALYSIS
O.D. Anderson & M.R. Perryman (eds.)

ON SOME SERIAL DEPENDENCE FORMULAE FOR ARUMA MODELS

O.D. ANDERSON
9 Ingham Grove, Lenton Gardens, Nottingham NG7 2LQ, England

This note establishes a simple method for obtaining close approximations to the moments of serial correlations obtained from sufficiently warmed up finite simulations from any member of a general homogeneously nonstationary class of linear models, excluding the once-integrated mixed autoregressive moving average processes.

1. INTRODUCTION

The general autoregressive integrated moving average process of order (p,d,q), which we shall abbreviate to ARIMA(p,d,q), is defined by a stochastic sequence, $\{Z_i\}$, satisfying

$$(1-\phi_1 B-\ldots-\phi_p B^p)(1-B)^d Z_i = (1-\theta_1 B-\ldots-\theta_q B^q)A_i \tag{1}$$

where $(\phi_1,\ldots,\phi_p)$ and $(\theta_1,\ldots,\theta_q)$ are two sets of real parameters, with the first subject to the stationarity condition, namely that the polynomial $1-\phi_1\zeta-\ldots-\phi_p\zeta^p$ in the complex variable ζ has no zero within or on the unit circle; and B is the backshift operator, such that B^j operating on any X_i, for instance A_i or Z_i, produces X_{i-j}. $\{A_i\}$ is a white noise sequence of independent but identically distributed normal zero-mean random variables, all with variance σ_A^2 say.

When the factor $(1-B)^d$ in (1) is replaced by any general homogeneous nonstationary operator of the form

$$U_d(B) = (1-u_1 B-\ldots-u_d B^d)$$

all of whose zeros lie precisely on the unit circle, but not necessarily all (or any) taking the value plus unity, we obtain the more general so-called ARUMA(p,d,q) class of models - see, for instance, Anderson (1980a).

Putting d = 0 in either the ARIMA or ARUMA specification retrieves the stationary ARMA(p,q) processes.

Given any series, $\{z_i: i=1,\ldots,n\}$ of length n, its sampled autocovariance, at lag k, is defined as

$$c_k^{(n)} = n^{-1}\sum_{i=1}^{n-k}(z_i-\bar{z})(z_{i+k}-\bar{z}) \qquad (k=0,1,\ldots,n-1) \tag{2}$$

where $\bar{z} = (z_1+\ldots+z_n)/n$ is the mean of the observed series, and the expectation of (2) is denoted by $E\left[c_k^{(n)}\right]$.

Associated with this is

$$E_k^{(n)} = E\left[c_k^{(n)}\right]/E\left[c_o^{(n)}\right] \tag{3}$$

which, for ARIMA models with d > 1 and all other proper ARUMA processes, has been shown to usually closely characterise the k-th serial correlation, defined by

$$r_k^{(n)} = c_k^{(n)}/c_o^{(n)} \qquad (k=0,1,\ldots,n-1).$$

For instance, see Anderson (1980b).

2. MORE THAN ONCE INTEGRATED ARMA MODELS

Anderson (1980a), equation (5), gave that for a general ARIMA(p,d,q), with d > 1,

$$E_k^{(n)} \simeq (1-\frac{k}{n})\{1-\frac{2k}{n}-\frac{(2k^2+1)}{n^2}\}/(1-\frac{1}{n^2}) \qquad (0 \le k \le n-1) \tag{4}$$

whilst, on page 916, he noted that the right of this was also the exact formula appropriate to a deterministic series having the form

$$z_i = z_1 + (i-1)m \qquad (i=1,\ldots,n) \tag{5}$$

where m defines an arbitrary but constant (non-zero) trend. Our first task will be to explain this observation.

Using an obvious condensation of notation, (1) may be rewritten as

$$(1-B)^d z_i = \phi_p^{-1}(B)\theta_q(B)A_i. \tag{6}$$

Now, from its definition (3), $E_k^{(n)}$ is clearly independent of σ_A^2, so we may put $\sigma_A^2 = 0$ (or, more strictly, consider the limit as $\sigma_A^2 \to 0$), when we are dealing with $A_i = 0$. Then (6) reduces to

$$(1-B)^d z_i = 0$$

and the solution of this difference equation is

$$z_i = \sum_{j=0}^{d-1} c_j i^j$$

for some set of arbitrary constants $c_o,\ldots,c_{d-1}$.

So, for large N and $t > 0$, we quickly get that

$$Z_{N+t} = O(N^{d-3}) + c_{d-2}N^{d-2} + c_{d-1}\{1 + \frac{(d-1)t}{N}\}N^{d-1}. \tag{7}$$

Then, for $\overline{Z} = (Z_{N+1} + \ldots + Z_{N+n})/n$, (7) gives

$$\overline{Z} = c_{d-2}N^{d-2} + c_{d-1}\{1 + \frac{(d-1)(n+1)}{2N}\} + O(N^{d-3})$$

and thus, for $\{z_{N+1},\ldots,z_{N+n}\}$,

$$c_k^{(n)}(0) = c_{d-1}^2(d-1)^2N^{2(d-2)}\sum_{t=1}^{n-k}(t - \frac{n+1}{2})(t+k - \frac{n+1}{2}) + O(N^{2d-5}) \tag{8}$$

where $c_k^{(n)}(0)$ indicates that we are here looking at $c_k^{(n)}$ for a process with $\sigma_A^2 = 0$ (which of course is deterministic).

Simplifying (8) then yields

$$c_k^{(n)}(0) = c_{d-1}^2(d-1)^2N^{2(d-2)}(n-k)(n^2-2kn-2k^2-1)/12 + O(N^{2d-5}).$$

So, as $N \to \infty$,

$$E_k^{(n)}(0) = \frac{c_k^{(n)}(0)}{c_o^{(n)}(0)} \to \frac{(n-k)(n^2-2kn-2k^2-1)}{n(n^2-1)}.$$

That is, given sufficient warming up of the process,

$$E_k^{(n)} \simeq (1 - \frac{k}{n})\{1 - \frac{2k}{n} - \frac{(2k^2+1)}{n^2}\}/(1 - \frac{1}{n^2})$$

since the $E_k^{(n)}$ are independent of σ_A^2, which is the result that we set out to establish.

Also note that, in addition to this providing a very easy way of obtaining a close approximation for $E_k^{(n)}$, for a warmed up ARIMA(p,d,q) process with $d > 1$, it demonstrates that the behaviour is like that for a determinstic series

$$z_i = \sum_{j=0}^{d-1} c_j i^j \qquad (i=N+1,\ldots,N+n)$$

which can be rewritten in the form

$$z_i = z_{N+1} + \sum_{j=1}^{d-1} c_j\{(N+i)^j - (N+1)^j\}$$

that has N sufficiently large so that, in fact,

$$z_i \simeq z_{N+1} + (i-1)(d-1)c_{d-1}N^{d-2}$$

where the approximation becomes equality when d = 2.

That is, for d > 1, warmed up ARIMA(p,d,q) processes have local behaviour like that of a linear trend, as was hypothesised on page 916 of Anderson (1980a). Basically this reflects the elementary fact that the curvature of a polynomial rapidly decreases as one goes out beyond its last turning point - that is, locally, the polynomial soon approximates closely to a straight line.

3. HIGHER MOMENTS

The same approach can evidently be used to obtain ratios of higher order covariance moments for the ARIMA(p,d,q) class, with d > 1. For instance, clearly

$$\frac{E\left[c_k^{(n)}(0)\,c_h^{(n)}(0)\right]}{E\left[c_k^{(n)}(0)\right]E\left[c_h^{(n)}(0)\right]} = 1 \qquad (0 \leq h,\ k \leq n-1) \tag{9}$$

and due to the lack of dependence on σ_A^2, the zeros can be dropped on the left of (9).

Now a closer approximation than $E_k^{(n)}$ for $E\left[r_k^{(n)}\right]$ is provided by

$$E_k^{(n)}\left\{1 - \frac{E\left[c_k^{(n)}c_o^{(n)}\right]}{E\left[c_k^{(n)}\right]E\left[c_o^{(n)}\right]} + \frac{E\left[c_o^{(n)2}\right]}{E^2\left[c_o^{(n)}\right]}\right\}$$

and also

$$\mathrm{Var}\left[r_k^{(n)}\right] \simeq E_k^{(n)2}\left\{\frac{E\left[c_k^{(n)2}\right]}{E^2\left[c_k^{(n)}\right]} - \frac{2E\left[c_k^{(n)}c_o^{(n)}\right]}{E\left[c_k^{(n)}\right]E\left[c_o^{(n)}\right]} + \frac{E\left[c_o^{(n)2}\right]}{E^2\left[c_o^{(n)}\right]}\right\}.$$

So, for the ARIMA(p,d,q) class with d > 1, we get that $E_k^{(n)}$ is also the second approximation to $E\left[r_k^{(n)}\right]$ and

$$\mathrm{Var}\left[r_k^{(n)}\right] \simeq 0 .$$

4. OTHER ARUMA MODELS

Finally, the same approach can be applied to all other ARUMA models, which do not degenerate to either ARMA(p,q) or ARIMA(p,1,q).

For instance, when $U_d(b) = (1+B)$, the equivalent deterministic process is $Z_{N+t} = (-1)^t Z_N$ and we get, after sufficient warming up, that

$$E\left[r_k^{(n)}\right] \simeq E_k^{(n)}(0) \simeq (-1)^k\left(1 - \frac{k}{n}\right) \qquad (k=0,1,\ldots,n-1)$$

whilst, for $U_d(B) = (1-2B\cos\omega + B^2)$, we have $Z_i = c_1\cos(i\omega+c_2)$ and hence, when $2\pi/\omega << n$,

$$E\left[r_k^{(n)}\right] \simeq E_k^{(n)}(0) \simeq \left(1 - \frac{k}{n}\right)\cos k\omega \qquad (k=0,\ldots,n-1)$$

and, for all such processes,

$$\mathrm{Var}\left[r_k^{(n)}\right] \simeq 0 .$$

Thus we conclude that for all ARUMA models, except ARMA and ARIMA(p,1,q), the observed serial correlations, for finite lengthed series from warmed up realisations, will almost certainly be closely characterised by the appropriate (deterministic) $E_k^{(n)}(0)$, which are all simple expressions.

RECOGNITION OF PRIORITY

I am grateful to Dr R.M.J. Heuts of Tilburg University for noting that the $E\left[c_k^{(n)}\right]$ formula, given an ARMA(p,q) model, quoted in O.D. Anderson (1980a), equation (3), is in fact implicit in a pair of formulae, numbered (51) and (52), on page 448 of T.W. Anderson (1971).

ACKNOWLEDGEMENT

This paper was written whilst the author was a visiting professor at Charles University, Prague, November-December 1980.

REFERENCES

ANDERSON, O.D. (1980a). Serial dependence properties of linear processes. *J. Opl Res. Soc.* 31, 905-917.

ANDERSON, O.D. (1980b). Serial dependence properties of ARUMA models. *Cahiers du CERO* 22, 309-323.

ANDERSON, T.W. (1971). *The Statistical Analysis of Time Series.* Wiley, New York and London.

APPLIED TIME SERIES ANALYSIS
O.D. Anderson & M.R. Perryman (eds.)

A STATISTICAL INTERPRETATION OF THE R AND S ARRAY APPROACH TO ARMA MODEL INDENTIFICATION

J. Bee Bednar and Brenda Roberts

Department of Mathematics
The University of Tulsa
Tulsa, Oklahoma 74104
U.S.A.

ABSTRACT

This paper develops and evaluates an R and S array based algorithm for estimation of a biased AIC for autoregressive moving average (ARMA) models. The algorithm is obtained by first relating the Yule-Walker or Levinson method for solving the Yule-Walker equations to R and S array recursions and then interpreting combinations of the R and S arrays in a statistical sense. It is shown that, given the final moving average coefficient, the AIC can be calculated directly from the R and S arrays. Moreover, the log absolute value of this coefficient is exactly the bias in the estimation algorithm. Because the coefficient becomes zero as the model fit improves, the bias performs a desirable function with regard to model determination.

In addition to an AIC approach to model selection the R and S array relationship to Levinson recursion leads quite naturally to a new set of arrays which provide additional methodologies for model selection. At least one of these arrays measures the transition between a predictor space of dimension equal to the assumed number n of past observations needed to accurately predict and a space of dimension less than n. Thus a measure of the degree of linear dependence of the observed stochastic process on past values is obtainable.

INTRODUCTION

H. Akaike (1969, 1970) introduced the concept of final prediction error (FPE) and demonstrated its effectiveness in differentiating between competing models. The success of FPE lead in 1971 to his now famous information criterion (AIC). This information criterion is determined by the data size, the number of independent model parameters, and the maximum likelihood estimate for the innovation variance of the model.

For autoregressive (AR) models, AIC is very nearly optimal. It is also completely automatic. For more general ARMA processes automation has been difficult. Computational maximization of the likelihood function involves solution of a nonlinear system of equations for each model considered. Needless to say this has made extension of AIC to the ARMA case difficult.

More recently, Gray, Kelly and McIntire (1978) and Woodward and Gray (1979) have developed a model identification methodology based on a recursive algorithm of Pye and Atchison (1973). The technique primarily exploits the presence of certain patterns in the S array of an ARMA process. Woodward and Gray (1979) showed that this pattern is almost equivalent to consistent patterns in a "generalized partial autocorrelation array" which results in model identification procedure reminiscent of the partial autocorrelation approach of Box and Jenkins (1976).

By extending the Woodward and Gray results, Bednar (1980a, 1980b) has developed a relationship between the Pye and Atchison R and S array algorithm and traditional recursive methods for solving the Yule-Walker equations. This relationship provides the basis for a computationally convenient algorithm for obtaining a biased AIC estimate. How R and S recursion relates to Yule-Walker and how these two concepts relate to AIC are the chief contributions of this paper.

YULE-WALKER AND THE AIC

Consider the ARMA (p, q) process

$$z_t = f_{1p}^{q} z_{t-1} + \cdots + f_{pp}^{q} z_{t-p} + a_t - g_{1q}^{p} a_{t-1} - \cdots - g_{qq}^{p} a_{t-q} \tag{2.1}$$

where a_t is a white Gaussian innovation with zero mean and variance σ_a^2. With

$$r_k = E(z_{t-k} z_t)$$

and

$$s_k = E(z_{t-k} a_t)$$

one can easily show [Box and Jenkins (1976), 474] that

$$r_k = f_{1p}^{q} r_{k-1} + \cdots + f_{pp}^{q} r_{k-p} + s_k - g_{1q}^{p} s_{k-1} - \cdots - g_{qq}^{p} s_{k-q} \tag{2.2}$$

Letting k = q in (2.2) and using the fact that $s_k = 0$ when $k > 0$ yields

$$\sigma_a^2 g_{qq}^{p} = \sum_{i=1}^{p} f_{ip}^{q} r_{q-i} - r_q \tag{2.3}$$

This last equation forms the basis for the algorithm to be presented in subsequent paragraphs. In succinct form, an estimate of σ_a^2 produces an estimate of the AIC when the number, p+q, of model parameters is known. Since p+q is known, an estimate of $f_{1p}^{q}, f_{2p}^{q}, \ldots, f_{pp}^{q}$ and g_{qq}^{p} would appear to be all that is necessary to estimate σ_a^2 and hence the AIC. The f's in equation (2.3) are easy to estimate. One need only take k=q+i i=1,2,..,p in (2.2) to obtain the p equations

$$r_{q+i} = - \sum_{k=1}^{p} f_{kp}^{q} r_{q+i-k} \tag{2.4}$$

for them. Unfortunately the same is not true for g_{qq}^{p}. There does not appear to be a straightforward procedure to estimate g_{qq}^{p}. On the other hand there does appear to be a path around this difficulty.

To see that (2.3) and (2.4) do indeed form a basis for maximum likelihood model selection, consider the likelihood function L(Θ) of a model with parameter vector Θ. For an assumed Gaussian density and data of length N,

$$L(\Theta) = \left(\frac{1}{2\pi\sigma_a^2} \right)^{N/2} \exp \left\{ - \frac{1}{2\sigma_a^2} \sum_{i=1}^{N} a_i^2 \right\}$$

which for large N gives the approximate likelihood

$$L(\Theta \mid a_{1g}\cdots,a_{p+q}) = \left(\frac{1}{2\pi\sigma_a^2} \right)^{\frac{N-p-q}{2}} \exp \left\{ - \frac{1}{2\sigma_a^2} \sum_{i=p+q+1}^{N} a_i^2 \right\}$$

Taking the partial derivative of the log likelihood function $l(\Theta)=\log L(\Theta)$ produces

$$\frac{\partial l(\Theta)}{\partial f_{l,p}^{q}} = - \frac{1}{2\sigma_a^2} \sum_{n=p+q+1}^{N} 2\left\{ \sum_{j=0}^{p} f_{jp}^{q} z(n-j)z(n-l) - z(n-l) \sum_{i=1}^{q} g_{iq}^{p}\, a(n-i)\right\}$$

Equating this partial derivative to zero gives

$$\sum_{n=p+q+1}^{N} \sum_{j=0}^{q} f_{jp}^{q}\, z(n-j)\, z(n-l) = \sum_{n=p+q+1}^{N} \sum_{i=1}^{q} g_i^{p}\, a(n-i)\, z(n-l) \qquad (2.5)$$

or equivalently

$$\sum_{j=0}^{p} f_{jp}^{q} \sum_{n=p+q+1}^{N} z(n-j)\, z(n-l) = \sum_{i=1}^{q} g_{iq}^{p} \sum_{n=p+q+1}^{N} a(n-i)\, z(n-l)$$

Replacing the sum over n by expectations of both sides of this last equation yields

$$\sum_{j=0}^{p} f_{jp}^{q}\, r_{q+j-l} = \sum_{i=1}^{q} g_{iq}^{p}\, s_{-i+q+l} \qquad (2.6)$$

which is essentially a restatement of equation (2.2). Thus, $l(\Theta)$ is approximately maximized when the autoregressive parameter vector satisfies (2.4) and, in this case $\sigma_a^2 g_{qq}^{p}$ is estimated by the right-hand side of (2.3).

By definition, AIC is given by
AIC = -2(maximum log likelihood)+2(number of independently adjusted parameters)

Since

$$L(\Theta) = \left(\frac{1}{2\pi\sigma_a^2} \right)^{N/2} \exp \left[- \frac{1}{2\sigma_a^2} \sum_{i=1}^{N} a_i^2 \right]$$

$$\begin{aligned} AIC(p,q) &= N \log 2\pi\sigma_a^2 + \frac{1}{\sigma_a^2} \sum_{i=1}^{N} a_i^2 + 2(p+q+1) \\ &\simeq N \log 2\pi\sigma_a^2 + N + 2(p+q+1) \end{aligned}$$

or, neglecting constants

$$AIC(p,q) = N \log \sigma_a^2 + 2(p+q)$$

From (2.3), an estimate of AIC is

$$\mathrm{AIC}(p,q) = N \log \left| r_q - \sum_{i=1}^{p} f_{ip}^{q} r_{q-1} \right| + N \log \left| g_{qq}^{p} \right| + 2(p+q)$$

Because of considerations which will become apparent in the next section, it will be more convenient to estimate

$$\overline{\mathrm{AIC}}(p,q) = \mathrm{AIC}(p,q) + N \log \left| g_{qq}^{p} \right|$$

It will also be clear that this change will have little effect on the ability to automatically determine the model orders. Since computation of f_{ip}^{q}, $i=1,2,\ldots,p$ completely determines $\overline{\mathrm{AIC}}$ (p,q), and since estimates of f_{ip}^{q} can be obtained from the Yule-Walker equations, attention is now focused on how $\overline{\mathrm{AIC}}$ (p,q) can be obtained without actually solving the Yule-Walker equations.

RELATIONSHIP OF LEVINSON RECURSION TO THE R AND S ARRAYS

Solution of the equations

$$r_{q+i} = \sum_{j=1}^{p} f_{jp}^{q} \; r_{q+i-j}$$

for all possible p and q has been shown by Levinson (1947), Durbin (1960) and Trench (1965) to be a simple, fast recursive procedure. To emphasize the fact that p and q will represent the chosen ARMA orders, the recursive algorithm for a fixed moving average order j and autoregressive order k satisfies

$$f_{ik}^{j} = f_{i,k-1}^{j} - f_{kk}^{j} \, b_{k-i-1,k-1}^{j} \qquad 0 \leq i \leq k-1 \tag{3.1}$$

$$b_{ik}^{j} = b_{i,k-1}^{j} - b_{kk}^{j} \, f_{k-i-1,k-1}^{j} \qquad 0 \leq i \leq k-1 \tag{3.2}$$

$$f_{kk}^{j} = \left[r_{j+k+1} - \sum_{i=0}^{k-1} f_{i,k-1}^{j} \; r_{j+k-i} \right] / \left(r_j d_{k-1}^{j} \right) \tag{3.3}$$

$$b_{kk}^{j} = \left[r_{j-k-1} - \sum_{i=0}^{k-1} b_{i,k-1}^{j} \; r_{j-k+i} \right] / \left(r_j d_{k-1}^{j} \right) \tag{3.4}$$

$$d_{k-1}^{j} = \left(1 - f_{k-1,k-1}^{j} \; b_{k-1,k-1}^{j} \right) d_{k-2}^{j} \tag{3.5}$$

with initial conditions

$$f_{00}^{j} = r_{j+1} / r_j \;, \; b_{00}^{j} = r_{j-1} / r_j \;, \; d_{-1}^{j} = 1 \tag{3.6}$$

Moreover, the determinant $R_{kk} = \left| \left[r_{j+(m-n)} \right] \right|$ where $1 \leq m \leq k$, $1 \leq n \leq k$, and j is fixed satisfies

$$R_{kk} = d_{k-1}^{j} \cdot d_{k-2}^{j} \cdot \cdots \cdot d_{0}^{j} \tag{3.7}$$

Although of only passing interest here Luenberger (1969) demonstrates that the rank of R_{kk} is related to the dimension of the predictor space so that a transition from a full rank R_{kk} to a singular $R_{k+1,k+1}$ represents a measure of when the true autoregressive order has been achieved. Clearly,

$$E^{j}_{k-1,k-1} = (1 - f^{j}_{k-1,k-1} \; b^{j}_{k-1,k-1}) \tag{3.8}$$

is a potentially effective measure of this change. When $E^{j}_{k-1,k-1}$ is near zero the rank of the matrix is no longer full and k is approaching the true autoregressive order. Bednar (1980b) has studied this array and its use in model identification.

Woodward and Gray (1979) were apparently the first to recognize that the R and S array recursions

$$R_{n+1}(r_m) = R_n(r_{m+1}) \left[\frac{S_n(r_{m+1})}{S_n(r_m)} - 1 \right] \tag{3.9}$$

$$S_n(r_m) = S_{n-1}(r_{m+1}) \left[\frac{R_n(r_{m+1})}{R_n(r_m)} - 1 \right] \tag{3.10}$$

of Pye and Atchison (1973) can be used to calculate f^{j}_{kk}. They show that

$$f^{j}_{kk} = - \frac{S_k(r_{-k+j+1})}{S_k(r_{-k-j})} \tag{3.11}$$

and use f^{j}_{kk} as elements of the generalized partial autocorrelation array (GPAC) to analyze sample autocovariances.

Bednar (1980a, 1980b) has shown that the GPAC array patterns used by Woodward and Gray can be weakened to obtain approximate to as opposed to exact model fits. In this analysis it has been further demonstrated that

$$b^{j}_{kk} = - \frac{S_k(r_{-k-j+1})}{S_k(r_{-k+j})} \tag{3.12}$$

$$S_k(r_{-k+j+1}) = (-1)^k \, (1 - \sum_{i=0}^{k} f^{j}_{ik}) \tag{3.13}$$

$$S_k(r_{-k-j}) = (-1)^k \, (1 - \sum_{i=0}^{k} b^{j+1}_{i,k}) \tag{3.14}$$

$$R_{k+1}(r_{-k+j+1}) \; S_k(r_{-k+j+1}) = (-1)^k \, (r_{j+k+1} - \sum_{i=0}^{k-1} f^{j}_{i,k-1} \; r_{j+k-i}) \tag{3.15}$$

$$R_{k+1}(r_{-k-j}) \; S_k(r_{-k-j}) = (-1)^k \, (r_{j-k} - \sum_{i=0}^{k-1} b^{j+1}_{i,k-1} \; r_{(j+1)-k+i}) \tag{3.16}$$

In addition, one has

$$b^{j}_{kk} = \begin{cases} \begin{matrix} 0 \\ f_{kk} \end{matrix} & j = 0 \\ (f^{j-1}_{kk})^{-1} & j = 1,2,\ldots \end{cases} \tag{3.17}$$

and

$$b_{ik}^{j+1} = f_{k-i-1,k}\,(-\,b_{kk}^{j+1}) \qquad i = 0,1,2,\ldots k-1 \tag{3.18}$$

These relationships are important not only because they characterize recursive solutions to Toeplitz systems in terms of R and S arrays but also because they permit direct estimation of $\overline{\text{AIC}}$ (p,q). Clearly, any of the quantities specifying the recursions of equations (3.1)-(3.5) can now be calculated from the R and S arrays. The converse is also true but it is equation (2.3) which actually links AIC to Levinson recursion and then to the R and S arrays. This is a relatively straightforward procedure once the appropriate trick has been found. Estimation of AIC from the R and S arrays is the subject of the next section.

AIC AND THE R AND S ARRAYS

To estimate AIC, one must have an estimate of $\sigma_a^2 g_{jj}^k$. From (3.16)

$$S_k(r_{-k-j})\;R_{k+1}(r_{-k-j}) = (-1)^k\left[r_{j-k} - \sum_{i=0}^{k-1} b_{i,k-1}^{j+1}\, r_{j+1-k+1}\right]$$

$$= (-1)^{k-1}\left[r_{j-k} - \sum_{i=0}^{k-2} b_{i,k-1}^{j+1}\, r_{j+1-k+i} - b_{k-1,k-1}^{j+1}\, r_j\right]$$

Replacing b's by f's via (3.17) and (3.18), the right-hand side of the preceding equation becomes

$$(-1)^k\left[r_{j-k} + \frac{\sum\limits_{i=0}^{k-2} f_{k-i-2,k-1}^{j}\, r_{j+1-k+i} - r_j}{f_{k-1,k-1}^{j}}\right]$$

so that

$$S_k(r_{-k-j}\; r_{k+1}(r_{-k-j})\; f_{k-1,k-1}^{j}$$

$$= (-1)^k\left[f_{k-1,k-1}^{j}\, r_{j-k} + \sum_{i=0}^{k-2} f_{k-i-2,k-1}^{j}\, r_{j+1-k-1} - r_j\right]$$

or

$$S_k(r_{-k-j})\;R_{k+1}(r_{-k-j})\; f_{k-1,k-1}^{j} = (-1)^k\left[\sum_{i=0}^{k-1} f_{i,k-1}^{j}\, r_{j-i-1} - r_j\right] \tag{4.1}$$

From (2.3)

$$\sigma a^2\, g_{jj}^{k} = (-1)^k\, S_k(r_{-k-j})\; R_{k+1}(r_{-k-j})\; f_{k-1,k-1}^{j} \tag{4.2}$$

Using Woodward and Gray's formula for $f_{k-1,k-1}^{j}$ one finally obtains

$$\left| \sigma_a^2 \, g_{jj}^k \right| = \left| \frac{S_k(r_{-k-j}) \; R_{k+1}(r_{-k-j}) \; S_{k-1}(r_{-k+j+2})}{S_{k-1}\,(r_{-k-j+1})} \right| \tag{4.3}$$

Actually (4.3) is valid for all k>1. From the initial conditions of equation (3.6) together with the definition of the first column of the R array, one has

$$\left| \sigma_a^2 \, g_{jj}^1 \right| = \left| \frac{S_1(r_{-1-j}) \; R_2(r_{-1-j}) \; R_1(r_{j+1})}{R_1(r_j)} \right| \tag{4.4}$$

The right-hand sides of equations (4.3) and (4.4) contain the computational aspects of $\overline{\text{AIC}}$, the left-hand sides the interpretive. Noting that for a purely autoregressive process, j=0 , $g_{jj}^k=1$, and hence $\sigma_a^2 g_{jj}^k=\sigma_a^2$, it is quite easy to see that

$$\overline{\text{AIC}}\ (k,0) - \text{AIC}\ (k,0)$$

so that $\overline{\text{AIC}}$ (k,0) is Akaike's information criterion by the Yule-Walker method.

For the ARMA case, j>1 and g_{jj}^k need not be unity. It must, however, be zero for all k>p if p is the true autoregressive order and j-1=q the true moving average order. Although not quite as easy to see, $\sigma_a^2 g_{jj}^k$ must also be zero for all j>q+1. Since the logarithm of zero is negative infinity these zeros suggests that the location (p,q+1) of the first minimum in the $\overline{\text{AIC}}$ (k,j) table can be used as an effective measure of the orders for the ARMA process under consideration.

One could, of course, use standard likelihood methods to estimate g_{jj}^k and correct $\overline{\text{AIC}}$ accordingly. This will not be done in this paper. Instead, numerical examples will be synthesized to demonstrate the usefulness of $\overline{\text{AIC}}$. It is felt that these simulations demonstrate that $\overline{\text{AIC}}$ is a reasonable and effective alternative to more computationally intensive methods for process order estimation. Attention is now focused on these simulations.

NUMERICAL EXAMPLES

A practical review of the remarks of the preceding paragraphs is in order. What has been shown is that $\sigma_a^2 g_{jj}^k$, where g_{jj}^k is the highest order moving average term in a (k,j) order ARMA process, can be estimated by a suitable combination of elements of the R and S arrays of Pye and Atchison. It is clear from the analysis that $\sigma_a^2 g_{jj}^k$ satisfies the pattern described in Table I. Taking the logarithm of the tabular entries of Table I, multiplying by N, and adding twice the number of independent parameters will produce a new table dominated by $-\infty$'s replacing each of the zeros in the pth column and q+i rows, i=1,2,..., as well as those in the q+1 st row beginning in the p th column. For the sake of compactness this $\overline{\text{AIC}}$ pattern will not be given in tabular form for the theoretical case. It will be given for the case of an exact autocorrelation of an ARMA (1,1) process.

Table II below gives $\overline{\text{AIC}}$ using the true autocorrelation for the process

$$z_t - .87z_{t-1} = a_t - .48a_{t-1}$$

TABLE I

Theoretical Pattern of $\sigma_a^2 g_{jj}^k$

Autoregressive Order

Moving Average Order	1	2	. . .	p-1	p	p+1	. . .
0	$\sigma_a^2 g_{00}^1$	$\sigma_a^2 g_{00}^2$	. . .	$\sigma_a^2 g_{00}^{p-1}$	$\sigma_a^2 g_{00}^p$	$\sigma_a^2 g_{00}^{p+1}$	. . .
1	$\sigma_a^2 g_{11}^1$	$\sigma_a^2 g_{11}^2$	. . .	$\sigma_a^2 g_{11}^{p-1}$	$\sigma_a^2 g_{11}^p$	$\sigma_a^2 g_{11}^{p+1}$	. . .
.	.	.		.	.	.	
.	.	.		.	.	.	
.	.	.		.	.	.	
q	$\sigma_a^2 g_{qq}^1$	$\sigma_a^2 g_{qq}^2$	. . .	$\sigma_a^2 g_{qq}^{p-1}$	$\sigma_a^2 g_{qq}^p$	$\sigma_a^2 g_{qq}^{p+1}$	. . .
q+1	$\sigma_a^2 g_{q+1,q+1}^1$	$\sigma_a^2 g_{q+1,q+1}^2$	. . .	$\sigma_a^2 g_{q+1,q+1}^{p-1}$	0	0	. . .
q+2	$\sigma_a^2 g_{q+2,q+2}^1$	$\sigma_a^2 g_{q+2,q+2}^2$	. . .	$\sigma_a^2 g_{q+2,q+2}^{p-1}$	0	0	0 0 0
.	.	.		.	.	0	0 0 0
.	.	.		.	.	0	0 0 0
.	.	.		.	.	0	0 0 0

This example shows quite clearly the theoretical behavior expected of $\overline{AIC}$. The minimum value $-.15 \times 10^{53}$ occurs in column one beginning in row 3 (corresponding to q = 2). The model is correctly identified as ARMA (1,1).

Table III below contains $\overline{AIC}$ for the true autocorrelation for the process

$$z_t - .19z_{t-1} + .75z_{t-2} = a_t - .5a_{t-1}$$

This process is interesting because, as shown by Bednar (1980b), realizations of this process can lead to erroneous values in the R and S arrays. These values occur because of a near zero in the lag 1 autocorrelation value. Note that $\overline{AIC}$ in Table III does not exhibit the true theoretical behavior. Table IV contains the AIC estimates for a 255 point realization of the process under consideration. The minimum value of $-.18406 \times 10^4$ still properly specifies that the process under consideration is ARMA (2,1). Several additional simulations with this and other models were run with similar results.

The $\overline{AIC}$ table for the metal series given by Makridakis (1978) as analyzed by Woodward and Gray (1979) is apparently well-modeled by a ARMA (13,1). Both FPE and Parzen's CAT criterion (1974) model this series as an ARMA (2,0). Both Akaike and Parzen content that an ARMA (2,0) is almost certainly not the correct model because of poor agreement between the estimated and the sample spectrum. To get the supposedly correct order, Woodward and Gray must first difference the data and then analyze the residual series. The absolute minimum of a 15-column, 7-row $\overline{AIC}$ array, partially displayed in Table V, occurs at a row and column pair consistent with ARMA (14,3) model. Similar magnitude values ($\sim -.8 \times 10^3$) also present in the $\overline{AIC}$ array indicate potential ARMA (3,6), ARMA (5,5), ARMA (13,2) and ARMA (15,3) fits to the observed data.

TABLE II

$\overline{\text{AIC}}$ Array for the Exact Autocorrelation of

$$A_t - .87\ Z_{t-1} = a_t - .48\ a_{t-1}$$

Moving Average Order	Autoregressive Orders 1	2	3	4	5	6
0	-.581E 02	.567E 02	.466E 02	.460E 02	.518E 02	.257E 05
1	-.179E 03	.203E 04	-.175E 03	-.173E 03	.389E 03	.243E 05
2	-.150E 53	-.150E 53	-.150E 53	-.150E 53	-.150E 53	-.150E 53
3	-.150E 53	-.150E 53	-.150E 53	-.150E 53	-.150E 53	-.150E 53
4	-.150E 53	.519E 02	.262E 05	-.150E 53	-.150E 53	-.150E 53
5	-.150E 53	-.150E 53	-.150E 53	-.150E 53	-.150E 53	-.150E 53
6	-.150E 53	-.192E 03	.260E 05	.260E 05	.260E 05	-.150E 53

TABLE III

$\overline{\text{AIC}}$ Array for the Exact Autocorrelation of

$$Z_t - .19\ Z_{t-1} + .75\ Z_t = a_t - .5\ a_{t-1}$$

Moving Average Order	Autoregressive Orders 1	2	3	4	5	6
0	.200E 01	-.127E 04	-.904E 01	-.275E 02	-.301E 02	-.217E 03
1	.115E 04	.950E 03	.181E 04	-.166E 03	-.198E 04	.475E 03
2	-.372E 02	-.250E 04	-.240E 03	.189E 04	-.245E 04	-.515E 02
3	.171E 03	-.213E 04	-.229E 04	-.236E 03	-.691E 02	.914E 02
4	-.689E 02	-.150E 53	-.150E 53	-.150E 53	-.150E 53	-.150E 53
5	.312E 02	-.231E 04	-.150E 53	-.150E 53	-.150E 53	-.150E 53
6	-.924E 02	-.150E 53	-.150E 53	-.150E 53	-.150E 53	-.150E 53

TABLE IV

$\overline{\text{AIC}}$ Array for 255 Point Realization of

$$Z_t - .19\ Z_{t-1} + .75\ Z_{t-2} = a_t - .5\ a_{t-1}$$

Moving Average Order	Autoregressive Orders 1	2	3	4	5	6
0	.192E 01	-.118E 04	-.455E 02	-.754E 02	-.984E 02	.699E 02
1	.966E 03	.605E 03	.687E 03	-.434E 03	-.123E 04	-.133E 03
2	-.611E 02	-.184E 04	-.422E 03	.731E 03	-.241E 03	-.379E 03
3	.244E 03	-.116E 04	-.155E 04	.253E 03	.105E 04	-.330E 03
4	-.106E 03	-.962E 03	-.116E 04	-.101E 04	.326E 02	.700E 03
5	.103E 03	-.452E 03	-.667E 03	-.824E 03	-.952E 03	-.232E 03
6	-.120E 03	-.593E 03	.207E 03	-.705E 03	-.109E 04	-.667E 03

TABLE V

$\overline{\text{AIC}}$ for the Makridakis Metal Series

Moving Average Order	Autoregressive Order 11	12	13	14	15
0	-.273E 03	-.665E 03	-.966E 02	-.199E 01	.101E 03
1	-.827E 02	.329E 03	-.239E 03	-.189E 02	-.299E 03
2	.649E 03	-.516E 03	-.435E 03	-.234E 03	-.299E 03
3	.134E 03	.475E 03	-.775E 03	-.352E 03	.478E 01
4	.303E 03	-.103E 03	-.321E 03	-.822E 03	-.745E 03
5	.115E 03	.852E 02	-.139E 03	-.397E 03	-.125E 03
6	.100E 03	-.596E 02	-.156E 03	-.139E 03	-.598E 03

The only columns which exhibit close to the theoretical behavior indicate autoregressive orders of 1,2,3,13, and 14. Spectral comparisons between the (13,1) and (14,3) indicate only minor differences. Thus either of these models appear satisfactory.

CONCLUSIONS AND FUTURE RESEARCH

Simulations run using minimum AIC as an indicator of the model orders of ARMA processes have always chosen an order pair consistent with the known orders. Only in a few instances have the selected orders been incorrect and in these cases, AIC has tended to over-estimate the required number of parameters. It seems that AIC is an effective statistic in the choice of the model orders. Computationally, AIC, through the use of the R and S arrays, is much more efficient then exact likelihood. The AIC statistic is not as accurate as exact likelihood but inaccuracies tend to be biased towards higher order errors rather than under-estimation of the number of model parameters. Over-all, AIC performs well as an automatic order selector.

Because of its close ties with the recursions needed to solve the Yule-Walker equations, it is tempting to speculate on procedures which would permit generalizations of AIC (or AIC) to other model situations. Recent work by Friedlander, et.al. (1979)would appear to offer a natural generalization of the ideas presented in this paper to "α stationary time series." At the very least, order selection can be done more as a least squares operation than a sample covariance problem. This is a step in the refinement direction. The much more general question of whether or not full linear or ridge regressions can be placed within the present framework is not clear, but AIC is really an analysis of the residuals in the regression problem so that there is at least hope. Extension of the ideas presented in this paper to vector time series does not appear to represent great difficulty.

REFERENCES

AKAIKE, H. (1976). Canonical correlation analysis of time series and the use of an information criterion. In *System Identification: Advances and Case Studies*, Academic Press, Inc.

AKAIKE, H. (1969). Fitting autoregressive models for prediction. In *Ann. Inst. Statist. Math 21*, 243-247.

AKAIKE, H. (1970). Statistical predictor identification. In *Ann. Inst. Statist. Math 22*, 203-217.

BEDNAR, J.B. (1980a). On the approximation of FIR by IIR digital filters. Submitted to *IEEE ASSP*.

BEDNAR, J.B. (1980b). On the relationship between Levinson recursion and the R and S arrays for ARMA model identification. Cities Service Technical Center Research Memorandum GP 80-09. Submitted to *Comm. Stat. - Simula. and Comput*.

BOX, G.E.P. and JENKINS, G.M. (1976). *Time Series Analysis: Forecasting and Control*, Holden-Day, Inc.

DURBIN, J. (1960). The fitting of time series models. In *Rev. Inst. Int. Stat. 28*, 233-243.

FRIEDLANDER, B. *et al* (1979). New inversion formulas for matrices classified in terms of their distance from Toeplitz matrices. In *Linear Algebra and Its Applications 27*, 31-60.

GRAY, H.L., KELLY, G.D. and McINTIRE, D.D. (1978). A new approach to ARMA modeling. In *Comm. of Stat. - Simula. and Comput. B7(1)*, 1-77.

LEVINSON, N. (1947). The Wiener RMS error criterion in filter design and prediction. In *J. Math. Phy. 25*, 261-278.

LUENBERGER, D.G. (1969). *Optimization by Vector Space Methods*, John Wiley and Sons.

MAKRIDAKIS, S. (1978). *Interactive Forecasting*, Holden-Day, Inc.

PARZEN, E. (1974). Some recent advances in time series modeling. In *IEEE Transactions on Automatic Control AC-19*, 723-730.

PYE, W.D. and ATCHISON, T.A. (1973). An algorithm for the computation of the higher order G-transformation. In *SIAM J. Numerical Analysis 10*, 1-7.

TRENCH, W.F. (1964). An algorithm for the inversion of finite Toeplitz matrices. In *SIAM 7(3)*, 515-521.

WOODWARD, W.A. and GRAY, H.L. (1979). On the relationship between the R and S arrays and the Box-Jenkins method of ARMA model identification. Tech Report No. 134, Dept. of Stat., ONR Contract, SMU, Dallas, Texas.

YULE, G.U. (1927). On a method of investigating periodicities in disturbed series, with special reference to Wolfer's sunspot numbers. In *Phil. Trans. A226*, 267-298.

APPLIED TIME SERIES ANALYSIS
O.D. Anderson & M.R. Perryman (eds.)

A NEW LOOK AT MULTIPLICATIVE MODELS

Michael S. Broida

Department of Production and Decision Sciences
Miami University
Oxford, Ohio
U.S.A.

Simplistic time series and forecasting techniques such as the Shiskin or X-11 multiplicative model and Winters' Method have been "passed over" by many statisticians and forecasters in favor of powerful and complex techniques such as Box-Jenkins. Forecasters generally agree that defects in the X-11 model include inability to deal with cycle, lack of precise confidence intervals for the forecast, and, often, poor forecasts. The record of X-11 for dealing with seasonality and trend is rather good.

INTRODUCTION

Time series techniques, including moving average, exponential smoothing, X-11 (Shiskin's Method or Decomposition), Winters' technique, and, more recently, Box-Jenkins, have been used for years with varying success. Generally, the techniques are most useful when relationships over time are relatively stable. In the past, simple time series techniques, possibly excluding Box-Jenkins, are relatively weak when some form of economic cycle is a major factor in the analysis.

One of the "most elementary" techniques often taught to sophomore business students is the Decomposition or X-11 model. In fact, the technique is so common that it is often not given serious consideration by analysts as a useful model. In fact, generally, when used as devised by Julius Shiskin for use at the Census Bureau, the model in fact did not give good medium range forecasts. A medium range forecast is defined as 3 months to 2 years for the purposes of this paper. However, with very little extra work, the model has proved to be the backbone of a relatively accurate short and medium range forecasting tool.

There are two recommended changes from the basic model developed by Shiskin. First, some logical estimate of the cycle can be included in the forecast. This is generally not done with the versions of the technique taught in statistics courses. Second, if the model is first employed as a screening or filtering device (prior to any forecast attempt), then the effects of a few bizarre, unusual or "one time" events can be modified so that we have a consistent data set. The construction of a consistent data set generally goes hand in hand with a more accurate forecast. Examples of unusual events which upset the ability to forecast include windfall or extraordinary income, prolonged wildcat or other strikes, or perhaps a tornado, earthquake, flood, or other disaster.

Often, the unusual data point is only a reflection of calendar variation. If sales are being analyzed for a firm doing a predominantly weekend business, we know that some months have 5 weekends. For example, September of 1979 had 5 weekends, while September of 1980 had only 4. Thus, receipts in 1975 could appear to be as much as 25% "too high", everything else equal. Companies paying their employees every two weeks will find that there are at least two months every year where they have three payrolls, and the next year the three payroll months will occur at different points in time. These variations can also be factored out of the analysis, and incorporated into the forecast as the concluding step.

Last, it is often the case that only four to six years of quarterly or monthly data is available. If older data is available, it is often too incompatible to be of real use. When these "short" sets are the only data available for forecasting, Box-Jenkins and similar techniques are usually not employed. However, this four to six year span is ideal for X-11 type techniques.

X-11 AS A SCREENING TOOL

In order to forecast using a time series technique, any time series technique, it is necessary to establish a consistent data base and to have all unusual data points modified to reflect some sort of a constant condition. Ordinarily, the analyst would automatically attempt to modify data points which are not good indicators of what we want to measure, but often only the most extreme points are discovered by visual inspection of the raw data. For example, during a study of Miami University water consumption, we found that for April, 1978 we had 73.5 million gallons of water used, almost twice the next peak high over an 8 year period ending March, 1981. Interviews determined that during one weekend in April, 1978 students had a "flush in." All showers, faucets, and toilets were run in an attempt to drain the town. (No single motive has been established.) Expert opinion has estimated that over 35 million gallons of water were wasted. The simple average of April, 1977 and April, 1979 yields an estimate of 36.5 million gallons as the actual water demanded excluding the abnormal demand, which corresponds closely with the expert opinion. In any case, the figure of 73.5 million cannot be used as raw data. Even an educated guess of actual demand will produce superior forecasts. That is, a new, artificial number must replace the actual figure if the actual figure includes an event which is extraneous to the forecast. Other abnormal, but less spectacular, points probably will not be uncovered until a preliminary analysis.

The easiest preliminary analysis involves simply running the raw data (with corrections for spectacular abnormal points) through the decomposition. Three separate sources of information are available to the analyst. First, the residual plot (of the deseasonalized or deseasonalized-decycled data) will show if any outliers are present; that is, points which break the pattern. Second, the centered moving average as a percentage and the seasonal index for that point in time should not differ radically. Any points in time which have a large difference should be considered as candidates for further study. Last, the irregulars should be close to 1.0 (or 100 if an index is used). Any irregulars which deviate radically from 1.0 should be considered as suspicious.

Unfortunately, exact confidence intervals are not available to allow for an easy decision. For example, if a certain irregular for time k is measured as 1.05, we cannot be certain whether that is pretty small or whether it is unacceptably large. Judgment alone must be used to compile a list of all data points which are associated with violating any one of the three measures indicated.

Once we have compiled a list of "inconsistent" data points, we should use whatever research means available to discover whether an abnormality occurred which is outside of the factors we wish included in the forecast. In the absence of an explicit reason, it is recommended that a point should not be changed; that is, we should not smooth the data for smoothing sake alone. The discovery of a windfall, strike, tornado, etc., which caused a deviation in the data, still leaves the problem of what to do about it.

ADJUSTING THE DATA

The single most important task for the analyst is to be sure that the data base being considered for forecasting purpose is consistent enough in order that a forecast could be made. It is quite common that the original numbers would have to be in some way altered before the forecast could continue. Suppose, for example, we are studying income tax receipts for a community. Exhibit 1 shows a set of hypothetical figures that we will be using in the discussion. Notice that generally the first quarter has the fewest receipts as compared to any other time during the year. The best quarter tends to be the third quarter which would be the late summer period. Over the course of the four years as illustrated, notice the slow, steady growth of tax receipts. Every first quarter is slightly higher than the one before. Every third quarter except for 1974 is higher than the one before. And every fourth quarter is higher than the one before. The total tax receipts for the year would tend to be increasing over the four year period.

Year	Quarter	
1973	1	100,000
	2	105,000
	3	120,000
	4	110,000
1974	1	106,000
	2	110,000
	3	92,000
	4	115,000
1975	1	110,000
	2	114,000
	3	131,000
	4	118,000
1976	1	112,000
	2	120,000
	3	136,000
	4	122,000

Exhibit 1
Income Tax Receipts, Hypothetical Example, Quarterly Data

The one number which appears to break the pattern is the third quarter of 1974. The tax receipts collected at that point in time are far lower than an analyst would expect. It turns out that during the third quarter of 1974 the major industry in our town underwent a lengthy strike and because of this strike tax receipts fell off drastically. Now what would happen if this raw data was used to make a forecast? First, the regression line will not be quite accurate because of the third quarter of 1974. This low data point will tend to lower the regression line and in this case would show a lower rate of growth of tax receipts than really exists. All of the seasonal indexes will also be thrown off. The first, third, and fourth will be thrown off slightly. The third seasonal index will be thrown off considerably. In fact, the seasonal index for quarters 1, 2, and 4 will be slightly overestimated and time period 3 will be considerably underestimated because the data point for quarter 3, 1974 will be averaged in with all the others that show that summer in normally a good season for collecting tax money. What is necessary is for the analyst to remove the actual figure of $92,000 and to replace it with an educated guess which would reflect how much tax receipts would have been collected if it had not been for that strike, that is, if the strike had not taken place. There is no technique for coming up with a number like this. Some analysts would simply take the third quarter of the year on each side, that would be 1973 which has the value of $120,000 and 1975 which has the value of $131,000, and use a simple average. Perhaps $125,500 would be reasonable. If we were to insert this newly created number in place of the one which actually occurred, we will end up with a consistent data base that will show us what our income tax receipts will be in the absence of the crippling strike in this major industry.

If we were to look at the typical seasonal index as compared with specific seasonal index, we will find that 1974 will have a specific index well under the typical third quarter; 1973, 1975, and 1976 will have a specific seasonal index somewhat above the average or typical third quarter. We observe this one data point so far below all the others when we examine the seasonal indexes or when we look at the error terms. Sometimes we find there is a data error.

As stated earlier, the data must be consistent. Let us suppose that we are looking at property tax receipts. Let us also suppose that in the middle of the time period we are studying, a 10% rollback property tax occurred. This rollback would be a 10% reduction in the amount of taxes collected which would be mandated by the state. What we would find is that the first half of the time period we are analyzing, tax receipts might be proceeding at a somewhat constant rate, possibly increasing. All of a sudden, there will be a break. The data points, when graphed, would suddenly drop sharply downward and from that point on perhaps they would begin moving upward slowly through time. If our intention is to forecast property tax receipts for the next few years, before the analysis can be made mathematically, it is necessary that we create a consistent data set. It would be necessary for the analyst to go back to the first years, the years prior to the rollback, and to take every data point for those years and to adjust them downward by 10%. That would be to pretend the rollback had taken place several years before it actually did. Even though the numbers we derive for analysis never existed, since the rollback is in effect at the end of our analysis and is going to continue in the near future, it is critical that all numbers present prior to the rollback be adjusted downward, that is, to pretend the rollback had been in existence through the whole time period. We now have a consistent data set which can be analyzed and from which forecasts can be made.

Other events which can impair our forecast would be when conditions change in some unforeseeable way which drastically alters whatever series of numbers we are studying. For example, consider electric power usage by month. If we were to use 1973, 1974, and 1975, and 1976 figures in an attempt to predict usage for 1978 and 1979, we very well may have come up with very good estimates of what electrical demand would have been had the electricity been available. Due to the lengthy coal strike in the winter of 1978 which forced facilities to curtail electricity and to strongly urge (and even mandate) conservation, it is possible that instead of growth of electrical usage as may have occurred had it been freely available, we actually may have had less electricity used in the early part of 1978 as compared to the same parts of the previous years. Our forecasts which may have been made prior to the coal strike would never have included a "correction factor" to account for the strike. Rather, we would be forecasting what the demand would have been had the electricity been available. Likewise, if we attempt to forecast water usage, and if some prolonged drought requires that conservation be mandated, then all of the figures we forecasted would represent simply what would have been demanded had it been available rather than in fact what is actually going to be used.

Another important source of unexplained variation is the calendar itself. August, 1978 had 23 working days, but August, 1977 had only 21. Some months have 3 paydays, other months have only 2. A month that has 3 paydays this year may have only 2 paydays next year. Many figures that we study are drastically affected, either because of the number of working days involved or the number of recreation days involved. A trend line would tend to reflect some sort of an average, that is, if an average August had 22 working days, that figure would, to some extent, be used as a basis. Since August, 1978 happened to have 23 working days our simulation would tend to slightly underestimate some actual figure for August, 1978 and since August, 1977 had only 21 working days, our simulation would slightly overestimate August, 1977. If August, 1976 had exactly 22 working days our simulated figures might be very close for that particular year.

Over the course of a quarter we can define that the number of working days and the number of recreation days is considerably closer to a constant than for any particular month. Quite often attempting to forecast on the quarterly basis will provide better period by period forecasts. If we are after only a yearly forecast, either data base, monthly or quarterly, might prove reasonably satisfactory. In fact, the Census II method, developed by Julius Shiskin, is similar to decomposition but it makes certain refinements, including working day corrections.

FORECASTING WITH THE X-11 MODEL

We have now decomposed the time series into the four model components - trend, seasonal variations, cyclical variations, and irregular movements. Having performed this decomposition, we can use this information to forecast future values of the time series. It should be noted that the forecasting procedure to be described is based on the fundamental assumption that the time series will behave in the future in the same way as it has behaved in the past. Of course, this assumption is the assumption that is made in all other forecasting procedures and thus it should not be disturbing.

The most common use of the information provided by the decomposition method involves the generation of forecasts using the trend and seasonal factors only. Forecasters usually restrict their attention to these two components because the other components, cyclical variations and irregular movements, cannot be accurately predicted for future time periods. Obviously, irregular movements by their very nature cannot be predicted for future time periods. However, theoretically, the cyclical variations in the time series ought to be predictable. That is, the cyclical factors we have derived could be treated in the same manner as the seasonal factors we derived. The cyclical factors calculated for time periods in the same portion of different cycles could be averaged, just as the seasonal factors for the various quarters in different years were averaged. These averages could then be normalized so that the cyclical factors add to the number of periods in a cycle. This procedure, although theoretically possible to implement, is very difficult if not impossible to use. First, this procedure can be used only if a well defined, repeating cycle of a reasonably constant duration can be identified. In many situations this is not the case. Second, even if such a cycle does exist, it may not be possible to analyze the cycle because of lack of data. In order to obtain accurate cyclical factors, data for several cycles must exist. Since cyclical fluctuations by definition have a duration of from 2 to 10 years or longer, more than 30 years of data may be required for the calculation of cyclical factors which can be used in forecasting. Many times these data requirements cannot be met. Because of these problems the cyclical variations in a time series usually cannot be predicted accurately for future time periods. Thus forecasts are usually made using the trend and seasonal factors only.

Let us consider making forecasts for future observations in our illustrative time series. We obtain a forecast by calculating a trend value for the future period and by subsequently multiplying this trend value by the appropriate seasonal factor for the period. The trend value for our time series is given by the equation:

$$\text{Trend} = 22.61 + 0.59t$$

The seasonal factors for our time series are as follows:

QUARTER 1: 0.46

QUARTER 2: 1.22

QUARTER 3: 0.68

QUARTER 4: 0.64

The trend value for the first quarter in year 5 (period 17) is:

$$22.61 + 0.59(17) = 32.64$$

THE MIDDLETOWN FORECASTING MODEL

The Middletown forecasting model is a modified version of the time series decomposition technique. The model is modified in two ways. First, all calculations for cycle are omitted entirely. Cycles cannot be adequately measured over short time periods. Since the Middletown model was designed to operate favorably with a 4 to 5 year data base, no estimate of the cycle has even been considered.

Even in models where cycle is calculated, the cycle itself is never used in the forecast directly. The second modification of the X-11 model is the most important of the two and does in fact use an estimate of the cycle in an attempt to improve the forecast. To demonstrate the second modification, refer to Exhibit 2, a hypothetical manager's report. Columns 3 and 4 (the first data columns) represent the current period's position (which is the original data base) and the expected period's position (which is our simulation using trend and seasonal index) to try to reproduce the current period. Generally speaking, if the current and the expected positions are relatively close we are probably going to be making good forecasts. The year to date, Column 5, simply accumulates the current period's position from the beginning part of the year, which would be January for monthly data, to the end of the year, which would be December for monthly data. Thus, the 12th, 24th, 36th. The forecast for the first quarter in year 5 is then $(0.46)(32.64) \cong 15$. The forecasts for quarters 2, 3, and 4 in year 5 (periods 18, 19, and 20) are as follows:

$$(1.22)(22.61 + 0.59(18)) \cong 41$$

$$(1.68)(22.61 + 0.59(19)) \cong 57$$

$$(0.64)(22.61 + 0.59(20)) \cong 22$$

Forecasts for subsequent periods can be generated in a similar manner.

Notice in this example the cycle value shown in Table E Column 4 are all relatively close to base 1 (or base 100 if they are converted to index numbers). The largest deviation occurs for year 1 quarter 3 with a cycle value of 1.04, indicating that, due to weather, economic conditions, or fads, we are in a peak of a boom period with demand about 4% above average. However, with all of the numbers so close to base 1 (or 100), our interpretation is that there is virtually no cycle present.

It should also be noted that unless eight to ten (or more) years of data is present, it is very difficult to distinguish between cycle and trend. Our trend line is likely to include the cycle in its entirety if we have only four or five years of data. However, if we do estimate cycle, we can use 5%-6% as a yardstick to determine whether cycle exists. That is, index numbers below 95 or above 105 will indicate the presence of cycle. In this example none of the cyclical factors meet this criteria; thus we have no evidence of cycle.

The year to date figures for the expected column reflect the same information in terms of the forecasted amounts. Generally speaking, the year to date figures (actual vs. expected) should be relatively close together toward the end of the year in contrast with an individual monthly forecast compared with an individual month. This is because a lot of internal smoothing should automatically take place; better forecasts are made for a year than for a specific month. There are many reasons why we might misforecast a particular month, but over the course of a whole year the errors would tend to balance out. For example, if one was studying golf course usage, one June may have been very rainy and less golfers than would

Manager's Report

Rate of growth used is \$196. per period or \$28296. per year

PERIOD	YEAR	CURRENT PERIOD'S POSITION	EXPECTED PERIOD'S POSITION	YEAR TO DATE	EXPECTED YEAR TO DATE	VARIATION BETWEEN ACTUAL AND EXPECTED -FOR PERIOD- DOLLARS	%	VARIATION BETWEEN ACTUAL AND EXPECTED -YEAR TO DATE- DOLLARS	%	UPDATED EXPECT YEAR TO DATE
Jan.	1973	114504.	142140.	114504.	142140.	-27636.	-21.5	-27636.	-21.5	
Feb.	1973	109963.	117223.	224467.	259363.	- 7260.	- 6.4	-34896.	-14.4	
Mar.	1973	142485.	132568.	366952.	391931.	9917.	7.2	-24979.	- 6.6	
Apr.	1973	116809.	104392.	483761.	496323.	12417.	11.2	-12562.	- 2.6	
.	.	.	.	.	.	.	.	.	.	
.	.	.	.	.	.	.	.	.	.	
.	.	.	.	.	.	.	.	.	.	
June	1977	129265.	126088.	798242.	775329.	3177.	2.5	22913.	2.9	
July	1977	141606.	143003.	939848.	918332.	- 1397.	- 1.0	21516.	2.3	
Aug.	1977	120003.	131112.	1059851.	1049443.	-11109.	- 8.8	10408.	1.0	
Sept.	1977	159553.	132624.	1219404.	1182067.	26929.	18.4	37337.	3.1	
Oct.	1977	125675.	158581.	1345079.	1340647.	-32906.	-23.2	4432.	0.3	
Nov.	1977	115315.	115002.	1460394.	1455648.	313.	0.3	4746.	0.3	
Dec.	1977	131957.	128279.	1592351.	1583926.	3678.	2.8	8425.	0.5	
Jan.	1978		155934.		155934.					156761.
Feb.	1978		128581.		284514.					286024.
Mar.	1978		145392.		429906.					432187.
Apr.	1978		114473.		544380.					547267.
May	1978		116521.		660901.					664407.
June	1978		128342.		789243.					793430.
July	1978		145556.		934799.					939757.
Aug.	1978		133449.		1068247.					1073913.
Sept.	1978		134984.		1203231.					1209613.
Oct.	1978		161399.		1364629.					1371867.
Nov.	1978		117043.		1481671.					1489530.
Dec.	1978		130551.		1612222.					1620773.
Jan.	1979		158693.		158693.					159534.
.	.		.		.					.
.	.		.		.					.
.	.		.		.					.

Exhibit 2
Hypothetical Water Usage Example

have been expected were able to play, but over the course of the whole summer golfers somewhat make up for it in July and August. In this example you might not entirely offset the poor June's weather, that is, you might never have the actual figures add up to what is expected. But over the course of a whole season, or year, a lot of smoothing will occur. It's also the case that in some years there are 5 weekends in a month and in other years there are only 4. If the set of numbers being studied was weekend oriented, then you would tend to find that some months would be very difficult to predict unless you were aware of some of these calendar discrepancies. But over the course of a whole year there tends to be a much closer number of weekends per year.

Columns 7 and 8 show the variation period by period between actual and predicted. Column 7 shows the actual number of "dollars", Column 8 shows the amount of error as a percentage. Normally, the numbers in Column 8 should vary with plus and minus signs rather randomly and should be as close to zero as possible. Any error over 10% should be regarded as a data point that is going to require additional investigation. Errors of plus 10% or minus 10% are problems. Columns 9 and 10 show the variation between the year to date figures, actual and expected. For January of each year the numbers are identical to the numbers shown in Columns 7 and 8. Therefore, it's the November and December figures that are of more interest to us. If smoothing takes place, the actual dollar error as well as the percentage error should both be decreasing to very nearly zero. If this error series calculates to be less then 2% consistently, there is evidence of little or no cycle and we have a forecasting tool that should be adequate for our purposes. As we examine the last months of each year one should normally expect the signs to be somewhat random. If they are consistently plus or consistently minus as you get into the December portion of the year, there is some evidence of cycle, whether it be mild or fairly strong.

The modification utilizes techniques to introduce any evidence of cycle into the analysis. In Columns 4 and 6 there are forecasts which utilize the trend line using the detrended data times the correct seasonal factors. Columns 4 and 6, when extended into the future, would be called a raw forecast. Column 4 is a month by month or period by period forecast. Column 6 is the year to date figures. Once again, the yearly forecast is expected to be considerably more accurate than any particular month. Column 11 is called updated expected year to date. Another heading for this column could be revised forecast. How these revisions are made is the second modification of the standard decomposition technique. Turning to Exhibit 2, notice that the last number in Column 10 which is the percentage error for the final December, is listed as +.5, that is, 1/2 of 1%. Notice that not only is this last error term positive, but the preceding several are all positive. A positive percentage error indicates that the actual year to date figures that would be the actual amount of money collected (or number of units or number of gallons, etc.), is higher than our simulated number. This would seem to indicate that our data is in a boom period of a cycle and our trend line is underestimating what is actually happening. As a matter of fact, our trend line together with its seasonal facts should be underestimating reality by approximately that 1/2 of 1% as reported as being the last number in Column 10. The technique works as follows. Where we have a positive error we add 1.00 to the last percentage error, in this example creating the number 1.005. This number will become a multiplicative factor. Every number in the expected period's position or expected year to date can be multiplied by 1.005, that is, the individual forecast will be raised by exactly 1/2 of 1%. The reasoning is as follows. If in fact the data is in an upturn, what we are doing is moving our trend line so that it moves approximately through the last known data points, that is, our trend line together with the seasonal factors are matched at the end. If the data continues the upturn, the actual volume will still be underestimated but by considerably less than if the raw forecasts had been used. If the data levels off, the forecasts will be virtually exactly correct, and if the data had been at a peak and is now beginning to downturn, the updated forecasts might overestimate what will actually happen in the next year or two, but are likely to be as close to actual events as the raw

Manager's Report

PERIOD	YEAR	CURRENT PERIOD'S POSITION	EXPECTED PERIOD'S POSITION	YEAR TO DATE	EXPECTED YEAR TO DATE	VARIATION BETWEEN ACTUAL AND EXPECTED -FOR PERIOD- DOLLARS	VARIATION BETWEEN ACTUAL AND EXPECTED -FOR PERIOD- %	VARIATION BETWEEN ACTUAL AND EXPECTED -YEAR TO DATE- DOLLARS	VARIATION BETWEEN ACTUAL AND EXPECTED -YEAR TO DATE- %	UPDATED EXPECT YEAR TO DATE
Jan.	1975	160583.	144899.	160583.	144899.	15684.	10.3	15684.	10.3	
Feb.	1975	100194.	119495.	260777.	264393.	-19301.	-17.6	- 3616.	- 1.4	
Mar.	1975	169545.	135133.	430322.	399527.	34412.	22.6	30796.	7.4	
Apr.	1975	82753.	106408.	513075.	505935.	-23655.	-25.0	7141.	1.4	
May	1975	123904.	108324.	636979.	614258.	15580.	13.4	22721.	3.6	
.	.	.	.	.	.	.	.	.	.	
.	.	.	.	.	.	.	.	.	.	
.	.	.	.	.	.	.	.	.	.	
Jan.	1977	152191.	150416.	152191.	150416.	1775.	1.2	1775.	1.2	
Feb.	1977	118793.	124038.	270984.	274454.	- 5245.	- 4.3	- 3470.	- 1.3	
Mar.	1977	113695.	140262.	384679.	414716.	-26567.	-20.9	-30037.	- 7.5	
Apr.	1977	131864.	110441.	516543.	525157.	21423.	17.7	- 8614.	- 1.7	
May	1977	106922.	112422.	623465.	637580.	- 5500.	- 5.0	-14115.	- 2.2	
June	1977	110418.	123834.	733883.	761414.	-13416.	-11.5	-27531.	- 3.7	
July	1977	153896.	140451.	887779.	901865.	13445.	9.1	-14086.	- 1.6	
Aug.	1977		134573.		1026438.					1010015.
Sept.	1977		139370.		1165808.					1147155.
Oct.	1977		157707.		1323515.					1302339.
Nov.	1977		99732.		1423247.					1400475.
Dec.	1977		130935.		1554182.					1529315.
Jan.	1978		127328.		127328.					125291.
Feb.	1978		150695.		278032.					273575.
Mar.	1978		146581.		424604.					417810.

Exhibit 3

forecasts are. There are some cases where the raw forecasts could do better than the updated forecasts, but these would be extremely rare.

Now let us suppose that the last percentage error term is negative, such as shown in Exhibit 3. Exhibit 3 shows July, 1977 to have an aggregate error of -1.6. This is -1.6%. If we add 100% to this (1 - .016 as a decimal), we will create a multiplicative factor of .984. In this situation every element of the raw forecasts will be adjusted downward by roughly 1.6% or multiplied by .984. The updated forecasts, therefore, will be lower than the raw forecasts.

To restate the logic, where there are consistent negative percentage errors this is an indication that the simulation is higher than the actual data at that point in time. This could indicate that the data is in the down portion of a cycle where the trend line is somewhat on more of an average. Again, what we want to do is to run the trend line through the last known data points. That is, a somewhat parallel shift of position. Again, if the data continues in this downturn we will still be overestimating what will actually occur, but will be much closer than the raw forecasts. If the data levels off, the forecast will be virtually perfect. If the data was in a trough and begins an upturn we may underestimate what will actually happen but are as likely to have as good a forecast with the update as the raw forecast. In this way the cycle which is most pertinent to the data, that is, the cyclic activity if the last point in time can be directly incorporated into the analysis.

One word of warning is in order here. In order for the update technique to be useful it is necessary that at least half of the year has expired. In Exhibit 2 our correction factor of 1/2 of 1% is a December figure. In Exhibit 3 our correction factor of -1.6% is a July figure. If the last reported error term had been a January, February, March, April, or May figure, the error terms will often not have had a chance to stabilize and are not usable for making an update or a correction to the raw forecast.

APPLIED TIME SERIES ANALYSIS
O.D. Anderson & M.R. Perryman (eds.)

INTEREST RATE FORECASTING AND PORTFOLIO ANALYSIS USING THE STATE SPACE FORECASTING SYSTEM

Dr. Alan V. Cameron
State Space Systems Inc.
2091 Business Center Drive, Suite 100
Irvine, CA. 92715
USA

This paper analyzes the relationships between several interest rate series and other leading indicators and develops efficient portfolios of investments in several of these areas. The series analyzed include -

- Certificate of Deposit Rates
- Eurodollar Deposit Rates
- Commercial Paper Rates
- Foreign Currency Rates of Return
- T-Bill Interest Rates
- T-Bond Rates
- Prime Rates.

1.0 INTRODUCTION

Forecasting of interest rates is an important function for almost all business and government organizations. Because of the changes occurring in the money markets, classical methods are not necessarily sufficient to meet these demanding forecasting problems. The State Space Forecasting approach is able to identify and use lead, lag and feedback relationships between the interest rate series. The models which have been found can significantly improve the accuracy of interest rate forecasts and related portfolio and business decisions which depend on these forecasts.

Analysis of many of these individual interest rate series shows that they are close to "random walk" processes. This result implies that based on the single series alone, the best forecast for all future values is given by the current interest rate and that interest rates are equally likely to rise as fall.

Analysis of groups of interest rate series and related economic series shows however, that there are important relationships between the series. These leading, lagging and feedback relationships which are identified by the State Space program are very useful in improving forecast accuracy and developing efficient portfolios of investments.

These financial application areas are of particular interest because they tie together information and concepts from -

- State Space Forecasting - where forecasts for interacting time series are developed using available historical data, leading indicators and alternative methods. (Reference 1, Cameron & Mehra 1979).

- Modern Portfolio Theory - where portfolios of investments are developed to produce the highest level of return for the least risk using forecasts for each individual investment and the correlation matrix of the forecasts. (Reference 3,Bilson 1981).

- Evaluation of Alternative Forecasting Methods - where the value of additional forecasting analysis and professional forecasting services can be assessed by their impact on portfolio performance.

- Preparation of Composite Forecasts - where forecasts prepared by alternative approaches, e.g. econometric forecasts or forecasts from professional services, can be combined to improve both forecast and portfolio performance.

This paper is divided into several sections. Section 2 discusses forecasts for Short and Long Term Interest Rates. Section 3 discusses Prime Interest Rate Forecasts. Section 4 discusses Actual and Predicted Accuracy for State Space Models. Section 5 discusses State Space Portfolio Analysis methods and results.

2.0 FORECASTING SHORT AND LONG TERM INTEREST RATES

What are the relationships between short and long term interest rates? How will long term interest rates react to Administration policies of reduced growth in Money Supply? In this Section we will show how the State Space Forecasting program can be used to quantify these important relationships.

In his paper, " Efficient Markets, Interest Rates and Monetary Policy", Donald J. Mullineaux (1981 Reference 2) has recently provided an enlightening discussion of these questions from both the conventional economists viewpoint and the " efficient markets" viewpoint.

If the short and long term markets are highly efficient , then the yields available are likely to be close to "random walks". On the other hand if the conventional view is correct there should be significant leading and lagging relationships between money supply and short and long term interest rates.

Using State Space, we can easily analyze the available data and begin to quantify any leads, lags and feedbacks that exist between interest rates and money supply.

The following shows a summary of the results obtained with the State Space Forecasting program in analyzing the relationships between -

- Money Supply (M1B)
- Short Term Interest Rates - 90 Day T-Bills (FYGM3)
- Long Term Interest Rates - 10 year T-Bonds (FYGT10)
- Commercial and Industrial Loans Outstanding (FCLBW)

Logarithms of monthly data for each of these 4 series from the CITIBASE data base were used (from January 1974 to May 1981). The table shows results for the single interest rate series first, then bivariate models with money supply and finally a four series model of the interest rates, money supply and loan demand.

MODEL #	SERIES MODELED	ORDER OF STATE VECTOR	ONE PERIOD AHEAD R-SQUARED TEST	
			ORIGINAL DATA	DIFFERENCED DATA
1	LFYGM3	2	95.2%	24.6%
2	LFYGT10	2	96.8%	21.5%
3	LFM1B	3	99.8%	4.6%
	LFYGM3		96.4%	43.4%
4	LFM1B	3	99.8%	2.7%
	LFYGT10		97.3%	32.9%
5	LFCLBW	4	99.6%	37.3%
	LFM1B		99.8%	7.2%
	LFYGM3		96.7	49.3%
	LFYGT10		97.0%	30.6%

As the table indicates significant leading and lagging relationships exist between these series.

3.0 PRIME RATE FORECASTS

Most corporations and businesses are effected by changes in the prime interest rate. The most direct effect is increased loan costs because loans have interest rates based on the prime rate. For example, SBA loans may be issued with a prime plus 2% interest rate and business credit lines normally pay rates based on the prime. If a business has a $12 million loan based on the prime rate and the rate increases unexpectedly by 2%, the business is faced with unexpected interest charges of $20,000 per month or $240,000 per year.

In this section, we will analyze the relationships between the prime rate, the 90-day T-Bill rate and the money supply. As a first step, using the monthly average prime rate data from the CITIBASE data base for 1/78 to 5/81, the State Space Forecasting program explains 44% of the monthly changes in the (logarithms of) the prime rate, as shown below.

STATE SPACE FORECASTS
V2.2

41 OBSERVATIONS, 1 SERIES

DIFFERENCING PERFORMED

SERIES	REGULAR
LFYPR	1

THE FOLLOWING ARE THE ELEMENTS OF THE STATE VECTOR
LFYPR(T)
LFYPR(T+ 1)

VARIANCE EXPLAINED

(R SQUARED TEST	ORIGINAL DATA	DIFFERENCED DATA
LFYPR	94.8336%	44.6246%

Using monthly T-Bill average interest rates as a leading indicator, a first order autoregressive model developed by the program explains 65% of the month to month variance in the prime rate. Finally using both money supply (FM1B) and T-Bill rates together as leading indicators, the State Space program explains 74% of the month to month changes in the prime rate as summarized below.

R-SQUARED TEST	ORIGINAL DATA	DIFFERENCED DATA
LFM1B	99.380%	19.310%
LFYGM3	94.352%	60.515%
LFYPR	97.583%	74.094%

This is a very significant improvement in forecast accuracy. Instead of being faced with a typical change in the prime rate of 1% per month a business could be now faced with a possible uncertainty of only 0.3% per month in the prime rate.

4.0 COMPARING ACTUAL AND PREDICTED FORECAST ACCURACY FOR STATE SPACE MODELS

One of the primary measures of forecast accuracy calculated by the State Space Forecasting program is the R-squared statistic . This basic statistic calculates the percentage of the variance in the original data which has been explained by the one step ahead forecasts. For example, if the variance in the original data was 100 and the variance in the residual forecast errors was 10, then the R-squared statistic would be 90%. This indicates that 90% of the original variance has been explained by the one-step ahead forecasts and 10% of the variance remains unexplained in the residual forecast errors. Note that this is the usual definition of the R-squared test used in forecasting applications, but it differs to some extent from the R-squared test used in simple regression analysis. Regression typically calculates R-squared using the variance of the "fit residuals", i.e. the residual errors obtained in estimating the current value of one series based on the current & prior values of other series. State Space and other forecasting packages use the one step ahead forecast residuals to determine the R-squared test.

If differencing is performed, the State Space program also calculates an R-Squared statistic based on the variance in the differenced data. In both cases, the R-squared statistic depends on the variance in the one step ahead residual forecast errors calculated over the historical data.

In a complete forecast and model analysis, a more precise measure of forecast accuracy can be obtained by "ex-ante" analysis, i.e. an analysis of forecast accuracy when each forecast uses no information whatsoever from the future. Normally State Space models and R-squared statistics use correlation functions based on all the historical data, so some loss of accuracy can be expected in the "ex-ante" analysis.

To illustrate this situation, a two time series model of the logarithms of Money Supply (M1B) and 90 day T-Bill Interest rates was analyzed using monthly average data from January 1973 to February 1981. In the usual "ex-post" format, the State Space model explains 40.7% of the variance in the one month ahead interest rate changes.

To generate the "ex-ante" forecasts for these series, the model development was repeated 24 times using first 97 terms, then 96, 95,... etc. to generate new models and forecasts for each set of data.

The MAD for the "ex-ante" forecast errors was 0.769 versus 0.705 for the "ex-post" MAD (i.e. a 9% increase). The "ex-ante" forecast MAD was 25% less than the"random walk" MAD of 1.015. The standard deviation of 1.105 for the "ex-ante" forecast errors is higher than the 0.916 for the "ex-post" forecast errors (i.e. a 20% increase), but is still 25% less than the "random walk" forecast errors.

The direction of the interest rate change was correctly predicted in 75% of the 24 months analyzed by both the "ex-ante" and the "ex-post" analysis. These relatively small changes between the "ex-post" and "ex-ante" forecast performance provide additional support for the validity of the models that have been developed.

5.0 STATE SPACE PORTFOLIO ANALYSIS

In this section, we will outline a procedure using State Space Forecasting to choose a combination of investments which maximizes an investment portfolio return and minimizes portfolio variance.

This application is based on the important new approach developed by John Bilson in his recent paper " International Currency Management" presented at the Quebec Forecasting Conference, May 1981 (Reference 3). The procedure and results outlined below extend Bilson's results by incorporating capabilities of the State Space Forecasting program. The procedure has the following steps :

1. Prepare the foreasts and covariance matrix for the forecasts for each of the securities using a joint State Space Forecasting model based on historical data for each security.

2. Calculate the baseline investment portfolio and the portfolio rate of return and standard deviation using the forecasts based on the prior step and modern portfolio theory.

3. Build a composite forecasting system by incorporating leading indicators and alternative independent forecasts in the State Space model.

4. Recalculate the optimal portfolio rate of return and standard deviation using the composite forecasts.

5. Calculate the improvement in portfolio performance provided by the alternative composite forecasts.

The advantages of this approach are -

- the inclusion of lead, lag and feedback effects between the individual investments.

Classical portfolio theory incorporates only the covariance matrix (at lag 0) and does not include leading, lagging and feedback relationships between the investments.

- the inclusion of alternative forecasts prepared by professional services into the forecasting and portfolio analysis.

The State Space Forecasting system is used to build a composite forecast from the independent components of all available data (including alternative forecasts or leading indicators). Traditional restrictions on the weighting of composite forecast coefficients are avoided by using the full State Space model to build the composite forecasts. (Reference 3, Bilson).

- the economic evaluation of the forecasting system performance and the value of additional analysis and information.

This State Space approach allows the assessment of the incremental value of leading indicators and/or alternative forecasts on overall portfolio performance. Isolated improvements in forecast accuracy for one investment may have little value if the investment is already diversified in the portfolio. Information and analysis which reduces the non-diversified risk elements in portfolios are more valuable. (Reference 3, Bilson).

A PORTFOLIO OF MONEY MARKET INVESTMENTS

The following analysis shows two portfolios developed using the State Space approach for monthly T-Bills, Certificates of deposit, Eurodollar Deposit rates and Commercial paper. The first portfolio was developed using only historical data for these five investments. The second portfolio used two leading indicators, (M1B) money supply and loan demand.

The use of the leading indicators improved the rate of return for these investments and also reduced the risk. The ratio of the Expected Return (E) to the standard deviation (S) increased to 26.69 compared to 17.64 for the original portfolio. The proportion of the portfolio funds invested in each security also changed as a result of including the leading indicator information. The expected return from the second portfolio is 15.74% per year.

STATE SPACE PORTFOLIO ANALYSIS

DATE: 08/02/81 TIME : 01:52EDT

RISK FREE INTEREST RATE 0.00%

SECURITY NAME	EXPECTED (E) RETURN	STANDARD DEVIATION(S)	E/S	PORTFOLIO COMPOSITION
FYGM3	14.30%	0.82	17.34	82%
FYCD3M	15.90%	1.39	11.47	-14%
FYUR90	16.80%	1.06	15.87	0%
FYCP60	15.70%	0.97	16.19	32%
	------	----	-----	---
TOTAL	14.52%	0.82	17.64	100%

STATE SPACE PORTFOLIO ANALYSIS

DATE: 08/02/81 TIME: 01:45EDT

RISK FREE INTEREST RATE 0.00%

SECURITY NAME	EXPECTED RETURN(E)	STANDARD DEVIATION(S)	E/S	PORTFOLIO COMPOSITION
FYGM3	15.00%	0.58	25.72	71%
FYCD3M	17.20%	1.18	14.59	-9%
FYUR90	18.00%	0.78	23.05	15%
FYCP60	17.10%	0.71	23.94	23%
	------	----	-----	---
TOTAL	15.74	0.59	26.69	100%

FORECASTS AND PORTFOLIOS FOR FOREIGN CURRENCIES

In this section we will analyze foreign currency data using the State Space forecasting program and we will develop efficient portfolios from combinations of investments in the foreign currencies. (Reference 3, Bilson)

As a first step the following data was available for the analysis - daily data on spot and futures prices for 120 trading days for the foreign currencies -

Canadian dollars	(CDDO)
Swiss Francs	(SFDO)
German Marks	(DMDO)
United Kingdom Pounds	(UKDO)
Japanese Yen	(JYDO)

The baseline portfolio was developed using only the historical data for these five currencies. The portfolio developed is shown below. This portfolio has an expected return of 1.4% for the day and a standard deviation of 1.55 in the rate of return, i.e. making the ratio of return over risk (E/S) equal to 0.90. The portfolio composition is made up of short positions in Canadian dollars and British Pounds and long positions in Swiss Francs, German Marks and Japanese Yen. The proportion of the portfolio invested in each currency is also calculated , e.g. the amount of Canadian dollars short should be approximately twice the long position in Swiss Francs, etc.

STATE SPACE PORTFOLIO ANALYSIS

DATE: 08/01/81 TIME : 20:30 EDT

RISK FREE INTEREST RATE 0.00%

SECURITY NAME	EXPECTED RETURN	STANDARD DEVIATIONS(S)	E/S	PORTFOLIO COMPOSITION
CDDO	0.03%	0.22	0.13	-208%
SFDO	0.50%	0.68	0.74	111%
DMDO	0.50%	0.66	0.76	157%
UKDO	0.30%	0.71	0.42	-106%
JYDO	0.30%	0.47	0.64	146%
TOTAL	1.40%	1.55	0.90	100%

For the next step, one month ahead futures prices for these five currencies were also included in the forecasting model. These futures prices could be regarded as leading indicators for the spot prices. The State Space model which was developed uses the current and lagged correlations between the series to improve the forecasts. Following Bilson's idea, the futures prices could also be regarded as elements of our State Space composite forecasting model for the spot prices.

This second portfolio which is shown below has a reward/risk ratio (E/S) of 1.17 compared to 0.90 for the baseline portfolio, i.e. an improvement of 30%. The expected return for this example is 1.37% versus 1.40% for the baseline, i.e. a small decline due to changes in the forecasts with the leading indicators. The improvements in the forecasts produce reductions in the standard deviations for each of the currencies, except for the Canadian dollar whose risk is unchanged. The changes in the cross correlations for the forecasts and the changes in the forecasts themselves combine to produce the revised portfolio composition shown.

STATE SPACE PORTFOLIO ANLYSIS

DATE: 08/02/81 TIME: 01:35EDT

RISK FREE INTEREST RATE 0.00%

SECURITY NAME	EXPECTED RETURN (E)	STANDARD DEVIATION(S)	E/S	PORTFOLIO COMPOSITION
CDDO	0.03%	0.22	0.13	-177%
SFDO	0.47%	0.66	0.71	60%
DMDO	0.49%	0.63	0.77	47%
UKDO	0.34%	0.69	0.50	-66%
JYDO	0.48%	0.46	1.05	236%
TOTAL	1.37%	1.17	1.17	100%

These results are already good for a speculative portfolio. A risk/reward of 1 implies that 95% of the actual results will be between the range of -1 to +3 for each unit invested in the portfolio.

As a next step for this type of analysis, several potential improvements are possible. Bilson's original work using a classical composite forecasting scheme, reports the results of a portfolio model which has been improved in two important ways -

More than one maturity is allowed in the portfolio to allow for cross- currency and cross-maturity spreading.

Actual professional forecasts were used rather than the spot rate which was the alternative forecast in the earlier model.

With these changes the portfolio performance improves dramatically. Bilson reports that over a three year period the portfolio had an average risk/ reward ratio of approximately 3.6 for each month, accumulating to a total of approx. 130 for the period.

SUMMARY

This paper has shown that there are useful leading indicators for interest rates & other financial time series which can be easily identified using the State Space Forecasting program.

Once these forecasting procedures have been developed, the forecasts & the associated covariance matrices can then be combined in a Modern Portfolio Theory program to produce actual decision strategies for investments. The actual examples described in the paper illustrate the significant potential of these financial applications.

REFERENCES

1. Cameron, A.V. and Mehra, R.K., "State Space Forecasting - A Handbook on Business and Economic Forecasting for Single and Multiple Time Series ", 1980, State Space Systems Inc. 300 pages.

2. Mullineaux, D.J., " Efficient Markets, Interest Rates and Monetary Policy ", Federal Reserve Bank of Philadelphia, Business Review, May-June 1981, pp. 3-10.

3. Bilson, John, F.O. " International Currency Management, " International Forecasting Symposium, Quebec, Canada, May 1981.

4. Mehra, Raman K. and Lainiotis, D.G. eds., "System Identification - Advances & Case Studies", Academic Press 1976.

5. Cameron, A.V. and Mehra, R.K., "A Multidimensional Identification and Forecasting Technique Using State Space Models", TIMS/ORSA Miami Conference, November 1976.

6. Cameron, A.V., "State Space Forecasts & Models for GNP & Money Supply", Time Series Analysis, Anderson, O.D. & Perryman, M.R., eds., North-Holland, 1981.

APPLIED TIME SERIES ANALYSIS
O.D. Anderson & M.R. Perryman (eds.)

DEFENSE EXPENDITURE AND ECONOMIC GROWTH IN DEVELOPING COUNTRIES: A TEMPORAL CROSS-SECTIONAL ANALYSIS

M.W. Luke Chan
McMaster University and Institute for Policy Analysis, University of Toronto

Cheng Hsiao
Institute for Policy Analysis, University of Toronto, and Princeton University

C.W. Kenneth Keng
Ontario Hydro and University of Toronto

The role of defense expenditure in economic growth is a subject of intense debate by economists. Conclusions of empirical studies based on conventional econometric modelling methodology depends critically on the ways in which the macro-econometric models are specified. In this paper, we rely on the Wiener-Granger notion of causality to investigate the causal relationship between economic growth, investment, and defense expenditures. Both conventional methods and a modified Akaike's Final Prediction Error criterion together with hypothesis testing and diagnostic checkings are used for system identifications. Several empirical results are reported and discussed.

INTRODUCTION

It is usually assumed by economists that defense expenditure can exert two different effects on economic growth. Defense expenditure by developed countries reduces the resources available for investment and thus causes a reduction in economic growth. On the other hand, defense expenditure by developing countries may have positive contributions towards economic growth. It is argued that defense expenditure can provide various useful inputs towards the civilian economies. These include manpower training and the supply of many dual use infrastructures, such as transportation and communication.

Until recently, the study of defense expenditure and economic growth, despite its obvious importance, has received little formal attention by economists. Traditionally, the analysis of defense expenditure and its influences centers around the construction of a macroeconomic model and examines the defense expenditure multiplier (see, for example, Fishman (1969), and Lynk (1978)). It is quite obvious that the specification of the model will somehow dictate the outcome of the analysis. Benoit (1973, 1978), using quantile analysis, correlation analysis, and regression analysis, found significant positive links between defense expenditures and economic growth for the 44 developing countries included in his study. On the other hand, Smith (1977) found that advanced capitalist countries with high military expenditures have much lower investment, lower growth, and higher rates of unemployment.

The policy implications associated with the empirical studies, especially Benoit's study, for the developing countries are obvious and serious. However, correlation analysis cannot be used to justify the existence of causality and the conclusion of any regression analysis is valid only when the model is correctly specified. It is felt that an empirical analysis of defense expenditure and economic growth which does not rely on a priori constraints will help to shed some light on this very important issue. Using the data published by Benoit

(1973) for developing countries, and Granger's (1969) definition of causality, we will develop a more sophisticated lead-lag analysis to carry out our study.

In the second section of this paper, we will discuss the basic concepts of causality. Section three is devoted to the formulations and discussions of the cross-sectional time series model for the analysis of causality between defense expenditure and economic growth. The last section is reserved for concluding remarks.

BASIC CONCEPTS

Analysis of economic data has been approached with two different philosphies, that of time series analysis and that of classical econometrics. If the endogenous and exogenous variables have been distinguished, and if prior information is available with which to identify structural parameters and the dynamic properties of the structural equations, we can represent the economy by a dynamic simultaneous equations model. But if theories are inexact, the time series techniques have the advantage of avoiding spurious and false restrictions. For instance, Sims (1980) has persuasively argued the advantages of treating all variables as endogenous and estimating unconstrained vector autoregressive (AR) models in the first stage; then formulating and testing hypotheses with economic content in the second stage. In this paper we use the time series approach to analyze the impact of defense expenditure on economic growth.

Causality is a central concept in the discussion of economic laws. In experimental science, such as physics or biology, the concept has a clear meaning. For instance, Basmann (1963) presents the definition of causality as: "Assume that (the) mechanism under investigation can be isolated from all systematic, i.e., non-random external influences; assume that the mechanism can be started over repeatedly from any definite initial condition. If, every time the mechanism is started up from approximately the same initial condition, it tends to run through approximately the same sequence of events, then the mechanism is said to be causal."

However, in economics, we only observe certain facts and cannot perform controlled experiments. To make inferences based on observed data, economists often hypothesize a model which generates the observations of real world data. In this context, causality is often defined in terms of the characteristics of a model. For instance, Simon (1952) essentially argues that causality is an asymmetrical relation between certain variables, or subsets of variables, in a self-contained structure. Zellner (1978) takes up the philosophical definition (Feigle (1953)) and argues that causality should be defined as "predictability according to well thought of economic laws". While we agree that a proper definition of causality should be in terms of economic laws, there is a practical difficulty in applying this definition to model construction with time series techniques. The definition essentially assumes that in order to establish a causal ordering we must have a priori knowledge. But an empirical model building strategy such as that suggested by Sims (1980) is recommended when economists disagree about the set of laws governing economic relationships. If a model is specified according to a set of incorrect laws, the estimation is biased, and the model may thereby become useless as a framework within which to do formal statistical tests. In this sense, a definition of causality not relying on economic theory may provide useful insights to many problems.

Granger (1969) has suggested a definition of causality that makes no mention of economic laws (also see Caines and Chan (1975), Pierce and Haugh (1977), Sims (1973)). It is based on the stochastic nature of the variables. Its central feature is the direction of the flow of time. Granger's definition is at variance with the philosophical definition in certain important aspects (Zellner

(1978)). In certain cases it may even obscure conventional causal ordering (Sims (1977)). However, Sims has also demonstrated that these possibilities exist only under special, rather restrictive assumptions. Geweke (1978) has shown that in the complete dynamic simultaneous equation model exogenous variables cause endogenous variables in the sense of Granger. It does seem that there will be a large class of applications where the causal ordering arising from the most plausible behavioral structure will be consistent with a Granger ordering. Therefore, we shall adopt Granger's definition in this paper despite the disagreement about its appropriateness. However, we must caution readers that this definition is not infallable. The statistical analysis based on this concept should be used as a complement to, and not substitute for, economic analysis.

In developing his definition of causality, Granger (1969) considers a stationary stochastic process A_t. Let $\{x_t, y_t\}$ be components of A_t. Let $\bar{A}_t$, $\bar{X}_t$, $\bar{Y}_t$ represent the set of past values of A_t, x_t and y_t before time t, respectively. Let $A_t - x_t$ as the set of elements in A_t without the element x_t. The set $\bar{A}_t - \bar{X}_t$ is defined analoguosly to $\bar{A}_t$. Denote by $\sigma^2(y_t|\bar{A}_t)$ the minimum mean square linear prediction error of y_t given information set $\bar{A}_t$. We then have the following definitions.

Definition 1 (Causality): If $\sigma^2(y_t|\bar{A}_t) < \sigma^2(y_t|\bar{A}_t-\bar{X}_t)$, i.e., the prediction of y using past x is more accurate than without using past x, in the mean square error sense, we say that x causes y, denoted by x => y.

Definition 2 (Feedback): If $\sigma^2(y_t|\bar{A}_t) < \sigma^2(y_t|\bar{A}_t-\bar{X}_t)$ and $\sigma^2(x_t|\bar{A}_t) < \sigma^2(x_t|\bar{A}_t-\bar{Y}_t)$, we say that feedback occurs, denoted by x <=> y.

Granger (1969) also introduces the notion of instantaneous causality. In this paper we shall ignore it since, as discussed in Caines (1976), Hsiao (1979b), etc., it is more natural to treat the notion of instantaneous causality as contemporaneous correlation.

Given these definitions we may write down the causal ordering of economic variables in terms of constraints on the time series model. We know that under fairly general conditions (Masani (1966)) a regular full rank stationary process, say $\{y, x, z\}$ possesses an autoregressive (AR) representation,

$$(1)\quad \begin{bmatrix} y_t \\ x_t \\ z_t \end{bmatrix} = \begin{bmatrix} \psi_{11}(L) & \psi_{12}(L) & \psi_{13}(L) \\ \psi_{21}(L) & \psi_{22}(L) & \psi_{23}(L) \\ \psi_{31}(L) & \psi_{32}(L) & \psi_{33}(L) \end{bmatrix} \begin{bmatrix} y_t \\ x_t \\ z_t \end{bmatrix} + \begin{bmatrix} u_t \\ v_t \\ w_t \end{bmatrix}$$

In equation (1), L denotes the lag operator, $Ly_t = y_{t-1}$. Typical elements of $\psi_{jk}(L)$ is given by $\sum_{\ell=1}^{\infty} \psi_{jk\ell} L^{\ell}$. The $\{u_t, v_t, w_t\}$ are white noice innovations with constant variance covariance matrix Ω. If z does not cause y, $\psi_{13\ell} = 0$ for all ℓ (e.g. see Skoog (1976)).

Thus, we may identify the causal relation among economic variables by fitting a vector AR model for these variables, and checking if $\psi_{jk}(L) \equiv 0$. However, Hsiao (1979 a,b) has shown that the test of $\psi_{jk}(L) \equiv 0$ is quite sensitive to the order of lags chosen, yet most applied work arbitrarily selected the order of lags for $\psi_{jk}(L)$. Furthermore, shortages of degree of freedom and multicollinearity in economic time series data often make it difficult to interpret the statistical results (e.g. Ando (1977), Klein (1977)). However, if we have temproal cross sectional data, we will have many more degrees of freedom than single time series observations for an agent. This will also reduce the collinearity problem, and thus make the statistical inferences more robust. In the following section we shall explore the feasibility and robustness of testing causal relations using temporal cross-sectional data.

A TIME SERIES MODEL OF GROSS DOMESTIC PRODUCT, INVESTMENT AND DEFENSE EXPENDITURE

In this section we will fit a time series model of the civilian gross domestic product (GDP), investment (INV) and defense expanditure (DEF) to identify the impact of defense expenditure on economic growth. The data we used in this analysis is based on the published information gathered by Benoit (1973, pp. 265-310). The original data set contains 44 developing countries for the period of 1950 to 1967. Among these 44 countries we only use those 22 countries which have complete annual observations from 1950 to 1967.

We first take the first difference of the logarithm of each variable. There are two reasons for doing so. One is that we are interested in the relationship between the rate of growth of each variable. The other is that the data contain 22 different countries. It is difficult to imagine that every country is a random draw from the same universe. If the difference in each country's behavior can be attributed to some unobserved characteristics which stay constant over time (e.g., see Anderson and Hsiao (1981 a,b), Balesfra and Nerlove (1966), Mundlak (1978)), we eliminate this country specific effect by differencing the data.

One major difficulty in fitting (1) lies in the decision about the maximum order of lags in $\psi_{jk}(L)$. There have been extensive investigations of these topics which are closely related to this subject (e.g. Akaike (1969, 70), Anderson (1963), Parzen (1976)). Under the assumption that the first differenced temporal cross-sectional data are repeated measurements from the same autoregressive vectored-valued process, we can derive the asymptotic normality property of the ordinary least squares estimator (e.g. see Anderson (1978). Therefore, with straightforward modifications we can still use those model selection criteria which were proposed for single time series.

In this paper we use two procedures to identify a vector AR model. One is the conventional approach. We first determine the order of the unrestrictd AR model, then test for $\psi_{jk} = 0$ in the second stage. The other is a procedure suggested by Hsiao (1979 a,b) using the concept of the final prediction error (Akaike (1969)). This procedure combines the determination of the order of an autoregressive form with some second stage hypothese testing. We first report results based on the standard procedure. Suppose the maximum order of dependence is Q. Here we let Q = 5 because the maximum allowable order for using the FIML option in TSP78 computer package at the University of Toronto is a fifth order one and also because previous empirical studies indicated that the effect of these variables on each other after five years was negligible. Let H_q stand for the hypothesis that $\psi_Q = 0, \ldots, \psi_{q+1} = 0$, $\psi_q \neq 0$. The likelihood ratio test statistics for H_5 is 5.96, for H_4 is 12.00, and for H_3 is 21.66. Each H_q has 9 degrees of freedom. The critical value of a chi square distribution with 9 degrees of freedom for the 10% significance level is 14.68 and for the 1% level is 21.67. We therefore choose q=3 as the order of the unconstrained vector autoregressive process.

To proceed to test the causal relationship among the variables, we have to test (i) $\psi_{13}(L)=0$; (ii) $\psi_{12}(L)=\psi_{13}(L)=0$; (iii) $\psi_{13}(L)=\psi_{23}(L)=0$; (iv) $\psi_{12}(L)=\psi_{13}(L) = \psi_{23}(L)=0$; (v) $\psi_{13}(L)=\psi_{23}(L)=\psi_{32}(L)=0$ and so on. Treating an unconstrained third order on AR model as the maintained hypothesis, we may either test each of these hypotheses against the maintained hypothesis or we may perform the following sequential test:

$H_1 : \psi_{13}(L) \neq 0;$

$H_2 : \psi_{13}(L) = 0, \quad \psi_{12}(L) \neq 0;$
$H_3 : \psi_{13}(L) = 0, \quad \psi_{23}(L) \neq 0;$
$H_4 : \psi_{13}(L) = \psi_{23}(L) = 0, \quad \psi_{12}(L) \neq 0;$
$H_5 : \psi_{13}(L) = \psi_{23}(L) = 0, \quad \psi_{32}(L) \neq 0;$

.
.
.

The advantage of testing each of the hypotheses against the maintained hypothesis is that the size of the test of each of the hypothees is the same. The diadvantage is that the otherwise "significant coefficients" (or insignificant coefficients) might be contaminated by other insignificant coefficients (or significant coefficients), hence making it difficult to reject (or accept) the null hypothesis.

The advantage of testing each hypothesis sequentially by treating the previous null hypothesis as the new maintained hypothesis is that the test is more sharply focused. The disadvantage is that the test of ψ_{jk} might either be accepted or rejected depending on the order in which it is tested. To balance the desirability of the sensitivity to a non-zero constraint without favoring a particular $\psi_{jk} = 0$, we let $y = \nabla \ln$ GDP, $x = \nabla \ln$ INV, $z = \nabla \ln$ DEF and use the FIML method to estimate models with different constraints. The results are listed in Table 1.

Table 1: The log-likelihood Value of Various 3rd Order Autoregressive Models

Model	Log-Likelihood Value	Model	Log-Likelihood Value
1. all $_{jk} \neq 0$	461.449	21. $\psi_{21}=\psi_{31}=\psi_{32}=0$	457.406
2. $\psi_{13}=0$	458.84	22. $\psi_{21}=\psi_{31}=\psi_{13}=0$	455.766
3. $\psi_{23}=0$	458.82	23. $\psi_{21}=\psi_{31}=\psi_{23}=0$	455.710
4. $\psi_{12}=0$	455.523***	24. $\psi_{21}=\psi_{31}=\psi_{12}=0$	450.586
5. $\psi_{21}=0$	459.015	25. $\psi_{13}=\psi_{23}=\psi_{21}=0$	455.123
6. $\psi_{31}=0$	460.407	26. $\psi_{13}=\psi_{23}=\psi_{32}=0$	456.121
7. $\psi_{32}=0$	460.407	27. $\psi_{13}=\psi_{23}=\psi_{31}=0$	455.766
8. $\psi_{13}=\psi_{23}=0$	457.146	28. $\psi_{12}=\psi_{13}=\psi_{23}=\psi_{32}=0$	450.577***
9. $\psi_{12}=\psi_{32}=0$	454.212*	29. $\psi_{12}=\psi_{13}=\psi_{21}=\psi_{31}=0$	448.959**
10. $\psi_{21}=\psi_{31}=0$	457.927	30. $\psi_{13}=\psi_{21}=\psi_{31}=\psi_{32}=0$	455.241
11. $\psi_{12}=\psi_{13}=0$	453.580**	31. $\psi_{13}=\psi_{23}=\psi_{31}=\psi_{32}=0$	455.625
12. $\psi_{21}=\psi_{23}=0$	456.789	32. $\psi_{21}=\psi_{23}=\psi_{31}=\psi_{32}=0$	455.188
13. $\psi_{31}=\psi_{32}=0$	459.866	33. $\psi_{12}=\psi_{21}=\psi_{31}=\psi_{32}=0$	449.911**
14. $\psi_{12}=\psi_{13}=\psi_{23}=0$	451.886**	34. $\psi_{12}=\psi_{13}=\psi_{24}=\psi_{31}=\psi_{32}=0$	450.074*
15. $\psi_{12}=\psi_{21}=\psi_{31}=0$	450.586**	35. $\psi_{12}=\psi_{13}=\psi_{21}=\psi_{31}=\psi_{32}=0$	448.269**
16. $\psi_{12}=\psi_{31}=\psi_{32}=0$	453.691*	36. $\psi_{13}=\psi_{21}=\psi_{23}=\psi_{31}=\psi_{32}=0$	453.512
17. $\psi_{12}=\psi_{32}=\psi_{21}=0$	450.458**	37. $\psi_{12}=\psi_{13}=\psi_{21}=\psi_{23}=\psi_{31}=\psi_{32}=0$	446.656**
18. $\psi_{12}=\psi_{23}=\psi_{31}=0$	452.135**		
19. $\psi_{12}=\psi_{13}=\psi_{32}=0$	452.268**		
20. $\psi_{12}=\psi_{32}=\psi_{23}=0$	451.793**		

* significant at 10% level
** significant at 5% level — against Model 1
*** significant at 1% level

Among tests with just one $\psi_{jk} = 0$, the only significant case is $\psi_{12} = 0$ (model

4). Among two or more combinations of $\psi_{jk} = 0$, we see that those combinations which are significant are the ones which contain $\psi_{12} = 0$. Either comparing models 21 (among those models with three $\psi_{jk} = 0$, it has the highest likelihood value), 36, and 37 against model 1 or comparing them sequentially, they all demonstrate that the only causal factor for economic growth in this model is investment expenditure. Defense expenditure does not cause gross domestic product or investment. Nor is defense expenditure caused by either.

The above procedure restricts the maximum lag order of each variable to be identical. There is no economic justification for making this assumption. To check whether this arbitrary assumption would induce bias in our results, we allow each variable to enter into each equation with a different lag length. We use Akaike's (1969) final prediction error (FPE) criterion to determine the order of lags for each ψ_{jk} in each equation.

The FPE is defined as the (asymptotic) mean square prediction error,

$$(2) \quad \text{FPE of } \hat{y}_t = E(y_t - \hat{y}_t)^2$$

where $\hat{y}_t$ is the predictor of y_t,

$$(3) \quad \hat{y}_t = \hat{\psi}_{11}^m(L)y_t + \hat{\psi}_{12}^n(L)x_t + \hat{\psi}_{13}^r(L)z_t + \hat{a}$$

The superscripts m, n, and r denote the order of lags in $\psi_{11}(L)$, $\psi_{12}(L)$, and $\psi_{13}(L)$. $\hat{\psi}_{11}^m(L)$, $\hat{\psi}_{12}^n(L)$, $\hat{\psi}_{13}^r(L)$, and $\hat{a}$ are least squares estimates obtained when we treat the observations from $-Q + 1$ to 0 as fixed $\{T: t=-M+1,\ldots, 0,1,\ldots T\}$; $m, n, r \leq Q$. Akaike [1969] defines the estimate of FPE in this case as

$$(4) \quad \text{FPE}_y(m,n,r) = \frac{T+m+n+r-1}{T-m-n-r-1} \; \frac{\sum_{t=1}^{T} (y_t-\hat{y}_t)^2}{T}.$$

The model chosen for y is the one which yields the smallest FPE.

The use of the FPE criterion to identify a model is equivalent to a combination of first stage model fitting with some second stage hypotheses testing. Shibata (1976) has derived the asymptotic distribution and risks of Akaike's statistic. It is noted that the probability of selecting too low an order using Akaike's method approaches zero very quickly as sample size increases. The probability of selecting too high an order does not approach zero, but it dies out fairly quickly. This inconsistency of Akaike's criterion can be eliminated by modifying the criterion with a more severe penalty for overparameterization (e.g., see Bhansali and Downham (1977)). However, as their Monte Carlo studies illustrate, although a consistent method can be derived by generalizing Akaike's FPE criterion, the probability of choosing too low an order is not insignificant in finite samples. We would think that the cost of selecting too high an order is smaller than the cost of selecting too low an order in this kind of analysis. Also the use of the criterion fits in nicely with Granger's (1969) definition introduced above and seem to have good properties in finite samples (e.g., see Geweke and Messe (1979), Quandt and Trussell (1979)). Thus, we shall use the FPE criterion as a fundamental criterion in selecting the order of an AR process.

The FPE criterion is mainly developed for the study of a single sequence of observations. Here, we have temporal cross-sectional data. As mentioned before, under the assumption that they are repeated measurements from the same autoregressive vector-valued process, it is a straight-forward manner to show that the FPE formula is now computed by

$$(5) \quad \text{FPE}_y(m,n,r) = \frac{NT+m+n+r+1}{NT-m-n-r-1} \; \frac{\sum_{i=1}^{N} \sum_{t=1}^{T} (y_{it}-\hat{y}_{it})^2}{NT},$$

where N denotes the number of cross-sectional units, and T denotes the number of time series observations used to obtain the least squares estimate of

$$\hat{y}_{it} = \hat{\psi}^{m}_{11}(L)y_{it} + \hat{\psi}^{n}_{12}(L)x_{it} + \hat{\psi}_{13}(L)z_{it} + \hat{a}, \quad i=1,\ldots,N \quad t=1,\ldots,T, \tag{6}$$

by treating the first Q observations of each cross-section unit as fixed constants.

Once we succeed in modifying the FPE formula, the identification of a time series model based on this criterion is straightforward. If Q is the maximum order of $\psi_{jk}(L)$, then one way to select the order of $\psi_{jk}(L)$ is to let the order of each $\psi_{jk}(L)$ vary between 0 and Q. For a system of p variables, this means there will be $(Q+1)^p$ combinations of $\psi_{jk}(L)$ for the j-th equation. In the case where p=3 and Q=15, we will have to compute 4096 FPE's for each of these three equations. To reduce the computation burden to less than (usually substantially than) $[p+(p-1)](Q+1)$ (in this case 80), we use a sequential procedure suggested by Hsiao (1979a) to identify the system.[1]

The FPE's obtained from treating each variable as a one-dimensional AR process are presented in Table 2 with maximum order of lag Q assumed to be equal to 7. The smallest FPE's for GDP, INV, and DEP are 5, 3 and 1, respectively. We next treat each of these variables as output variables and use other variables as input variables. Holding the order of the autoregressive operator of the controlled variable to the one specified above, we compute FPE's of the controlled variable by varying the order of lags of the manipulated variable from 1 to 7. The orders which give the smallest FPE's are presented in Table 3.

Table 2: FPE's of One-dimensional Autoregressive Processes for GDP, DEF and INV*

Time Order of Lags	FPE of ∇GDP x 10^{-2}	FPE of ∇INV x 10^{-1}	FPE of ∇DEF x 10^{-1}
1	0.1365	0.2054	0.4465
2	0.1363	0.2079	0.4506
3	0.1377	0.2008	0.4562
4	0.1336	0.2026	0.4620
5	0.1326	0.2037	0.4678
6	0.1331	0.2015	0.4611
7	0.1337	0.2041	0.4649

*Note: ∇GDP, ∇INV, ∇DEF are defined as the first difference of the logarithm of real civilian gross domestic product, real defense expenditure, and real investment expenditure, respectively.

We then choose the specification of the output variable as the one with the smallest FPE and add a second input variable to the equation. The results are presented in Table 4.

After checking the omitted variable effect as described in Hsiao (1979a), we tentatively choose the following specification:

$$\begin{bmatrix} (1-L)\log GDP \\ (1-L)\log INV \\ (1-L)\log DEF \end{bmatrix} = \begin{bmatrix} \psi^{5}_{11}(L) & \psi^{3}_{12}(L) & \psi^{1}_{13}(L) \\ 0 & \psi^{3}_{22}(L) & 0 \\ 0 & 0 & {}_{33}(L) \end{bmatrix} \begin{bmatrix} (1-L)\log GDP \\ (1-L)\log INV \\ (1-L)\log DEF \end{bmatrix} + \begin{bmatrix} u \\ v \\ w \end{bmatrix} \tag{7}$$

Table 3: The Optimal Lag Order of a Single Manipulated Variable and the FPE's of the Controlled Variable

Controlled Variable*	Manipulated Variable	Optimal Order of Manipulated Variable	FPE x 10^{-2}
∇GDP(5)	∇DEF	1	0.1331
∇GDP(5)	∇INV	3	0.1313
∇DEF(1)	∇GDP	5	4.6157
∇INV(3)	∇GDP	2	0.2009

*Note: Numbers in brackets are the orders of autoregressive operators in the controlled variables.

Table 4: The Optimal Order of Second Manipulated Variable and the FPE of the Controlled Variable

Controlled Variables*	First manipulated variable*	Second manipulated variable	Optimal Order of 2nd manipulated variable	FPEXx10^{-2}
∇GDP(5)	∇INV(3)	∇DEF	1	0.1312
∇DEF(3)	∇GDP(0)	∇INV	1	4.5176
∇INV(3)	∇GDP(0)	∇DEF	1	0.2019

*Numbers in brackets indicate the orders of lags.

Specification (7) would imply that both investment and defense expenditure are causing gross domestic product in the Weiner-Granger sense. However, a likelihood ratio test of (7) against the specification:

$$(8)\quad \begin{bmatrix} (1-L)\log GDP \\ (1-L)\log INV \\ (1-L)\log DEF \end{bmatrix} = \begin{bmatrix} \psi_1^5(L) & \psi_{12}^3(L) & 0 \\ 0 & \psi_{22}^3(L) & 0 \\ 0 & 0 & \psi_{33}^1(L) \end{bmatrix} \begin{bmatrix} (1-L)\log GDP \\ (1-L)\log INV \\ (1-L)\log DEF \end{bmatrix} + \begin{bmatrix} u \\ v \\ w \end{bmatrix}$$

has a chi-square value of 1.014 with one degree of freedom. This statistic is not significant at 25% level. Therefore, instead of choosing (7) as the maintained hypothesis, we choose (8) as the maintained hypothesis.

To further check the adequacy of specification (8), a sequence of likelihood ratio tests are carried out by deliberately over-fitting and under-fitting (8). The results are reported in Tables 5 and 6. They do not seem to indicate any serious problem with our specification. We therefore choose (8) as a proper AR model for the gross domestic product, investment, and defense expenditure. Based on this specification, we find that changes in investment cause changes in gross domestic product, but changes in defense expenditure do not cause changes in investment or changes in domestic product.

Table 5: Likelihood Ratio Tests of (8) Against Overfitted Models

Model Order	1	2	3	4	5	6	7	8
11	5	5	5	5	5	5	5	5
12	3	3	3	3	3	3	3	3
13	1	3	0	0	0	0	0	2
21	0	0	2	0	0	0	2	2
22	3	3	3	3	3	3	3	3
23	0	0	0	2	0	0	2	2
31	0	0	0	0	2	0	0	2
32	0	0	0	0	0	2	0	2
33	1	1	1	1	1	1	1	1
Degree of Freedom	1	3	2	2	2	2	4	10
Likelihood Ratio Statistic	1.014	1.121	4.166	1.210	0.838	0.347	5.050	8.477

Table 6: Likelihood Ratio Tests of (8) Against Underfitted Model

Model Order	1	2	3	4	5	6	7	8
11	5	5	5	5	5	4	3	3
12	3	3	2	1	0	3	2	1
13	0	0	0	0	0	0	0	0
21	0	0	0	0	0	0	0	0
22	2	1	3	3	3	1	2	1
23	0	0	0	0	0	0	0	0
31	0	0	0	0	0	0	0	0
32	0	0	0	0	0	0	0	0
33	1	1	1	1	1	1	1	1
Degree of Freedom	1	2	1	2	3	3	4	6
Likelihood Ratio Statistic	7.758*	7.872**	7.736*	7.759**	10.177**	10.637**	19.711*	20.069*

* significant at 2.5% level
** significant at 1 % level

CONCLUSION

As Sims (1980) has remarked, estimation of economic models as unrestricted reduced forms by treating all variables as endogenous provides an opportunity to drop the burden of supposed a priori knowledge econometricians have been used to caryying. In this paper, we follow this idea by constructing a time series model for gross domestic product, investment and defense expenditure using temporal cross-sectional data. Contrary to what Benoit (1973, 1978) has observed, we find that defense expenditure has no impact on economic growth. This, of course, casts some doubt on Benoit's policy implication (1978, p. 280) of, "... deliberately utilize military forces and equipment for secondary uses on behalf of civilian economic objectives would also appear to have important potential economic benefits,...".

In this paper we have followed the literature on defense expenditure and economic growth in assuming that the temporal cross-sectional observations are repeated measurements of the same underlying unknown structure. However, the pooling of time series and cross-sectional data involves a lot of complicated issues (see, for example, Anderson and Hsiao (1980)). For this reason, results presented here may in some ways be considered as preliminary. The exploration of causality testing using panel data which takes account of individual specific and time specific effects is a non-trivial one and should be addressed in future work.

FOOTNOTES

*This work was supported by National Science Foundation grant SES80-07576 and Social Sciences and Humanities Research Council of Canada grant 410-80-0080.

[1]We modify the procedure described in step 5' of Hsiao (1979a) by letting ψ_{jk} vary between 0 and Q rather than between 0 and the order chosen by the step-wise procedure because the omitted variable effect can go either direction.

REFERENCES

Akaike, H. (1969), "Statistical Predictor Identification", Annals of the Institute of Statistical Mathematics, 21, 203-217.

Akaike, H. (1970), "Autoregressive Model Fitting for Control", Annals of the Institute of Statistical Mathematics, 22, 163-180.

Anderson, T.W. (1963), "Determination of the Order of Dependence in Normally Distributed Time Series", Time Series Analysis (ed. M. Rosenblatt), New York, John wiley, 425-446.

Anderson, T.W. (1978), "Repeated Measurements on Autoregressive Processes", Journal of the American Statistical Association, 70, 371-378.

Anderson, T.W. and C. Hsiao (1981a), "Estimation of Dynamic Models with Error Components", Journal of the American Statistical Association, 76, 598-606.

Anderson, T.W. and C. Hsiao (1981b), "Formulation and Estimation of Dynamic Models Using Panel Data", forthcoming in the Journal of Econometrics.

Balestra, P. and M. Nerlove (1966), "Pooling Cross-Section and Time Series Data in the Estimation of a Dynamic Model: The Demand for Natural Gas", Econometrica, 34, 585-612.

Benoit, Emile (1973), Defense and Economic Growth in Developing Countries, New York, Lexington Books.

Benoit, Emile (1978), "Growth and Defense in Developing Countries", Economic Development and Cultural Change, July, pp. 271-280.

Bhansali, R.J. and D.Y. Downham (1977), "Some Properties of the Order of an Autoregressive Model Selected by a Generalization of Akaike's FPE Criterion", Biometrika, 64, 547-51.

Caines, P.E. (1976), "Weak and Strong Feedback Free Processes", IEEE Trans. Automatic Control, AC-21, 1137-1139.

Caines, P.E. and C.W. Chan (1975), "Feedback between Stationary Stochastic Process", IEEE Trans. Automatic Control, AC-20, 498-508.

Feigl, H. (1953), "Notes on Causality", in H. Feigl and M. Brodbeck (editors), Readings in the Philosophy of Science, New York: Appleton-Century-Crofts, Inc., 408-418.

Fishman, L. (1962), "A Note of Disarmament and Effective Demand" Journal of Political Economy, LXX No. 2, pp. 183-186.

Geweke, J. (1978), "Testing the Exogeneity Specification in the Complete Dynamic Simultaneous Equation Model", Journal of Econometrics, 7, 163-186.

Geweke, J. and R. Messe (1979), "Estimating Distributed Lag of Unknown Order", paper presented in Econometric Society 1979 North American Summer Meeting, Montreal.

Granger, C.W.J. (1969), "Investigating Causal Relations by Econometric Models and Cross-Spectral Methods", Econometrica, 37, 424-38.

Hsiao, C. (1979a), "Autoregressive Modelling of Canadian Money and Income Data", Journal of the American Statistical Association, 74, 553-560.

Hsiao, C. (1979b), "Causality Tests in Econometrics", Journal of Economic Dynamics and Control, 1, 321-346.

Lynk, Edward L. (1979), "Defense Expenditure: Changes and Their Implications", Institute of Social and Economic Research, Department of Economics and Related Studies, University of York Discussion Paper 25.

Masani, P. (1966), "Recent Trends in Multivariate Prediction Theory", Multivariate Analysis I (ed. P.R. Krishnaiah), New York, Academic Press Inc. 351-382.

McElroy, F.W. (1978), "A Simple Method of Causal Ordering", International Economic Review, 19, 1-24.

Mundlak, Y. (1978), "On the Pooling of Time Series and Cross-Section Data", Econometrica, 46, 69-81.

Parzen, E. (1976), "Multiple Time Series: Determining the Order of Approximating Autoregressive Schemes", in Multivariate AnalysisVI, (P.R. Krishnaiah, Ed.), Academic Press, New York.

Pierce, D.A., and L.d. Haugh (1977), "Causality in Temporal Systems: Characterizations and Survey", Journal of Econometrics, 5, 265-293.

Quandt, R.E. and T.J. Trussell (1979), "Some Empirical Evidence on Model Selection Rules", Research memo. no. 248, Econometric Research Program, Princeton University.

Shibata, R. (1976), "Selection of the Order of an Autoregressive Model by Akaike's Information Criteria", Biometrika, 63, 117-126.

Simon, H.A. (1953), "Causal Ordering and Identifiability", in Studies in Econometric Method, edited by W.C. Hood and T.C. Koopmans, Cowles Commission Monograph 14, New York.

Sims, C.A. (1977), "Exogeneity and Causal Orderings in Macroeconomic Models", in New Methods of Business Cycle Research, Minneapolis Federal Reserve Bank.

Skoog, G. (1976), "Causality Characterizations: Bivariate, Trivarte and Multivariate Propositions", Staff report no. 14, Federal Reserve Bank of Minneapolis.

Smith, R. (1977), "Military Expenditure and Capitalism", Cambridge Journal of Economics, 1(1), 61-76.

Zellner, A. (1978), "Causality and Econometrics", Paper presented at a joint University of Rochester and Carnegie-Mellon University Conference.

APPLIED TIME SERIES ANALYSIS
O.D. Anderson & M.R. Perryman (eds.)

IDENTIFICATION AND TESTING OF OPTIMAL LAG STRUCTURES AND CAUSALITY IN ECONOMIC FORECASTING

Dr. Richard A. Fey
Nem C. Jain

Dynamics Associates
Division of Interactive Data Corporation/Chase Econometrics
1033 Massachusetts Avenue
Cambridge, MA 02138
USA

The objective of this paper is to analyze the determination of lag structures in forecasting models. In view of the value and widespread use of lags in forecasting models, comparatively little research has been devoted to the problem of choosing among alternative lag structures. We propose the use of the information criterion developed by Akaike to determine the optimal lag structure. This criterion maximizes the log likelihood after correction for degrees of freedom and therefore introduces the concept of parsimony into the determination of lag structure. We further investigate this problem using Akaike's state space methodology to analyze the interrelationships among various economic variables. These results enable us to investigate causality and feedback among economic variables in the sense defined by Granger. We then apply these results to the analysis of a simple macrodynamic economic system of the determination of income and the price level. Specifically, we investigate whether the feedback from income to price levels, reported by Nelson, remains when the exogenous variables driving the system are explicitly included in the model.

INTRODUCTION

The objective of this paper is to analyze the determination of lag structures among economic variables using techniques of system identification and parameter estimation developed in the control literature. The practitioner of forecasting is confronted with the bewildering array of alternative lag structures. Frequently, the quality of the data are not good enough to discriminate among the various lag distributions using the usual statistical tests. These difficulties are further compounded by the possibility of feedbacks among the economic variables, requiring the further step of identifying whether the system is simultaneous or recursive. Recursive systems have the appealing property that causal inferences may be drawn from the structure of the model, placing still greater burdens on model identification.

Using the venerable St. Louis equation, we show that the impact of fluctuations in the nominal supply of money affect fluctuations in nominal income with a much shorter lag than previously reported. On the other hand, our results are generally consistent with the hypothesis that fluctuations in the nominal supply of money are exogenous to income and have a relatively greater impact on income than do fluctuations in high employment government expenditures. While these results are generally in accord with the monetarist position, the main conclusion of this paper is that our findings do not support the hypothesis that changes in income cause changes in price levels as reported by Nelson.

DETERMINATION OF OPTIMAL LAG STRUCTURES

The problem of determining the length of a lag structure is essentially analogous to determining the order of an autoregressive (AR) representation. A plausible approach to this problem involves the use of maximum likelihood. For a given set of data estimate a sequence of AR models beginning with one lag and continuing with higher orders of the AR process until the likelihood function for the given sample attains a maximum. It has been shown that the resulting model is over specified in the order of the AR process. Akaike (1976) identified an alternative criterion called AIC to measure the deviation of a parametric stochastic model from the true structure. AIC is defined as:

$$\begin{aligned} AIC = &- 2 \cdot \text{(maximum log likelihood)} \\ &+ 2 \cdot \text{(number of free parameters in the model)}. \end{aligned}$$

For a Gaussian autoregressive model of order M, this criterion can be simplified to:

$$AIC = - N \cdot \log|c| + 2Mr^2$$

where N = number of observations
r = number of variables constituting Y
$|c|$ = determinant of the covariance of the r dimensional white noise residual process.

This criterion clearly utilizes the important principle of parsimony in statistical model building and provides a consistent criterion to compare increasing model goodness of fit versus increasing number of model parameters. It is a common practice in econometric modeling to develop models having long lag structures. To overcome the problems of multicollinearity and loss of observations, distributed lag structures as suggested by Koyck or Almon are frequently estimated without having any firm idea about the optimum lag structure or the possibility of any feedback between the variables. This may lead to model specification error and biased parameter estimates or at least inefficient estimates. With the AIC criterion, we have an objective basis to evaluate the lag structures in autoregressive representations.

The St. Louis equation and the rational expectations hypothesis itself depend on particular assumptions about the recursive structure. Given the linear system

$$\begin{bmatrix} a_{11}(B), & a_{12}(B) \\ a_{21}(B), & a_{22}(B) \end{bmatrix} \cdot \begin{bmatrix} Y_T \\ P_T \end{bmatrix} = \begin{bmatrix} b_{11}(B), & b_{12}(B) \\ b_{21}(B), & b_{22}(B) \end{bmatrix} \cdot \begin{bmatrix} Z_{1,T} \\ Z_{2,T} \end{bmatrix} + \begin{bmatrix} \eta_{1,T} \\ \eta_{2,T} \end{bmatrix}$$

where $a_{ij}(B)$ are polynomials and b_{ij} are matrix polynomials in the backshift operation B, Y is the change in the log of the nominal income, P is the change in the log of the price levels, Z_i are vector exogenous variables, and η_1, and η_2, are uncorrelated across equations. Nelson (1979) suggests that the "monetarist" position as well as the rational expectations hypothesis imply the structural assumption that $a_{12}(B) = b_{12}(B) = b_{21}(B) = 0$. The implication of these restrictions is that changes in income cause changes in price levels, but changes in prices do not cause changes in income and further that the group of exogenous variables affecting income are independent of the group of exogenous variables affecting price levels. Nelson develops a series of tests of these zero restrictions based solely on time series analysis of income and prices.

By using the bivariate ARMA model, Nelson fails to explicitly specify the exogenous variables of the system. Consequently, the estimated innovations of price and income from univariate models may differ from those implied by a theoretical model including exogenous variables. Specifically, if the exogenous variable(s) affected price and income through significantly different lag structures, then this unexplained residual could be the source of the significant cross correlation reported between the innovation of price at time T and the innovation of income at time T-1. Similarly, the two-way regressions of the Sims test will suffer from biased coefficients as a result of the exclusion of the exogenous variables.

We will test the validity of Nelson's results using a more general time series method that enables us to explicitly include some plausible exogenous variables, estimate the entire linear system and directly test the zero restrictions and their causal implications.

THE STATE SPACE MODEL

In the previous section, we have seen how the AIC criterion allows us to compare models with different structures. But AR models are not in general parsimonious as suggested by Box and Jenkins, who instead advocate building mixed ARMA models to achieve parsimony. Developing unique ARMA models for a single time series may not be difficult, but in a multiple time series case, a number of different ARMA models are possible. This poses a serious problem of model identification. The problems can be resolved by developing an equivalent representation called the state-space model. Suppose

$$Y_T - \phi_1 Y_{T-1} \cdots - \phi_p Y_{T-p} = U_T - \Theta_1 U_{T-1} \cdots - \Theta_q U_{T-q} \tag{1}$$

is an ARMA representation of an observed Y_T, where U_T is an independent multivariate normal random vector. Using B as the backward shift operator and $\phi(B)$ and $\Theta(B)$ as matrix polynominals in B with $\phi(0) = \Theta(0)=I$, the above equation can be written as

$$\phi(B)Y_T = \Theta(B)U_T$$

$$Y_T = \phi^{-1}(B)\ \Theta(B)U_T.$$

Since any mixed model can be expressed as a linear filter, it follows that:

$$Y_T = \sum_{s=0}^{\infty} \psi_s\ U_{T-s}.$$

State-space representation of this model is obtained by taking conditional expectations where $Y_{T+1/T}$ is the conditional expectation of Y_{T+1} at time T:

$$Y_{T+i/T} = \sum_{s=i}^{\infty} \psi_s\ U_{T+i-s} \tag{1A}$$

and

$$Y_{T+i/T+1} = Y_{T+i/T} + \psi_{i-1}U_{T+1}. \tag{2}$$

Also from equation 1, we get

$$Y_{T+p/T} - \phi_1 Y_{T+p-1/T} - \cdots - \phi_p Y_T = 0$$

and

$$Y_{T+p/T} = \phi_1 Y_{T+p-1/T} + \cdots + \phi_p Y_T.$$

For equation 2, we write:

$$Y_{T+p/T+1} = (\phi_1 Y_{T+p-1/T} + \cdots + \phi_p Y_T) + \psi_{p-1} U_{T+1}. \quad (3)$$

Thus, from 2 and 3 we can write the system of equations in 1 as

$$\begin{bmatrix} Y_{T+1} \\ Y_{T+2/T+1} \\ \cdot \\ \cdot \\ \cdot \\ Y_{T+p/T+1} \end{bmatrix} = \begin{bmatrix} 0 & 1 & 0 & \cdots & 0 \\ 0 & 0 & 1 & \cdots & 0 \\ & & \cdot & & \\ & & \cdot & & \\ & & \cdot & & \\ \phi_p & \phi_{p-1} & \phi_{p-2} & \cdots & \phi_1 \end{bmatrix} \cdot \begin{bmatrix} Y_T \\ Y_{T+1/T} \\ \cdot \\ \cdot \\ \cdot \\ Y_{T+p-1/T} \end{bmatrix} + \begin{bmatrix} 1 \\ \psi_1 \\ \cdot \\ \cdot \\ \cdot \\ \psi_{p-1} \end{bmatrix} U_{T+1} \quad (4)$$

or in terms of vectors,

$$Z_{T+1} = F \cdot Z_T + G \cdot U_{T+1} \quad (5)$$

$$Y_T = H \cdot Z_T \quad (6)$$

The n-dimensional state vector Z_T is the set of present and past information sufficient to predict future values. U_T is a p-dimensional vector of one-step-ahead prediction errors which is Gaussian white noise with covariance matrix S. As such, the state space model is an alternate representation equivalent to an ARMA model in the sense that both will produce an output vector Y_T with identical properties. However, there are several important differences in practice.

Multi-input ARMA models are nonunique unless complex restrictions are imposed on the coefficients. In addition, multi-input ARMA model identification is tedious at best. These problems are easily solved for state vector models. In equation 5 the transition matrix F is stable if all eigenvalues lie within the unit circle and hence the processes Z_T and Y_T are asymptotically stationary processes. However, multiplication of equations 5 and 6 by any nonsingular matrix will yield a different state space representation that is also valid. In effect, the coordinates of the system have been changed. If we define an nxn nonsingular matrix T such that TFT^{-1} and HT^{-1} have a "canonical" form, then the state vector model is always uniquely determined. Equations 5 and 6 become

$$Z'_{T+1} = F' \cdot Z'_T + T \cdot G \cdot U_{T+1} \quad (7)$$

$$Y_T = H \cdot T^{-1} \cdot Z'_T. \quad (8)$$

This unique identification property makes state vector models particularly useful for analysis of recursive structures. Model identification in the canonical state vector model is achieved by selecting the first maximum set of linearly independent elements within the sequence of

$$Y_1(T/T), Y_2(T/T), \ldots, Y_r(T/T), \ldots, Y_1(T+k/T), Y_2(T+k/T), \ldots, Y_r(T+k/T).$$

To achieve this, canonical correlation analysis is performed between the sets of present and past values, and present and future values. Non-zero canonical correlation values between the two sets determine the number of linearly independent elements in the present and future sets. The amount of past information to be used in the correlation analysis is decided by fitting a sequence of AR models and computing the AIC for each model. The optimum lag into the past is chosen as the order of the model having the minimum value of AIC. The test for the smallest canonical correlation is based on the Bartlett's chi-square statistic minus twice the associated degrees of freedom. For negative values of this criterion, the canonical correlation value is judged to be zero.

Estimates of the F matrix are obtained during the canonical correlation analysis. The optimum AR model, used to define the amount of past information to be included, also provides estimates of the impulse-response function and the innovation covariance matrix used to estimate the stationary Kalman filter gain matrix G where:

$$\hat{Z}_{T+1/T} = F \cdot Z_T \qquad \text{one-step-ahead prediction} \quad (9)$$

$$U_{T+1} = Y_{T+1} - H \cdot \hat{Z}_{T+1/T} = Y_{T+1} - H \cdot F \cdot Z_T \qquad \text{innovation} \quad (10)$$

$$Z_{T+1/T+1} = \hat{Z}_{T+1/T} + G \cdot U_{T+1} \qquad \text{updating equations} \quad (11)$$

The matrix, H, is the matrix of regression coefficients of Y and Z and takes the form (I,O) under the assumption of non-singularity of the covariance matrix of Y.

The goodness of fit of the model is evaluated in terms of conventional R^2 and by the relative goodness of fit criterion defined as:

$$TIC = - \log \frac{|c|}{|c_0|} + 2 \cdot (\text{number of free parameters in the model})$$

where $|c|$ and $|c_0|$ are the determinants of the residual and data covariance matrices, respectively. This gives a measure of the lack of fit of the identified model. By changing the sign of the criterion, we can interpret TIC as the extent of agreement of the fitted model to the true one. Thus, the higher the value of TIC, the better is the fitted model.

The state vector model is particularly well suited to the present study. First, it provides a general model framework that allows simultaneity, recursiveness, or feedback among the variables included without prior restrictions on model structure other than linearity. The identified structural relationships among the variables are uniquely determined by the past and present data and are optimal in the sense defined by Akaike. The model identification and estimation procedure takes advantage of recent developments in the system control literature to yield unique and parsimonious model structures.

In the next section we use the canonical state vector model to analyze a simple macroeconomic system using the state space approach. The method has been programmed on the Chase Econometrics/Interactive Data Corporation time-sharing system as part of its high level modeling language designed specifically for economic, financial and marketing applications.

A direct application of the concepts of causality and exogeneity developed by Granger (1969) and Sims (1972) leads to four tests:

1. The coefficients of the F matrix

2. By comparing the amount of variance, R^2 (Y/X), explained by a model involving a variable, say X, with the variance, R^2 (Y/Y), explained by a model having all variables of the previous model except X. The hypothesis that X does not cause Y can be tested as an F ratio defined as:

$$F(K_2 - K_1, N - K_2 - 1) = \frac{(R^2(Y/X,Y) - R^2(Y/Y)) \,/\, (K_2 - K_1)}{(1 - R^2(Y/X,Y)) \,/\, (N - 1 - K_2)}$$

where K_1 = number of parameters in model without X.

K_2 = number of parameters in model with X.

N = sample size.

3. By comparing the change in AIC. If the AIC improves by adding the variable X, then X causes Y.

4. By comparing the error variance of Y from a model containing X with the error variance of the model without X.

EXAMPLE - A SIMPLE MACROECONOMIC SYSTEM

Using the state space model of the previous section and Granger's definitions of feedback and causality, we first look at univariate models of money supply (M1), income (GNP), and prices (deflator); next we consider two bivariate models, one of money and income and the other of income and prices. Since these series are nonstationary, we are dealing with changes in the logs of these series and in keeping with Nelson's reasons for avoiding Korean war distortions in the early fifties and price control distortions in the early seventies, we use the sample period from 1955 to 1970 for this analysis. The results are shown in Table 1. The one-step-ahead error variance for income reduces to .6912 when modelled with money as compared to .7983 when modelled alone, thus suggesting causality in the Granger's sense from money to income. Similar analysis for causality from income to money does not support feedback from income to money. These results are further confirmed by the coefficients of the F matrix for the bivariate model and the significant increase in R^2 from 14.23% to 25.74% and the AIC criterion from 7.516 to 42.503. Similarly, to test causality for income to prices we model the bivariate relationship between income and prices as:

$$\begin{bmatrix} Y_{T+1} \\ P_{T+1} \\ P_{T+2} \end{bmatrix} = F \cdot \begin{bmatrix} Y_T \\ P_T \\ P_{T+1} \end{bmatrix} + G \cdot \begin{bmatrix} U_{Y_{T+1}} \\ U_{P_{T+1}} \\ \end{bmatrix}$$

$$F = \begin{bmatrix} 0.4194 & -0.1082 & 0.0 \\ 0.0 & 0.0 & 1.000 \\ -0.0115 & 0.2676 & 0.6049 \end{bmatrix}$$

$$G = \begin{bmatrix} 1.000 & 0.0 \\ 0.0 & 1.000 \\ 0.2624 & 0.0688 \end{bmatrix} .$$

where $U_{y,T+1}$ and $U_{p,T+1}$ are the innovations of income and price respectively at time T+1. The state space model when written in the transfer function form becomes:

$$INCOME_T = .4194 * INCOME_{T-1} - .1082 * PRICE_{T-1} + U_{1,T}$$

$$PRICE_T = .6049 * PRICE_{T-1} + .2394 * PRICE_{T-2}$$
$$+ .2624 * INCOME_{T-1} + .0996 * INCOME_{T-2} + U_{p,T}$$
$$+ .5361\ U_{p,T-1}$$

This representation clearly shows feedback from INCOME to PRICES but very weak feedback from PRICES to INCOME. Very negligible drop in error variance for INCOME from .7983 to .7951 also confirms that PRICE is not a significant variable to affect INCOME. The error variance of price decreases from .08325 to .0735, thereby further supporting the hypothesis of feedback from INCOME to PRICES. These equations also confirm Nelson's finding that INCOME and PRICE have different autoregressive structures and that the impact of income on prices is small but persistent.

We now introduce an exogenous variable, nominal supply of money, into the income model. The results are shown in Table 1. This model shows that money causes income and price and is exogenous to them. It further supports the bivariate model results that income and price affect each other, but this time the importance of prices affecting income has increased considerably. This interdependence between income and prices is in clear contrast to the bivariate model results of Nelson in that there appears to be a strong feedback from income to prices and a somewhat weaker impact of prices on income. The exogeneity of money is clearly supported by the absence of significant feedbacks in the F matrix and by the fact that the R^2 and the one-step-ahead error variance show negligible improvement over the univariate model for money. Given the very substantial impact of money on income, and relatively modest impact on prices, the bivariate results showing income causing changes in prices appear to stem from the specification bias resulting from excluding money, with income apparently becoming a proxy variable for money. These results are not significantly altered if output, measured as differences in logarithms of real GNP, is substituted for income.

Now consider the impacts of fiscal policy on income and prices. Table 1 includes the results of a three-variable model using the high employment budget deficit as our indicator of fiscal policy. This model appears to explain prices nearly as well as the previous model but does not explain income as well. However, the tests clearly show that the deficit is exogenous to income and prices and that changes in the high employment deficit cause changes in income and prices as in the previous model.

A model with four variables shows further improvements over the three-variable model. The AIC increases to 140.332 from 71.237, R^2 for money increases from 50.36% to 60.39% and error variance of money decreases from .2081 to .1660. Income and price variables also show marginal improvements in terms of higher R^2 values and lower error variances. The F and G matrices of this model suggest the following causal chain:

- The deficit is exogenous to money, income, and price, and it causes changes in money
- Money is exogenous to income and prices and causes both price and income but is not exogenous to the deficit
- Income has a persistent feedback from price
- Price, though driven by money and its own history, has a transient feedback from income as shown by the third element in the fourth row of the G matrix.

These results clearly do not agree with the recursive structure between income and prices described by Nelson. Indeed when the exogenous variables are explicitly included in the model, a strong case can be made for two-way interdependence with a

TABLE 1

Model Type	Variables	R^2 in %	Error Variance	AIC	F Matrix D(T)	M(T)	I(T)	P(T)	P(T+1)	G Matrix U_D(T+1)	U_M(T+1)	U_I(T+1)	U_P(T+1)
Univariate	Money(T+1)	50.04	.2094	41.029		.7069							
Univariate	Income(T+1)	14.23	.7983	7.516			.3854						
Univariate	Price(T+2)	46.66	.08325	32.337				.2297	.6657				.2199
Bivariate	Money(T+1) Income(T+1)	50.46 25.74	.2077 .6912	42.503		.7314 .3817	-.0538 .2120						
Bivariate	Income(T+1) Price(T+2)	14.58 52.88	.7951 .0735	47.766			.4194 -.0115	-.1082 .2676	0 .6049			1 .2624	0 .0688
3 Variables	Money(T+1) Income(T+1) Price(T+2)	50.36 28.96 56.72	.2081 .6613 .0676	71.237		.7570 .4350 .1892	-.0342 .2528 .0888	-.0996 -.2069 .2015	0 0 .6488		1 0 .1398	0 1 .1162	0 0 .1978
3 Variables	Deficit(T+1) Income(T+1) Price(T+2)	80.621 21.131 54.686	11.45 .7341 .9797	127.272	.8933 -.2777 -.1630		.0060 .3744 -.0354	.0077 -.1959 .2284	0 0 .5816	1 0 .021		0 1 .1679	0 0 .1988
4 Variables	Deficit(T+1) Money(T+1) Income(T+1) Price(T+2)	80.97 60.39 29.45 57.03	11.24 .1660 .6567 .0671	140.322	.9422 -.3964 -.0857 -.0810	.0985 .5354 .3871 .1524	-.0238 .0136 .2572 -.0835	.0008 -.1744 -.2231 .2051	0 0 0 .6092	1 0 0 .0047	0 1 0 .0887	0 1 0 .1186	0 0 0 .1885

significant negative impact of price changes on income. These results are quite the opposite of what one would expect under the rational expectations hypothesis. There appears to be some specification bias in the bivariate model. However, the four-variable model still leaves much of the determination of income unexplained and may require different specifications including either higher lag terms or additional driving variables to reflect the impact of investment on income.

An interesting implication of the above model for current economic policy is that changes in the money supply appear to have much larger impacts on income than on inflation. Consequently, the use of monetary policy to control inflation must cause more severe reductions in income growth. Moreover, there is a substantial negative impact of the deficit on money, indicating substantial monetization of debt by the Federal Reserve.

REFERENCES

Anderson, L.C. and Carlson, K.M., "A Monetarist Model for Economic Stabilization, "Monthly Review Federal Reserve Bank of St. Louis, 52 (April 1970).

Akaike, H., "Fitting Autoregressive Models for Prediction," Annals of the Institute of Statistical Mathematics, 21, 243-247, 1969.

Akaike, H., "Statistical PredictorIdentification," Annals of the Institute of Statistical Mathematics, 22, 203-217, 1970.

Akaike, H., "Canonical Correlation Analysis of Time Series and the Use of an Information Criterion," Mehra and Lainiotis, Eds., System Identification: Advances and Case studies. Academic Press, New York, 1976.

Box, G.E.P. and Jenkins, G.M., Time Series Analysis, Forecasting and Control, San Francisco, Holden-Day, 1970.

Cameron, A.V., "State Space Models and Forecasts for Money and Supply and Income Data," State Space Newsletter, Volume 2, #10, 1979.

Cooper, J.P. and Nelson, C.F., "The Ex Ante Prediction Performance of the St. Louis and FRB-MIT-Penn Economic Model and Some Evidence on Composite Predictions," Journal of Money, Credit and Banking 7, February, 1975.

Granger, C.W.H., "Investigating Causal Relations by Econometric Models and Cross Spectral Methods," Econometrica 37, 424-38, 1969.

Hsiao, C., "Autoregressive Modelling of Canadian Money and Income Data," Journal of American Statistical Association, page 553, September 1979.

Mehra, R.K. and Lainiotis, D.G., Eds., System Identification: Advances and Case Studies, Academic Press, New York, 1976.

Nelson, C.R., "Recursive Structure in U.S. Income, Prices and Output," Journal of Political Economy, Volume 87, #6, 1307-1327, 1979.

Pierce, D.A. and Haugh L.D., "Causality in Temporal Systems: Characterization and a Survey," Journal of Econometrics, 5, 205-293, 1977.

Sims, C.A., "Money, Income and Causality", American Economic Review 62, 540-552, 1972.

APPLIED TIME SERIES ANALYSIS
O.D. Anderson & M.R. Perryman (eds.)

NONSTATIONARY SECOND-ORDER MOVING AVERAGE PROCESSES

Marc Hallin

Université Libre de Bruxelles
Institut de Statistique C.P. 210
1050 Bruxelles - BELGIUM

We study how to obtain all possible second-order moving average (MA(s)) models with time-dependent coefficients corresponding to a given (time-dependent) autocovariance function. Two different approaches are considered. The first one relies on a factorization of infinite autocovariance matrices, and leads to a necessary and sufficient condition for a process being an MA(2) process. The second approach shows how a generalization of the stationary factorization property of the characteristic polynomial associated with MA autocovariance functions can be used for theoretical model-building purposes.

In a preceding paper [5] we studied the relationship between nonstationary autocovariance (or autocorrelation) functions and the possible MA(1) representations for it, i.e. the set of all possible moving average models (with time-dependent coefficients) corresponding to a given (time-dependent) autocovariance structure. We gave a necessary and sufficient condition for a (time-dependent) autocovariance function to be an MA(1) autocovariance function, and showed that, for such autocovariances, a one-parameter family of MA(1) models (with time-dependent parameters) exists. An explicit formulation of these models was given, involving continued fractions.

It was quite natural to look for a MA(q) version of these results. Unfortunately, the continued fraction methods on which they rely, and, more particularly, the concept of *chain sequence* could not be readily extended to the case of higher order processes.

Two alternative approaches are described here. The first one is based on a Cholesky factorization of the finite segments of the (infinite) covariance matrix of the process. We merely use it (section 1) to obtain a necessary and sufficient condition for a process to be an MA(2) process. For model-building purposes, however, we prefer the second one (section 3), which relies on a generalization of the classical stationary factorization property of the characteristic polynomial associated with (stationary) MA(2) autocovariance functions, and provides better insight into the structure of the solution.

1. DEFINITIONS. NOTATIONS. A FIRST FACTORIZATION APPROACH.

Let $\{\varepsilon_t;\ t \in \mathbb{Z}\}$ denote a real second order white noise :

$$E(\varepsilon_u) = 0 \quad E(\varepsilon_u . \varepsilon_v) = \delta_{uv} \quad u, v \in \mathbb{Z} .$$

Let

$$A_t(L) = a_{to} + a_{t1}L + a_{t2}L^2 \quad t \in \mathbb{Z} , \tag{1}$$

where L denotes the lag operator, be a linear difference operator of order two ($a_{to}.a_{t2} \neq 0,\ t \in \mathbb{Z}$) such that $a_{to} > 0$, and consider the process $\{z_t\ ;\ t \in \mathbb{Z}\}$ generated (in a quadratic mean sense) by

$$z_t = A_t(L)\ \varepsilon_t = a_{to}\ \varepsilon_t + a_{t1}\ \varepsilon_{t-1} + a_{t2}\ \varepsilon_{t-2} \qquad t \in \mathbb{Z} : \tag{2}$$

z_t is a second-order *moving average* (MA(2)) *process*, (1) an MA(2) *model* for it; z_t is generally nonstationary, although time-dependent models may generate stationary processes (cf. [5] for an example). The autocovariances of z_t are

$$\gamma_{to} = \mathrm{Var}(z_t) = \qquad a_{to}^2 + a_{t1}^2 + a_{t2}^2 \tag{3a}$$

$$\gamma_{t1} = \mathrm{Cov}(z_t, z_{t-1}) = a_{t1}\ a_{t-1,o} + a_{t2}\ a_{t-1,1} \tag{3b}$$

$$\gamma_{t2} = \mathrm{Cov}(z_t, z_{t-2}) = a_{t2}\ a_{t-2,o} \tag{3c}$$

As a function of t, we refer to the triple $\gamma_t = (\gamma_{to}, \gamma_{t1}, \gamma_{t2})$ as z_t's *autocovariance function*. Also, z_t's autocovariance matrix is the infinite positive definite band matrix Γ with bandwidth 5

$$\Gamma = \begin{pmatrix} \ddots & 0 & \gamma_{t+3,2} & \gamma_{t+2,1} & \gamma_{t+1,o} & \gamma_{t+1,1} & \gamma_{t+1,2} & 0 & & & \\ & & 0 & \gamma_{t+2,2} & \gamma_{t+1,1} & \gamma_{to} & \gamma_{t1} & \gamma_{t2} & 0 & & \\ & & & 0 & \gamma_{t+1,2} & \gamma_{t1} & \gamma_{t-1,o} & \gamma_{t-1,1} & \gamma_{t-1,2} & 0 & \ddots \end{pmatrix}$$

(the heavy line indicates Γ's diagonal).

The corollary to the following theorem extends a stationary result by T.W. Anderson ([2], pp. 224-225) and O.D. Anderson ([1], itself a seasonal version of the preceding one); we state it here for MA(2) processes - it obviously holds in the general MA(q) case.

Theorem 1. (Factorization of the covariance matrix). An infinite symmetric band matrix Γ with bandwidth 5 is definite positive iff it admits a factorization $\Gamma = AA'$, with A a real matrix of the form

$$A = \begin{pmatrix} \ddots & & \ddots & \ddots & \ddots & & \\ & 0 & a_{to} & a_{t1} & a_{t2} & 0 & \\ & & 0 & a_{t-1,o} & a_{t-1,1} & a_{t-1,2} & 0 \\ & & & \ddots & \ddots & \ddots & \ddots \end{pmatrix}$$

Corollary 1. A process $\{z_t;\ t \in \mathbb{Z}\}$ (with mean zero) is MA(2) iff it is 2-dependent (in a quadratic mean sense, i.e. iff its autocovariances vanish after a lag 2 : $\mathrm{cov}(z_t, z_{t-2}) \neq 0$, $\mathrm{cov}(z_t, z_{t-k}) = 0\ \forall\ k > 2$).

Corollary 2. An MA(2) process generally admits a three-parameter family of MA(2) representations.

Proof. Denote by Γ_{tk} the principal minor of Γ corresponding to rows t, t-1, ..., t-k. Suppose that Γ is definite positive : so is $\Gamma_{t,k+2}$. Hence, $\Gamma_{t,k+2}$ is of the form $B_{t,k+2}\ B'_{t,k+2}$, where $b_{t,k+2}$ is a real upper triangular matrix (cf. [8]). $\Gamma_{t,k+2}$ being a band matrix, $B_{t,k+2}$ is also a band matrix, with the same bandwidth (but the lower diagonals are all zero). Γ_{tk}, as an upper left corner submatrix of $\Gamma_{t,k+2}$, can be written as

$$\Gamma_{tk} = (B^{(k+1)}_{t,k+2})\ (B^{(k+1)}_{t,k+2})' ,$$

where $B^{(k+1)}_{t,k+2}$ is the rectangular matrix consisting of the first k+1 rows of $B_{t,k+2}$; put $A_{t,k+2} = B^{(k+1)}_{t,k+2}$.

So far, we established a factorization property for the finite segments of Γ. Consider t as an arbitrarily fixed value : the matrices A_{tk} are generally not uniquely defined. It is easy to see that, requiring $a_{t+i,o} \geq 0$ $(i = 1,2,\ldots,k)$, if suitable values of a_{t1}, a_{t2} and $a_{t-1,1}$ are chosen, $A_{tk} = A_{tk}\ (a_{t1}, a_{t2}, a_{t-1,1})$ is completely determined; denote by $\mathcal{D}_{tk}$ the set of possible values for $(a_{t1}, a_{t2}, a_{t-1,1})$: $\mathcal{D}_{tk}$ is a nonempty bounded (since $a^2_{t1} + a^2_{t2} \leq \gamma_{to}$, $a^2_{t-1,1} \leq \gamma_{t-1,o}$) subset of $\mathbb{R}^3$ such that $\mathcal{D}_{tk} \supseteq \mathcal{D}_{t-\ell,k+m}$. More precisely, $\mathcal{D}_{t,k+1}$ is the set of points $(a_{t1}, a_{t2}, a_{t-1,1})$ in $\mathcal{D}_{tk}$ such that the system

$$\begin{aligned} \gamma_{t-k+1,2} &= a_{t-k+1,2}\ x \\ \gamma_{t-k,1} &= a_{t-k,1}\ x + a_{t-k,2}\ y \\ \gamma_{t-k-1,o} &= x^2 + y^2 + z^2 \end{aligned} ,$$

where $a_{t-k+1,2}$, $a_{t-k,1}$ and $a_{t-k,2}$ are the appropriate entries in $A_{tk}\ (a_{t1}, a_{t2}, a_{t-1,1})$, admits a real solution (x,y,z) with $x \geq 0$; this is possible iff

$$(\gamma_{t-k+1,2}/a_{t-k+1,2})^2+(\gamma_{t-k,1}-a_{t-k,1}\,\gamma_{t-k+1,2}/a_{t-k+1,2})^2/a_{t-k,2}^2 < \gamma_{t-k-1,o}\ .$$

Hence, the $\mathcal{D}_{tk}$'s are compact subsets of $\mathbb{R}^3$, and

$$\mathcal{D}_t = \bigcap_{2<k\in\mathbb{N}} \mathcal{D}_{t,\pm k} \neq \emptyset\ .$$

Finally, define the infinite matrix A as the matrix with finite principal minors $A_{tk} = A_{tk}\,(a_{t1}, a_{t2}, a_{t-1,1})$, where $(a_{t1}, a_{t2}, a_{t-1,1}) \in \mathcal{D}_t$: A is of the required form, which completes the direct proof. As for the converse, if $\Gamma = AA'$, it is trivially positive definite.

Proof of Corollary 1. If Γ is the autocovariance matrix of a 2-dependent process z_t, $\gamma_{t2} > 0$ $(t \in \mathbb{Z})$; hence $0 \notin \mathcal{D}_t$, and $a_{to} > 0$, $a_{t2} \neq 0\ \forall\ t \in \mathbb{Z}$. Developing $\Gamma = AA'$ yields (3), which shows that the coefficients a_{tj} define a MA(2) model for z_t.

Proof of Corollary 2. Corollary 2 immediatly follows from the preceding proof, since $\mathcal{D}_t \in \mathbb{R}^3$.

2. ADJOINT DIFFERENCE OPERATORS AND SYMBOLIC PRODUCTS.

Consider again the linear difference operator of order two $A_t(L)$ defined in (1); the *adjoint operator* $A_t^*(L)$ is defined (cf. [9]) as

$$A_t^*(L) = a_{to}^* + a_{t1}^*L + a_{t2}^*L^2\ , \tag{4}$$

with

$$a_{ti}^* = a_{t+2-i,2-i}\ ,$$

i.e.

$$a_{to}^* = a_{t+2,2} \qquad a_{t1}^* = a_{t+1,1} \qquad a_{t2}^* = a_{to}\ .$$

Define also the autocovariance difference operator $\Gamma_t(L)$, of order four, as

$$\Gamma_t(L) = \gamma_{t+2,2} + \gamma_{t+1,1}L + \gamma_{to}L^2 + \gamma_{t1}L^3 + \gamma_{t2}L^4\ .$$

Finally, recall that the *symbolic product* of two difference operators is obtained by applying usual noncommutative multiplication rules, the lag operator L operating on any time index appearing on its right (*example* : $a_{t2}L^2\ .\ b_{to}L = a_{t2}\,b_{t-2,o}L^3$).

The following theorem can be obtained in a very simple way by developing the symbolic product $A_t(L).A_t^*(L)$:

Theorem 2. (Factorization of the autocovariance operator). $A_t(L)\varepsilon_t$ is a MA(2) model for a process with autocovariance function $\gamma_t = (\gamma_{to}\ \gamma_{t1}\ \gamma_{t2})$ iff

$$\Gamma_t(L) = A_t(L).A_t^*(L)\ . \tag{5}$$

Hence, the adjoint model operator is a symbolic divisor of the autocovariance operator, the symbolic quotient being the model operator itself. What happens in the stationary case ? $\Gamma_t(L)$ reduces to a fourth degree polynomial (with constant coefficients) in L (the operator's characteristic polynomial); if we only consider time-independent MA(2) models, $A_t(L)$ also turns into a characteristic polynomial $A(L)$, of degree two. The adjoint of $A(L)$ is nothing else than $A^*(L) = A(L^{-1})$. (5) gives

$$\Gamma(L) = A(L).A(L^{-1}) \ ;$$

but the symbolic product $A(L).A(L^{-1})$ is equivalent, here, to an ordinary product of polynomials $A(L)\ A(L^{-1})$, and Theorem 2 reduces to the well-known property of the characteristic polynomials associated with moving average autocovariance functions.

3. FACTORIZING THE MA(2) AUTOCOVARIANCE OPERATOR.

It results from the preceding section that the theoretical model-building problem, i.e. the problem of obtaining all the possible MA(2) models for a given MA(2) autocovariance function, is equivalent to that of obtaining all the symbolic divisors (of order two) of the autocovariance operator which are also the adjoint of the corresponding symbolic quotient. Just as the various possible MA(2) models (with constant coefficients) for a given stationary autocovariance function can be determined by choosing appropriatedly two out of the four roots of the characteristic equation $\Gamma(x) = 0$, so the various MA(2) models for a given nonstationary MA(2) autocovariance function will be obtainable by choosing appropriatedly two solutions of the fourth order homogeneous difference equation $\Gamma_t(L)\ \psi_t = 0$.

(a) *Obtention of a symbolic divisor* $B_t(L) = b_{to}+b_{t1}L+b_{t2}L^2$ of $\Gamma_t(L)$.

Recall that the solutions of an homogeneous linear difference equation of order q constitute a vectorial space of dimension q. Conversely, any q-dimensional vectorial space of sequences determines an homogeneous equation of order q, hence, up to multiplicative constants, a difference operator of order q. The homogeneous equation $\Gamma_t(L)\psi_t = 0$ determines a solution space of dimension four. $B_t(L)$ is a symbolic divisor of $\Gamma_t(L)$ iff any solution of $B_t(L)\psi_t = 0$ is also a solution of $\Gamma_t(L)\psi_t = 0$, hence iff its two-dimensional solution space is a subspace of $\Gamma_t(L)\psi_t = 0$'s four-dimensional one.

Now, let ψ_t^1 and ψ_t^2 be any couple of linearly independent (non proportional) solutions of $\Gamma_t(L)\psi_t = 0$; they characterize a second-order homogeneous equation $B_t(L)\psi_t = 0$, with

$$\frac{b_{t1}}{b_{to}} = - \begin{vmatrix} \psi_t^1 & \psi_t^2 \\ \psi_{t-2}^1 & \psi_{t-2}^2 \end{vmatrix} \Big/ \begin{vmatrix} \psi_{t-1}^1 & \psi_{t-1}^2 \\ \psi_{t-2}^1 & \psi_{t-2}^2 \end{vmatrix} \tag{6a}$$

$$\frac{b_{t2}}{b_{to}} = \begin{vmatrix} \psi_t^1 & \psi_t^2 \\ \psi_{t-1}^1 & \psi_{t-1}^2 \end{vmatrix} \Big/ \begin{vmatrix} \psi_{t-1}^1 & \psi_{t-1}^2 \\ \psi_{t-2}^1 & \psi_{t-2}^2 \end{vmatrix} . \tag{6b}$$

Furthermore, any symbolic divisor $B_t(L)$ of $\Gamma_t(L)$ can be obtained this way. If $B_t(L)$ is of the form $A_t^*(L)$, this gives us the values of $(a_{t+1,1}/a_{t+2,2}) = (b_{t1}/b_{to})$ and $(a_{to}/a_{t+2,2}) = (b_{t2}/b_{to})$, which is insufficient to determine the model operator $A_t(L)$, even up to multiplicative constants.

(b) *Further requirement : the quotient operator admits the divisor as its adjoint.*

Up to this point, we have not required anything from the quotient operator.

(b1) If we add (3c) to (6a) and (6b), the model is completely determined; indeed, we obtain

$$a_{to} = \left(\gamma_{t+2,2} \begin{vmatrix} \psi_t^1 & \psi_t^2 \\ \psi_{t-1}^1 & \psi_{t-1}^2 \end{vmatrix} \Big/ \begin{vmatrix} \psi_{t-1}^1 & \psi_{t-1}^2 \\ \psi_{t-2}^1 & \psi_{t-2}^2 \end{vmatrix} \right)^{1/2} , \tag{7a}$$

hence

$$a_{t2}^2 = \gamma_{t2} \begin{vmatrix} \psi_{t-3}^1 & \psi_{t-3}^2 \\ \psi_{t-4}^1 & \psi_{t-4}^2 \end{vmatrix} \Big/ \begin{vmatrix} \psi_{t-2}^1 & \psi_{t-2}^2 \\ \psi_{t-3}^1 & \psi_{t-3}^2 \end{vmatrix} , \quad \text{sign}\,(a_{t2}) = \text{sign}\,(\gamma_{t2}) \tag{7b}$$

and

$$a_{t1}^2 = \gamma_{t+1,2} \begin{vmatrix} \psi_{t-1}^1 & \psi_{t-1}^2 \\ \psi_{t-3}^1 & \psi_{t-3}^2 \end{vmatrix}^2 \Big/ \begin{vmatrix} \psi_{t-1}^1 & \psi_{t-1}^2 \\ \psi_{t-2}^1 & \psi_{t-2}^2 \end{vmatrix} \begin{vmatrix} \psi_{t-2}^1 & \psi_{t-2}^2 \\ \psi_{t-3}^1 & \psi_{t-3}^2 \end{vmatrix} ,$$

$$\text{sign}(a_{t1}) = - \text{sign} \begin{vmatrix} \psi_{t-1}^1 & \psi_{t-1}^2 \\ \psi_{t-2}^1 & \psi_{t-2}^2 \end{vmatrix} \begin{vmatrix} \psi_{t-1}^1 & \psi_{t-1}^2 \\ \psi_{t-3}^1 & \psi_{t-3}^2 \end{vmatrix} . \tag{7c}$$

(b2) (3b) requires that ψ_t^1 and ψ_t^2 satisfy

$$\gamma_{t+1,2} \begin{vmatrix} \psi_{t-1}^1 & \psi_{t-1}^2 \\ \psi_{t-3}^1 & \psi_{t-3}^2 \end{vmatrix} + \gamma_{t1} \begin{vmatrix} \psi_{t-2}^1 & \psi_{t-2}^2 \\ \psi_{t-3}^1 & \psi_{t-3}^2 \end{vmatrix} + \gamma_{t2} \begin{vmatrix} \psi_{t-2}^1 & \psi_{t-2}^2 \\ \psi_{t-4}^1 & \psi_{t-4}^2 \end{vmatrix} = 0 \quad t \in \mathbb{Z}. \tag{8}$$

Now, (8) is satisfied on $\mathbb{Z}$ iff it is satisfied for one arbitrary value t_o in

$\mathbb{Z}$. Indeed, if ψ_t^1 and ψ_t^2 are two solutions of $\Gamma_t(L)\psi_t = 0$, then $\begin{vmatrix} \psi_t^1 & \psi_t^2 \\ \psi_{t_o-2}^1 & \psi_{t_o-2}^2 \end{vmatrix}$ is also a solution :

$$\gamma_{t+2,2}\begin{vmatrix} \psi_{t_o}^1 & \psi_{t_o}^2 \\ \psi_{t_o-2}^1 & \psi_{t_o-2}^2 \end{vmatrix} + \gamma_{t+1,1}\begin{vmatrix} \psi_{t_o-1}^1 & \psi_{t_o-1}^2 \\ \psi_{t_o-2}^1 & \psi_{t_o-2}^2 \end{vmatrix} + \gamma_{t1}\begin{vmatrix} \psi_{t_o-3}^1 & \psi_{t_o-3}^2 \\ \psi_{t_o-2}^1 & \psi_{t_o-2}^2 \end{vmatrix}$$

$$+ \gamma_{t2}\begin{vmatrix} \psi_{t_o-4}^1 & \psi_{t_o-4}^2 \\ \psi_{t_o-2}^1 & \psi_{t_o-2}^2 \end{vmatrix} = 0 \ ;$$

hence

$$\gamma_{t_o1}\begin{vmatrix} \psi_{t_o-2}^1 & \psi_{t_o-2}^2 \\ \psi_{t_o-3}^1 & \psi_{t_o-3}^2 \end{vmatrix} + \gamma_{t_o2}\begin{vmatrix} \psi_{t_o-2}^1 & \psi_{t_o-2}^2 \\ \psi_{t_o-4}^1 & \psi_{t_o-4}^2 \end{vmatrix}$$

$$= \gamma_{t_o+1,1}\begin{vmatrix} \psi_{t_o-1}^1 & \psi_{t_o-1}^2 \\ \psi_{t_o-2}^1 & \psi_{t_o-2}^2 \end{vmatrix} + \gamma_{t_o+2,2}\begin{vmatrix} \psi_{t_o}^1 & \psi_{t_o}^2 \\ \psi_{t_o-2}^1 & \psi_{t_o-2}^2 \end{vmatrix}$$

In the stationary case $\Gamma_t(L) = \Gamma(L)$, if we admit time-independent models $A_t(L) = A(L)$ only, (8) reduces to a property of suitably chosen roots of equations of the type $A(x)\ A(1/x) = 0$.

(b3) Let us show, now, that (8) implies that condition (3a) is automatically satisfied; (3a) can be written as

$$\gamma_{t+2,2}\begin{vmatrix} \psi_t^1 & \psi_t^2 \\ \psi_{t-1}^1 & \psi_{t-1}^2 \end{vmatrix}\begin{vmatrix} \psi_{t-2}^1 & \psi_{t-2}^2 \\ \psi_{t-3}^1 & \psi_{t-3}^2 \end{vmatrix} + \gamma_{t2}\begin{vmatrix} \psi_{t-1}^1 & \psi_{t-1}^2 \\ \psi_{t-2}^1 & \psi_{t-2}^2 \end{vmatrix}\begin{vmatrix} \psi_{t-3}^1 & \psi_{t-3}^2 \\ \psi_{t-4}^1 & \psi_{t-4}^2 \end{vmatrix} + \gamma_{t+1,2}\begin{vmatrix} \psi_{t-1}^1 & \psi_{t-1}^2 \\ \psi_{t-3}^1 & \psi_{t-3}^2 \end{vmatrix}^2$$

$$- \gamma_{to}\begin{vmatrix} \psi_{t-1}^1 & \psi_{t-1}^2 \\ \psi_{t-2}^1 & \psi_{t-2}^2 \end{vmatrix}\begin{vmatrix} \psi_{t-2}^1 & \psi_{t-2}^2 \\ \psi_{t-3}^1 & \psi_{t-3}^2 \end{vmatrix} = 0 \ . \qquad (9)$$

From (8), we know that

$$\gamma_{t+1,2}\begin{vmatrix}\psi^1_{t-1} & \psi^2_{t-1}\\ \psi^1_{t-3} & \psi^2_{t-3}\end{vmatrix} = -\gamma_{t1}\begin{vmatrix}\psi^1_{t-2} & \psi^2_{t-2}\\ \psi^1_{t-3} & \psi^2_{t-3}\end{vmatrix} - \gamma_{t2}\begin{vmatrix}\psi^1_{t-2} & \psi^2_{t-2}\\ \psi^1_{t-4} & \psi^2_{t-4}\end{vmatrix} ;$$

since ψ^1_t and ψ^2_t are solutions of $\Gamma_t(L)\psi_t = 0$,

$$\gamma_{t+2,2}\begin{vmatrix}\psi^1_{t} & \psi^2_{t}\\ \psi^1_{t-1} & \psi^2_{t-1}\end{vmatrix} + \gamma_{to}\begin{vmatrix}\psi^1_{t-2} & \psi^2_{t-2}\\ \psi^1_{t-1} & \psi^2_{t-1}\end{vmatrix} + \gamma_{t1}\begin{vmatrix}\psi^1_{t-3} & \psi^2_{t-3}\\ \psi^1_{t-1} & \psi^2_{t-1}\end{vmatrix} = \gamma_{t2}\begin{vmatrix}\psi^1_{t-1} & \psi^2_{t-1}\\ \psi^1_{t-4} & \psi^2_{t-4}\end{vmatrix} .$$

Replacing in (9), we obtain

$$\begin{vmatrix}\psi^1_{t-1} & \psi^2_{t-1}\\ \psi^1_{t-4} & \psi^2_{t-4}\end{vmatrix}\begin{vmatrix}\psi^1_{t-2} & \psi^2_{t-2}\\ \psi^1_{t-3} & \psi^2_{t-3}\end{vmatrix} + \begin{vmatrix}\psi^1_{t-1} & \psi^2_{t-1}\\ \psi^1_{t-3} & \psi^2_{t-3}\end{vmatrix}\begin{vmatrix}\psi^1_{t-4} & \psi^2_{t-4}\\ \psi^1_{t-2} & \psi^2_{t-2}\end{vmatrix} + \begin{vmatrix}\psi^1_{t-1} & \psi^2_{t-1}\\ \psi^1_{t-2} & \psi^2_{t-2}\end{vmatrix}\begin{vmatrix}\psi^1_{t-3} & \psi^2_{t-3}\\ \psi^1_{t-4} & \psi^2_{t-4}\end{vmatrix} = 0 ,$$

which is an easily verifiable determinantal identity.

Summing up, if ψ^1_t and ψ^2_t are two linearly independent solutions of $\Gamma_t(L)\psi_t=0$ such that (8) is satisfied for some t_o in $\mathbb{Z}$, then the coefficients given in (7) define a factorization $\Gamma_t(L) = A_t(L)\, A^*_t(L)$. How many are these factorizations ? The selections of ψ^1_t and ψ^2_t involve four initial conditions each. But we are only interested in the vectorial space they are spanning, which depends on four parameters only (if t_o is an arbitrary moment in $\mathbb{Z}$, we always can fix $\psi^1_{t_o-1} = \psi^2_{t_o-2} = 1$ and $\psi^1_{t_o-2} = \psi^2_{t_o-1} = 0$). From these four remaining parameters, one can be dropped using (8). There exists thus a three-parameters family of possible factorizations; parametrizations for this family are

$$\psi^1_{t_o-3} , \psi^2_{t_o-3} , \psi^1_{t_o-4} \quad \left(\text{with} \begin{pmatrix}\psi^1_{t_o-1} & \psi^2_{t_o-1}\\ \psi^1_{t_o-2} & \psi^2_{t_o-2}\end{pmatrix} = \begin{pmatrix}1 & 0\\ 0 & 1\end{pmatrix}\right) ; \tag{10a}$$

or

$$a_{t_o1} , \quad a_{t_o2} , \quad a_{t_o-1,1} \quad (\text{as in } \textit{Theorem 1}) ; \tag{10b}$$

or

$$\frac{\begin{vmatrix}\psi^1_{t_o-1} & \psi^2_{t_o-1}\\ \psi^1_{t_o-3} & \psi^2_{t_o-3}\end{vmatrix}}{\begin{vmatrix}\psi^1_{t_o-1} & \psi^2_{t_o-1}\\ \psi^1_{t_o-2} & \psi^2_{t_o-2}\end{vmatrix}} , \quad -\frac{\begin{vmatrix}\psi^1_{t_o-1} & \psi^2_{t_o-1}\\ \psi^1_{t_o-4} & \psi^2_{t_o-4}\end{vmatrix}}{\begin{vmatrix}\psi^1_{t_o-1} & \psi^2_{t_o-1}\\ \psi^1_{t_o-2} & \psi^2_{t_o-2}\end{vmatrix}} , \quad \frac{\begin{vmatrix}\psi^1_{t_o-2} & \psi^2_{t_o-2}\\ \psi^1_{t_o-4} & \psi^2_{t_o-4}\end{vmatrix}}{\begin{vmatrix}\psi^1_{t_o-1} & \psi^2_{t_o-1}\\ \psi^1_{t_o-2} & \psi^2_{t_o-2}\end{vmatrix}} . \tag{10c}$$

(b4) If we want, however, the operators $A_t(L)$ to have real coefficients only, we have a last condition :

$$\gamma_{t+2,2} \begin{vmatrix} \psi_t^1 & \psi_t^2 \\ \psi_{t-1}^1 & \psi_{t-1}^2 \end{vmatrix} \begin{vmatrix} \psi_{t-1}^1 & \psi_{t-1}^2 \\ \psi_{t-2}^1 & \psi_{t-2}^2 \end{vmatrix} > 0 \quad \forall\, t \in \mathbb{Z} . \tag{11}$$

From *Corollary 2* to *Theorem 1*, we know that there exists a nonempty subset of "initial parameters" ((10a), (10b) or (10c)) such that (11) holds. In the MA(1) case, we obtained [5] an explicit formulation for the corresponding set of parameters (a real interval); a similar explicit formulation is probably obtainable here in terms of *generalized continued fractions* (cf. [3]).

REFERENCES

[1] Anderson, O.D. : On the individual moving average inequality. Metrika (1978) 241-245.

[2] Anderson, T.W. : Statistical analysis of time series. J. Wiley, N.Y. (1971).

[3] De Bruin, M.G. : Convergence of Generalized C-Fractions. Journal of Approx. Th. (1978) 177-207.

[4] Hallin, M. : Invertibility and generalized invertibility of time series models; Addendum to "Invertibility and ...". J.R.S.S (B), (1980) 210-212 and (1981) 103.

[5] Hallin, M. : Nonstationary first-order moving average processes : the model-building problem. Time Series Analysis, O.D. Anderson and M.R. Perryman eds. North-Holland, Amsterdam (1981), 189-206.

[6] Hallin, M. and Ingenbleek, J.-F. : Nonstationary Yule-Walker equations. Submitted for publication (1981).

[7] Hallin, M. and Ingenbleek, J.-F. : The model-building problem for nonstationary multivariate autoregressive processes. Time Series Analysis : Theory and Practice 1. O.D. Anderson ed., North-Holland, Amsterdam (to appear).

[8] Marcus, M. and Minc, H. : A survey of matrix theory and matrix inequalities. Allyn and Bacon, Boston (1964).

[9] Miller, K.S. : Linear difference equations. Benjamin, N.Y. (1968).

APPLIED TIME SERIES ANALYSIS
O.D. Anderson & M.R. Perryman (eds.)

INTERNATIONAL INTEREST RATE INTERDEPENDENCE: A TIME SERIES ANALYSIS

Stephen E. Haynes

Department of Economics
University of Oregon
Eugene, Oregon

This paper investigates interest rate interdependence using a formal model of international capital flows. The model is tested with data from the U.S. and Canada for both fixed and flexible exchange rate periods. Cross spectral and cross correlation estimates support the model where only the Canadian interest rate is endogenous, and indicate that (1) adjustment occurs more rapidly under fixed exchange rates; (2) interdependence results from capital flows; and (3) risk alone does not explain why long-run interdependence is less-than-perfect.

INTRODUCTION

In comparison to closed economy studies of the term structure of interest rates, empirical research into the interdependence of interest rates between countries (hereafter referred to as IRI) is in a rather rudimentary stage of development. The few exploratory studies represent correlation research since they lack reference to formal models.[1] Other related research uses formal models, but does not focus on IRI and frequently relies on restrictive assumptions concerning the response of interest rates to capital movements. For example, at one extreme, the literature on both covered interest rate arbitrage and capital flow estimation typically ignores IRI because interest rates are assumed exogenous, i.e., independent of capital movements.[2] At the other extreme, many of the "Monetary" capital flow papers assume that capital movements correlate interest rates perfectly.[3]

This study investigates the pattern and source of IRI using a formal model of international capital markets. Developed by Haynes & Pippenger (1979, 1982), the model predicts that capital flows influence domestic and foreign interest rates, generating both direct and inverse links of IRI. Special cases of the model, consistent with existing capital flow formulations, imply exogenous interest rates, perfect IRI, or the endogeneity of only one interest rate.

Empirical tests of the model and its cases examine short- and long-term interest rate data from the United States and Canada for both fixed and flexible exchange rate periods. The tests focus on three issues. First, the cross spectral technique of model discrimination complemented with cross correlation analysis is used to choose among alternative models. Estimates show direct long-run IRI which is less-than-perfect, but no inverse IRI, and support the model where the Canadian but not the U.S. interest rate is endogenous. Furthermore, estimates indicate that adjustment occurs more rapidly under fixed exchange rates than under floating rates, a finding consistent with the hypothesis that exchange rate uncertainty associated with a floating regime retards investment.

The second issue concerns the source of the interdependence. Bohi (1972) has suggested that long-run IRI stems from the covariation of business cycles between nations, not from capital flows, implying that the capital flow model is misspecified. Evidence examined herein fails to support his argument.

MODEL

As derived in Haynes & Pippenger (1979, 1982), the model is the following:

$$r_1(t) = e_1(t) - \lambda_1 A_1(t) \tag{1}$$

$$r_2(t) = e_2(t) + \lambda_2 A_2(t) \tag{2}$$

$$A_1(t) = A_2(t) = A(t) \tag{3}$$

$$A(t) = \frac{1}{b + apD - aD^2} [r_1(t) - r_2(t) + z(t)] \tag{4}$$

where $A_1(t)$ is the excess stock supply of securities in country one (the foreign country);
$A_2(t)$ is the excess stock demand for similar securities in country two (the domestic country);
$r_1(t)$ and $r_2(t)$ are the observed interest rates in countries one and two, respectively;
$e_1(t)$ and $e_2(t)$ are the autarky interest rates in countries one and two, respectively, and are assumed to follow independent random walks;[4]
$z(t)$ is an error term;
p is investors' rate of discount, assumed constant;
λ_1, λ_2, a, and b are positive constants;
and D is the differential operator ($Dx(t)$ equals $dx(t)/dt$).

Equations (1) and (2) describe intranational behavior; equation (3) is a two-country market clearing condition; and equation (4) describes how wealth maximizers invest internationally in the presence of transaction costs (a positive) and risk costs (b positive). Other returns (costs) of international investment are captured by error term $z(t)$, and are assumed to have a transitory influence on investors' behavior. Thus, $z(t)$ is assumed equal to $Dy(t)$, where $y(t)$ is white noise.

Three special cases of the model can be identified. First, in the absence of transaction and risk costs (a and b equal zero) and ignoring $z(t)$, equation (2) reduces to "$r_1(t) = r_2(t)$." This static case is similar to Monetary capital flow models in that both predict perfect capital market integration and interest rate interdependence. Second, assuming significant transaction and risk costs (a and b positive) and exogenous interest rates (λ_1 and λ_2 equal zero), the model becomes a popular single equation portfolio-stock formulation.[5] Third, by assuming the endogeneity of the foreign interest rate $r_1(t)$, but not the domestic interest rate $r_2(t)$ (λ_1 positive and λ_2 equals zero), the model becomes one where the foreign country is small relative to the domestic country. Therefore, instead of imposing upon the data an arbitrary assumption such as perfect interest rate interdependence at one extreme, or exogenous interest rates at the other extreme, the model is sufficiently general to permit the data to determine the pattern of interest rate endogeneity.

As is apparent from the reduced-form equations of the model (equations (A) and (B) in Appendix 1), the model predicts both direct and inverse links between $r_1(t)$ and $r_2(t)$. Variations in autarky interest rates $e_1(t)$ and $e_2(t)$ lead to the direct link. For example, an increase in $e_1(t)$ generates a rise in $r_1(t)$, hence also in $A(t)$, a capital flow which increases $r_2(t)$ (and moderates the rise in $r_1(t)$). Variations in $z(t)$ cause the inverse link between $r_1(t)$ and $r_2(t)$. For example, an increase in $z(t)$ leads to a rise in $A(t)$, a capital flow which then decreases $r_1(t)$ but increases $r_2(t)$.

The net link between $r_1(t)$ and $r_2(t)$ depends upon the variation of $e_1(t)$ plus $e_2(t)$ relative to that of $z(t)$. If the variation in $e_1(t)$ plus $e_2(t)$ is large (small) relative to that of $z(t)$, interest rates $r_1(t)$ and $r_2(t)$ are directly (inversely) correlated. If the variation in $e_1(t)$ plus $e_2(t)$ approximates that of $z(t)$, interest rates could appear uncorrelated since the direct and inverse links may cancel.[6] Thus, even with significant capital mobility and endogenous interest rates, IRI could appear insignificant.

EMPIRICAL TESTS

United States and Canada data are used to investigate the three issues concerning IRI summarized in the introduction.[7] In the following tests, Canada is country one, the foreign country, and the United States is country two, the domestic country. Three samples of monthly end-of-period data are examined: January 1955 to April 1962; May 1962 to May 1970; and June 1970 to November 1978. The first and third are periods of floating exchange rates; the second is a period of fixed exchange rates. For each period, IRI is examined using both three-month treasury bill rates and long-term government bond yields. In addition, some comparable weekly data are examined.

Pattern of IRI

The first issue is the empirical examination of IRI from the perspective of the model and its three cases. Time domain correlation analysis of the interdependence may prove ineffective in disentangling the direct and inverse links since, if present, the links may tend to cancel. Spectral analysis, however, can disentangle those direct and inverse links of IRI which are predicted by the model.[8] To see this, consider the exogenous variables of the model--autarky interest rates $e_1(t)$ and $e_2(t)$ and error term $z(t)$. The interest rates are assumed to perform random walks, implying that their variation is concentrated at low frequencies, i.e., the long run. In contrast, the term $z(t)$ is assumed to be significant primarily at high frequencies, i.e., the short run. Since changes in $e_1(t)$ and $e_2(t)$ lead to direct IRI, and changes in $z(t)$ cause inverse IRI, the model predicts direct IRI at low frequencies and inverse IRI at high frequencies.[9]

Such decomposition of IRI can be seen formally by examining the theoretical cross spectral statistics derived from the reduced-form equations of the model (see Appendix 1). Theoretical coherence square (K^2) and phase (Ph) between the Canadian and U.S. interest rates predicted by the model and its cases are plotted in Fig. 1. For the general model, significant coherence square with a phase of unity (and zero) is predicted at low frequencies, reflecting the direct IRI; and significant coherence square with a phase of 0.5 is predicted at high frequencies, reflecting the inverse IRI. The static case (a, b, and $z(t)$ equal zero) implies perfect and direct IRI, i.e., a coherence square of unity and phase of unity (and zero) for all frequencies. For the single equation portfolio stock case (λ_1 and λ_2 equal zero), insignificant coherence square is predicted for all frequencies. For the case where the foreign nation is small relative to the domestic one (λ_1 positive and λ_2 equal zero), significant coherence square with a phase of unity (and zero) is predicted at low frequency; and coherence square declines and phase lies in the first quandrant, i.e., between zero and 0.25, as frequency increases.

Empirical coherence square and phase between changes in the Canadian and U.S. short-term interest rates are recorded in Table 1.[10] Estimates between long-term interest rates are similar (they are omitted for brevity). Comparison of the theoretical cross spectral statistics in Fig. 1 to corresponding empirical ones permits discrimination among the model and its cases. At low frequency, empirical coherence square is significant and phase does not differ significantly from unity (and zero). As frequency increases (i.e., as period decreases), coherence square tends to decline and the confidence interval on most phase estimates includes some

Table 1. Empirical Coherence Square (K^2) and Phase (Ph) Between Changes in Canadian and U.S. Three-Month Treasury Bill Interest Rates

(cycles/ month) FREQ.	(months) PERIOD	Jan. 1955–April 1962 $K^2_{CAN,US}$	Jan. 1955–April 1962 $Ph_{CAN,US}$	May 1962–May 1970 $K^2_{CAN,US}$	May 1962–May 1970 $Ph_{CAN,US}$	June 1970–Nov. 1978 $K^2_{CAN,US}$	June 1970–Nov. 1978 $Ph_{CAN,US}$
0.00	∞	0.837*	0.983±.05	0.487*	0.984±.12	0.473*	0.035±.13
0.04	23.81	0.564*	0.000±.10	0.433*	0.011±.14	0.566*	0.041±.10
0.08	12.05	0.446*	0.057±.13	0.225	0.058	0.483*	0.043±.13
0.12	8.00	0.307*	0.099±.21	0.121	0.102	0.185	0.149
0.17	5.99	0.061	0.175	0.115	0.134	0.325*	0.265±.19
0.21	4.80	0.154	0.261	0.041	0.058	0.245	0.246
0.25	4.00	0.088	0.332	0.027	0.887	0.258	0.301
0.29	3.42	0.162	0.789	0.049	0.012	0.131	0.306
0.33	3.00	0.507*	0.870±.11	0.097	0.015	0.087	0.001
0.37	2.67	0.291	0.002	0.174	0.956	0.161	0.004
0.42	2.39	0.129	0.026	0.303*	0.953±.23	0.225	0.013
0.46	2.18	0.276	0.936	0.152	0.952	0.172	0.983
0.50	2.00	0.585*	0.961±.09	0.004	0.820	0.020	0.925

NOTE: Asterisk (*) identifies estimates which are significant at the 5 percent level; plus and minus (±) value forms a 95 percent confidence interval where coherence is significant.

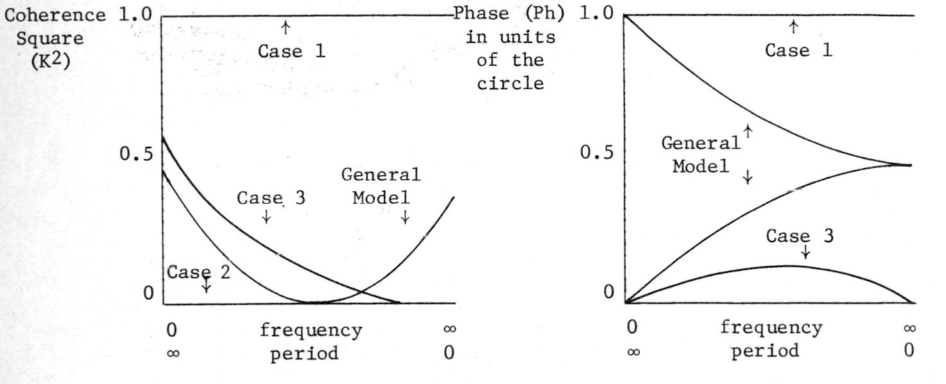

Notes: General Model is equations (1), (2), (3), and (4); Case 1 is statics (a, b, and z(t) equal zero); Case 2 is single equation portfolio stock (λ_1 and λ_2 equal zero); and Case 3 is small foreign nation (λ_2 equals zero). Also, phase for General Model remains between zero and 0.5 if λ_2 [autospectrum of $e_1(t)$] < λ_1[autospectrum of $e_2(t)$], and between 0.5 and unity, otherwise; and phase for Case 3 remains between zero and 0.25.

Fig. 1: Theoretical Coherence Square (K^2) and Phase (Ph) from the Foreign to the Domestic Interest Rate

value between zero and 0.25. Such evidence generally supports the model where λ_1 is positive but λ_2 is zero.[11]

In the presence of lagged adjustment, a more direct method of investigating the endogeneity of interest rates involves examination of the cross correlation estimates between changes in the two interest rates.[12] In particular, if λ_1 is positive but λ_2 is zero and adjustment is lagged, changes in the U.S. rate should lead changes in the Canadian rate, but not vice versa.

Table 2 records cross correlation estimates between changes in the Canadian and U.S. interest rates using the monthly data. For the two floating exchange rate periods and for short- and long-term interest rates, changes in the U.S. rate significantly lead (at the five percent level) changes in the Canadian rate by one and/or two months, and the sign of the link is positive. For the fixed exchange rate period, a similar pattern is present for short-term interest rates at the ten percent level. Interpreted from the perspective of the capital flow model, this evidence indicates that λ_1 is positive.

Based on short-term interest rates, many of the cross correlation estimates where the U.S. rate leads the Canadian rate by four or more months are negative,

Table 2. Empirical Cross Correlation Estimates Between Changes in Canadian and U.S. Interest Rates: Monthly Data

	(months)	Jan. 1955– April 1962		May 1962– May 1970		June 1970– Nov. 1978	
	LAG	BILLS	BONDS	BILLS	BONDS	BILLS	BONDS
	12	-.109	0.012	-0.136	0.061	0.145	0.082
	11	0.021	0.030	-0.207	0.046	0.114	0.164
	10	-0.036	0.022	-0.175	-0.102	0.057	-0.115
	9	-0.188	-0.014	-0.135	0.071	0.119	0.042
	8	-0.057	-0.020	-0.132	-0.102	-0.106	-0.020
	7	-0.321*	-0.058	0.007	0.046	-0.150	-0.122
	6	0.206	0.024	-0.092	0.107	-0.067	-0.139
U.S.	5	0.039	0.069	-0.052	-0.026	0.068	0.042
Rate	4	-0.072	-0.039	-0.030	0.005	0.023	0.042
Leading	3	0.110	0.009	0.093	-0.117	0.135	0.085
	2	0.364*	0.030	0.182	0.112	0.388*	0.142
	1	0.072	0.297*	0.172	0.071	0.387*	0.247*
	0	0.402*	0.392*	0.388*	0.632*	0.349*	0.544*
	1	0.018	0.189	0.075	0.020	0.050	0.115
	2	0.128	-0.191	-0.075	-0.173	0.160	0.057
	3	0.123	0.132	0.018	0.015	0.130	-0.150
	4	0.095	0.201	-0.051	-0.199	-0.013	-0.042
Canadian	5	-0.077	0.088	0.060	-0.005	-0.073	-0.135
Rate	6	-0.046	-0.092	-0.023	0.046	-0.125	-0.142
Leading	7	0.059	-0.055	-0.066	0.087	-0.095	-0.092
	8	0.011	0.062	-0.130	-0.117	-0.242*	0.072
	9	-0.001	0.044	-0.012	0.020	-0.041	0.055
	10	0.066	-0.170	0.000	0.010	-0.011	0.062
	11	-0.164	-0.138	-0.097	-0.061	0.139	0.092
	12	0.158	-0.088	-0.030	0.122	-0.009	-0.002

NOTE: Asterisk labels estimates significant at the 5 percent level.

with one estimate significant at the five percent level and four significant at the ten percent level. Although weak, such evidence (in combination with the positive estimates at lags one and/or two) suggests that λ_1 declines over time, i.e., that the supply response captured by equation (1) becomes more elastic as the time horizon lengthens.

Cross correlation estimates where the Canadian rate leads the U.S. rate show no significant positive links, suggesting that λ_2 is zero. A rough pattern of positive estimates at lags one through four and negative estimates thereafter, however, indicates that capital might have a weak influence on the U.S. interest rate, and that the influence diminishes over time.

Table 3 records cross correlation estimates between changes in the U.S. and Canadian treasury bill rates for most of the latter two periods using weekly data.[13] By indicating that the U.S. rate leads the Canadian rate out to approximately eight weeks, but that the Canadian rate does not lead the U.S. rate, such estimates are consistent with corresponding ones based on monthly data in Table 2. To summarize, by supporting the endogeneity of the Canadian but not the U.S. interest rate, these cross correlation estimates corroborate the spectral findings.

Table 3. Empirical Cross Correlation Estimates Between Changes in Canadian and U.S. Treasury Bill Rates: Weekly Data

	May 1962-May 1970		June 1970-May 1974	
(weeks) LAG	U.S. RATE LEADING	CANADIAN RATE LEADING	U.S. RATE LEADING	CANADIAN RATE LEADING
0	0.237*	0.237*	0.160*	0.160*
1	0.233*	0.051	0.181*	0.051
2	0.100*	-0.018	0.135	-0.039
3	0.085	0.066	0.161*	-0.020
4	0.109*	0.087	0.224*	-0.010
5	0.132*	0.110*	0.218*	0.013
6	0.126*	0.053	0.054	-0.038
7	0.056	-0.016	0.176*	-0.000
8	0.006	-0.041	0.158*	0.141
9	0.099	-0.004	0.077	0.028
10	0.115*	-0.004	0.068	0.057
11	0.008	0.039	0.014	0.051
12	0.091	0.040	0.138	0.005
13	0.110*	0.025	0.095	0.025
14	-0.015	-0.025	0.160*	-0.040
15	-0.048	-0.010	0.048	0.026
16	0.028	-0.032	0.070	-0.013
17	-0.046	0.012	0.056	-0.012
18	-0.068	0.029	0.064	-0.023
19	-0.062	-0.001	0.027	-0.056
20	-0.012	-0.005	-0.073	0.028
21	-0.004	0.003	0.074	-0.014
22	-0.038	-0.015	0.090	-0.013
23	-0.020	-0.029	-0.062	-0.014
24	0.029	0.000	0.054	0.070

NOTE: Asterisk labels estimates significant at the 5 percent level.

A final observation about the pattern of IRI concerns the adjustment lags under the two exchange rate regimes. Cross correlation estimates in Table 2 suggest that the lag of the response of the Canadian interest rate to changes in the U.S. rate is longer in the two floating exchange rate periods than in the fixed rate period. Also, estimates in Table 3 show that the strongest response occurs at the four-week lag in the floating rate period but contemporaneously in the fixed rate period. Such evidence regarding more rapid adjustment under fixed exchange rates is consistent with the hypothesis that a floating exchange rate regime and resulting exchange rate uncertainty retard international investment.

Source of IRI

In an interesting exploratory paper, Bohi (1972) postulated that high frequency IRI between nations arises from capital flows, while low frequency IRI results from the covariation of national business cycles. To test IRI arising solely from capital flows, Bohi examined high frequency IRI among short-term interest rates from ten industrialized nations. Upon finding minimal significant interdependence at high frequencies (a finding consistent with those herein), Bohi (1972, p. 600) concluded that "either domestic monetary conditions are more important to interest rate behavior than external monetary conditions; their monetary authorities have succesfully insulated their money markets from foreign influence; or capital is not very mobile across national frontiers."

To test Bohi's contention that low frequency IRI results from business cycle covariation and not from capital flows, empirical coherence square and phase describing IRI were reestimated after conditionalizing the link on the industrial production indexes of both the U.S. and Canada. Such conditional statistics are very similar to corresponding bivariate ones.[14] Contrary to the assertion of Bohi, this finding suggests that significant IRI at low frequencies results from the international movement of capital, hence that autarky interest rates $e_1(t)$ and $e_2(t)$ of the model are statistically independent, as assumed.

Explanation for Less-Than-Perfect IRI

The model provides three explanations why IRI may not be perfect.[15] As equation (4) shows, they are transaction costs (a positive), risk (b positive), and the significance of capital flows drive by forces other than interest rates ($z(t)$ nonzero). At low frequency, however, the model implies that risk alone prevents perfect IRI. This follows because, at low frequency, transaction costs correlated with the speed of adjustment become insignificant and the variation in $z(t)$ is assumed zero. As discussed next, the low frequency risk explanation can be tested by examining the third cross spectral statistic, the gain.

The empirical low frequency gain from the U.S. interest rate to the Canadian one, and vice versa, are recorded in Table 4. In five of the six samples, the gain from the U.S. rate to the Canadian rate exceeds the gain in the other direction, and in three samples the two gains differ by a factor of two. Also, for the single sample where the gain from the Canadian to the U.S. rate is larger, their difference is minimal. Such apparent assymetry in arbitrage can only be explained by the model if the response of interest rates to capital flows differs, i.e., if λ_1 and λ_2 differ in magnitude. Evidence presented above suggests the endogeneity of the Canadian but not the U.S. interest rate, i.e., that λ_1 is positive and λ_2 is zero.

Comparison of the theoretical to the empirical gain statistics at low frequency under the assumption that λ_1 is positive and λ_2 is zero provides a test of the risk explanation of less-than-perfect IRI. With λ_2 equaling zero, the theoretical low frequency gain from the Canadian to the U.S. interest rate unambiguously exceeds the gain from the U.S. to the Canadian interest rate.[16] However, this prediction is inconsistent with the evidence reported in Table 4, which suggests that risk is not the sole impediment to perfect low frequency IRI.

Table 4. Low Frequency Gain Estimates Between Changes in Canadian and U.S. Interest Rates

	January 1955–April 1962	May 1962–May 1970	June 1970–November 1978
BILLS			
Gain from the U.S. to Canadian Rate	1.289±.22	1.189±.39	0.668±.33
BILLS			
Gain from the Canadian to the U.S. Rate	0.649±.16	0.409±.21	0.709±.35
BONDS			
Gain from the U.S. to Canadian Rate	1.151±.22	0.813±.17	0.8821.16
BONDS			
Gain from the Canadian to the U.S. Rate	0.609±.19	0.746±.14	0.700±.13

NOTE: Plus and minus (±) number forms a 95 percent confidence interval.

Relaxation of the variance assumption of error term z(t) does lead to consistency between the theoretical and empirical gain statistics. In particular, if the model were modified by assuming that z(t) approximates white noise, i.e., that the variance of z(t) is evenly distributed over all frequencies, the term z(t) could also impede perfect low frequency IRI. Given such a modification (and assuming the insignificance of risk), the theoretical gain from the U.S. to the Canadian interest rate must exceed the gain in the other direction.[17] Such a prediction is consistent with the empirical gain statistics, implying that error term z(t), unlike risk, could represent the sole explanation why the significant low frequency IRI is less-than-perfect.[18]

CONCLUSION

Most empirical studies of interest rate interdependence between nations represent correlation analysis since formal models are neither specified nor tested. Most related research using models, such as the literature on covered interest rate arbitrage and international capital flow estimation, is founded on restrictive assumptions concerning the response of interest rates to capital flows, hence is not useful in modeling interest rate interdependence resulting from capital flows.

This paper, an extension of one test in Haynes and Pippenger (1982), investigates interest rate interdependence from the perspective of a formal model of international capital flows. The model predicts both direct and inverse interdependence resulting from capital flows; special cases of the model are consistent with existing models, and imply exogenous interest rates, perfect interest rate interdependence, or the endogeneity of one but not both interest rates.

The cross spectral technique of model discrimination, complemented with cross correlation analysis, is used to choose among alternative models. Thus, the data and not ad hoc theory determine the pattern and source of interest rate interdependence. Empirical examination of short- and long-term interest rate data from the United States and Canada over both fixed and floating exchange rate periods shows only a long-run direct link, and supports the model where the Canadian but not the U.S. interest rate is endogenous. The evidence also indicates that (1) the adjustment process occurs more rapidly under fixed (in comparison to floating) exchange rates; (2) the observed interdependence stems from capital flows and not from the covariation of national business cycles; and (3) risk by itself does not explain why long-run interdependence is less-than-perfect.

APPENDIX 1

The structural equations of the model are equations (1), (2), (3), and (4). Reduced form equations for $r_1(t)$ and $r_2(t)$ are equations (A) and (B), respectively:

$$r_1(t) = \frac{1}{(b+apD-aD^2+\lambda_1+\lambda_2)}[(b+apD-aD^2+\lambda_2)\, e_1(t) + \lambda_1 e_2(t) - \lambda_1 z(t)] \tag{A}$$

$$r_2(t) = \frac{1}{(b+apD-aD^2+\lambda_1+\lambda_2)}[\lambda_2 e_1(t) + (b+apD-aD^2+\lambda_1)\, e_2(t) + \lambda_2 z(t)] \tag{B}$$

The following autospectra and cross spectra are derived from equations (A) and (B).

The autospectrum of X(t) is defined as $E(dX(iw)\cdot\overline{dX(iw)})$, and the cross spectrum from X(t) to Y(t) is $E(dY(iw)\cdot\overline{dX(iw)})$; where E is the expectations operator; dX(iw) and dY(iw) are the Fourier transforms of X(t) and Y(t), respectively; $\overline{dX(iw)}$ is the complex conjugate of dX(iw); w is frequency in cycles per time period; and i equals $\sqrt{-1}$.

For the general model, the frequency response function from $r_1(t)$, the input, to $r_2(t)$, the output, is the cross spectrum from $r_1(t)$ to $r_2(t)$, $V_{r_1r_2}(iw)$, divided by the autospectrum of $r_1(t)$, $V_{r_1}(w)$:

$$\frac{V_{r_1r_2}(iw)}{V_{r_1}(w)} =$$

$$\frac{\lambda_2(b+aw_2+\lambda_2)V_{e_1}(w)+\lambda_1(b+aw^2+\lambda_1)V_{e_2}(w)-\lambda_1\lambda_2V_z(w)+i[\lambda_1 awpV_{e_2}(w)-\lambda_2 awpV_{e_1}(w)]}{[(b+aw^2+\lambda_2)^2+(awp)^2]V_{e_1}(w)+\lambda_1{}^2V_{e_2}(w)+\lambda_1{}^2V_z(w)}$$

where $V_{e_1}(w)$, $V_{e_2}(w)$, and $V_z(w)$ are the autospectra of $e_1(t)$, $e_2(t)$, and $z(t)$, respectively.

The coherence square between $r_1(t)$ and $r_2(t)$, $K_{r_1r_2}{}^2(w)$, is

$$K_{r_1r_2}{}^2(w) = \frac{|V_{r_1r_2}(iw)|^2}{V_{r_1}(w)V_{r_2}(w)} = \frac{G}{H\cdot I}$$

where

$$G = [\lambda_2(b+aw^2+\lambda_2)V_{e_1}(w)+\lambda_1(b+aw^2+\lambda_1)V_{e_2}(w)-\lambda_1\lambda_2V_z(w)]^2+[\lambda_1 awpV_{e_2}(w)-\lambda_2 awpV_{e_1}(w)]^2;$$

$H = [(b+aw^2+\lambda_2)^2+(awp)^2]V_{e_1}(w)+\lambda_1{}^2V_{e_2}(w)+\lambda_1{}^2V_z(w)$;

$I = [(b+aw^2+\lambda_1)^2+(awp)^2]V_{e_2}(w)+\lambda_2{}^2V_{e_1}(w)+\lambda_2{}^2V_z(w)$; and

$|V_{r_1r_2}(iw)|$ is the modulus of $V_{r_1r_2}(iw)$.

The gain from $r_1(t)$ to $r_2(t)$, $G_{r_1r_2}(w)$, is

$$G_{r_1r_2}(w) = \frac{|V_{r_1r_2}(iw)|}{V_{r_1}(w)} = \frac{\sqrt{G}}{H}$$

and the gain from $r_2(t)$ to $r_1(t)$, $G_{r_2r_1}(w)$, is

$$G_{r_2r_1}(w) = \frac{|V_{r_1r_2}(iw)|}{V_{r_2}(w)} = \frac{\sqrt{G}}{I}$$

The phase from $r_1(t)$ to $r_2(t)$, $Ph_{r_1r_2}(w)$, is

$$Ph_{r_1r_2}(w) = \tan^{-1}\left[\frac{-QV_{r_1r_2}(iw)}{CV_{r_1r_2}(iw)}\right] =$$

$$\tan^{-1}\left[\frac{\lambda_1 awpV_{e_2}(w)-\lambda_2 awpV_{e_1}(w)}{\lambda_2(b+aw^2+\lambda_2)V_{e_1}(w)+\lambda_1(b+aw^2+\lambda_1)V_{e_2}(w)-\lambda_1\lambda_2V_z(w)}\right]$$

where $QV_{r_1r_2}(iw)$ is the quadrature of the cross spectrum; and $CV_{r_1r_2}(iw)$ is the co-spectrum of the cross spectrum.

Given the assumptions above the behavior of $e_1(t)$, $e_2(t)$, and $z(t)$, the theoretical statistics in Fig. 1 are obtained. Patterns for the three cases are obtained by setting the appropriate parameters to zero.

FOOTNOTES

[1]Examples include papers by Scott (1970), Cooper (1971), Argy & Hodjera (1973), Bisignano (1975), Laffer (1975), Genberg (1976) and Beenstock & Longbottom (1981), based on standard time domain methods of correlation analysis; Bohi (1972), based on cross spectral analysis; Fase (1976), based on principle components analysis; and White & Woodbury (1980), based on factor analysis. For the most part, these papers assume that observed IRI results from capital flows. For examples of studies that model and test the determination of the Eurodollar or Eurobond rate assuming the exogeneity of the U.S. interest rate, see Hendershott (1967) and Solnik & Grail (1975); for a summary of this research, see Dufey & Giddy (1978).

[2]For an arbitrage paper which does not assume the exogeneity of interest rates, see Pippenger (1978). For survey articles describing empirical

research on international capital flows, see Hodjera (1973) and Bryant (1975).

3 For example, see Kouri & Porter (1974) and Hodjera (1976).

4 Granger & Rees (1968) find that yields on British securities tend to follow random walks. It is assumed here that autarky interest rates behave in a similar fashion.

5 For single equation capital flow studies that use this Koyck stock model, see Arndt (1968), Miller & Whitman (1970, 1972, 1973), and Kreicher (1981). By also setting a to zero, the model collapses to the static single equation stock model used by Branson & Hill (1971), Hodjera (1971), and Hodgson & Holmes (1977).

6 As can be seen from the theoretical coherence square between $r_1(t)$ and $r_2(t)$ (discussed below), the net link between interest rates also depends on the parameters of the model.

7 The tests reported here extend one implication in Haynes & Pippenger (1982). The main objective of that study was to distinguish stock, flow, and combined stock-flow models of international investment by examining spectral estimates of the link between capital flows and interest rate differentials. Monthly data examined herein are from OECD Main Economic Indicators (1973) and subsequent issues; weekly data are from the arbitrage series published by the U.S. Board of Governors, Federal Reserve Bulletin. All data are end-of-period.

8 See Jenkins & Watts (1968) and Hause (1971) regarding cross spectral analysis. The frequency domain technique is used because of its power in model discrimination; for examples, see Pippenger & Phillips (1973) and Haynes (1979). The three cross spectral statistics have time domain analogs: the coherence square is the coefficient of determination (R^2) decomposed by frequency; the gain is similar to the absolute value of a regression coefficient decomposed by frequency; and the phase combines lead-lag properties with the sign of the link, for differing frequencies.

9 Thus, the model is assumed identified in the frequency domain by its error variance properties--shifts in eqs. (1) and (2) "trace out" eq. (4) at low frequencies, while shifts in eq. (4) "trace out" eqs. (1) and (2) at high frequencies. For support of such identification patterns based on spectral estimates of the link between long-term capital flows and bond yield differentials, as well as between bond yields (based on U.S. and Canadian data from 1963 to 1970), see Haynes & Pippenger (1982).

10 To use cross spectral analysis, data must be covariance stationary; see Jenkins & Watts (1968, pp. 155-157). With the data used here, first differencing is sufficient. The gain statistics are ignored here because in this test they do not provide additional discriminatory information.

11 Coherence square and some phase estimates also support the model where λ_1 is zero and λ_2 is positive. Cross correlation estimates described next, however, support only the model where λ_1 is positive and λ_2 is zero (assuming adjustment is lagged).

12 Examination of autocorrelation and spectral density estimates generally suggest that each interest rate series follows a random walk. See Jenkins & Watts (1976) and Pierce & Haugh (1977) regarding prewhitening time series data (first differencing herein) and relating cross correlation and regression estimates of distributed lag models.

13 The end-of-week arbitrage data from the Federal Reserve Bulletin was reported only from the early 1960s to May 1974.

[14]An analogous test conditionalizing on capital flows would prove inconclusive since (because of eq. (4)) the model does not predict zero interdependence when A(t) is held constant.

[15]The model assumes fixed exchange rates, hence technically should not be tested with data from periods of floating exchange rates. A comparison of results herein to those in Pippenger (1978), however, suggests that (at least at low frequency) the bias resulting from not controlling for movements in exchange rates may not be serious. Using U.S. and Canadian data, Pippenger estimated the cross spectrum between weekly changes in three-month treasury bill differentials and the forward premium from January 2, 1959 to December 31, 1964, reporting a low frequency coherence square of 0.908 (based on 30 lags). As reported above in Table 1, the low frequency coherence square between monthly changes in the treasury bill rates for the first period (which roughly compares to Pippenger's sample) is 0.837 (based on 12 lags). The small difference between the estimates suggests that fluctuations in the forward premium have little impact on the strength of long-run IRI.

[16]This prediction also assumes equality between the autospectra of $e_1(t)$ and $e_2(t)$, i.e., that $V_{e_1}(w)$ equals $V_{e_2}(w)$. To derive the prediction, compare the two theoretical gains between $r_1(t)$ and $r_2(t)$ in Appendix 1 after setting λ_2, w, and $V_z(w)$ to zero, and equating $V_{e_1}(w)$ and $V_{e_2}(w)$.

[17]Modifying the behavior of z(t) does not alter the basic pattern of the theoretical cross spectral statistics between interest rates plotted in Fig. 1. To derive this second gain prediction, compare the two theoretical gains between $r_1(t)$ and $r_2(t)$ in Appendix 1 after setting λ_2, w, and b to zero. If both risk and z(t) were significant at low frequency, i.e., if b and $V_z(w)$ were positive, the relative magnitude of the two theoretical gain statistics becomes ambiguous. Hence, risk in combination with z(t) could also prevent perfect IRI.

[18]Another explanation why IRI may be less than perfect is based on Fisherian capital theory. If expected inflation rates in the U.S. and Canada are less-than-perfectly correlated, then nominal interest rates in the two nations would not be perfectly correlated even if real rates were equal. Yet, since partial coherence square between interest rates conditional on the observed inflation rates of the two nations (calculated from Consumer Price Indexes) does not differ significantly from bivariate estimates, the Fisherian explanation appears invalidated. Finally, extending the model to explicitly capture Monetary policy may help reconcile the empirical and theoretical gain statistics.

BIBLIOGRAPHY

Argy, V. and Z. Hodjera. "Financial Integration and Interest Rate Linkages in the Industrial Countries," International Monetary Fund Staff Papers, 20 (March 1973):1-77.

Arndt, S. "International Short-term Capital Movements: A Distributed Lag Model of Speculation in the Foreign Exchange," Econometrica, 36 (January 1968): 59-70.

Askari, H., A. Raymond, and G. Weil. "Long-term Capital Mobility under Alternative Exchange Systems," Canadian Journal of Economics, 10/1 (February 1977):69-78.

Beenstock, M. and J. A. Longbottom. "The Term Structure of Interest Rates in a Small Open Economy," Journal of Money, Credit, and Banking, 13/1 (February 1981):44-59.

Bisignano, J. "The Interdependence of National Monetary Policies," Federal Reserve Bank of San Francisco Business Review, (Spring 1975):41-8.

Bohi, D. R. "The International Interdependence of Interest Rates," Kyklos, 25/3 (1972):597-600.

Box, E. P. and G. M. Jenkins. Time Series Analysis: Forecasting and Control, San Francisco: Holden-Day, 1976.

Branson, W. H. and R. D. Hill. "Capital Movements among Major OECD Countries: Some Preliminary Results," Journal of Finance, 26/2 (May 1971):269-86.

Bryant, R. C. "Empirical Research on Financial Capital Flows," in International Trade and Finance: Frontiers for Research, P. Kenen, ed. Cambridge: Cambridge University Press, 1975.

Cooper, R. N. "Towards an International Capital Market?" in North American and Western European Economic Policies, Proceedings of a Conference held by the International Economic Association, ed. by C. Kindleberger and A. Shondield. New York: St. Martina Press, 1971:192-208.

Dufey, G. and I. H. Giddy. The International Money Market, Englewood Cliffs, New Jersey: Prentice-Hall, 1978.

Fase, M. M. G. "The Interdependence of Short-term Interest Rates in the Major Financial Centres of the World: Some Evidence for 1961-1972," Kyklos, 29/1 (1976):63-96.

Gernberg, A. H. "Aspects of the Monetary Approach to Balance of Payments Theory, An Empirical Study of Sweden," in The Monetary Approach to the Balance of Payments, ed. by J. Frenkel and H. Johnson. Toronto: University of Toronto Press, 1976:298-325.

Granger, C. W. J. and H. J. B. Rees, "Spectral Analysis of the Term Structure of Interest Rates," Review of Economic Studies, 35/1 (January 1968):67-76.

Hause, J. C. "Spectral Analysis and the Detection of Lead-Lag Relationships," American Economic Review, 61/1 (March 1971):213-17.

Haynes, S. E. "Spectral Analysis as a Tool for Model Discrimination: An Application to International Capital Movements," Economics Letters, 2/3 (1979):263-267.

Haynes, S. E. and J. Pippenger, "International Capital Movements: A Synthesis of Alternative Models," Economics Letters, 3/2 (1979):179-185.

__________. "Discrimination among Alternative Models of International Capital Markets," Journal of Macroeconomics, 4/1 (Winter 1982).

Hendershott, P. H. "The Structure of International Interest Rates: The U.S. Treasury Bill Rate and the Eurodollar Deposit Rate," The Journal of Finance, 22 (September 1967):455-65.

Hodgson, J. S. and A. B. Holmes, "Structural Stability of International Capital Mobility: An Analysis of Short-term U.S.-Canadian Bank Claims," Review of Economics and Statistics, 59/4 (November 1977):465-73.

Hodjera, Z. "International Short-term Capital Movements: A Survey of Theory and Empirical Analysis," International Monetary Fund Staff Papers, 20 (November 1973):683-740.

__________. "Alternative Approaches in the Analysis of International Capital Movements: A Case Study of Austria and France," International Monetary Fund Staff Papers, 23 (November 1976):598-623.

Jenkins, G. M. and D. G. Watts. Spectral Analysis and Its Applications. San Francisco: Holden-Day, 1968.

Kouri, P. J. K. and M. G. Porter. "International Capital Flows and Portfolio Equilibrium," Journal of Political Economy, 82/3 (May/June 1974):443-68.

Kreicher, L. L. "International Portfolio Capital Flows and Real Rates of Interest," Review of Economics and Statistics, 63 (Feb. 1981):20-8.

Laffer, A. B. "The Phenomenon of Worldwide Inflation: A Study of International Market Integration," in D. Meiselman and A. Laffer, eds., The Phenomenon of Worldwide Inflation, Washington, American Enterprise Institute for Public Policy Research, 1975.

Miller, N. and M. V. N. Whitman. "A Mean-Variance Analysis of United States Long-term Portfolio Foreign Investment," Quarterly Journal of Economics, 84 (May 1970):175-96.

__________. "The Outflow of Short-term Funds from the United States: Adjustments of Stocks and Flows," in International Mobility and Movement of Capital, F. Machlup et al., eds. National Bureau of Economic Research, New York: Columbia University Press, 1972.

__________. "Alternative Theories and Tests of U.S. Short-term Foreign Investments," The Journal of Finance, 28/5 (December 1973):1131-50.

Office of Economic Cooperation and Development, Main Economic Indicators 1955-1971, Paris, 1973.

Pierce, D. and L. Haugh. "Causality in Temporal Systems: Characterizations and a Survey," Journal of Econometrics, 5/3 (May 1977):265-94.

Pippenger, J. "Interest Arbitrage between Canada and the United States: A New Perspective," Canadian Journal of Economics, 11/2 (May 1978):183-93.

Pippenger, J. and L. Phillips. "Stabilization of the Canadian Dollar: 1952-1960," Econometrica, 41/5 (September 1973):797-816.

Scott, I. O. "The Euro-dollar Market and Its Public Policy Implications," Economic Policies and Practices, paper no. 12, prepared for the Joint Economic Committee, 91st Congress, 2nd Session. Washington: U.S. Government Printing Office, 1970.

Solnik, B. and J. Grail. "The Eurobonds, Determinants of the Demand for Capital: International Interest Rate Structure," Journal of Bank Research, 5/4 (Winter 1975):218-30.

White, B. B. and J. R. Woodbury, III. "Exchange Rates Systems and International Capital Market Integration," Journal of Money, Credit, and Banking, 12/2 (May 1980):175-83.

APPLIED TIME SERIES ANALYSIS
O.D. Anderson & M.R. Perryman (eds.)

VARIANCE OF PREDICTORS AND CHOICE OF TIME SERIES MODELS

Agnar Höskuldsson

Technical University of Denmark
Copenhagen
Denmark

The variance of predictors from autoregressive models is computed, and it is shown that in time series analysis we have the same dilemma between fit and prediction as in linear regression. A new approach to the modelling of time series is suggested by studying the Cholesky factorizations of the covariance matrix and its inverse. In a simulation experiment, we show that actual prediction variances are generally larger than those claimed by time series analysts.

1. INTRODUCTION

Denote by X_t , t=1,2,, ...,n, a univariate time series. In an autoregressive moving average model the time series X_t is supposed to satisfy the following model

$$(1)\quad X_t = a_1 X_{t-1} + a_2 X_{t-2} + \ldots + a_m X_{t-m} + Z_t + b_1 Z_{t-1} + \ldots + b_q Z_{t-q}$$

Here $a_1, \ldots, a_m$ and $b_1, \ldots, b_q$ are constants, and Z_t is the innovation process of X_t and thus the Z_t's are mutually uncorrelated and $Var(Z_t)=s^2$, i.e. a constant variance. From applied point of view the model (1) is a natural one; we seek to explain X_t by a linear combination of the preceeding values, $a_1 X_{t-1}+\ldots+a_m X_{t-m}$, and of "lack of fit" from preceeding values, $b_1 Z_{t-1}+\ldots+b_q Z_{t-q}$, and a residual Z_t. But some persons within natural sciences do not like to include the moving average part in the model (1), because one is explaining an observed data value with the aid of lack of fit for some other data values.

Box and Jenkins (1970) have suggested a methodology for analysing models of the type (1). Their methodology, which has been widely accepted by many practitioners of time series analysis, can be summarized in the following items:

1. Carry out sufficient differencing of the time series X_t in order to remove trend and seasonality to ensure stationarity of the resulting series.

2. A maximum order m of the autoregressive model is suggested if the partial correlations show a "cut-of" at a lag larger than m, and similarly for a moving average model and the autocorrelations. Also, if the q-conditioned partial correlations show a "cut-off" at a lag larger than m, an (m,q) model (1) is suggested.

3. Predictions of the time series X_t at time N+1, N+2 etc are carried out by using (1) recursively. The variance of the error of one lag prediction, $X_{N+1} - \hat{X}_{N+1}$ is estimated by the residual variance s^2 of the model.

Apart from these guidelines a careful study of the residuals is suggested. From theoretical point of view these guidelies are insufficient. Differencing of series can cause problems, especially for multiple series. For the choice of order it is better to use Akaike's AIC (Akaike 1976) and Parzen's CAT (Parzen 1974) criteria. Alternatively a test of the order can be used (see e.g. McLeod 1978 and Hosking 1978, 1980). The use of residual variance for the one-lag predictor is also unsatisfactory. The parameters in the model are estimated from the data and uncertainties are therefore included in the estimates.

Although the variances of the estimates are asymptotically of order 1/N, they can by no means be neglected in small or moderately small samples, as shown in the present paper.

2. A VIEW ON TIME SERIES ANALYSIS

The time series X_t,t= ,...,N, are supposed to be normally distributed with mean and covariances

$$u_t = E(X_t)$$

$$R(t,s) = E((X_t-u_t)(X_s-u_s)),\ R = (R(t,s)) \text{ a } N\times N \text{ matrix.}$$

Here we will suppose that $u_t=0$, t=1,...,N, although it is a basic item, how the mean value function should be studied. The likelihood function for (X_t) is

$$(2)\ L = ((2\pi)^N \operatorname{Det}(R))^{-\frac{1}{2}} \exp(-\tfrac{1}{2} X^T R^{-1} X)$$

Here $X^T = (X_1,\ldots,X_N)$ and Det(R) is the determinant of R. If we consider the Cholesky decomposition of R, $R=FF^T$, with F a lower triangular matrix, the exponent of the likelihood function can be written as $-\frac{1}{2}(F^{-1}X)^T(F^{-1}X)$. Therefore the transformation $Z = F^{-1}X$ transforms X into independent variables $Z^T=(Z_1,Z_2,\ldots,Z_N)$. The idea of the model (1) is also to transform X to a sequence of independent variables. The model (1) is supposed to be valid for $X_{m+1},X_{m+2},\ldots, X_N$. Therefore the rows of F^{-1} from the m-th and onwards include the same parameters as in the model. The matrix F^{-1} is thus basic for the analysis of the time series; the variables (X_t) are mutually independent if and only if F^{-1} is diagonal and graphic analysis of F^{-1} is possible.

The transformation X=FZ can be written componentwize as

$$(3)\ X_t = F(t,1)Z_1 + \ldots + F(t,t)Z_t\ ,\ t=1,\ldots,N$$

and $Z=F^{-1}X$ as

$$(4)\ Z_t = F\_(t,1)X_1 + \ldots + F\_(t,t)X_t\ ,\ t=1,\ldots,N.$$

where $F=(F(t,s))$ and $F^{-1} = (F\_(t,s))$. The formulas (3) and (4) suppose that (Z_t) have unit variance. Let us consider closer the autoregressive model only,

$$(5)\ X_t = a_1X_{t-1} + \ldots + a_m X_{t-m} + Z_t\ .$$

Lemma 1. Suppose that X_t satisfies the autoregressive model (5). The t-th row of F^{-1} for t>m, is given by

(6) $F\_(t,.) = (0 \ldots 0 - a_m \; -a_{m-1} \ldots -a_1 \; 1)/F(t,t)$

If X_t is stationary, then

(7) $F(m+1,m+1) = F(m+2,m+2) = \ldots = F(N,N)$

Proof. (6) follows from (4) and (5), since the coefficients in (4) and (5) are unique. (7) is proved in appendix.

From the point of view of analysing the model (5), stationarity is not needed. In fact one can expect it to be a valuable extension of traditional analysis to allow the variance of X_t and thus of Z_t (i.e. $F(t,t)^2$) change by time, e.g. to let the variance of X_t be computed from a part of the sample value (e.g. not including values far away form time t).

The inverse F^{-1} can easily be computed in steps. Let

$$F_s = F_{11}, \quad F_{s+1} = \begin{bmatrix} F_{11} & 0 \\ F_{21} & F_{22} \end{bmatrix}$$

Then the inverse of F_{s+1} is

$$(8) \quad F_{s+1}^{-1} = \begin{bmatrix} F_{11}^{-1} & 0 \\ -F_{22}^{-1} & F_{21}F_{11}^{-1} \, F_{22}^{-1} \end{bmatrix}$$

Thus if the inverse of the s×s matrix F_s has been computed, the inverse of F_{s+1}, adding one row to F_s, can easily be computed; only the last row of F_{s+1}^{-1} need to be computed.

The partitioning of the matrix F has a simple interpretation in terms of conditional means and covariances. Let $X = (X^{(1)}, X^{(2)})$ and the covariance matrix R be partitioned in the same way as F.

The conditional mean of $X^{(2)}$ given $X^{(1)}$ is computed as $R_{21}R_{11}^{-1}X^{(1)}$, and $R_{21}R_{11}^{-1} = F_{21}F_{11}^{-1}$.

The conditional covariance of $X^{(2)}$ given $X^{(1)}$ is $R_{22} - R_{21}R_{11}^{-1}R_{12} = F_{22} \, F_{22}^T$

These formulas given an interpretation of the computations done in (8).

3. SOME RESULTS FROM LINEAR REGRESSION

The question of prediction in linear regression analysis has been studied intensively during recent years. The reason for this interest is the dilemma that by increasing the amount of independent variables in the regression the fit is improved, but the variance of predictors increases.

Intuitively, by introducing more variables, more uncertainty is introduced in the coefficients of the equation thus increasing the uncertainty of the prediction.

It may be instructive to look at the basic formulas. In matrix form a linear regression model has the form Y=XB+E, where X is the design matrix and $B=(b_1,\ldots,b_p)$

the regression coefficients. A predictor y at $x=(x_1,\dots,x_p)$ is computed as $y=b_1x_1+\dots+b_px_p$.

The estimate for the variance of the predictor is

$$\mathrm{Var}(y) \simeq s^2\, x^T(X^TX)^{-1}\, x$$

By including more regressors $(n-p)s^2$ always decreases (or more precisely does not increase), but $x^T(X^TX)^{-1}\,x$ always increases (does not decrease). Therefore if the variance of the predictors is to be minimized, the possible values x of the regressors, which will be used in the predictions, should be incorporated in the regression analysis. Since it is only safe to predict for x-values within the region given by the n rows of X, several authors have suggested some short of averaging over the rows of X. One such criteria is Mallow's C_p,

$$C_p = \Sigma(y_i-\hat{y}_i)^2\,/s^2 + 2p-n$$

Here $\Sigma(y_i-\hat{y}_i)^2$ is the residual sum of square for a given set of p regressors and s^2 is an estimate of the variance usually based on the full model. If C_p is minimized, the sum of the variance of the predictors based on the rows of X is approximately minimized. Values C_p close to p indicate low bias in the regression model. Thus when selecting regressors on the basis of C_p, one balances between small values of C_p, which mean small prediction variance, and values of C_p close to p, which mean a small bias.

This dilemma between fit and prediction is of course present in time series analysis.

4. THE VARIANCE OF PREDICTORS OF AN AUTOREGRESSIVE PROCESS

Here we shall show an approximate formula for the variance of the predictors of an autoregressive model. A one-lag predictor has the form

$$(9)\quad Y = a_1X_N + a_2X_{N-1} + \dots + a_m\, X_{N-m+1}$$

where $\underline{a} = (a_1,\dots,a_m)$ are the estimates computed from the Yule-Walker equations, $\underline{a}=R_m^{-1}\underline{r}_m$

with

$$R_m = (r_{i-j}, i,j=1,\dots,m), \underline{r}_m^T = (r_1,\dots,r_m)$$

$$r_0 = 1, r_k = N^{-1}\sum_{j=1}^{N-k} X_j\, X_{j+k} \qquad k=1,\dots,m.$$

We will suppose that the series (X_t) has been normalized so that it has mean (average) zero and variance 1. We shall utilize that $\underline{a}$ is asymptotically normally distributed with mean α and covariance matrix $s^2R_m^{-1}/N$, where s^2 is the residual or innovation variance.

<u>Lemma 2.</u> The predictor y has the variance

$$\mathrm{Var}(y) = s^2 \cdot \frac{m}{N} + 1-F(m+1,m+1)^2$$

where $F(m+1,m+1)$ is the $(m+1,m+1)$-th element of F in the Cholesky decomposition of $R=FF^T$.

Proof: If we use the general formula

$$Var(y) = E_a(Var(y|\underline{a})) + Var_a(E(y|\underline{a}))$$

where the last term in our case is zero, we get

$$Var(y) = E_a(\underline{a}^T R_m \underline{a})$$

$$= tr(R_m \, Cov(\underline{a})) + \underline{a}^T R_m \underline{a}$$

$$= s^2 \cdot \frac{m}{N} + 1-F(m+1,m+1)^2 .$$

For the second equality the estimated values of $\underline{a}$ have been inserted for the theoretical ones, and for the last equality see appendix.

This lemma shows us that Var(y) always increases with the order, since $F(m,m)^2$ always decreases. Usually, if an m-th order autoregression is used, the variance s^2 is estimated by the residual variance $s^2(m)$. In this case $s^2(m) = F(M+1,m+1)^2$ (see appendix) and the variance of the predictor is,

(10) $$v(m) = Var(y) = 1-F(m+1,m+1)^2 \quad (1- m/N)$$

The increase in variance of the predictor is

(11) $$v(m) - v(m-1) = (F(m,m)^2- F(m+1,m+1)^2)(1- m/N) + F(m,m)^2/N$$

$$= (a_{m+1}^{(m+1)})^2 \; F(m,m)^2(1- m/N) + F(m,m)^2/N$$

Here $\underline{a}^{(m+1)}$ are the estimates of the (m+1)-th order autoregression.

From the last equation we see that in the case the process is of order m, the variance of the predictor does not increase beyond m ($a_k^{(k)}=0$ for $k \geq m+1$).

In practice it is common to view the prediction being conditioned on the sample. The conditional variance of the predictor is given by the following lemma.

Lemma 3. The variance of the predictor (9) conditioned on $X=(X_N, X_{N-1},\ldots,X_{N-m+1})$ is asymptotically.

(12) $$Var(Y|X_N, \ldots, X_{N-m+1}) = s^2 \cdot X^T R_m^{-1} X/N$$

Proof: Follows from the asymptotic distribution of $\underline{a}$.

In (12) it is $X^T R_m^{-1} X$, which changes by the order m. From the last formula in appendix it follows, that it always increases by m.

We see thus that we have the same dilemma in time series analysis as in linear regression concerning fit and prediction, and like in linear regression the choice of order of the model will always be a subjective balance between the observed reduction in the residual variance and the increase (11) in the prediction variance.

5. THE ANALYSIS OF THE F^{-1} MATRIX

The matrix F^{-1} transforms the series X_t into independent variables, i.e. into white noise. We shall here suggest some methods, which can be used in choosing time series models. Numerous writers have suggested many criteria for studying time series, but we believe that practitioners should look at the matrix F^{-1} and choose their model from it. Of course not the whole F^{-1} should be computed, but F_s^{-1}, the first s rows, with s well beyond a possible order of the model.

The form of F^{-1}

From the point of view of matric calculus it is more convenient to work with the autoregressive coefficients in a reverse order. For that purpose let $\underline{d}^{(m)}$ be the autoregressive coefficients in a reverse order,

$$d_1^{(m)} = a_m^{(m)} , \; d_2^{(m)} = a_{m-1}^{(m)} , \; \ldots , \; d_m^{(m)} = a_1^{(m)} .$$

The coefficients $\underline{d}^{(m)}$ may be computed from

$$(13) \quad (F(m+1,1), \ldots ,F(m+1,m)) = d^{(m)^T} F_m$$

or

$$(14) \quad \underline{d}^{(m)^T} = (F(m+1,1), \ldots ,F(m+1,m)) \; F_m^{-1}$$

This follows from (5) and (3) by identifying the coefficients to the Z_i's.

The following lemma shows the form of F^{-1}.

Lemma 4. The (m+1)-th row of F^{-1} is

$$(15) \quad (- \underline{d}^{(m)^T} \; 1 \; \underline{0})/F(m+1,m+1))$$

Proof. Follows from (14) and (8).

F(m+1,m+1) is the residual standard deviation. From (15) we see that, if the (m+1)-th row of F^{-1} is multiplied by F(m+1,m+1), the autoregressive parameters are obtained with opposite sign.

A plot of $d^{(i)}$ versus $d^{(i+1)}$

As noted in lemma 1, an autoregressive process of order m is characterized by that

$$\underline{d}^{(s)^T} = (\underline{0}^T \; \underline{d}^{(m)^T}), \; s>m, \; \underline{0} \text{ zero vector}$$

This means that, if $(0,\underline{d}^{(i)^T})$ is plotted against $\underline{d}^{(i+1)}$, and the points lie on the straight line through origin with slope 1, then the order is approximately i. Of course the rows of F^{-1} may be plotted instead; the slope of the straight line is then different from 1.

F^{-1} for partial residuals

Suppose that the process is an autoregressive process of order m. Suppose we fit an autoregressive model of order s, with s<m. Then the residual process will be an autoregressive process of order m-s. That can be easily verified by writing the autoregressive model for the residual proces y_t,

$$(16)\quad y_t = X_{s+t}-(a_1^{(s)}X_{s+t-1}+ \dots + a_s^{(s)}X_t) \quad t=1, \dots, N-s$$

and identifying the coefficients to the X's. We shall show, how to determine the F^{-1} matrix associated with the residual process (16). Let

$$Y^T = (x_1, \dots, x_s, y_1, \dots, y_{n-s})^T, \quad G = \begin{bmatrix} I_s & 0 & 0 \\ -\underline{d}^{(s)^T} & 1 & 0 \\ 0 & -\underline{d}^{(s)^T} & 1 \\ \dots & & \end{bmatrix}$$

Here I_s is a unit matrix of order s. The rows below I_s are the s-th order autoregressive coefficients with opposite sign.

Denoting by F_Y^{-1} the F^{-1} matrix of Y, we have

$$Z = F_Y^{-1}\, Y = F_Y^{-1}\, GX = F^{-1}\, X$$

This gives

$$F_Y^{-1}\, G = F^{-1}$$

or

$$F_Y^{-1} = F^{-1}G^{-1}$$

The first s rows of F_Y^{-1} are the same as those of F^{-1} and may be excluded from the analysis. The diagonal elements of F_Y^{-1} are the same as those of F^{-1}. The matrix G^{-1} is easily computed, but is not considered here. - As noted by Kleiner (1979), a common practice is to fit an autoregressive model up to some order, which gives a relatively flat spectrum for the residual series. In our terms this means that F_Y^{-1} should be diagonal. The closeness to independence, i.e. the diagonality of F_Y^{-1}, of the residual series can be investigated in many ways, but is not considered here.

The need for a moving average component

If one has fitted an autoregressive model of order m, it may be of interest to investigate the need for including a moving average component of some order. We shall show how this can be achieved. -

Let q in the model (1) be equal to m, and let $\underline{e}$ be equal to $\underline{b}$ in reverse order, $e_1=b_m, \dots, e_m= b_1$. Let D_m be a diagonal matrix with F(1,1), ...,F(m,m) in the diagonal. In the same way as in (13) one can show that

$$(17)\quad (F(m+1,1), \dots, F(m+1,m)) = \underline{d}^T F_m + \underline{e}^T D_m$$

Similarly, if $F_{2,m+1}$ is the m×m matrix containing the rows 2 to m+1 of F, and $D_{2,m+1}$ a diagonal matrix containing (F(2,2) ,...,F(m+1,m+1)) in its diagonal, we have

$$(18)\quad (F(m+2,2), \dots, F(m+2,m+1)) = \underline{d}^T F_{2,m+1} + \underline{e}^T D_{2,m+1}$$

Both F_m and $F_{2,m+1}$ can be inverted. Therefore, if (17) is multiplied by F_m^{-1} from right, and similarly for (18), we can eliminate $\underline{d}$ by subtracting the two resulting equations.

$$(F(m+1,1), \ldots, F(m+1,m))F_m^{-1} - (F(m+2,2), \ldots, F(m+2,m+1))F_{2,m+1}^{-1}$$

$$= e^T D_m F_m^{-1} - e^T D_{2,m+1} F_{2,m+1}^{-1}$$

Multiplying this equation by $F_{2,m+1}$, we obtain

$$(F(m+2,2),\ldots,F(m+2,m+1))-(F(m+1,1),\ldots,F(m+1,m))F_m^{-1} F_{2,m+1}$$

$$= (F(m+2,2),\ldots,F(m+2,m+1)) - \underline{d}^{(m)} F_{2,m}$$

$$=-\underline{e}^T(D_m F_m^{-1} F_{2,m} - D_{2,m+1})$$

If the process is an autoregressive one of order m, the term with $\underline{e}$ should be zero. The matrix in paranthesis at e^T contains zero at and above the diagonal and is therefore singular. The coefficient $e_1=b_m$ can be set equal zero; the other coefficients can easily either be computed or put equal to zero. The resulting value $\underline{e}$ can be put into equation (17) and it solved for $\underline{d}$. The difference of this solution from $\underline{d}^{(m)}$ tells us, if there is a need for including a moving average component to the model.

6. A SIMULATION EXPERIMENT

Asymptotic formulas in time series analysis are not always reliable for small samples. In order to illustrate the variation of the basic figures we have carried out a simple simulation experiment. 200 realisations of the series

$$X_t = 1.2\, X_{t-1} - 0.6\, X_{t-2} + 0.8\, X_{t-3} - 0.4\, X_{t-4} + Z_t$$

for t=1,...,30 were simulated with Z_t being normally distributed with standard deviation equal to 0.2. (In fact 61 values of each series were generated and the first 30 dropped). For each of the 200 series three predictors were computed,

$y_1 = X_{31}$ the 31-st value of the simulated series

y_4 = predictor based on 4-th order estimated model

y_3 = predictor based on 3-d order estimated model.

The standard deviation of the three predictors in the sample of 200 values was

X_{31}	y_4	y_3
1.885	0.623	0.613

These figures reflect a difference in the standard deviations, which is larger than can be accounted for by the residual variance. The average values in the sample of 200 variances were as follows:

residual variances		Variances of predictors		Conditional variances of lemma 3	
$s^2(3)$	$s^2(4)$	v(3)	v(4)	3. order	4. order
0.0780	0.0733	0.930	0.936	0.196	0.211

These figures show that the decrease of the residual variance is of the same size as the increase in the variance of the predictors, indicating that a third order model is equally good as a fourth order model. The conditional variances of the predictors are much larger than residual variances, which shows that they cannot be neglected like most program packages in time series analysis do.

7. CONCLUDING REMARKS

In this paper the author has advocated for more nuanced statistical analysis of time series than is commonly practiced. It is suggested that the correlation matrix, of order well above possible model order, is factorized by its Cholesky decomposition. The Cholesky part is inverted and the inverted matrix studied graphically, the Cholesky part for the residuals computed and the need for including a moving average part in the model is investigated. An analysis of time series by testing for the order and by computations of criteria like AIC and CAT should be supplemented by this analysis of the Cholesky part and by computation of the variance of predictors. In that way the practitioners of time series analysis are told more properly what features of the model (1) are in fact present in the data.

APPENDIX

We shall here show some useful formulas, when analysing stationary time series. We suppose $E(X_t)=0$ and $Var(X_t)=1$.

$$s^2(m) = \inf_a E(X_t-(a_1X_{t-1} + \dots + a_m X_{t-m}))^2$$

$$= 1- \underline{a}^{(m)^T} R_m \underline{a}^{(m)} \qquad \text{for } \underline{a}^{(m)} = R_m^{-1} \underline{r}_m$$

$$= 1- \underline{r}_m^T \underline{a}^{(m)}$$

$$= 1- \underline{r}_m^T R_m^{-1} \underline{r}_m$$

$$= 1- R2(m)$$

$$= Det(R_{m+1})/Det(R_m)$$

$$= F(m+1,m+1)^2$$

$$= F(m,m)^2 \ (1- (a_{m+1}^{(m+1)})^2)$$

Here R2(m) is the multiple correlation coefficient between X_t and $(X_{t-1},\dots,X_{t-m})$. All equalities follow easily from matrix calculus except perhaps the last one, which can easily be deducted from the following general equation:

$$X^T R^{-1} X = X_1^T R_{11}^{-1} X_1 + (X_2 - F_{21} F_{11}^{-1} X_1)^T F_{22}^{-1^T} F_{22}^{-1} (X_2 - F_{21} F_{11}^{-1} X_1).$$

Here the partition of X,F and $R=FF^T$ is similar to the one in (8). This equation is used in the well-known decomposition of a normal density into marginal and conditional distribution.

REFERENCES

Akaike, H. (1976): Canonical Correlation Analysis of Time Series and the use of an Information Criterion. In R.K. Mehra and D.G. Lainotis (Eds.): System Identification: Advances and case studies. Academic Press. New York. pp. 27-96.

Box, G.E.P. and Jenkins, G.M. (1970): Time Series Analysis: Forecasting and Control. San Francisco. Holden Day.

Hosking, J.R.M. (1978): A unified derivation of the asymptotic distributions of goodness-of-fit statistics for autoregressive time series models. J.R. Statist. Soc. B, 40, pp. 341-349.

Hosking, J.R.M. (1980): Lagrange multiplier tests of time series models. J.R. Statist. Soc. B, 42, pp. 170-181.

Kleiner, B., Martin, R.D. and Thomson, D.J. (1979): Robust estimation of power spectra. J.R. Statist. Soc. B, 41, pp. 313-351.

McLeod, A.I. (1978): On the distribution of residual autocorrelations in Box-Jenkins Models. J.R. Statist. Soc. B, 40, pp. 296-302.

Parzen, E. (1974): Some recent advances in time series modelling. IEEE Trans. AC Vol 19, pp. 723-730.

APPLIED TIME SERIES ANALYSIS
O.D. Anderson & M.R. Perryman (eds.)

TRANSFER FUNCTION FORECASTS OF ECONOMIC ACTIVITY

William L. Huth

Management Science Department
Northeastern University
Boston, Massachusetts
U.S.A.

This paper considers the Business Conditions Digest (BCD) composite measures of economic activity as realizations of a stochastic time series model. The composite index of "roughly" coincident indicators is an aggregation of four component series that move in phase with economic activity. The leading indicator composite includes twelve time series whose movements tend to foreshadow those of the coincident index. The BCD indicators are selected from a broad set of possible candidates on the basis of both theoretical and empirical considerations. That is, a series is examined for selection not only in the light of its statistical properties and empirical performance in relation ot NBER reference cycles but also its relevance with respect to the various theories of business cycle dynamics. Given that the coincident index mirrored economic activity and that the leading index contained information about the future of that activity it was hypothesized that the series could be related by a combined transfer function-noise model to provide forecasts of future economic activity.

This paper considers the Business Conditions Digest (BCD) composite measures of economic activity as realizations of a stochastic time series model. The U.S. economy is represented as a dynamic process converting inputs into outputs. The composite index of "roughly coincident indicators is an aggregation of four component time series whose movements are in phase with economic activity. The coincident index is, thus, a proxy for the output series of a dynamic economic system. The leading indicator composite includes twelve time series whose movements are conjectured to foreshadow those of the coincident index.

The current system of indicators has evolved from early attempts at understanding, measuring, and forecasting the business cycle. The Harvard ABC curves were an early, quasi-quantitative, approach to measuring cyclic economic activity. These curves were used as a sequential feedback system to forecast the next step in economic activity. Work in earnest began on the present indicator system under the auspices of the National Bureau of Economic Research (NBER). The initial NBER sponsored examinations of business cycles were by Wesley Mitchel (1927) and later jointly with Arthur Burns (1947). The system of composite indexes was first developed by Moore and Shiskin (1967) and recently altered by Zarnowitz and Boschan (1977). As stated by the BEA in the Handbook of Cyclical Indicators (1977).

> ...the primary purpose of the composite indexes is to indicate changes in the direction of aggregate economic activity, many users have also come to view them as indicators of the current and future levels of economic activity, and the coincident index has come to be considered as a monthly approximation of economic activity.

The BCD component series are selected from a broad set of some 300 candidate time series on the basis of theoretical and empirical considerations. An individual series is examined for selection as a composite component not only in the light of its statistical properties and empirical performance in relation to NBER defined reference cycles but, also, with respect to the various theoretical explanations of business cycle dynamics.

Cyclical economic activity has long been the focus of a prodigous amount of retrospection and analysis. In the Bible, one finds mentioned periods of "feast and famine" and Jacob's mystical dream of the seven fat and lean cows. Sir William Petty, in the 17th century, referred to cyclic activity as "dearths and plenties." Later classical treatments invoked the vehicle of theory to explain the periodic wax and wain phenomenon observed in the economic functioning of a society. These explanations of variability in economic activity sprang from refutations of Jean Baptise Say's conjecture that the process of production generated - pari passu - consumption. Malthus, Carlyle's "dismal scientist", noted the existence of a lagged relation between production and consumption which J. S. Mill developed into an inventory theory of the business cycle. In addition to excesses of production over consumption, theoretical business cycle explanations have been constructed on the basis of other factors, including the investment-savings relation, financial factors, i.e., money and credit, and expectations.

Investment-savings theories of the cycle have been developed by Marx, Jevons, Von Hayek, Keynes, Schumpeter, and Samuelson to name a few. The Schumpeterian treatment was volumnous and noteworthy for the development of a taxonomy of superimposed cycles. The major vehicle for cyclic motion in Schumpeter's work was large swings in investment occuring as the result of innovative activity. The culmination of investment theories of the cycle was stated, in the form of difference equations, by Samuelson in his paper that dealt with interactions between the investment accelerator and the consumption multiplier. Accelerator-multiplier interaction in the mathematical form of difference equations illustrated endogenous cyclic motion in economic activity. The monetary theory of the cycle was pioneered by Hawtrey in the late 19th century and expounded upon by, among others, Fisher and Friedman. In the monetary theory variability was initiated via credit contractions and expansions. Fisher eluded to monetary causes of cyclic activity in his paper, "The Business Cycle Largely a Dance of the Dollar." Friedman's business cycle theory has centered around change in the rate of change of monetary aggregates and the corresponding cyclical behavior of the velocity of money.

Each theory of the business cycle has its justifications and shortcomings. No one theory has been able to successfully explain cyclic economic activity. However, the information concerning the generation of cyclical variability must be considered in the development of quantitative models that attempt to predict the direction and magnitude of change in industrial activity. These theoretic business cycle explanations were considered in the selection of the dozen series that make up the index of leading indicators shown in Table 1. In addition to compatibility with business cycle theory the series were also examined on the basis of their performance with respect to the NBER reference cycles. Measures of relative performance included statistical adequacy, timing, smoothness, and currency.[1] The individual series are then combined into a single index that is considered a summary measure of ex ante change in economic activity.[2]

[1]For the interested reader, a discussion of the selection progress is given in Zarnowitz and Boschan (1977) pp. 171-174.

[2]The method of composite index construction is given in the Handbook of Cyclical Indicators (1977) pp. 73-76.

The composite index of coincident indicators is compiled from the individual series given in Table 2. The coincident index, as stated earlier, can be considered as an appropriate measure of monthly economic activity and, as such, its movements are foreshadowed by those of the leading index.

Given that the coincident index mirrored economic activity and that the leading index contained information about the future of that activity it was hypothesized that the series could be related by a combined transfer function-noise model to provide forecasts of future economic activity.

Table 1

Composite Index of Leading Indicators

Component Series

1. Average Work Week of Production Workers
2. Manufacturing Layoff Rate
3. Index of Net Business Formation
4. New Orders of Consumer Goods and Materials
5. Contracts and Orders for Plant and Equipment
6. Index of Private Housing New Building Permits
7. Net change of Inventories on Hand and on Order
8. Vendor Performance
9. Percent Change in Sensitive Prices (WPI)
10. Index of Stock Prices
11. Money Balance (M1A)
12. Percent Change in Total Liquid Assets

Table 2

Composite Index of Coincident Indicators

Component Series

1. Employees on Non-agricultural Payrolls
2. Index of Industrial production
3. Personal Income Excluding Transfer Payments
4. Manufacturing and Trade Sales

The data for the leading index (LEAD) and the coincident index (COIN) were gathered from the Handbook of Cyclical Indicators (1977). Each series consists of 300 monthly observations covering the 25 year period from 1950 through 1974. All correlations and estimates that follow were computed via the SIBYL/RUNNER Interactive Forecasting System (1979).

The model fitted in this section is from the general family of linear difference equations of the form

$$Y_t = \delta_1 Y_{t-1} + \delta_2 Y_{t-2} + \ldots + \delta_r Y_{t-r} + \omega_o x_{t-b} - \omega_1 x_{t-b-1} - \ldots - \omega_s x_{t-b-s} + a_t \qquad (1)$$

Where:

Y_t represents the output series, COIN.
X_t represents the input series, LEAD.
a_t represents the noise model and is given by the general ARMA model

$$a_t = \phi_1 a_{t-1} + \phi_2 a_{t-2} + \ldots + \phi_p a_{t-p} + a'_t - \theta_1 a'_{t-1} - \ldots - \theta_q a'_{t-q} \quad (2)$$

of order p,q. In equation (1) the parameters r, s, and b refer to, respectively, the number of lagged values in the output series, the input series, and the delay lag for the input series. The value of b is of considerable importance in that given an input series movement the corresponding output series change will occur in b periods given that the input series does indeed lead the output series.

Equation (1) represents the output series, Y_t, as a weighted sum of its previous values and previous values of the input series, X_t. Since the I/O series COIN and LEAD are each univariate time series they can each be thought of as realizations of some integrated auto-regressive moving average process. That is,

$$Y_t = \phi_1 Y_{t-1} + \ldots + \phi_p Y_{t-p} + a_t - \theta_1 a_{t-1} \ldots - \theta_q a_{t-q} \quad (3)$$

and

$$X_t = \phi'_1 X_{t-1} + \ldots + \phi'_p X_{t-p} + e_t - \theta'_1 e_{t-1} - \ldots - \theta'_q e_{t-q} \quad (4)$$

are ARMA processes of p autoregressive terms and q moving average terms that generate the observed series COIN and LEAD. The series LEAD is given by its own ARMA (p,q) generating function whereas COIN is determined jointly by its own generating process and that of LEAD. Thus, in order to specify the parameters of (1) the series COIN and LEAD were first "prewhitened" by specifying their ARMA generating functions given by (3) and (4), the results of which are reported in the next section.

The first step in transforming the I/O series to white noise involved identification of the ARMA process that yields normally distributed residuals (white noise) for each series. Too fascilitate the determination of the ARMA models and their order, autocorrelations and partial autocorrelations were computed for the differenced series. These correlograms are shown in Figures 1-4. From the diagrams in Figures 1 and 2, the singly differenced series LEAD was identified as an ARMA (2,0) process and estimated as

$$X_t = .486 X_{t-1} + .163 X_{t-2} \quad (5)$$

where X_t is the differenced series. The mean square error (MSE) from the estimation was 0.898 and the residual autocorrelations are shown in Figure 5. The series COIN was, after an examination of Figures 3 and 4, identified as an ARMA (1,0) and estimated as

$$Y_t = .542 Y_{t-1} \quad (6)$$

where Y_t was the once differenced series. The MSE was .524 and the residual autocorrelations are displayed in Figure 6.

The two residual series were then used to generate the cross correlations portrayed in Figure 7. From the figure it was noted immediately that the value of b, i.e., the lag with the largest cross correlation coefficient, was zero, indicating that to an extent, the leading index was not foreshadowing the coincident index. Various other univeriate models were tentatively entertained and in each case the result, b = 0, was the same. The cross correlation coefficients for negative lags (the west side of the zero lag dotted line that bisects the diagram vertically) were not significantly different from zero denoting that, at least, the coincident index is not at leading indicator of the leading index. For positive time lags the cross correlations do indicate a somewhat nebulous lag relation and they appear to follow an AR (1) process after a delay of one period. Setting b = 0, r = 1, and s = 0 the transfer function

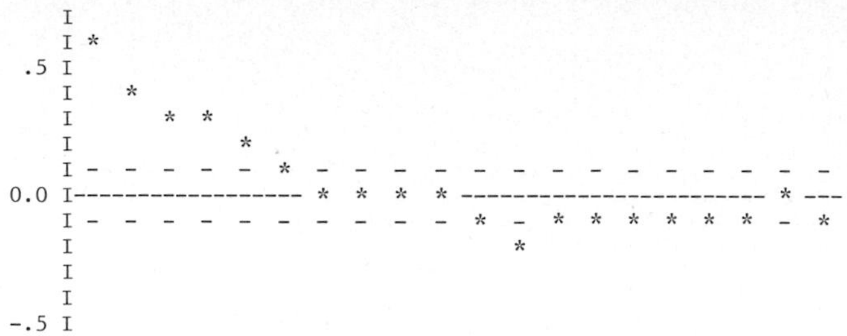

Figure 1: AUTOCORRELATIONS OF LEAD DIFFERENCED DATA

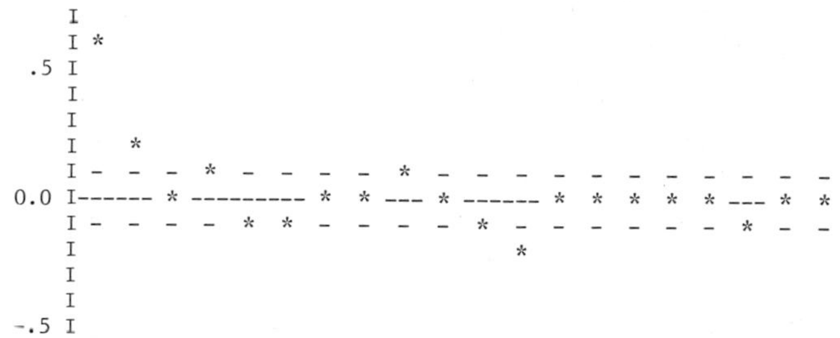

Figure 2: PARTIAL AUTOCORRELATIONS OF LEAD DIFFERENCED DATA

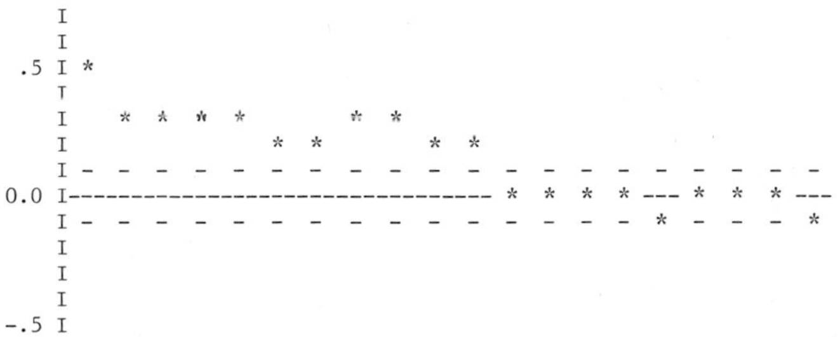

Figure 3: AUTOCORRELATIONS OF COIN DIFFERENCED DATA

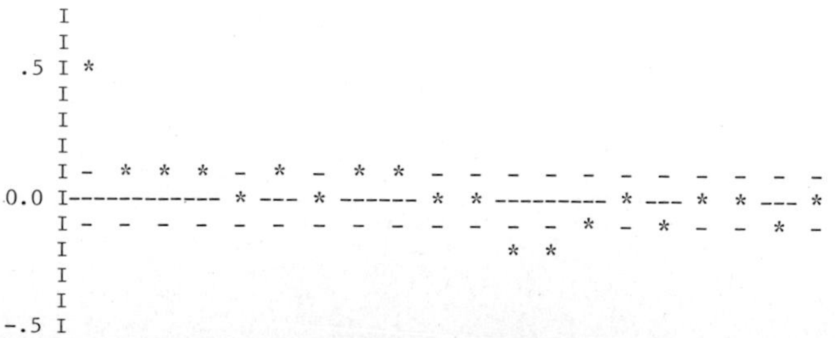

Figure 4: PARTIAL AUTOCORRELATIONS OF COIN DIFFERENCED DATA

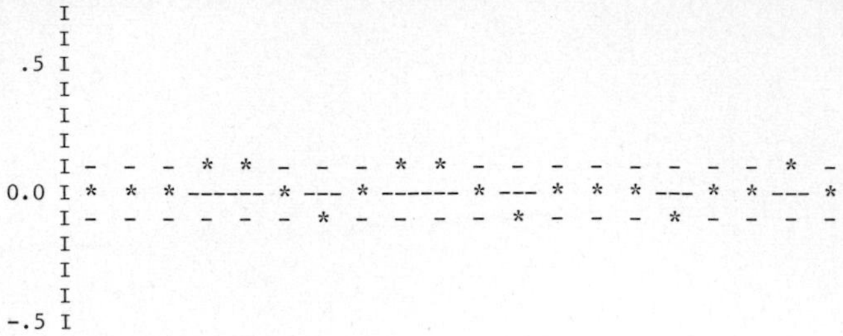

Figure 5: AUTOCORRELATIONS OF LEAD RESIDUALS

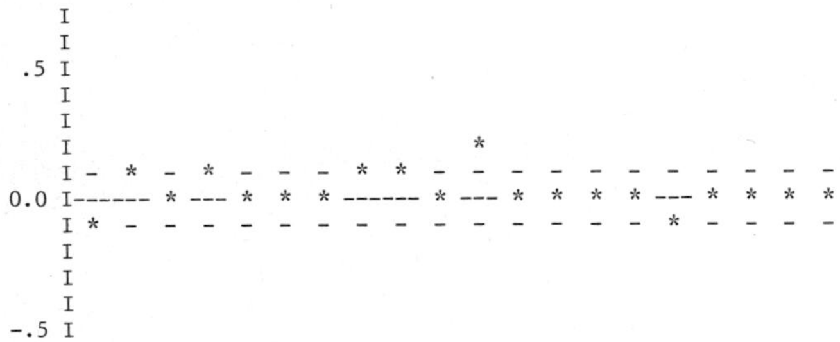

Figure 6: AUTOCORRELATIONS OF COIN RESIDUALS

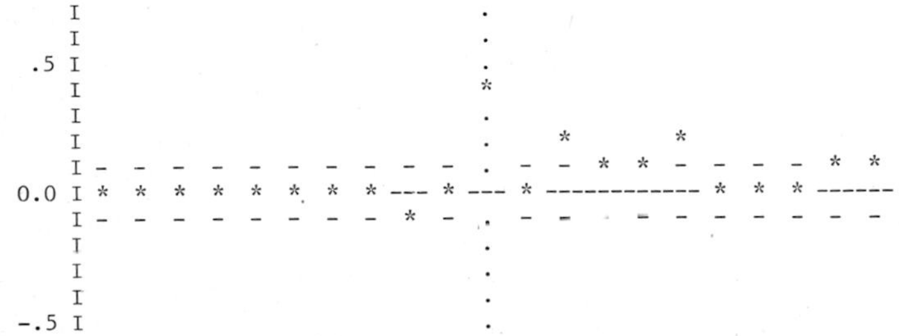

Figure 7: CROSS-AUTOCORRELATIONS WLD1.D VS. WCD1.DAT

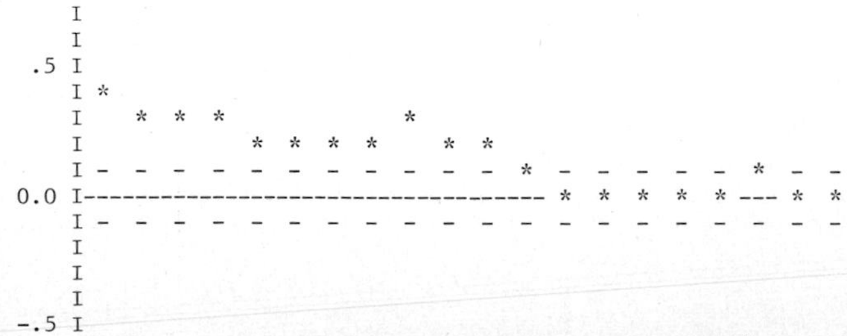

Figure 8: AUTOCORRELATIONS OF COIN NOISE

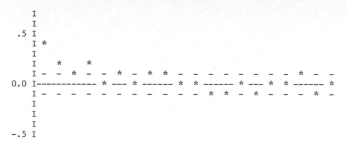

```
     I
     I
 .5  I
     I *
     I
     I    *     *
     I -  -  *  -  -  *  -  *  *  -  -  -  -  -  -  -  -  *  -  -
0.0  I------------ * --- * ------ *  * ------ * --- *  * ------ *
     I -  -  -  -  -  -  -  -  -  -  -  *  *  -  *  -  -  -  *  -
     I
     I
     I
-.5  I
```

Figure 9: PARTIAL AUTOCORRELATIONS OF COIN NOISE

```
     I
     I
 .5  I
     I
     I
     I                                  *
     I -  -  -  *  -  -  -  -  *  *  -  -  -  -  -  -  -  *  -  *
0.0  I *  *  * --- *  *  *  * --------- * ------ * --- * --- * ---
     I -  -  -  -  -  -  -  -  -  -  -  -  *  *  -  *  -  -  -  -
     I
     I
     I
 .5  I
```

Figure 10: AUTOCORRELATIONS OF COIN RESIDUALS

```
     I                                  .
     I                                  .
 .5  I                                  .
     I                                  .
     I                                  .
     I                                  *
     I *  -  *  *  -  -  *  -  -  -  .  -  -  -  -  -  -  -  -  -  -
0.0  I--- * ------ *  * --- * --- * ------ *  *  *  * --------- *  *
     I -  -  -  -  -  -  -  -  *  -  .  *  -  -  -  -  *  *  *  -  -
     I                                  .
     I                                  .
     I                                  .
-.5  I                                  .
```

Figure 11: CROSS-AUTOCORRELATIONS WLD1.D VS. RESIDUALS

$$Y_t = \delta_1 Y_{t-1} + \omega_o X_t + a_t \qquad (7)$$

was specified and initial estimates were computed using a regression relation between the impulse response parameters and the cross correlation coefficients. The noise model was identified as an ARMA (1,0) from the correlations given in Figures 8 and 9 and estimated as

$$a_t = .327 a_{t-1} + a'_t \qquad (8)$$

where a'_t is white noise as illustrated by the residual autocorrelations in Figure 10.

The combined transfer function noise model

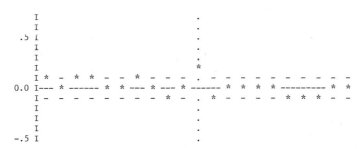

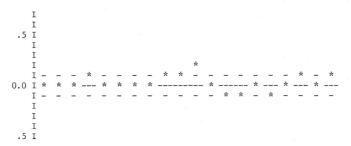

$$Y_t = \delta_1 Y_{t-1} + \omega_o X_t + \omega_1 a_{t-1} + a'_j \tag{9}$$

was estimated using Marquardt's non-linear least squares algorithm as

$$Y_t = .907\, Y_{t-1} + .169\, X_t + .327\, a_{t-1} \tag{10}$$

and each transfer function coefficient was significant at the .01 level. In replicating the sample history of the coincident index the MSE was 0.628 and the mean absolute percentage error (MAPE) was 0.719. In order to check the adequacy of the estimation, residual autocorrelations and cross correlations between the prewhitened input series and the residuals were computed and are displayed in Figure 10 and 11. From Figure 10, the residuals appear to be noise except for the value at lag 11. This value is significant at the .05 level but insignificant at the .10 level. The cross correlation diagram depicts two random series with no visible relationship.

In conclsuion the estimated transfer function-noise model performs well in replicating the sample history. The analysis has, however, shed an aura of doubt on the assumed relationship between the two composite indexes. Given the weight attached to the leading index as a harbinger of change in economic activity, further analysis of the lag structure between the two indexes would seem appropriate.

Bibliography

1. Applied Decision Systems, a Division of Temple, Barker & Sloane, Inc. (1979). SIBYL/Runner Interactive Forecasting: Users Guide. Lexington, MA.

2. Box, G. and Jenkins, G. (1976). Time Series Analysis: Forecasting and Control. Holden-Day, San Francisco, 337-418.

3. Makridakis, S. and Wheelwright, S. (1978a). Forecasting: Methods and Applications. Wiley, New York.

4. Makridakis, S. and Wheelwright, S. (1978b). Interactive Forecasting: Univariate and Multivariate Methods, 2nd ed. Holden-Day, San Francisco.

5. Mitchel, W. (1927). Business Cycles: The Problem and Its Setting. NBER, New York.

6. Mitchel, W. and Burns, A. (1947). Measuring Business Cycles. NBER, New York.

7. Moore, G. and Shiskin, J. (1967). Indicators of Business Expansions and Contractions. NBER, New York.

8. Moore, G. (1975). The Analysis of Economic Indicators. Scientific American 232, 17-23.

9. Nelson, C. (1973). Applied Time Series Analysis. Holden-Day, San Francisco.

10. U.S. Department of Commerce and the Bureau of Economic Analysis (1977). Handbook of Cyclical Indicators: A Supplement to Business Conditions Digest.

11. Zarnowitz, V. (1977). The Business Cycle Today. Ed: V. Zarnowitz, NBER, New York.

12. Zarnowitz, V. and Boschan, C. (1977). New Composite Indexes of Coincident and Lagging Indicators. In Handbook of Cyclical Indicators, 185-199.

13. Zarnowitz, V. and Boschan, C. (1977). Cyclical Indicators: An Evaluation and New Leading Indexes. In Handbook of Cyclical Indicators, 170-184.

APPLIED TIME SERIES ANALYSIS
O.D. Anderson & M.R. Perryman (eds.)

A MULTIPLE TIME SERIES INVESTIGATION OF THE RELATIONSHIPS AMONG A SET OF SOCIAL AND ECONOMIC VARIABLES

Kenneth J. Jones

Florence Heller School
Brandeis University
Waltham, Massachusetts

and

Priscilla Pitt Jones

Several Vector ARMA analyses were done in an attempt to uncover relationships between a pair of economic indicators, unemployment and leading indicator index, and several measures of social ill. Both monthly and yearly frequencies were used. In contrast to previous studies no cross domain relationships were found.

INTRODUCTION

The issue of how to assess causality appropriately in domains of time dependent or contemporaneous variables is currently of great interest in time series analysis (Pierce, 1977). Although the precise methodology of preference is not at this time agreed upon, this paper attempts to evaluate the plausibility of some claims of causal relations between indicators of social and economic conditions via multivariate or vector autoregressive-moving average (VARMA) modeling procedures (Jenkins, 1979; Tiao, *et al.*, 1979).

Research studies indicate relations between such social indicators as infant and neonatal mortality and unemployment (Brenner, 1973) and income (Fuchs, 1974); suicide rates and unemployment (Henry and Short, 1954); and unemployment and mental hospital admissions (Brenner, 1976). On the other hand, others have found contrary results (MacMahon, *et al.*, 1972) and attributed these to the existance of an irreducible minimum of social ill. In this paper, the hypothesis is entertained that while logically one may postulate a causal chain from economic condition indicators such as employment or robust economic activity to such social indicators as infant mortality, mental hospital admissions, homicides, and suicides, the chain may lead through so many unmeasured endogenous variables and be impinged upon by so many unmeasured exogenous variables that the linkages may be extremely weak.

The fact that causal relations have been found in prior studies may be more a defect of methodology than a true revelation. For the most part, these causal studies have depended upon regression analysis in some form or other. The fatal flaw of this approach has been pointed out numerous times (*e.g.* Granger and Newbold, 1977, p.202). The basic conclusion of these warnings is that when any two series are internally correlated, there is a built in cross-correlation effect which may be construed in a regression analysis context as manifesting a true "co-relation". If the cross-correlations which are found are of suitable lags, these may then be erroneously interpreted as indicating a causative relationship. In addition to the above problem, measurement of interval or aggregative protocols may give rise to spurious autocorrelations within series which further compound the problem (Kendall, 1973, Working, 1960).

In this paper various data designs are explored to suggest how two economic indicators may be related to several indicators of social ill. The index of leading indicators (JLEAD) of the U.S. Department of Commerce (1967 = 1.00)[1] and the number of unemployed males (UHHM) have been selected as the economic indicators. Depending on the analysis, either absolute numbers or rates of neonatal deaths (0-28 days), infant deaths (28 days to one year), homicides, suicides, and mental hospital admissions are employed as social indicators. The variables are measured on a monthly or yearly basis for the entire United States. Monthly data begin in 1965 and include January 1981 for a series of 193 observations. Yearly data start in 1947 and include 1980 yielding a series of 34 observations. The latter is quite short for this type of analysis. However, prior studies in which yearly observations have gone back to 1920 or 1900 with economic or social data almost surely have erred by assuming a constancy of data definitions which does not exist. Nevertheless, the sortness of the series imposes a considerable Type II error on the analysis of yearly data.

YEARLY DATA

Table 1 defines the data used for this study.

Table 1

Yearly Rates

Variable	Description
Neonatal	Deaths in the period 0 to 28 days per 1000 live births in the total U.S.
Homicide	Homicides per 100,000 population for U.S.
Suicide	Suicides per 100,000 population for U.S.
MHA	Mental hospital admissions per 100,000 population for U.S.
URM	Male unemployment rate (per cent)
JLEAD	Index of leading indicators, 1967 = 1.00

Rates were used in order to restrict the number of parameters to be estimated. If frequencies were used, it would have been necessary to include additional control variables such as population and births.

Plots of these rates are displayed in Figures 1 and 2. Although differencing is not strictly necessary to apply the multivariate ARMA model, it was decided to do this in order to minimize the number of parameters to be fitted. First differencing was decided upon for all six series. The auto- and cross-correlograms revealed that stationarity had been achieved and that an autoregressive process was indicated. As there was a dramatic drop in the chi-square value after the second full AR fit, an upper limit of two was indicated for the order of the AR processes. The computer programs used were those described in Tiao, *et al.* (1979).

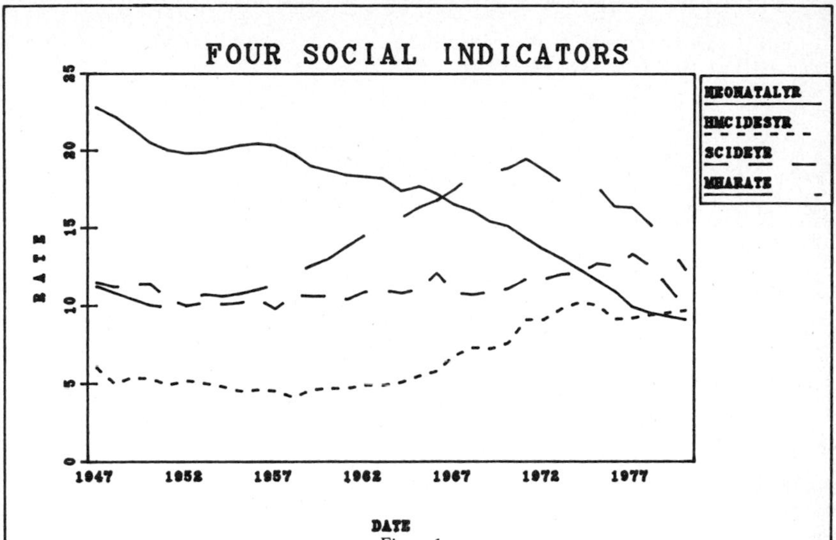

Figure 1

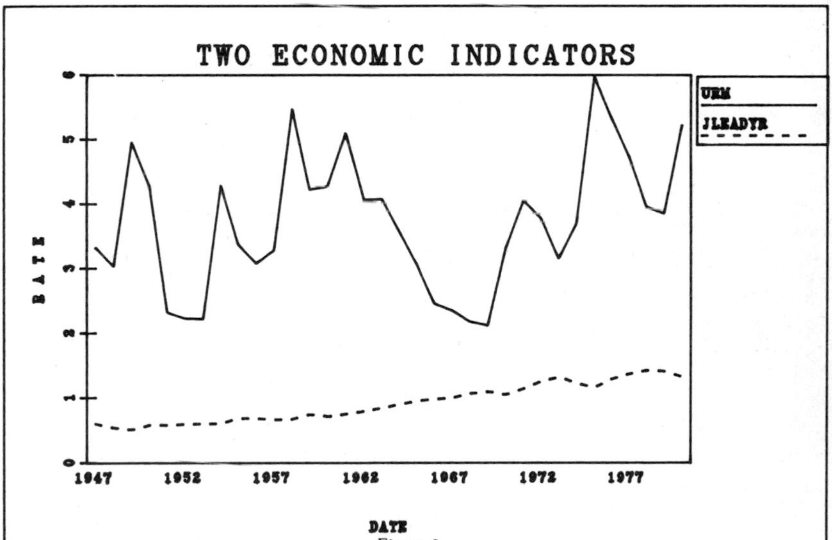

Figure 2

Figure 3 shows the final results for the model after the iterative fitting process. Several runs were made in which insignificant parameters were eliminated and in some cases examination of the residuals revealed cross-correlations which if sufficiently high (.5 or above) were "targeted" and eliminated. It was in this manner that the lag-4 and lag-5 parameters were included. The usual checks were done and these indicated that the residuals were white noise, there being no cross- or auto-correlation significant at the 5% level. To a lag of 8 only one correlation was above .30. Haugh's S test on the values to a lag of five gave no chi-square above the 5% level. The parameters also are correlated less than 0.5.

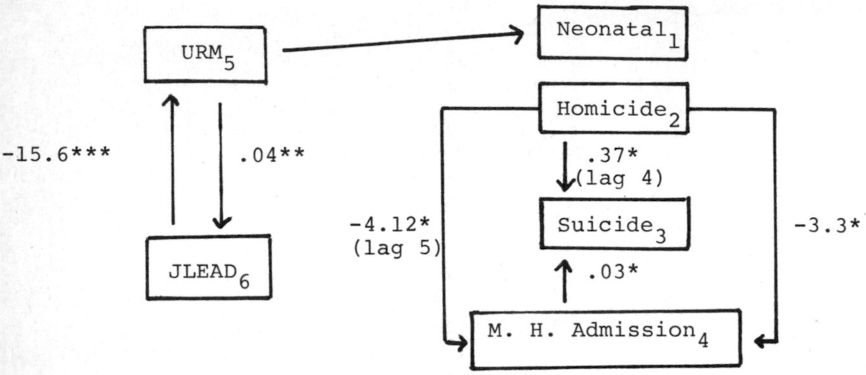

Figure 3

Estimates of the Phi Values for the Yearly Model for Rates

(N=34)

(unless otherwise noted, links are lag 1 relationship)

* = $p < .05$
** = $p < .01$
*** = $p < .001$
no star = $p < .20$

From Figure 3, it may be seen that although the analysis uncovers some links among the economic variables - the strongest of which implies that a leading indicator drop presages an increase in unemployment in the subsequent year which in turn has some small feedback effect - there is virtually no link to the social variables indicated by this model. The link to neonatal mortality is noted on the diagram, but if it is reliable, it is at least very weak - being significant at approximately the 20% confidence level. In the social variable domain, there is a complex of relationships, all of which are quite weak, but if real, suggesting a lagged relationship of mental hospital admissions to homicide rate.

It should be observed that Figure 3 notes only the off-diagonal coefficients of the resultant multivariate ARMA model. The model actually computed for Figure 3 also includes diagonal auto-regressive parameters of lag one for neonatal death rate, mental hospital admission rate, unemployment and leading indicators. To determine the amount of variance which would have been explained using these terms alone, a separate analysis was done. Table 2 gives the percent variance accounted for in the differenced series by each model.

From Table 2 it may be seen that the series which benefit most from the full model are variables three through six. However, Figure 3 reveals that cross-domain relationships are not responsible for these increases in explained variance. Virtually all of the explained variance in neonatal death rate is due to its own past history and the full model was unable to add to or replace any of this. In short, although it is possible to reduce these correlated series to a set of residuals which are white noise, in order to do so it is necessary to rely primarily on the past history of each series as the explanatory factors.

Table 2

Percent Variance for the Differenced Series

Variable	Full Model (17 parameters)	Diagonal Model (9 parameters)
1.Neonatal rate	20.1	19.3
2.Homicide rate	0.0	0.0
3.Suicide rate	23.2	0.0
4.M.H.Admission rate	66.9	46.3
5.Unemployment rate	51.0	1.0
6.Leading indicator index	44.9	10.5

Thus, the conclusion is that the system of economic and social variables is incomplete and there exist other causal agents which should be included. In addition, if these economic variables do relate to these social variables, they do so either through other modifying variables or at a level low enough to be undetectable with this sample size.

ANALYSIS OF THE MONTHLY DATA

In a attempt to reduce the large Type II error inherent in the analysis of short annual series, several monthly series were developed. These are illustrated in Figures 4 and 5. With the exception of the Index of Leading Indicators, all are in terms of frequencies. The denominators of the rate variables of the yearly analysis, births and

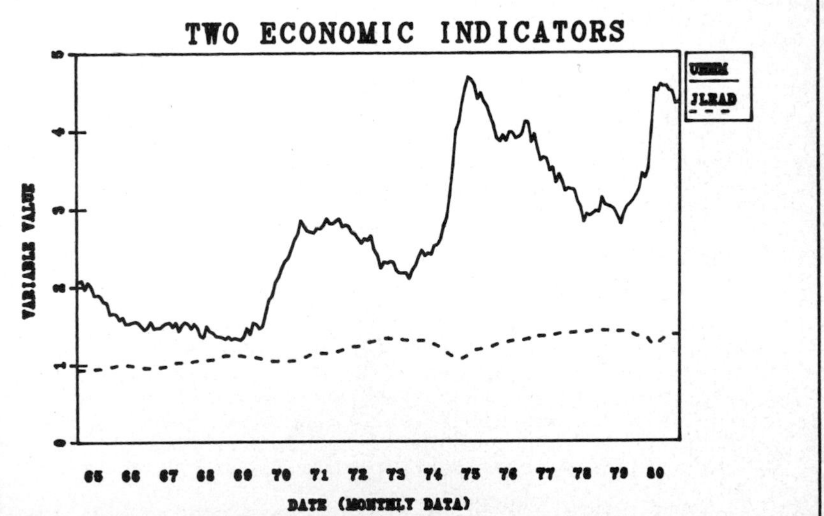

Figure 4

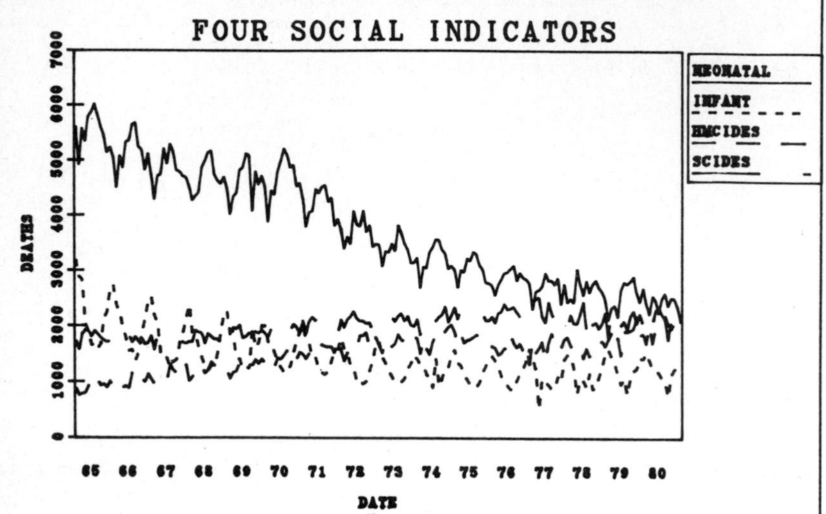

Figure 5

U.S. population, were included as additional explanatory series. Table 3 lists the series.

Table 3

Variables in the Monthly Model

Variable	Description	Differencing
1.Birth-1	U.S. births delayed 1 month	$(1-B^1)^1(1-B^{12})^1$
2.Infant	Deaths in the U.S. of infants from 29 days to 1 year	$(1-B^1)^1(1-B^{12})^1$
3.Neonatal	Deaths in the U.S. of infants from 0-28 days	$(1-B^1)^1(1-B^{12})^1$
4.Homicides	Homicides in the U.S.	$(1-B^1)^1(1-B^{12})^1$
5.Suicides	Suicides in the U.S.	$(1-B^1)^1$
6.NR	Population in the U.S.	$(1-B^1)^1(1-B^{12})^1$
7.UHHM	Male head of household unemployed	$(1-B^1)^1$
8.JLEAD	Index of Leading Indicators (1967=1)	$(1-B^1)^1$

Once again preliminary inspections of the data and correlograms were made and differencing parameters selected. It has been suggested that the operation of differencing may affect the multivariate relationships in the data and may not always be necessary in order to arrive at a proper fit. It is hoped that a future analysis of the undifferenced series will shed some light on this issue.

After several trials the model fitted was a seasonal ARMA $(4,0)*(0,1)^{12}$. In total 34 parameters were necessary to reduce the residual cross-correlations to white noise. All of the six seasonal terms, save one, were on the diagonal.

When the Haugh S test was done on the residual cross-correlations, no pair gave a significant chi-square for lag up to 12. Of the 768 correlations up to lag 12, only 7 were significant at the 1% level. None of these was greater than .25. It safely may be concluded that the model adequately reduces the series to white noise.

However, Figure 6 reveals - as in the prior analysis - a disappointing set of linkages.

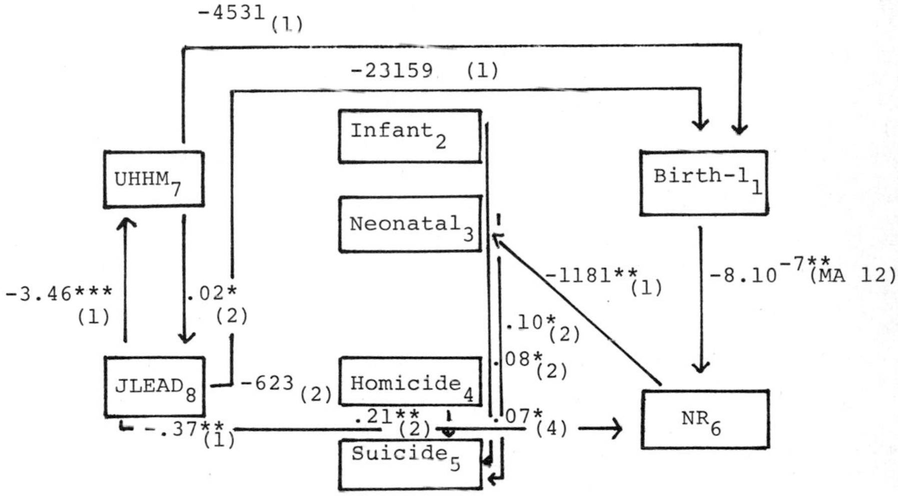

Figure 6

Parameters of the Monthly Model

* p<.05
** p<.01
*** p<.001
no star p<.20
(lag value)
All are AR parameters unless otherwise noted.

The links between the economic pair of variables and the social set are extremely weak. In fact, there is only one between JLEAD and homicides. There are others to the control variable population and births, but these are weak, in the wrong direction, or both. The links within the social set are weak and their pattern suggests the presence of unmeasured common causal variables rather than causal links. The economic pair themselves seem to reveal credible causal linkages similar to the yearly analysis with the same feedback relationship being manifest. The strong MA relationship between births and population is probably due to corrections made to monthly population estimates when birth data are completed at the end of the year.

One might well query the absence of links between neonatal and infant mortalities and births as well as homicides and suicides with population. One hypothesis is that the trend and seasonality of the social ill variables is largely attributable to the control variables and when these were differenced, this was lost.

Once again it is useful to determine whether the model presented in Figure 6 fits appreciably better than its diagonal form. Table 4 presents the variances explained by each.

Table 4

Percent Variance Explained by Each Model

	Model		
Variable	Full	Theoretical	Diagonal
1.Births	44.2	34.8	43.8
2.Infant	49.4	48.0	48.0
3.Neonatal	55.5	53.7	53.7
4.Homicides	50.6	50.1	49.3
5.Suicides	39.3	33.5	32.7
6.Population	40.3	37.4	33.8
7.Unemployment	27.5	24.4	8.9
8.Leading Indicator	38.5	37.2	36.9

In addition, a "theoretical" model was fitted. This was developed from logical considerations and prior studies asserting links between the economic variables and the social set. As a first try, lag one was chosen for all links. Also fitted were the diagonal elements of the full model. The parameters are noted in Figure 7. As is seen from the Figure none of the economic and social links is significant; nor is the variance explained in each variable appreciably greater than the diagonal model with the exception of unemployment. An inspection of the residuals gives little evidence that a choice of additional lags would improve the fit. Although the Haugh test indicated some pair-wise correlation, the pairs were not those of the theoretical model. In fact, these tests pointed toward the full model.

From these analyses, it is clear that there is considerable predictable variance in the eight variables, but that this is predictable mainly from the variable's own past history and with the exception of the economic variable diad little is added to the predictive equation by a knowledge of other variables. In particular, once again even with the significant increase in power given by the larger sample size, no relationship is elucidated between the economic set and the social set of variables. Although one can never really find for the null hypothesis, these analyses cast considerable doubt on the direct relationship of economic indices and the rise or fall of variables measuring the quality of life.

It is possible to ask what oversights could have led to the missing of relationships other than lack of statistical power. One could be the incompleteness of the model - that is, failing to include variables which act as suppressor variables. Suppressor variables when included partial out suppressor variance and allow the correlations between the economic and social variables to be manifest. Another could be a restriction of range problem. Thus, unemployment may not have an effect unless it moves radically or above a threshhold. A third problem could be that differencing somehow masked the effects. However, some preliminary analyses have been done using undifferenced data which tend to belie the existence of undiscovered relationships. A fourth problem involves the geographical aggregation of the data. Perhaps if a geographically more homogeneous data set were analyzed, such as a single state or community, the relationships might be more easily seen. Lastly, the data themselves in terms of their validity and accuracy may be challenged.

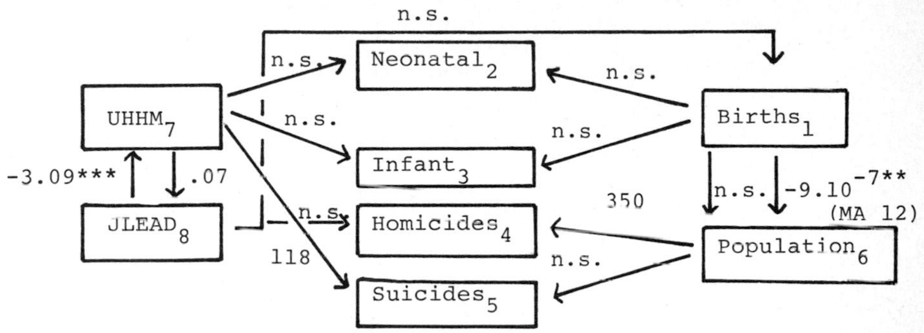

Figure 7

The Theoretical Model

*	p<.05
**	p<.01
***	p<.001
no star	p<.20
n.s.	p<1.0

The major conclusion is one of general lack of relationships in data which have in the past been found to be quite correlated when subject to regression type procedures. This is similar to other studies (e.g. Pierce, 1977) which have found that when ARMA analyses are applied to data corrected for autocorrelation many relationships tend to disappear.

FOOTNOTE:

[1] These variables were obtained from Data Resources (Lexington, Massachusetts) data banks under the codes given.

REFERENCES:

Brenner, M.H. (1976). Estimating the social costs of National Economic Policy. Washington: U.S. Government Printing Office.

_____ (1973). Fetal, infant and maternal mortality during period of economic instability. *Int. J. of Health Services.* 3,2, 69-82.

Brooks, C. H. (1978). Infant mortality in the SMSA's before Medicaid: test of a causal model. *Health Services Research.* 13,1.

Data Resources Inc. (1981). *Review of the U.S. Economy.*

Eisner, V. *et al.* (1968). Improvement in infant mortality and perinatal mortality in the United States 1965-1973. *Amer. J. Pub. Health*, 68,4.

Goldsmith. S.B. (1972). The status of health status indicators. *Health Services Report.* 87, 212-220.

Granger, C.W.J. and Newbold, P. (1977). *Forecasting Economic Time Series.* New York: Academic Press.

Grossman, M. and Jacibowitz, S. (1980). *Determinants of variations in infant mortality rates among counties of the United States: the roles of social policies and programs.* paper presented to World Congress on Health Economics. Netherlands.

Haugh, L.D. and Box, G.E.P. (1977). Identification of dynamic regression (distributed lag)model connecting two time series. *J. Amer. Stat. Assn.*, 72, 121-131.

Henry, A.F. and Short, F. (1954). *Suicide and Homicide.* New York: Free Press

Jenkins, G.M. (1979). Practical experiences with modelling and forecasting time series. Reprinted from *Forecasting* (1979). Ed. O.D. Anderson, North-Holland, Amsterdam, 43-166.

Kendall, M.G. (1973). *Time Series.* New York: Hafner.

Lee, K.S. *et al.* (1980). Neonatal mortality: an analysis of the recent improvements in the U.S. *Amer. J. of Pub. Health*, 70, 1.

MacMahon, B., Kovar, M.G., and Feldman, J.J. (1972). Infant mortality rate: socio-economic factors. *Vital Health Stat. Series*, 22, 14, DHEW, pub. no. HSM 72-1045.

National Center for Health Statistics. (1965-1978). *Vital Statistics of the United States. Vol. 1 Natality, Vol. 11 A Mortality.* USDHEW: Public Health Service.

Office of Health Research and Statistics. *Monthly Vital Statistics Report: Births, Marriages, Divorces, and Deaths.* DHEW: Public Health Service.

Pierce, D.S. (1977). Relationships and the lack thereof between economic time series with special reference to money and interest rates. *J. Amer. Stat. Assn.*, 77, 11-21.

Stockwell, E.G. (1962). Infant mortality and socioeconomic indicators: a changing relationship. *Millbank Mem. Fund Q.*, 11, 101-111.

Tiao, G.C., Box, G.E.P., Grupe, M.R., Hudak, G.B., Bell, W.R., and Chang, I. (1979). *The Wisconsin Multiple Time Series (WMTS-1) Program: A Preliminary Guide.* Madison: University of Wisconsin, Dept. of Statistics.

Working, H. Note on the correlation of first differences of averages in a random chain. *Econometrica,* 28, 916.

APPLIED TIME SERIES ANALYSIS
O.D. Anderson & M.R. Perryman (eds.)

TIME SERIES MODELING OF MACHINED SURFACES

S. G. Kapoor

Department of Mechanical and Industrial Engineering
University of Illinois at Urbana-Champaign
Urbana, IL 61801
U.S.A.

A combined deterministic-stochastic modeling approach is introduced to analyze machined surfaces. The deterministic component of the model is interpreted as the periodicity or trend present in the profile and the stochastic component represents the true surface roughness. The parameters of the stochastic models are then used to develop the characterization parameters in terms of physically meaningful characteristics such as profile heights, slopes, and curvatures. Data from the inside of a cylinder surface where the arc of a circle is part of the surface profile are analyzed to illustrate the proposed modeling and characterization approach.

INTRODUCTION

The machined surfaces often possess the deterministic component due to copying the tool on the workpiece or by a vibration of the tool and randomness by the detachment of material, geometry of the cutting action, and metallurgical or heat treatment effects. Therefore, these surfaces should be analyzed using statistical techniques. A number of statistical parameters such as center-line average (CLA) and root mean square (RMS) values have been proposed to characterize the surface finish. These parameters are simple to calculate and easy to implement on an instrument. However, it has been recognized for some time that the profiles differing markedly in their roughness and waviness values show identical CLA or RMS values. New techniques employing the height density function (Pesante (1964)), the autocorrelation function (Peklinik (1967)), or the spectrum (Kubo (1967)) have been proposed. The basic problem with these techniques has been the determination of a reliable characterization parameter. Furthermore, the parameters developed from these techniques suffer from a lack of physical interpretation.

The surface profiles are also analyzed using a purely deterministic approach. The finite Fourier analysis is used to represent both periodic and non-periodic parts of the profile. Raja and Radhakrishnan (1977) approximated turned and ground surface profiles by as much as 70 harmonics terms in a Fourier series. The surface profiles have recently been analyzed using stationary stochastic modeling approaches (Wu, Kapoor, and DeVries (1978), Pandit and Wu (1973), Kapoor and Wu (1978)). Characterization by the stochastic modeling method has an added advantage as the mathematical model is developed for the mechanics of profile generation together with the characterization parameters.

Despite being very useful, the method suffers a great disadvantage when the surface profiles possess a pronounced degree of periodicity or trends due to manufacturing processes. The characterization parameters developed from the model parameters are found to be inaccurate and do not truly reflect surface finish characteristics. It has been suggested by Whitehouse (1967) that the deterministic components of a surface profile should be filtered by using digital or analog methods. The problem with filtering is that occasionally it introduces spurious details and distorts the original trace thereby giving incorrect values of the characterization parameters.

The purpose of this paper is to present a method of analyzing machined surfaces using a combined deterministic-stochastic modeling approach. The deterministic component reveals the periodicity or trend present in the profile and the stochastic component represents the true surface roughness. Parameters of the stochastic models are then used to develop the characterization parameters in terms of physically meaningful characteristics such as profile heights, profile slopes, and curvatures as well as peaks and zero-crossings. The models and modeling strategy are discussed first. Data from the inside of a cylinder surface where the arc of a circle is a part of the surface profile are analyzed to illustrate the proposed modeling and characterization approach.

MODEL AND MODELING PROCEDURE

Let Y_t represent the surface profile heights at time t = 1,2,...,N as shown in Fig. 3. Any covariance stationary time series can be represented linearly into two parts.

$$Y_t = D(t,\alpha) + X_t \tag{1}$$

where $D(t,\alpha)$ is the deterministic function with parameter vector α, X_t is a stationary stochastic series and can be represented by an autoregressive moving average model (ARMA) of the form

$$X_t = \sum_{i=1}^{n} \phi_i X_{t-i} - \sum_{j=1}^{m} \theta_j a_{t-j} + a_t \tag{2}$$

The parameters ϕ and θ are referred to as the autoregressive and moving average parameters and a_t is a discrete white noise process with $E[a_t a_{t-k}] = \delta_k \sigma_a^2$ where δ_k refers to Kronecker delta function.

Modeling Procedure

The combined deterministic-stochastic modeling procedure involves three stages of modeling. The purpose of incorporating three stages is to minimize the chances of getting parameters estimates subject to local minima convergence.

In the first stage, a deterministic function is identified and parameters of the deterministic model are estimated by the least squares method, assuming uncorrelated errors. A successive modeling is done in order to obtain the adequate model. The adequacy of the model is checked by statistical tests such as F-test as discussed in Pandit (1980). Parameters of the adequate deterministic model are also used as initial values of the parameters of the final combined deterministic-stochastic model. In Phase II, the ARMA models are fitted to the residuals from Phase I. The ARMA (n,n-1) modeling strategy as proposed by Pandit (1980) is used to fit ARMA model. Similar to the deterministic case, the order of the adequate stochastic model and the parameter values are used as the starting values for the parameters to be estimated in Phase III.

Once Phases I and II are completed, a combined deterministic-stochastic model

$$Y_t = D(t,\alpha) + X_t$$

and

$$X_t = \sum_{i=1}^{n} \phi_i X_{t-i} - \sum_{j-1}^{m} \theta_j a_{t-j} + a_t \tag{3}$$

is fitted and the parameter values are estimated using the least squares method. Adequacy of the combined model is checked by examining the autocorrelation of the residuals.

DEVELOPMENT OF CHARACTERIZATION PARAMETERS

Since the stochastic part of the combined deterministic-stochastic model reveals the surface finish, parameters of the ARMA model are used to develop the characterization parameters for analyzing the surface finish of machined surfaces. Profile characteristics such as profile heights, slopes, and curvatures are considered as the basis for characterization. The zero, second, and fourth moments of the spectrum (m_0, m_2, and m_4) are physically termed as the variances of profile heights, slopes, and curvatures. Therefore, the power spectral density is derived from the ARMA model parameters and the spectral moments are then computed from the power spectral density function.

The power spectral density, $f(\omega)$ is obtained from the ARMA (n,m) model as follows:

$$f(\omega) = 2\pi\, \sigma_a^2\, F(j\omega)\, F(-j\omega) \qquad (4)$$

where

$$F(j\omega) = [1 - \sum_{k=1}^{m} \theta_k\, e^{-kj\omega\Delta}]/[1 - \sum_{i=1}^{n} \phi_i\, e^{-ij\omega\Delta}] \qquad (5)$$

where ω denotes frequency (cycles/mm); Δ denotes sampling interval; ϕ_i denotes the autoregressive parameters of the ARMA model; and θ_k denotes the moving average parameters.

The zero, second, and fourth spectral moments are next computed from

$$m_r = 2\int_0^{1/2\Delta} \omega^r\, f(\omega)\, d\omega \qquad (6)$$

where r = 0, 2, and 4. Note that the spectral moments m_1, m_3, etc., are zero and, therefore, are not considered in our analysis.

Having obtained the spectral moments, the geometrical statistical properties of a surface profile including the mean number of zero crossings and the average number of peaks as shown in Fig. 1 and, which also are the physical parameters that may have high correlation with both generating and functional parameters can be derived.

Mean Number of Zero Crossings

Under the assumption of normal distribution of profile heights, the mean number of zero crossings per unit length, z_c, is defined by Nayak (1971) in terms of the variance of profile heights and slopes as

$$z_c = 1/\pi\ (m_2/m_0)^{1/2} \qquad (7)$$

Physically, the mean number of zero crossings denotes the reciprocal of the average wavelength of the profile.

Average Number of Peaks or Maxima

By definition, the maxima in a curve are those points where the slope is zero and the curvature is negative. Therefore, the frequency of maxima of profile height, X_t, must be the frequency of negative zero crossings of the derivative process $\dot{X}(t)$. Hence, if the average number of maxima of $X(t)$ is N_m and z_c is the frequency of zero crossings of $\dot{X}(t)$, we have $N_m = Z_c \cdot \dot{X}(t)$. It can be shown that the average number of maxima, N_m of profile peaks is

$$N_m = 1/2\pi\ (m_4/m_2)^{1/2} \qquad (8)$$

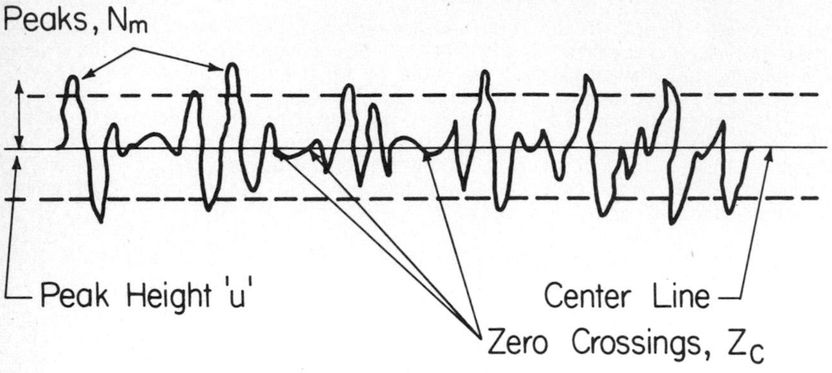

Figure 1 Surface Characterization Parameters

ANALYSIS OF SURFACE PROFILE OF A CYLINDRICAL SURFACE

Surface profile data were collected from the inside surface of a cylinder used in the automotive industry. A microcomputer based three-dimensional surface characterization system (Fig. 2) was employed to collect, digitize, and store the data on the main computer. A sampling interval of 50 μm was used to digitize the continuous record and the length of the profile trace was 12 mm.

It is obvious from the trace of the data (Fig. 3a) that there is an inherent curvature due to the cylindrical surface. This curvature along the orientation of the specimen to the stylus tip will add a deterministic trend to the data (Fig. 3b). To accomplish removal of the arc (the deterministic portion of the data) from the profile data, the following mathematical equation was proposed

$$(x - h)^2 + (y - k)^2 = R^2 \qquad (9)$$

where x denotes the sampling distance along the surface; y, the profile height at X due to deterministic trend; h and k, the offsets from the center of the circle to the first height location in the x- and y-directions, respectively; and R, the theoretical radius of the profile arc. The actual profile height, Y_t, consists of basically two parts. The first part refers to the height due to the curvature and offset, Y_t' and the other part shows the actual profile variation relating to the surface finish.

Using the three-stage modeling strategy, the determinisitc model was fitted first by employing the method of least squares, i.e.:

$$Y_t = k - [R^2 - (t - h)^2]^{1/2} + X_t \qquad (10)$$

where t is the observation number and represents the sampling distance, x/Δ (where Δ denotes the sampling interval), Y_t the profile height at t, and X_t the residual. The estimation results are given in the second column of Table 1. The residuals obtained from the least squares fitting are free from arc and shown in Fig. 3c.

The ARMA models are fitted in Stage 2 to the residuals X_t by using an (n,n-1) modeling strategy, and an ARMA (4,1) model was found adequate based on the

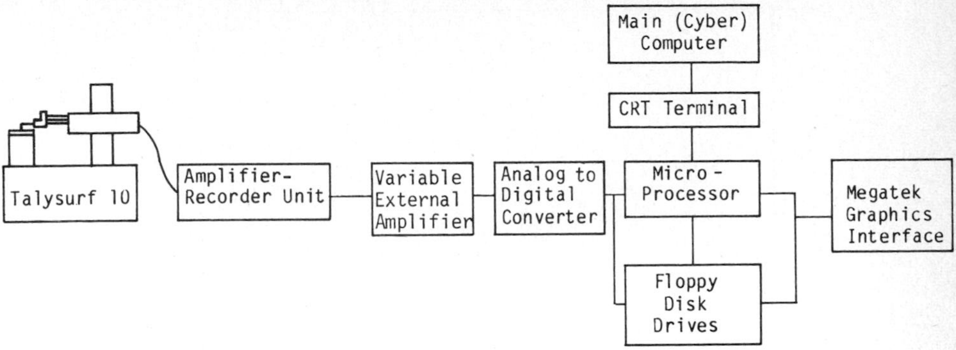

Figure 2 Block diagram of the micro-computer based three-dimensional surface characterization system

statistical tests. These parameters are listed in the third column of Table 1. The order of the model and the parameter values will be confirmed in the final estimation.

Finally, in the third stage, a combined deterministic-stochastic modeling was done. The combined model is mathematically represented as:

$$Y_t = k - [R^2 - (t - h)^2]^{1/2} + X_t \tag{11}$$

$$X_t = \phi_1 X_{t-1} + \phi_2 X_{t-2} + \phi_3 X_{t-3} + \phi_4 X_{t-4} + a_t - \theta_1 a_{t-1}$$

The order of the stochastic model was again found to be ARMA (4,1). The final estimates of the parameters are also listed in Table 1.

The parameters of the ARMA (4,1) model were used to obtain the spectrum and the spectral moments by employing Eqs. (4) through (6). Having obtained the spectral moments, namely the variance of profile heights, profile slopes, and profile curvatures, the mean number of zero crossings and the average number of maxima were computed. The results are listed in Table 2. It is seen that the average number of peaks (maxima) is 2.47/mm. The peak counts are found to be a useful index in determining the lubrication characteristics of the inside surface of the cylinder. The peak counts are also considered important when the performance of two contacting surfaces are of interest. The mean number of zero crossings physically indicate the average wavelength of the profile. A surface profile of large wavelength is considered to be good for contact applications.

CONCLUSIONS

The machining process by which parts are manufactured lends itself to the introduction of deterministic and stochastic variation in the surface profile. A combined deterministic-stochastic modeling approach is, therefore, proposed to develop a mathematical model and to deduce parameters for the characterization of surfaces. The modeling procedure involves three stages. In the first stage, a deterministic function is identified through the least squares technique. The ARMA models are fitted to the residuals from stage I, using the ARMA (n,n-1) modeling strategy. Finally, the combined deterministic-stochastic models are fitted to the original surface profile data.

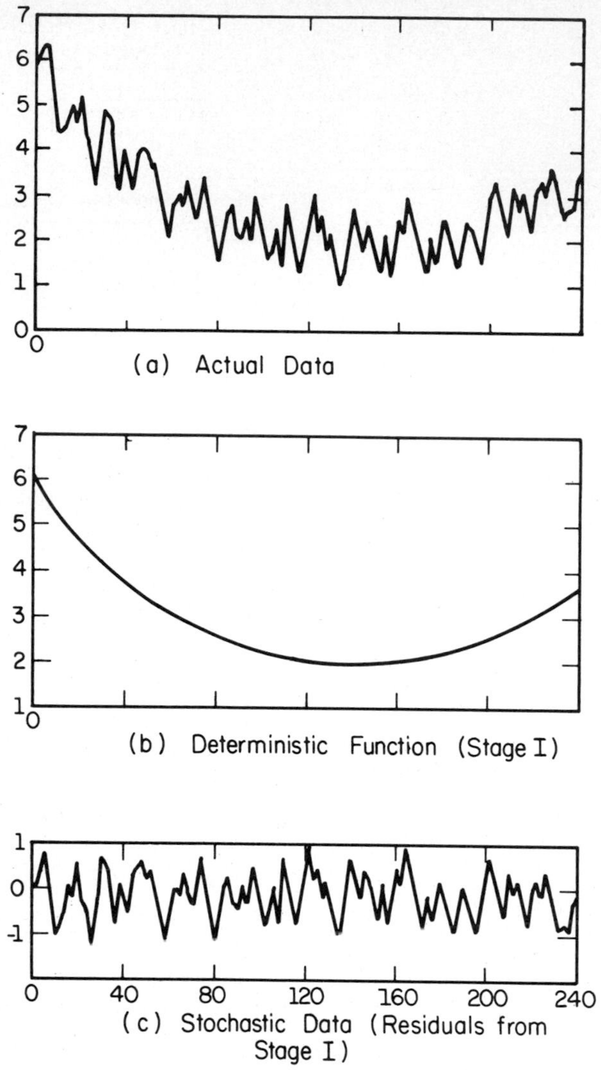

Figure 3 Surface Profile of a Cylindrical Surface

The surface profile data obtained from an inside surface of a cylinder used in an automotive industry are analyzed using the combined deterministic-stochastic modeling approach. The deterministic components of the models revealed the curvature of the cylindrical surface. The stochastic component representing the true surface roughness profile was used to develop characterization parameters in terms of the variance of profile heights, slopes, curvatures, the mean number of zero crossings and the average number of peaks.

TABLE 1

MODEL PARAMETERS

Parameters	Stage I	Stage II	Stage III
h	7.12		6.96
k	9.91		10.10
R	8.24		8.07
ϕ_1		1.61	1.79
ϕ_2		- .75	- .64
ϕ_3		.34	.54
ϕ_4		- .11	- .31
θ_1		.58	.49
RSS**	43.12	32.16	28.08
Q-Value*		15.8	13.6

*95% x^2 value (Table) = 28.9

**RSS = Residual Sum of Squares

TABLE 2

CHARACTERIZATION PARAMETERS

1.	Variance of Profile Heights, m_0	0.166 mm^2
2.	Variance of Profile Slopes, m_2	28.90
3.	Variance of Profile Curvatures, m_4	6960.7 mm^{-2}
4.	Mean Number of Zero Crossings, ZC	4.2 / mm
5.	Average Number of Maxima, Nm	2.47 / mm

REFERENCES

1. Kapoor, S. G., and Wu, S. M. (1978), "A Stochastic Approach to Paper Surface Characterization and Printability Criteria," J. Phy. D., Applied Physics, Vol., 11, pp. 83-97.

2. Kubo, M (1967), "Statistical Analysis of Surface Roughness Waveforms," Annals of CIRP, Vol. 15, pp. 381-385.

3. Nayak, P. R. (1971), "Random Process Model of Rough Surfaces," Trans. ASME, Vol. 93, No. 3, pp. 398-407.

4. Pandit, S. M. (1980), "Data Dependent Systems and Exponential Smoothing". In Analysing Time Series, (Proceedings of the International Conference held on Guernsey, Channel Islands, October 1979). Ed: O. D. Anderson, North-Holland, Amsterdam, pp. 217-238.

5. Pandit, S. M., and Wu, S. M. (1973), "Characterization of Abrasive Tools by Continuous Time Series," J. of Engr. for Ind., Trans. ASME, pp. 821-826.

6. Peklinik, J. (1967), "Investigation of Surface Topology," Annals of CIRP, Vol. 15, pp. 381-385.

7. Pesante, M. (1964), "Determination of Surface Roughness Topology by Means of Amplitude Density Curves," Annals of CIRP, Vol. 12, pp. 61-68.

8. Raja, J., and Radhakrishnan, V. (1977), "Analysis and Synthesis of Surface Profiles Using Fourier Series," Int. J. of Machine Tool Design and Res., Vol. 17, pp. 245-251.

9. Whitehouse, D. J. (1967), "Improved Type of Wave-Filter for Use in Surface Finish Measurement," Inst. Mech. Engr., (London), Proc., Vol. 182, pp. 306-318.

10. Wu, S. M., Kapoor, S. G., and DeVries, M. F. (1978), "Characterization of Coated Abrasive Wear by Statistical Invariants," Proc., NAMRC, pp. 346-351.

APPLIED TIME SERIES ANALYSIS
O.D. Anderson & M.R. Perryman (eds.)

The Rank Transformation as a Method for Dealing with Time Series Outliers

Gary D. Kelley
Texas Tech University
and
David E. Noel
Arizona State University

I. INTRODUCTION

The rank transformation is a statistical method whereby data to be analyzed are ranked from smallest to largest, and then the usual statistical procedure is applied on ranks. It has been shown that the rank transformation leads to robust and powerful statistical procedures in regression, experimental design, multiple cmparisons, discriminant analysis and cluster analysis, especially when the sampled population and/or the sample itself is contaminated with outliers (see Conover, 1980), Conover and Iman (1976 and1980), Randals et.al. (1978) and Worsley (1977).

At the present time, little work has been done regarding the application of the rank transformation to time series analysis. It is the intent of the authors of this paper to investigate in an heuristic manner, the rank transformation as applied to the initial phase of Box-Jenkins time series analysis: the selection of the appropriate autoregressive-moving average model. In particular, the paper will concentrate on the undesirable effects which outliers have on the modeling phase and how these effects can be alleviated by applying the rank transformation.

II. RESULTS

The Box-Jenkins modeling procedure calls for the examination of two sets of sample statistics computed from the time series to be analyzed: the sample autocorrelation function (acf) and the partial autocorrelation function (pacf) (Box-Jenkins, 1976). The behavior of these functions determine which autoregressive-moving averge model is selected to represent the given series. The results in this section show how the behavior of the acf and pacf are affected when these functions are computed on the ranks of the data instead of on the data themselves. Also, outliers will be introduced in the data and the acf and pacf will be computed both before and after ranking the data.

Table 1 shows the acf and pacf at lag k (k=1,2,...,10) for a randomly selected realization of length 50 generated from the AR(1) model $X_t = .85X_{t-1} + a_t$. The column labeled "Original Unranked" shows the acf and pacf of the generated series. The "Original Ranked" column shows the acf and pacf of the original series in which each observation has been replaced with its rank in the series. The two "Original" columns are seen to be quite similar. In the remaining columns labeled "2 Extremes Unranked" and "2 Extremes Ranked", the acf and pacf were computed for both unranked and ranked data after 2 randomly chosen observations from the original series were replaced with "outliers".

The magnitude of each outlier was randomly selected from four numbers $\pm$ 4s and $\pm$ 5s, where s is the standard deviation of the original series. The" Unranked" column shows marked deviation from the "Original Unranked," while the "Ranked"

Table 1. Shows the autocorrelation (acf) and the partial autocorrelation (pacf) at lag k computed from a series of length 50 from $X_t = .85X_{t-1} + a_t$.

		Original		2 Extremes	
	k	Unranked	Ranked	Unranked	Ranked
	1	.658	.671	-.002	.473
	2	.541	.564	-.008	.413
	3	.440	.456	.040	.402
	4	.318	.343	-.043	.226
acf	5	.291	.343	-.039	.192
	6	.155	.241	-.135	.018
	7	.182	.261	-.100	.066
	8	.160	.199	-.115	.015
	9	.112	.150	-.160	-.064
	10	.199	.169	-.020	.098
	1	.658	.671	-.002	.473
	2	.190	.217	-.008	.244
	3	.050	.021	.040	.191
	4	-.062	-.049	-.043	-.088
	5	.080	.137	-.039	-.011
pacf	6	-.146	-.085	-.138	-.196
	7	.142	.109	-.100	.084
	8	.013	-.052	-.124	-.014
	9	-.028	-.031	-.170	-.034
	10	.071	.068	-.049	.182

with extremes is not nearly as affected. It is difficult from a table such as this to observe how the original patterns of the acf and pacf are affected both by introducing outliers and by ranking. However, in the interest of conserving space, graphs of the functions are not included. By graphing these functions, the interested reader will be able to observe that the patterns of the original acf and pacf are preserved, even with the introduction of outliers in the series, after applying the rank transformation to the data.

Tables 2 and 3 show similar results for a realization of length 200 from the ARMA (1,1) model $X_t = .7X_{t-1} + a_t + .6a_{t-1}$. As before, the observations replaced with outliers were randomly chosen, as were the magnitudes of the outliers. The outliers ranged in magnitude from $\pm$ 4s to $\pm$ 9s. It can be seen (again, perhaps better graphically) that the acf and pacf computed on the ranked data reflect the original patterns which are crucial to Box-Jenkins model identification.

III. SUMMARY AND CONCLUSIONS

The procedure outlined in this paper differs from most other techniques for dealing with outliers in that no attempt is made to identify or replace the outliers prior to the actual analysis. It was shown that the rank transformation does not cause significant loss of information in the acf or the pacf, which is especially important when the series is contaminated with outliers.

For a more thorough treatment of the rank transformation as applied to time series analysis, the reader is referred to Noel (1982).

Table 2. Shows the autocorrelation at lag k computed from a series of length 200 from $X_t = .7X_{t-1} + a_t + .6a_{t-1}$.

	Original		2 extremes		5 extremes		10 extremes	
k	U	R	U	R	U	R	U	R
1	.765	.767	.669	.751	.458	.700	.206	.565
2	.412	.410	.371	.407	.167	.345	.094	.292
3	.183	.180	.202	.196	.082	.160	.105	.102
4	.023	.028	.059	.037	.026	.030	.016	-.047
5	-.100	-.085	-.075	-.094	-.057	-.093	-.060	-.141
6	-.191	-.184	-.174	-.194	-.091	-.172	.021	-.135
7	-.213	-.217	-.221	-.220	-.123	-.194	.029	-.190
8	-.176	-.176	-.195	-.190	-.033	-.180	-.021	-.144
9	-.110	-.101	-.110	-.120	-.062	-.121	-.002	-.062
10	-.026	-.023	-.037	-.040	-.008	-.041	.019	-.004
11	.030	.021	.038	.030	.050	.035	.049	.060
12	.058	.052	.087	.069	.094	.086	.169	.086
13	.085	.091	.074	.077	.068	.094	.112	.100
14	.123	.115	.060	.100	.052	.106	.017	.105
15	.116	.095	.057	.097	.035	.095	.003	.105

U = unranked series, R = ranked series

Table 3. Shows the partial autocorrelation for lag k computed from a series of length 200 from $X_t = .7X_{t-1} + a_t + .6a_{t-1}$.

	Original		2 extremes		5 extremes		10 extremes	
k	U	R	U	R	U	R	U	R
1	.767	.765	.669	.751	.458	.700	.206	.565
2	-.436	-.419	-.141	-.360	-.053	-.286	.054	-.039
3	.189	.162	.024	.134	.033	.106	.079	-.070
4	-.207	-.201	-.097	-.191	-.022	-.134	-.025	-.099
5	.005	-.017	-.106	-.037	-.077	-.095	-.074	-.082
6	-.172	-.120	-.089	-.119	-.041	-.037	.041	.012
7	.101	.058	-.054	.045	-.073	-.052	.029	-.134
8	-.041	-.031	.024	-.036	.078	.013	-.025	.026
9	.073	.051	.067	.077	-.086	.040	-.003	.037
10	-.008	.036	.011	-.009	.067	.026	.013	.088
11	-.020	-.041	.056	.048	.036	.041	.054	.045
12	.048	.032	-.003	-.044	.053	.016	.162	-.005
13	.027	.025	-.062	.015	-.007	-.024	.040	.040
14	.019	.085	.006	.070	.003	.063	-.046	.024
15	-.054	-.100	.016	-.052	.012	-.035	-.032	.035

U = unranked series, R = ranked series

References

Conover, W.J. Practical Nonparametric Statistics, 2nd. Edition. New York, Wiley, 1980.

Conover, W.J.and Iman, R.L. "On some alternative procedures using ranks for the analysis of experimental designs." Communications in Statistics, Vol. A5, No. 14 (1976), pp. 1349-1368.

Conover, W.J. and Iman, R.L. "The rank transformation as a method of discrimination with some examples." Communications in Statistics, Vol. A9, No. 5 (1980), pp. 465-487.

Conover, W.J. and Iman R.L. "Rank transformation as a bridge between parameteric and nonparametric statistics." The American Statistician, Vol. 35, No. 3 (1981), pp. 124-133.

Noel, D.E. An Examination of the Rank Transformation Method in Autoregressive Moving Average Time Series Analysis. DBA Dissertation, College of Business Administration, Texas Tech University (to be completed, 1982).

Randles, R.H., Broffitt, J.D., Ramberg, J.S. and Hogg, R.V. "Discriminant analysis based on ranks." Journal of the American Statistical Association, Vol. 73, No. 362 (1978), pp. 379-384.

Worsley, K.J. "A nonparametric extension of a cluster analysis method by Scott and Knott." Biometrics, Vol. 33, No. 3 (1977), pp. 532-535.

APPLIED TIME SERIES ANALYSIS
O.D. Anderson & M.R. Perryman (eds.)

RELATIONSHIPS BETWEEN TRANSFER FUNCTIONS, CAUSAL IMPLICATIONS AND ECONOMETRIC MODELING SPECIFICATION

Sergio G. Koreisha

Graduate School of Management
University of Oregon
Eugene, Oregon
USA

Construction of causally and structurally adequate simultaneous equations models can be accomplished by determining causal relations between potential variables and balancing these statistically derived inferences with economic theory to relate behavioral or technological forces among the variables. An appropriate lag structure for each of the equations can be determined by a two step multiple transfer function approach involving reduced form equations to obtain estimates of the determined endogenous variables which will not be correlated with the error terms of the various equations of the model. Testing the specification of already existing simultaneous equations models is done by constructing multiple transfer function models of the reduced form equations of the simultaneous equations models which permit incorporation of future cross correlations.

Causality inferences are made whenever relations between variables in econometric models are postulated. Typically assertions about which variables are endogenous and which are exogenous are not verified. Consequently, it is quite possible that many models currently being used are misspecified. Thus, an adequately specified model must be one which in some sense is causally correct. Testing for the direction of causality, for example, allows the researcher to avoid the specification and inconsistency problems discussed by Goldfeld and Blinder (1972) which result from incorrectly assuming that a policy variable is determined exogenously. Geweke (1978, p. 182) points out that "the exogeneity specification is the most important assertion about the way in which that aspect of the real world which the model purports to describe 'works.' While the recursiveness which the exogeneity specification asserts may be intended only as a conceptual and analytical simplification which would not be defended literally by the architect of the model in question, it is assumed to be approximately correct and therefore ought to be rejected if it is found to be at variance with the data. It makes no sense to discuss and test the detailed dynamics of an econometric model within a framework which can be rejected in its entirety by the data. Even if one were solely concerned with the predictive ability of such a model, proceeding on the basis of a system whose exogeneity specification was found to be unwarranted would be unwise because of the intimate link between exogeneity and forecasting ability shown by Sims (1972). Specification of a group of variables as exogenous is equivalent to a refusal to use knowledge of the endogenous variables in forecasting them — a restriction which will improve efficiency and therefore forecast accuracy if true, but increase forecast errors for both endogenous and exogenous variables if false."

In this article we will show how causality inferences can be integrated with transfer functions to check the specifications of econometric models. In Section 1 we present a definition of causality in terms of predictability. In Section 2 we propose the use of a new test, namely that of the causal transfer function test. In Section 3 we discuss how causality can be used not only to detect misspecification in simultaneous equations systems but also to formulate econometric relationships. In Section 4 we show how the theories discussed in previous sections can be put to work using Klein's Model I of the U. S. economy (1950). Finally, some concluding remarks are presented in Section 5.

1. DEFINITION OF CAUSALITY

Granger (1969) developed a definition of causality in terms of predictability which can be easily applied to temporal systems. "A variable X causes another variable Y, with respect to a given universe or information set that includes X and Y, if present Y can be better predicted by using past values of X than by not doing so, all other information contained in the past of the universe being used in either case."[1] This definition can be stated more formally as follows: Let $\{I_t, t=0,\pm1,\pm2,...\}$ be the given information set,[2] including at least (X_t, Y_t). Let $\bar{I} = \{I|:s<t\}$, $\bar{\bar{I}} = \{I|:s\leq t\}$, and similarly define $\bar{X}$, $\bar{\bar{X}}$, $\bar{Y}$, $\bar{\bar{Y}}$. Let $P_t(Y|B)$ denote the minimum Mean Square Error (MSE) single step predictor of Y_t given an information set B, and MSE(Y|B) be the resulting mean square error. Then

1. X causes Y if

$$MSE(Y|\bar{I})<MSE(Y|\bar{I}-\bar{X}), \tag{1}$$

2. X causes Y instantaneously[3] if

$$MSE(Y|\bar{I},\bar{\bar{X}})<MSE(Y|\bar{I}). \tag{2}$$

By assuming that the time series to be analyzed are jointly covariance-stationary,[4] and by considering only linear predictors and using expected squared forecast error as the criterion for predictive accuracy, Sims (1972) has proven some basic results which give empirical content to Granger's definition.

Following Sims (1972) we consider the jointly covariance-stationary pair of stochastic processes X and Y. If X and Y are jointly purely nondeterministic, then the processes can be expressed as moving average processes, namely as

$$\begin{aligned} X_t &= a(B)u_t+b(B)v_t \\ Y_t &= c(B)u_t+d(B)v_t \end{aligned} \tag{3}$$

where u and v are mutually uncorrelated white noise processes with unit variance.

Sims (1972, p. 544) has shown that "Y does not cause X in Granger's definition if, and only if, a(B) can be chosen to be identically zero." If X causes Y, then of the two orthogonal white noise processes which make up X and Y, one is "X itself 'whitened' and the other is the error in predicting Y from current and past X, whitened."

Granger (1969) showed that if there exists an autoregressive representation for X and Y given by

$$A(B)\begin{bmatrix} Y_t \\ X_t \end{bmatrix} = \begin{bmatrix} u_t \\ v_t \end{bmatrix} \tag{4}$$

where u_t and v_t are defined by (3), then the claim that Y does not cause X is equivalent to the upper right-hand element of the autoregressive polynomial matrix A(B) being zero. Building on this result Sims (1972, p. 545) proved a theorem which has become the basis of much of the current empirical work on causality:

> Theorem "When $\begin{bmatrix} X \\ Y \end{bmatrix}$ has an autoregressive representation, then Y can be expressed as a distributed lag function of current and past X with a residual which is not correlated with any values of X, past or future, if, and only if, Y does not cause X in Granger's sense."

Based on these definitions and the above theorem it is possible to develop causal tests for determining the adequacy of regression and simultaneous equation models.

2. CAUSAL TESTS

The first practical causal test was developed by Sims (1972). His methodology, as it is well known today, was quickly adopted and widely used by others to study relationships among socio and economic variables.[5] The Sims test, however, is sensitive to serially correlated errors and to the particular choice of lead/lag structures. Thus, the use of this test can lead to erroneous conclusions, (Feige and Pearce, 1978). A test which is less sensitive to the problems discussed above can be developed using Box-Jenkins transfer functions. Such a test is described below.

Using Granger's definitions, if one were to estimate transfer function models (TFMs) of the form

$$y_t = v(B^*)x_t \tag{5}$$

where

$$v(B^*) = (v_{-f}B^{-f}+\cdots+v_{-1}B^{-1}+v_0+v_1B+\cdots+v_gB^g) \tag{6}$$

instead of TFMs for which the v_j's for $j<0$ are assumed to be zero, then one could have a comprehensive test for determining whether or not causality directions inferred by the model can be justified.

If some values of v_j for $j<0$ in the causal transfer function model (5) are nonzero then clearly X could not cause Y. (If Y caused X one would expect that in a TFM for which Y is the explanatory variable, the values of v_j for $j\leq 0$ would be zero.) On the other hand, if only the v_j's for $j>0$ are nonzero, then X would be said to cause Y. (If X caused Y, one would expect that in a TFM for which Y is the explanatory variable, the values of v_j for $j\leq 0$ would be nonzero.) If none of the v_j's for $j<0$ is nonzero, but v_j either by itself or with other v_j's for $j>0$ are significant then this would be an indication of instantaneous causality between X and Y.

Pierce and Haugh (1977) developed a nomenclature for categorizing causality events based on combinations of the values which the transfer function weights can take.[6] Later in this section we present the causality events together with the necessary conditions for each of those events.

Determination of the causal relation between two variables using transfer functions is made with the use of innovations generated from ARIMA models of the two time series rather than with the actual series themselves. The reasons for doing this involve methodological issues, causal measurement-type principles, and problems with estimation of spurious correlations.

Working with the innovations permit us to use methodologies which are similar to the one prescribed by Box and Jenkins (1970) for formulating causal transfer function models with one main explanatory variable.

If we were trying to ascertain whether or not X caused Y, then a simple way to obtain a measurement of the influence of X on Y would be to filter out the systematic portion of the process governing Y, i.e., that portion which could be explained by the Y itself, and then to check if the remaining error process could be further explained by the X process. In other words, the outcome of the analysis of the transfer function equation

$$a_t = v'(B^*)x_t \tag{7}$$

where

$$a_t = \phi_y(B)\theta_y^{-1}(B)y_t \tag{8}$$

would determine the causality relationship between X and Y.

However, in order to minimize the possibility of estimating spurious correlations (Granger and Newbold, 1974; Granger and Newbold, 1977) between the variables it is recommended that

$$a_t = v(B^*)b_t \tag{9}$$

where

$$b_t = \phi_x(B)\theta_x^{-1}(B)x_t \tag{10}$$

be analyzed instead of (5). The justification for using (9) to minimize spurious correlation errors can be explained as follows: When two processes are appreciably correlated only over some narrow range of lags the covariance between two cross correlation estimates $r_{xy}(k)$ and $r_{xy}(k+1)$ can be approximated by

$$\mathrm{Cov}\left[r_{xy}(k), r_{xy}(k+\ell)\right] \approx (n-|k|)^{-1} \sum_{v=\infty}^{\infty} \rho_{xx}(v)\rho_{yy}(v+\ell) \tag{11}$$

Now suppose that y_t is generated by a white noise series and x_t is governed by an ARIMA process

$$x_t = \phi_x^{-1}(B)\theta_x(B)b_t \tag{12}$$

By construction, therefore, the two processes should not be cross correlated. However, from (11) it follows that

$$\mathrm{Cov}\left[r_{xa}(k), r_{xa}(k+\ell)\right] \approx (n-|k|)^{-1}\rho_{xx}(\ell) \tag{13}$$

and

$$\mathrm{Var}\left[r_{xa}(k)\right] \approx (n-|k|)^{-1} \tag{14}$$

so that

$$\rho\left[r_{xa}(k), r_{xa}(k+\ell)\right] \approx \rho_{xx}(\ell) \tag{15}$$

Thus, even though a_t and x_t are not cross correlated, the cross correlation function would be expected to vary about zero with standard deviation $(n-|k|)^{-1/2}$ in a systematic pattern typical of the behavior of the autocorrelation function $\rho_{xx}(\ell)$. Consequently, determination of causal relationships from models such as (5) could be misleading.

At the outset of this section we suggested that causal inferences could be made depending on the values of the causal transfer function weights v_j. Since estimates of v_j are obtained from estimates of cross correlations between prewhitened series adjusted by the ratio of respective sample standard deviations (a constant), we will treat for convenience discussions of statistical significance on the basis of estimated cross correlations.

Haugh (1976) has shown that if n observations are available for two jointly covariance-stationary linear processes x_t and y_t which have ARMA univariate representations where α and β are corresponding white noise innovation series, and, moreover, if the two series are independent, then $\sqrt{nP_{\alpha\beta}}$ and $\sqrt{nR_{\alpha\beta}}$ have the same asymptotic distribution N(0,I); where $P_{\alpha\beta} = (\rho_{\alpha\beta} = (\rho_{\alpha\beta}(k_1),\ldots,\rho_{\alpha\beta}(k_m))'$ and $k_1,\ldots,k_m$ are m different integers, and $R_{\alpha\beta}$ are the estimates of the cross correlations of the innovations.

Therefore, since asymptotically the set of lagged cross correlations between α and β follow a normal distribution, chi-square tests can be used to check for series independence. In fact Haugh has proposed two test statistics for determining series independence:

$$S_{MN} = n\sum_{k=-M}^{N} r^2_{\alpha\beta}(k) \tag{16}$$

and

$$S^*_{MN} = n^2\sum_{k=-M}^{N}(n-|k|)^{-1} r^2_{\alpha\beta}(k)\,. \tag{17}$$

Both statistics would be compared with a critical value from a chi-square distribution with (M+N+1) degrees of freedom. Statistic (16) is appropriate for large samples whereas as statistic (17) has been shown to be more appropriate for small samples (Haugh, 1976). The particular lag range [-M,N] which should be used in these statistics depends on the analysts own knowledge about the phenomenon being studied.

Tests for causality directions can, therefore, be made by limiting the range of k in either (16) or (17) to positive or negative integer values. Verification for possible instantaneous causality can be made by testing the significance of $r_{\alpha\beta}(0)$.

A nomenclature for categorizing causality events based on restrictions on the values of the estimated cross correlations of the whitened series has been developed by Pierce and Haugh (1977). These events and their respective conditions are listed in Table 1. The causality conditions in the table can, of course, be examined utilizing statistics (16) or (17). For example to test relationship I (X causes Y) the appropriate test would be to compare $n\sum_{k=1}^{M} r^2_{\alpha\beta}(k)$ with the desired critical value from χ^2 distribution with M degrees of freedom.

TABLE 1

Pierce-Haugh Conditions on Cross Correlations of

Whitened Series for Causality Events

(positive k = lag)

	Relationship	Restrictions on $r_{\alpha\beta}(k)$
I.	X causes Y	$r_{\alpha\beta}(k) \neq 0$ for some $k>0$
II.	Y causes X	$r_{\alpha\beta}(k) \neq$ for some $k<0$
III,IV.	Instantaneous Causality	$r_{\alpha\beta}(k) \neq 0$
V.	Feedback	$r_{\alpha\beta}(k) \neq 0$ for some $k>0$ and some $k<0$
VI.	X causes Y but not instantaneously	$r_{\alpha\beta}(k) \neq 0$ for some $k>0$ and $r_{\alpha\beta}(0) = 0$
VII.	Y does not cause X	$r_{\alpha\beta}(k) = 0 \quad \forall k<0$
VIII.	Y does not cause X at all	$r_{\alpha\beta}(k) = 0 \quad \forall k\leq 0$
IX.	Unidirectional Causality from X to Y	$r_{\alpha\beta}(k) \neq 0$ for some $k>0$ and $r_{\alpha\beta}(k)= 0$ for either all $k<0$ or all $k\leq 0$
X.	X and Y are related only instantaneously (if at all)	$r_{\alpha\beta}(k) = 0 \quad \forall k \neq 0$
XI.	X and Y are related instantaneously and in no other way	$r_{\alpha\beta}(0) \neq 0$ and $r_{\alpha\beta}(k) = 0 \quad \forall k \neq 0$
XII.	X and Y are independent	$r_{\alpha\beta}(k) = 0 \quad \forall k$

Adapted from Pierce and Haugh (1977).
Where α_t and β_t are defined as

$$\alpha_t = \phi_x(B)\theta_x^{-1}(B)x_t$$

$$\beta_t = \phi_y(B)\theta_y^{-1}(B)y_t$$

3. USAGE OF CAUSALITY IN THE DETECTION OF MISSPECIFICATION AND IN THE FORMULATION OF SIMULTANEOUS EQUATION SYSTEMS

Any simultaneous equation model (SEM) can be represented as

$$\underset{(m\times m)}{F(B)} \; \underset{(p\times 1)}{\underline{Y}_t} + \underset{(m\times n)}{G(B)} \; \underset{(n\times 1)}{\underline{X}_t} = \underset{(m\times m)}{H(B)} \; \underset{(m\times 1)}{\underline{U}_t} \tag{18}$$

where

$$\underline{Y}_t' = (y_{1t}, y_{2t}, \ldots, y_{mt})$$

$$\underline{X}_t = (x_{1t}, x_{2t}, \ldots, x_{nt})$$

$$\underline{U}_t = (u_{1t}, u_{2t}, \ldots, u_{mt})$$

and typical elements of F(B), G(B), and H(B) are given by

$$f_{ij} = \sum_{l=0}^{a_{ij}} f_{ijl} B^l, \quad g_{ij} = \sum_{l=0}^{b_{ij}} g_{ijl} B^l, \text{ and } h_{ij} = \sum_{l=0}^{c_{ij}} h_{ijl} B^l$$

respectively (Quenouille, 1957; Zellner and Palm, 1974).

Equation (18) can also be expressed as

$$\sum_{\ell=0}^{a} F_\ell B^\ell \underline{Y}_t + \sum_{\ell=0}^{b} G_\ell B^\ell \underline{X}_t = \sum_{\ell=0}^{c} H_\ell B^\ell \underline{U}_t \tag{19}$$

where F_ℓ, G_ℓ, and H_ℓ are matrices with all elements not depending on B; $a = \max_{ij} a_{ij}$, $b = \max_{ij} b_{ij}$, and $c = \max_{ij} c_{ij}$. The reduced form equations (RFE) which express the current values of the endogenous variables as functions of the lagged endogenous and current and lagged exogenous variables are[7]

$$\underline{Y}_t = -\sum_{\ell=1}^{a} F_0^{-1} F_\ell B^\ell \underline{Y}_t - \sum_{\ell=0}^{b} F_0^{-1} G_\ell B^\ell \underline{X}_t + \sum_{\ell=0}^{c} F_0^{-1} H_\ell B^\ell \underline{U}_t. \tag{20}$$

Consequently, because of the specific variable lag restrictions defined by the reduced form equations (20), it is possible to determine whether or not the choice of endogenous and exogenous variables and their lag structures were appropriate by constructing causal multiple transfer function (CMTF) models for each reduced form equation. If in analyzing the reduced form transfer function equations (RFTFEs) for which the values for v_{ij}, $j<0$, are not restricted to be zero, we find

1. equations containing nonzero coefficients for leading values of assumed to be exogenous variables, then clearly the variables exogeneity assumption cannot be justified.[8]

2. in all equations the coefficients of some of the variables assumed to be endogenous are nonzero only for lagged values of the variables, then the endogeneity assumption is not justifiable because these variables cause the other endogenous variables and are not simultaneously determined by the system at time t.

3. equations which contain nonzero coefficients for current and/or leading values of endogenous variables, then the system may possess some recursive elements.[9]

If one is interested in explaining the relationship (if any) between various economic variables, then causal transfer function models can be very useful in determining appropriate structural formulations. Determination of variable exogeneity or endogeneity can be made by calculating cross correlations between pairs of innovation series associated with each variable. Once directional causality (if any) can be ascertained, then structural formulations can be contemplated. For example, suppose that from the set of innovation series $\alpha = (\alpha_1, \alpha_2, \alpha_3, \alpha_4)$ we discovered that

a. α_1 caused α_4,

b. α_2 caused α_4,

c. α_3 was not causally related to any of the innovations, and

d. α_1 and α_2 were not causally related to each other.

Then, a likely model for estimation would be

$$\alpha_{4t} = v_1(B)\alpha_{1t}+v_2(B)\alpha_{2t}+\varepsilon_{4t}. \tag{21}$$

If the causal relation between α_3 and the other innovations or between α_1 and α_2 were different than above, then a simultaneous model might be more appropriate. Suppose, for example, that

a. α_1 caused α_4,

b. α_3 caused α_2,

c. causation between α_2 and α_4 is measured in both directions,

d. no causal relation was found between α_1 and α_3,

e. weak or no causal relation was found between α_3 and α_4 and between α_1 and α_2.

Then a potential model relating these variables would be

$$\begin{aligned} \alpha_{2t} &= v_3(B)\alpha_{3t}+v_4(B)\alpha_{4t}+\varepsilon_{2t} \\ \alpha_{4t} &= v_1(B)\alpha_{1t}+v_2(B)\alpha_{2t}+\varepsilon_{4t} \end{aligned} \tag{22}$$

with the appropriate lag structure still to be determined.

The analyst, however, must still check to see if the derived specification is identifiable. Hatanaka (1975) has shown that when lag lengths and shapes of lag distributions are not known a priori, the usual identification order condition in which lagged dependent variables are lumped together with strictly exogenous variables (Sargan, 1961) is not applicable. In this case the order condition takes an altered form in that one must cease to count repeat occurrences of the same variable with different lags in a single equation. In effect, this rule prevents lagged dependent variables from playing the same kind of formal role as strictly exogenous variables in identification.

Statistically derived models should not be thought as being the ultimate explanatory models. Economic principles should guide the construction of every model. Addition or deletion of variables should be considered when derived relationships do not fully harmonize with economic theory. Econometric model construction should be viewed as the art of balancing statistical analysis with economic principles.

Once a harmonious balance between economic theory and statistical causal relationships has been achieved, determination of variable lag structure and extimation of the parameters in the equations can be made using transfer function procedures such as those of Box and Jenkins (1970); Granger and Newbold (1977); Koreisha (1980); and Liu and Hanssens (1981). It should be noted, however, that unlike the case of single equation models which can use these procedures verbatim for determining appropriate lag structures, for simultaneous equations models such as (22), the procedure is more involved requiring a two-phase estimation procedure analogous to two-state least squares estimation:

Stage 1

a. Formulate the reduced form structure associated with the innovations of the SEM, i.e., current endogenous innovations as functions of lagged endogenous and current and lagged exogenous innovations.

b. Estimate transfer function reduced form equations associated with the SEM based on the innovations to obtain uncorrelated estimates for the endogenous innovations.

Stage 2

a. Substitute the uncorrelated estimates derived in Stage 1b. to estimate the structural equations of the innovations.

b. Transform the innovations specification into a specification based on the original variables using the procedures to be described in the next subsection.

In order for the parameter estimates derived in Stage 2a to be consistent, the structure associated with the lagged endogenous and the exogenous variables appearing in each of the structural equations should also have appeared in the reduced form equations for which the uncorrelated estimates for the current endogenous innovations were obtained. In fact we have to be even more restrictive. The length of the lag structure of any lagged endogenous or any current or lagged exogenous variable appearing in a particular structural equation cannot exceed the intersection of the length of the lag structure associated with these variables in the reduced form equations for which the uncorrelated estimates for the current endogenous innovations included in the structural equation were obtained. For example, suppose that two current endogenous innovations appear on the RHS of one of the structural equations. Furthermore, also suppose that the uncorrelated estimates, for these innovations obtained via the RFTFEs, included a variable x which in one reduced form equation had a lag structure going from t to t-k whereas in the other equation the lag structure went from only t to t-s, s<k. Then in order for the estimates in Stage 2a to be consistent (for that particular structural equation), the maximum lag associated with that variable x (if it is part of the structural equation) cannot be greater than t-s. Thus, the RFTFE's provide an upper limit on the length of the lag structure of variables in the structural equations.

In order to facilitate the explanation of the conversion of the innovations model to a specification based on the original variables it will be convenient to define a few terms. Let

α_{it} = innovation associated with current exogenous variable, $i\varepsilon\{1,\ldots,a\}$

β_{jt} = innovation associated with current endogenous variable j, $j\varepsilon\{1,\ldots,b\}$

$\hat{\beta}_{jt}$ = reduced form estimate of the innovation associated with the current endogenous variable j.

Values for the above innovation variables are derived from the following equations

$$\alpha_{it} = \phi_{x_i}(B)\theta_{x_i}^{-1}(B)\nabla^{d_1}x_{it} \tag{23}$$

$$\beta_{jt} = \phi_{y_j}(B)\theta_{y_j}^{-1}(B)\nabla^{d_2}y_{jt} \tag{24}$$

$$\hat{\beta}_{jt} = \sum_{i=1}^{a} \sum_{k=0}^{m} s_{ik}\alpha_{1,t-k} + \sum_{h=1}^{b} \sum_{\ell=1}^{n} s_{h\ell}\beta_{h,t-\ell} + u_{jt} \tag{25}$$

where m, n, s_{ik} and $s_{h\ell}$ are scalars determined from successive applications of cross correlation analysis between current residuals and the then included RHS innovations (current and lagged); d is the degree of differencing necessary to induce stationarity to the original series; and μ_{jt} is the error in the equations.

A typical structural innovation equation estimated in Stage 2a may be expressed as

$$\beta_{jt} = \sum_{i=a'}^{a''} \sum_{k=0}^{m'} \omega_{ik}\alpha_{i,t-k} + \sum_{h=b'}^{b''} \omega_{h0}\hat{\beta}_{ht} + \sum_{h=b'}^{b''} \sum_{\ell=1}^{n'} \omega_{h\ell}\beta_{h,t-\ell} + u_{jt} \tag{26}$$

where

$\{a',\ldots,a''\} \sqsubseteq \{1,\ldots a\}$; i.e., the set of exogenous variables appearing in that particular structural innovations equation

$\{b',\ldots,b''\} \sqsubseteq \{1,\ldots b\}$; i.e., the set of endogenous variables appearing in that particular structural innovations equation

and where m', n', ω_{ik}, ω_{h0}, and $\omega_{h\ell}$ will be determined from successive applications cross correlations analysis between the current residuals and the included RHS innovations; and μ_{jt} is error in the equation.

The only term in equation (26) which offers some complications in the conversion is the one which includes $\hat{\beta}_{ht}$ since for the estimates of the structural equations parameters in terms of the original variables to be consistent it is necessary to derive uncorrelated estimates of y_{ht}. The other terms can be converted directly through the filters (23) and (24).

From equation (24) we know that the consistent estimator $\hat{\beta}_{ht}$ has to be also a consistent estimator for $[\phi_{yh}(B)\theta_{yh}^{-1}(B)\nabla^d y_{ht}]$, i.e.,

$$\hat{\beta}_{ht} = \widehat{\phi_{yh}(B)\theta_{yh}^{-1}(B)\nabla^d y_{ht}} \tag{27}$$

Thus, a consistent estimator of y_{ht} can be expressed as[10]

$$\hat{y}_{ht} = \left[\phi_{yh}(B)\theta_{yh}^{-1}(B)\nabla^d\right]^{-1}\hat{\beta}_{ht}. \tag{28}$$

For example, let the prewhitening filter for y_{ht} be a simple autoregressive filter of order 1, i.e.,

$$(1-\phi B)y_{ht} = \beta_{ht} \tag{29}$$

thus,

$$\hat{\beta}_{ht} = \widehat{(1-\phi B)y_{ht}} \tag{30}$$

so that

$$\hat{y}_{ht} = \hat{\beta}_{ht} + \phi y_{h,t-1}. \tag{31}$$

After all the appropriate substitutions are made into (26) it is quite possible that the polynomials associated with lag orders of certain variables may be quite large and also that many of the actual estimates may turn out to be very small in magnitude. It is suggested that the derived equations be reestimated so that standard errors of the parameter coefficients may be obtained to determine the statistical significance of some of the higher lag order parameters and also of those with very small coefficients values.

Conversion of the parameterization of the innovations in terms of the original variables is made through the filters used in the prewhitening operation. We recommend that the ARIMA models fitted for the series x_{it} and y_{it} be as parsimonious in nature as possible since in later stages of the conversion process these models may force the estimation of equations containing very high lagged variables with corresponding coefficients which are small in magnitude.

With the lag structure now properly specified the causally "correct" model could, if desired, be subjected to further analysis. For instance, the parameters of the model could be more efficiently estimated using the joint estimation techniques derived by Palm and Zellner (1980).

Within the context of consistency discussed above, the polynomial components of all the variables in the model are not restricted to obey any set of rules. Zellner and Palm (1974), however, showed that the endogenous variables of a SEM when modeled as univariate ARIMA processes should all have the same autoregressive form. The implications of this result not only suggest that perhaps most existing SEMs could be misspecified, but also that the procedure described in this section for constructing econometric models using causal transfer functions may not be adequate. Prothero and Wallis (1976); Granger (1976); Wallis (1977); and Granger and Newbold (1977), however, have argued among other things that because of cancellation of common factors across equations and because of small values associated with many of the parameters particularly those of high lags, one cannot expect to observe common autoregressive processes for the endogenous variables.

We should add that Palm and Zellner (1980), more than likely, do not strongly adhere to the proposition that in practice we should observe a common autoregressive component for the endogenous variables since the procedure they proposed to jointly estimate all the parameters of transfer function equations permits the autoregressive components of each endogenous variable to be different.

Comparison with the Sims (1980) Procedure

Sims (1980) has proposed estimating profligately parameterized models as unrestricted reduced forms, treating all variables as endogenous. What he proposed entailed estimating an unconstrained vector of autoregressions, i.e.,

$$A(B)\underline{y}_t = C(B)\underline{x}_t + \underline{n}_t \tag{32a}$$

or

$$D(B)\begin{bmatrix}\underline{y}_t\\ \underline{x}_t\end{bmatrix} = \underline{n}_t \tag{32b}$$

where $\underline{n}_t$ is the error term,[11]

$$D(B) = \begin{bmatrix}A(B) & -C(B)\end{bmatrix}$$

and the order of the polynomials is set arbitrarily, and then using the equations for forecasting and for testing some specific hypothesis. Estimates of the autoregressive parameters are not reported because, as he explained, sampling problems often force the parameters to oscillate in sign.

Our approach, on the other hand, is more comprehensive since it allows the data rather than the analyst to specify the length of the lag structure for each equation individually, and also because it permits incorporation of moving average parameters on the RHS of (32b). The mathematical collaboration of these advantages are shown below.

Our method, after obtaining the proper reduced form estimates for the endogenous variables, entails estimating equations of the form

$$E(B)\underline{a}_{yt} = F(B)\underline{a}_{xt} + \underline{u}_t \tag{33}$$

where

$$\Phi_y(B)\underline{y}_t = \Theta_y(B)\underline{a}_{yt} \tag{34a}$$

$$\Phi_x(B)\underline{x}_t = \Theta_x(B)\underline{a}_{xt} \tag{34b}$$

and $\underline{u}_t$ is the error term.[12] The orders of the polynomials are <u>not</u> preset. They are individually determined from the data.

Thus, substituting (34) into (33) yields an equation of the form

$$E(B)\underline{\phi}_y(B)\Theta_y^{-1}(B)Y_t = F(B)_x(B)\Theta_x^{-1}(B)\underline{X}_t + \underline{u}_t \tag{35}$$

which can also be expressed as

$$\left[E(B)\underline{\Phi}_y(B)\Theta_y(B) - F(B)\underline{\Phi}_x(B)\Theta_x(B)\right]\begin{bmatrix}\underline{Y}_t\\ \underline{X}_t\end{bmatrix} = \Theta_x(B)\Theta_y(B)\underline{u}_t \tag{36}$$

and which in turn can be simplified to

$$G(B)\begin{bmatrix}\underline{y}_t\\ \underline{x}_t\end{bmatrix} = F(B)\underline{u}_t \tag{37}$$

where the definitions of the polynomials are obvious.

It should be emphasized that determining transfer function relationships among <u>stationary innovations</u> rather than the actual variables, as Granger and Newbold (1977) among others point out, greatly reduces the possibility of estimation errors due to spurious correlations. The time series used in Sims (1980) have not been individually filtered to determine the processes governing them. All regressions, however, were run on seasonally adjusted logged data which contained time trends.

4. APPLICATIONS

Analysis of Klein's Model of the U. S. economy using transfer function reduced form equations indicate possible misspecifications. Table 2 contains the original equations of the model based on yearly data covering the period 1920-1941. Table 3 contains the results of reduced form transfer function equations associated with Klein's model using prewhitened data. These equations were estimated using quarterly data covering 1947.2 to 1977.1 (120 observations).

TABLE 2

KLEIN'S MODEL I OF THE U. S. ECONOMY

Consumption Function: $C = aP+b(PW+GW)+cP_{-1}+d+u_1$

Investment Function: $I = eP+fP_{-1}+gK_{-1}+h+u_2$

Demand for Labor

Function: $PW = i(Y+T-GW)+j(Y+T=GW)_{-1}+kt+\ell tu_3$

Identity: $Y+T = C+I+G$

Identity: $Y = PW+GW+P$

Identity: $K = K_{-1}+I$

The variables are

C = Consumer Expenditures,
G = Government Expenditures,
I = Investment Expenditures,
P = Profits,
PW = Private Wage Bill,
GW = Government Wage Bill,
K = Capital Stock
T = Taxes,
t = Time,
Y = Income Net of Taxes.

For convenience we shall also define the two variables

W = Total Wage Bill (PW+GW)
X = Gross Private Income (Y+T-GW)

Analysis of the results associated with Table 3 indicate that the original set of structural equations are not appropriate for quarterly data. The exogenous variable taxes in the total wages equation appears with a leading coefficient, an indication that in this context it should be treated as an endogenous variable. (This finding corroborates the rational expectations hypothesis.) In all three reduced form equations we note that certain endogenous variables appear with significant leading coefficients, an indication of possible recursiveness in the system.

Using the prewhitened series associated with each of the Klein variables we calculated the cross correlations for all 28 possible pair combinations of the

TABLE 3

ESTIMATES OF THE PROFITS, TOTAL WAGE BILL, AND GROSS PRIVATE INVESTMENT REDUCED FORM TRANSFER FUNCTION EQUATIONS

(innovations)

$$
\begin{aligned}
P_t = & -\underset{(.084)}{.676} W_t - \underset{(.084)}{.354} W_{t-1} - \underset{(.037)}{.918} T_t - \underset{(.037)}{.337} T_{t-1} + \underset{(.056)}{.135} G_t \\
& + \underset{(.035)}{.139} I_{t-10} + \underset{(.041)}{.120} C_{t+6} + \underset{(.037)}{.093} C_{t+2} + \underset{(.037)}{.181} C_{t+1} \\
& - \underset{(.030)}{.054} K_{t+6} + \underset{(.022)}{.073} X_{t+7} + \underset{(.036)}{.864} X_t + \underset{(.040)}{.448} X_{t-1} + \underset{(.027)}{.136} X_{t-2}
\end{aligned}
$$

$$
\begin{aligned}
W_t = & - \underset{(.042)}{.097} P_t - \underset{(.037)}{.149} T_{t+8} + \underset{(.064)}{.189} G_{t-2} + \underset{(.034)}{.085} I_{t+8} - \underset{(.044)}{.142} C_{t+1} \\
& + \underset{(.049)}{.124} C_{t-6} + \underset{(.039)}{.086} K_{t-1} + \underset{(.029)}{.308} X_t
\end{aligned}
$$

$$
\begin{aligned}
X_t = & - \underset{(.030)}{.131} P_{t-3} + \underset{(.035)}{.203} P_{t-5} + \underset{(.029)}{.190} T_{t-4} + \underset{(.029)}{.253} T_{t-5} + \underset{(.051)}{.900} G_t \\
& - \underset{(.023)}{.534} I_{t-1} - \underset{(.025)}{.047} I_{t-3} - \underset{(.029)}{.144} I_{t-6} + \underset{(.035)}{.967} C_{t+1} - \underset{(.040)}{.377} C_t \\
& + \underset{(.025)}{1.001} K_t + \underset{(.051)}{.285} W_{t-7}
\end{aligned}
$$

eight variables. Table 4 contains the results of the Haugh-Type causal tests. Statistical significance were calculated by comparing the estimated cross correlation values (lead=8, lag=8) with $2(120)^{-1/2} = .18$. Figure 1 presents a summary of the causal relationship among all the variables in the Klein model.

These results provide further evidence that some of Klein's equations do not appear to be adequately specified on a quarterly basis. For instance, from Table 4 as well as from Figure 1 we observe that the demand for labor (W) does not seem to depend on lagged values of gross private income (X). The relationship appears to be instantaneous if it exists at all. Furthermore, W does not appear to be an endogenous variable, i.e., none of the other variables "cause" W (Pearce and Haugh relation VII). We also observe that the previous period capital stock (K_{-1}) does not appear to influence current investment decisions (I).

The amount of taxes collected, as expected, is determined by levels of profits, wages, consumption, income, investment, and capital stock, and taxes in turn "affect" the amount of government expenditures. The fact that the causal direction between taxes and all other economic variables except government expenditures always run towards taxes is a clear indication that taxes should not be considered as being determined exogenously among this set of variables. Also interesting is the fact that tax effects are felt contemporaneously by all the other variables included in the analysis.

TABLE 4

RESULTS OF THE CAUSAL CROSS CORRELATION TEST ON THE KLEIN VARIABLE INNOVATIONS

Innovation Pairs	Causal Direction ⟶ $120\Sigma_{i=-8}^{-1}$	⟵ $120\Sigma_{i=1}^{8}$	⟷ $120\Sigma_{i=-8}^{8}$	r(0)/SE	Haugh-Pierce Applicable Causal Relations*
P-W	15.38	15.86	32.70	1.22	I,VI,VII,VIII,IX
P-X	13.96	11.95	52.43	5.22	III,IV,X,XI
P-T	29.52	7.91	61.73	5.00	II,III,IV
P-G	23.69	6.25	33.83	2.00	II,III,IV
P-I	19.55	31.09	52.71	1.78	I,II,V,VI
P-C	20.89	20.21	42.55	1.22	I,II,V,VI
P-K	20.32	21.60	45.80	2.00	I,II,III,IV,V
W-X	3.73	9.12	65.12	3.67	III,IV,X,XI
W-T	26.86	4.90	46.45	3.89	II,III,IV
W-G	5.08	5.92	11.96	1.00	XII
W-I	10.85	10.15	49.82	5.44	III,IV,X,XI
W-C	17.70	7.50	44.40	4.44	II,III,IV
W-K	10.36	8.66	54.01	6.00	III,IV,X,XI
X-T	22.63	9.95	50.83	4.33	II,III,IV
X-G	7.49	6.14	13.64	.11	XII
X-I	19.69	20.53	87.85	7.00	I,II,III,IV,V,VI
X-C	20.28	6.58	53.36	5.22	II,III,IV
X-K	18.72	12.48	83.47	7.33	II,III,IV
T-G	23.68	3.74	34.33	2.67	II,III,IV
T-I	11.74	25.43	59.35	4.78	I,III,IV,VII,IX
T-C	13.16	20.17	48.04	3.89	I,III,IV,VII,IX
T-K	12.01	21.42	56.66	4.89	I,III,IV,VII,IX
G-I	7.43	17.95	29.71	2.11	I,III,IV,VII,IX
G-C	5.00	6.94	15.01	1.78	XII
G-K	7.09	14.47	30.31	3.00	III,IV,X,XI
I-C	22.60	32.88	55.52	.22	I,II,V,VI
I-K	21.54	10.18	137.75	10.44	II,III,IV
C-K	27.94	25.27	53.27	4.56	I,II,III,IV,V
α=.05	15.5	15.5	27.6		

*Assumes that left hand variable in the innovation pair column is the Y variable of Table 1.
Relationships are determined @ 5% level.

FIGURE 1. SCHEMATIC OF THE CAUSAL DIRECTIONS AMONG KLEIN'S VARIABLES

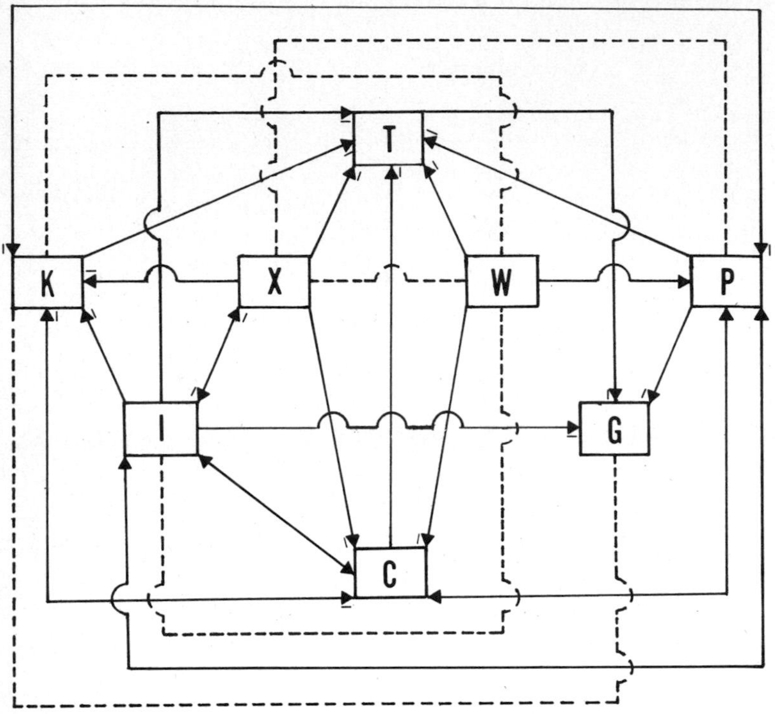

Legend:

---- instantaneous causality

⟶ causality direction indicated by arrow without instantaneous effects

⟶ causality direction indicated by arrow with instantaneous effects

Using the causality relationships expressed by Figure 1 we derived the following system of equations to represent a model of the U. S. economy based on Klein's variables:

$$\left.\begin{array}{lrcll}
\text{Capital Stock Identity:} & K_t & = & K_{t-1}+I_t & \text{(a)} \\
\text{Net Income Identity:} & Y_t & = & W_t+P_t & \text{(b)} \\
\text{Balance Identity:} & Y_t+T_t & = & C_t+I_t+G_t & \text{(c)} \\
\text{Consumption:} & a(B)C_t & = & b(B)P_t+c(B)W_t+u_{1t} & \text{(d)} \\
\text{Profit:} & d(B)P_t & = & e(B)I_t+f(B)W_t+g(B)K_t+u_{2t} & \text{(e)} \\
\text{Taxes:} & h(B)T_t & = & i(B)C_t+j(B)K_t+u_{3t} & \text{(f)} \\
\text{Government:} & k(B)G_t & = & \ell(B)T_t+u_{4t} & \text{(g)}
\end{array}\right\} \quad (38)$$

In the context of system (38) the total wage variable is surmized as being the exogenous variable. Given our sample period (1947-1977), this finding is not counterintuitive. During the past 30 years labor has become extremely powerful and has initiated innumerable changes in the economy. Labor has demanded increased benefits in the form of pension, insurance, wage, health plans, cost of living clauses, and partial compensation for lay-offs. Consistently, demands by labor have been met by industry.

FIGURE 2. SCHEMATIC OF THE CAUSAL RELATIONSHIPS AMONG THE VARIABLES IN SYSTEM (38) AND CAUSAL RELATIONSHIPS IMPLIED BY THE CROSS CORRELATION ANALYSIS

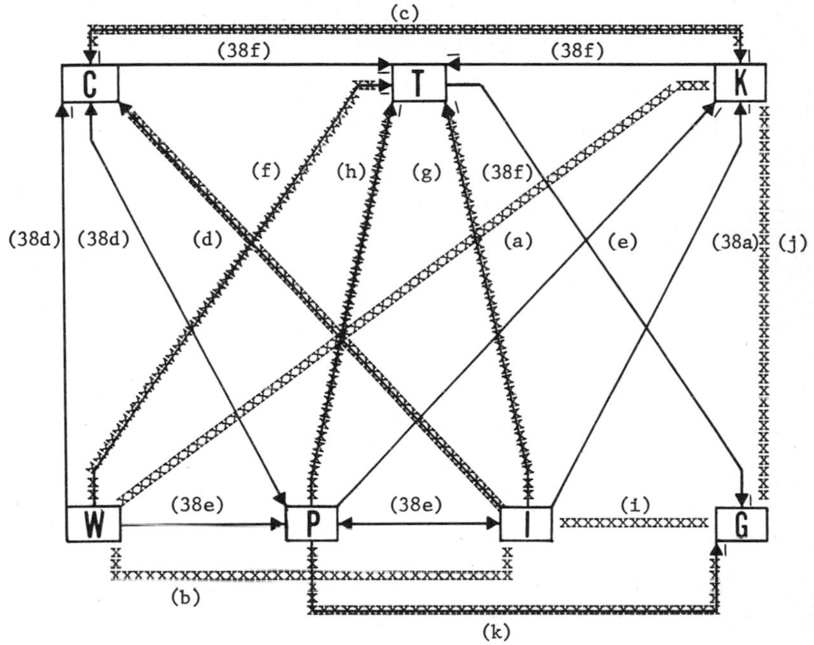

Legend:

causality direction indicated by arrow without instantaneous effects among variables formally modelled

causality direction indicated by arrow without instantaneous effects among variables not formally modelled, but implied by cross correlation analysis

causality direction indicated by arrow with instantaneous effects among variables formally modelled

causality direction indicated by arrow with instantaneous effects among variables not formally modelled, but implied by cross correlation analysis

xxx instantaneous causality measured among variables not formally modelled, but implied by cross correlation analysis

Numbers above the causal symbols represent the equation number relating the variables. A letter above the causal symbols indicates that that particular causal relation needs to be justified by relationships which will be formally structured.

System (38) satisfies the order and rank conditions for identification when the lag order of the variables is not known apriori. Figure 2 is a schematic representation of the causal relationships among the variables in system (38) and of the causal relationships implied by the cross correlation analysis. Table 5 summarizes the causal justification of the relationships implied by the cross correlation analyses but not formally modeled based on the results of the Haugh tests.

TABLE 5

CAUSAL JUSTIFICATION OF THE RELATIONSHIPS IMPLIED BY THE CROSS CORRELATION ANALYSIS BUT NOT FORMALLY INCORPORATED INTO (38)

	Implied Causal Relationship		Justification
1.	(a):	W xxx K	W ——≥ C, C ——≥ T, T ——> K
2.	(b):	W xxx I	W ——≥ C, C ——≥ T, T ≤—— K, K ≤—— I
3.	(c):	C≤xxx≥K	
	1.	C<xxx>K	C <——> P, P≤——≥ K
	2.	C xxx K	C ——≥ T, T≤—— K
4.	(d):	C<xxx>I	
	1.	C xxx I	C ——≥ T, T≤—— K, K≤—— I
	2.	C<xxx>I	C <——> P, P <——> I
5.	(e):	P≤——≥K	formally modelled
6.	(f):	W xxx≥T	W ——≥ C, C ——≥ T
7.	(g):	I xxx≥T	I ——≥ K, K ——≥ T
8.	(h):	P xxx≥T	P≤——≥ K, K ——≥ T
9.	(i):	I xxx≥G	I ——≥ K, K ——≥ T, T ——≥ G
10.	(j):	K xxx G	K ——≥ T, T ——≥ G
11.	(k):	P xxx≥G	P ≤——≥ K, K ——≥ T, T ——≥ G

N.B. Letters within parenthesis refer to the relationships associated with Figure 2

Table 6 contains the reduced form transfer function equations associated with the innovations of the variables in system (38). Table 7 contains the maximum consistent lag structure of the innovations structural equations inferred from the TFRFEs.

TABLE 6

REDUCED FORM TRANSFER FUNCTION EQUATIONS ASSOCIATED WITH THE INNOVATIONS OF THE VARIABLES IN SYSTEM (38)

$$c_t = \frac{\underset{(.100)}{.600}}{\underset{(.143)}{(1-.571B)}} w_t - \underset{(.154)}{.360}\, i_{t-1} - \underset{(.064)}{.173}\, i_{t-3} - \underset{(.060)}{.155}\, p_{t-5} + \underset{(.062)}{.126}\, p_{t-6}$$

$$+ \underset{(.140)}{.383}\, k_{t-1} - \underset{(.073)}{.164}\, t_{t-1} - \underset{(.060)}{.171}\, t_{t-7} + \underset{(.107)}{.273}\, g_{t-6} + n_{ct}$$

$$p_t = \underset{(.150)}{.256}\, w_t - \underset{(.162)}{.323}\, w_{t-7} - \underset{(.129)}{.313}\, c_{t-2} - \underset{(.207)}{.476}\, i_{t-1} + \underset{(.202)}{.231}\, k_{t-1}$$

$$+ \underset{(.092)}{.056}\, t_{t-1} + \underset{(.157)}{.116}\, g_{t-1} + n_{pt}$$

$$t_t = \frac{\underset{(.131)}{.888}}{\underset{(.042)}{(1-.765B)}} w_t - \underset{(.103)}{.081}\, c_{t-1} - \underset{(.074)}{.082}\, i_{t-1} - \underset{(.075)}{.228}\, k_{t-2} + \underset{(.082)}{.338}\, p_{t-1}$$

$$- \underset{(.078)}{.200}\, p_{t-9} - \underset{(.141)}{.368}\, g_{t-2} + n_{tt}$$

$$i_t = \underset{(.140)}{1.137}\, w_t + \underset{(.095)}{.669}\, c_{t-1} + \underset{(.096)}{.311}\, c_{t-4} + \underset{(.101)}{.313}\, c_{t-7} + \underset{(.077)}{.018}\, p_{t-1}$$

$$+ \underset{(.066)}{.216}\, k_{t-4} + \underset{(.073)}{.132}\, t_{t-6} - \underset{(.135)}{.339}\, g_{t-2} + n_{it}$$

TABLE 7

MAXIMUM LAG STRUCTURE IMPLIED BY THE RFTFE'S FOR THE INNOVATION STRUCTURAL EQUATIONS*

$$c_t = a + \omega_{w0} w_t - \omega_{w7} w_{t-7} - \omega_{p0} \hat{p}_t - \omega_{c2} c_{t-2} + e_{ct}$$

$$p_t = b + \omega_{w0} w_t - \omega_{i0} \hat{i}_t - \omega_{k0} \hat{k}_t - \omega_{k4} k_{t-4} - \omega_{p1} p_{t-1} + e_{pt}$$

$$t_t = c + \omega_{c0} \hat{c}_t - \omega_{c1} c_{t-1} - \omega_{k0} \hat{k}_t + e_{tt}$$

$$g_t = d + \omega_{t0} \hat{t}_t - \omega_{t1} t_{t-1} - \omega_{g2} g_{t-2} - \omega_{g3} g_{t-3} + e_{gt}$$

*Constant is optional.

As we pointed out earlier, the parameterization of the innovations structural equations inferred from the innovations RFTFEs does not necessarily have to be the final parameterization chosen for the innovations structural equations. Consistency in estimation requires only that the variables to be included in each innovations structural equation be selected from the set inferred from the innovations RFTFEs.

The second stage estimation results associated with the parameterization inferred from the RFTFEs are found in Table 8. The standard errors of the parameters have been adjusted so as to account for the 2 step estimation procedure. (See Maddala (1977), p. 239).

As we can see from Table 8 several parameters in the equations appear to be statistically insignificant. As a result simpler parameterizations for each of the equations were estimated. Table 9 contains the estimation results of the parameterization chosen to represent the innovations structural equations.

TABLE 8

ESTIMATION RESULTS OF THE PARAMETERIZATION OF THE INNOVATIONS

STRUCTURAL EQUATIONS INFERRED FROM THE RFTFE'S

$$c_t = \underset{(.537)}{-.002} + \underset{(.158)}{.512}\, w_t - \underset{(.177)}{.023}\, w_{t-7} + \underset{(.171)}{.289}\, c_{t-2} + \underset{(.287)}{.329}\, p_t + e_{ct}$$

$$c_t = \underset{(.148)}{.512}\, w_t - \underset{(.176)}{.023}\, w_{t-7} + \underset{(.167)}{.284}\, c_{t-2} + \underset{(.274)}{.329}\, p_t + e_{ct}$$

$$p_t = \underset{(.743)}{.468} + \underset{(.280)}{.160}\, w_t + \underset{(.199)}{.317}\, i_t - \underset{(.092)}{.311}\, k_t - \underset{(.101)}{.058}\, k_{t-4} + \underset{(.112)}{.072}\, p_{t-1} + e_{pt}$$

$$p_t = \underset{(.273)}{.182}\, w_t - \underset{(.098)}{.052}\, k_{t-4} + \underset{(.110)}{.071}\, p_{t-1} + \underset{(.196)}{.321}\, i_t - \underset{(.090)}{.305}\, k_t + e_{pt}$$

$$t_t = \underset{(.541)}{1.62} + \underset{(.151)}{.355}\, c_t - \underset{(.123)}{.088}\, c_{t-2} + \underset{(.056)}{.233}\, k_t$$

$$t_t = \underset{(.159)}{.343}\, c_t - \underset{(.128)}{.094}\, c_{t-2} + \underset{(.057)}{.274}\, k_t$$

$$g_t = \underset{(.391)}{.692} - \underset{(.075)}{.061}\, t_t + \underset{(.058)}{.105}\, t_{t-1} + \underset{(.094)}{.154}\, g_{t-2} - \underset{(.095)}{.017}\, g_{t-3} + e_{gt}$$

$$g_t = \underset{(.074)}{-.024}\, t_t + \underset{(.059)}{.119}\, t_{t-1} + \underset{(.094)}{.189}\, g_{t-2} + \underset{(.096)}{.015}\, g_{t-3} + e_{gt}$$

TABLE 9

ESTIMATION RESULTS OF THE PARAMETERIZATION CHOSEN TO REPRESENT THE INNOVATIONS STRUCTURAL EQUATIONS

$$c_t = \underset{(.142)}{.506}\, w_t + \underset{(.135)}{.297}\, c_{t-2} + \underset{(.218)}{.352}\, p_t + e_{ct}$$

$$p_t = \underset{(.189)}{.279}\, i_t - \underset{(.089)}{.296}\, k_t + \underset{(.253)}{.264}\, w_t + e_{pt}$$

$$t_t = \underset{(.535)}{1.62} + \underset{(.133)}{.304}\, c_t + \underset{(.054)}{.223}\, k_t + e_{tc}$$

$$g_t = \underset{(.050)}{.1091}\, t_{t-1} + \underset{(.094)}{.189}\, g_{t-2} + e_{gt}$$

The major change in parameterization occurred in the government expenditures equation: the innovation associated with the current endogenous variable taxes has been deleted together with the innovation g_{t-3}. This, of course, suggests that the government expenditures equation can be estimated outside the simultaneous equation system. Since the causality justifications for other equations in system (38) are not affected by the government expenditures equation, this change does not affect the causal adequacy of the remaining equations in the system. Furthermore, since none of the other non-identity structural equations includes the government expenditures variable the estimates obtained for the current endogenous innovations remain consistent. (The predetermined variables in the structural equations are included in the RFTFEs.) The fact that government expenditures together with wages are exogenous is actually not surprising considering our sample period (1947-1977). During this period the government was instrumental in initiating new programs and in dictating policy.

It should be noted, however, that for forecasting purposes the innovations' specification can be used without having to convert the entire structural equations specification to a specification based on the original variables. Once the innovations forecasts have been generated the conversion of those innovations can be made directly using the prewhitening filters as well as other pertinent innovation and original data.

When converting an innovations' specification to a specification based on the original variables one would expect the structural parameterization to be large since the conversion process involves products of several lagged order polynomials. The magnitude of the parameters, however, may be quite small in absolute value. Although standard errors are not available for the derived profligate specification one would suspect that the statistical significance of many of the parameters, especially high lagged parameters which are also small in magnitude, would be quite low.

A practical way to make the profligate specification more parsimonious is to qualitatively decide which of the variables, among the set obtained from the conversion process, to eliminate from the specification, and then to reestimate the new parameterization to obtain standard errors.[13] Candidates for exclusion are parameters which are small in magnitude in relation to the parameters associated with the same variable at different lags. (These parameters also tend to be the ones associated with high order lagged variables.)

In the reestimation of the equations one may, however, find that because of collinearity among the variables (as is often typical among economic time series), the parameter estimates as well as the standard errors may be significantly affected. One solution to this problem would be to follow Sims (1980), and not report the individual parameter values. Given that the data base is stable, the forecasting ability of the equations will not be affected by the multicollinearity, and several specific hypotheses on the explanatory powers of the model can be tested along the lines suggested by Sims (1980). We do not adhere to this philosophy. We prefer to drop some parameters and reestimate the equations.

After deleting several high order parameters we obtained the equations in Table 10 as our version of an adequate model in the causal sense, using Klein's variables.

TABLE 10

STRUCTURAL EQUATIONS OF THE CAUSALLY DERIVED MODEL

BASED ON THE ORIGINAL VARIABLES

$$C_t = \underset{(13.15)}{-46.96} + \underset{(.181)}{.740}\, W_t + \underset{(.180)}{.391}\, W_{t-2} + \underset{(.103)}{.518}\, P_t + e_1$$

$$P_t = \underset{(13.53)}{15.65} + \underset{(.069)}{.953}\, P_{t-1} + \underset{(.124)}{.274}\, P_{t-9} - \underset{(.125)}{.307}\, P_{t-10} - \underset{(.131)}{.234}\, I_t + \underset{(.124)}{.074}\, I_{t-1}$$

$$+ \underset{(.076)}{.139}\, K_t - \underset{(.076)}{.139}\, K_{t-1} - \underset{(.171)}{.303}\, W_{t-1} + \underset{(.158)}{.335}\, W_{t-2} + e_2$$

$$T_t = \underset{(.054)}{.843}\, T_{t-1} + \underset{(.098)}{.260}\, C_t - \underset{(.102)}{.223}\, C_{t-1} + \underset{(.061)}{.047}\, K_t - \underset{(.061)}{.045}\, K_{t-1} + e_3$$

The coefficients of the profits and taxes structural equations suggest that these equations should be viewed as representing changes in the processes governing these variables. (The equations have been expressed in nominal terms in order to facilitate comparing our results with Klein's specification using the same data base.) In fact, regressions in which all the variables were explicitly defined as changes, i.e., first differences, yielded results corroborating this supposition.[14]

It is interesting to observe that changes in profits 2-2½ years in the past influence current changes in profits. A possible explanation for this, in lieu of the fact that a lagged "change" in wages has a negative effect on changes in profits, is the fact that many union contracts are negotiated every 2-3 years. The fact that the coefficient associated with "changes" in investment is negative while the coefficient associated with changes in capital stock is positive supports the hypothesis that current investments have an immediate negative effect on current profits, but that they have, as hoped, positive effects in the long run, particularly since the formulation is based on quarterly data.

The results of the tax equation are basically what one would expect: the greater (lesser) the change in consumption and in capital stock the greater (lesser) the change will be in taxes collected. The constant has been omitted from this equation because it is highly correlated with K_t.

The consumption equation also seems very plausible since inherently there generally is a lag in spending patterns after a change in wages. We should point out that we also attempted to model the consumption equation in terms of changes, but found such a specification to be inferior given the fact that the sign of the coefficients associated with changes in profits was negative.

It should be reiterated that although our equations are plausible, we do not contend that they have ultimate explanatory power. Our purpose was to derive an adequate model with a very limited number of variables which would also provide insight into the lag structure of the equations.

Comparison with Klein's Original Solution

With the same set of quarterly data used to estimate our causally adequate model, we also estimated Klein's original specification in order to contrast the results. Table 11 contains the parameter estimates for the consumption, investment, and wages structural equations. (Klein's original estimates based on annual data can be found in Maddala, 1977.)

Although all but the constant and the time coefficients in the wages equation are statistically significant, there are major problems with this simultaneous equation specification. For instance, the fact that current wages are represented as a negative function of current income is highly suspect. One reason for such lack of harmony might be that the variable representing wages is an exogenous rather than an endogenous variable as assumed by Klein. Also disturbing is the negative sign associated with P_{t-1} in the consumption equation, given that the sign of the coefficient associated with P_t is positive and that the value of that coefficient was almost two times that of the coefficient associated with P_{t-1}. The negative and positive signs associated with P_t and P_{t-1}, respectively, in the investment equation together with almost the same coefficient value for both variables, on the other hand, suggests that current investments have an immediate negative effect on profit levels from one quarter to the next.

TABLE 11

ESTIMATES OF KLEIN'S STRUCTURAL EQUATIONS BASED ON QUARTERLY DATA

(1952.2 to 1977.4)

$$C_t = \underset{(13.8)}{-54.4} + \underset{(.385)}{1.18}\,P_t - \underset{(.360)}{.639}\,P_{t-1} + \underset{(.013)}{1.13}\,W_t + u_1$$

$$I_t = \underset{(20.9)}{77.98} - \underset{(.583)}{2.81}\,P_t + \underset{(.539)}{2.75}\,P_{t-1} + \underset{(.0006)}{.008}\,K_{t-1} + u_2$$

$$W_t = \underset{(8.70)}{5.26} - \underset{(.107)}{.139}\,X_t + \underset{(.103)}{.635}\,X_{t-1} - \underset{(.162)}{.198}\,t + u_3$$

Aside from specification differences, the causally derived and the Klein specifications differ in degrees of lag parameterization. The causal specification includes many more lagged variables. For example, our specification shows that events which occurred 2-2½ years in the past can still exercise some influence in current levels of profits, and that it takes some time (½ year) for changes in wages to noticeably influence consumption expenditures.

The aim in formulating a causally adequate model was not to construct the perfect model to explain the processes influencing the driving variables of the U. S. economy. What we showed is how a causally correct model could be formulated using economic and statistical theory, and how an appropriate structural parameterization could be derived using cross correlation analysis.

Our two stage estimation procedure is unique in that by estimating the transfer function reduced form equations associated with a structural specification for which the parameterization has not been determined, one obtains the maximum lag parameterization which will yield consistent estimates. Then, through a few refining steps, a consistent structural parameterization can be derived which can give a great deal of insight as to how the past is still influencing the present.

5. CONCLUDING REMARKS

In this article we established the relationship between transfer functions, causal implications, and econometric modeling specification. Starting with Granger's definition of causality in terms of predictability we have shown how causal transfer functions can be used to test whether or not a model is correctly specified and to establish new relationships among the variables under consideration. We have also applied the tests described here to Klein's Model of the U. S. economy using quarterly data. Causal transfer function models of the reduced form equations associated with Klein's structural equations were constructed to test the model's specification. We found that the model was misspecified and, using the causal relationships among the Klein variables, we constructed a model which was adequate.

FOOTNOTES

1. Granger (1969), p. 266.

2. The selected information set will consist of a plausible set of observed series underlying some theory or be based upon intuition about the problem.

3. If true delay between cause and effect is one day but if the stochastic processes are only observed monthly, then instantaneous causality will seem to occur.

4. Two stochastic processes $\{X(t):t\epsilon N\}$, $\{Y(t):t\epsilon N\}$ are said to be jointly covariance stationary if each is covariance stationary, and in attition,

$$K_{xy}(s+t',s) = K_{xy}(t')$$

where

$$K_{xy}(t,s) = E\ [X(t)-m_x(t)]\overline{[Y(s)-m_y(s)]}, \text{ and}$$

$m_x(\cdot)$ and $m_y(\cdot)$ = the mean functions of the processes X(t) and Y(s) respectively.

Notice that by convention the first argument of $K_{xy}(\cdot,\cdot)$ refers to the unconjugated process X(t), while the second refers to the index of the conjugated process Y(s). For more details see Dhrymes (1974), pp. 457-459.

5. See, for example Sargent (1976), Mehra (1978), and Michael (1978).

6. The actual categorization is based on prewhitened data and is not standardized by the ratio of standard deviations of the prewhitened series.

7. This assumes F_0 has full rank.

8. Note that the discussion deals with actual future values of the assumed to be exogenous variables, not their expected values.

9. We stress the term may possess recursive elements, since such results may also be confirmation of the variables' endogeneity. For example, suppose that one of the RFTFE's contained a current value of an endogenous variable on the RHS of the equation. Furthermore, assume that the current value of this endogenous variable also appeared on the RHS of the structural equation for which the left hand side (LHS) variable is the same as the LHS variable of the particular reduced form equation. Such a situation would confirm the instantaneous feedback expressed in the structural equation. However, if in the other equations this "endogenous" variable did not appear with leading coefficients, then there would be reason to doubt its endogeneity since it would appear that this variable "caused" all other endogenous variables.

10. See Maddala (1977), p. 149.

11. The error term is assumed to be white noise.

12. The error term $\underline{u}_t$ is permitted to be any ARIMA process.

13. See Haugh and Box (1977).

14. Although the coefficient associated with I_{t-1} is not statistically significant, the logic process used in constructing the profits equation suggested that the equation would make more sense if it were viewed as a equation based on changes in variables. The tax equation could also be viewed as a distributed lag equation. The choice of tax equation in Table 9 was based on the facts that in the equation construction process the lagged variables appeared to be significant; the SSR was less in terms of first differences; and collinearity was reduced.

REFERENCES

AMEMIYA, T. (1973) Generalized least squares with an estimated autocovariance matrix. *Econometrica 41*, 723-732.

BOX, G. and JENKINS, G. (1970) *Time Series Analysis: Forecasting and Control.* Holden Day, San Francisco.

DHRYMES, P.J. (1974) *Econometrics: Statistical Foundations and Applications.* Springer-Verlag, New York.

FEIGE, E.L. and PEARCE, D.K. (1979) The causal relationship between money and income: some caveats for time series analysis. *Journal of Economics and Statistics LXI*, 521-533.

GEWEKE, J. (1978) Testing and exogeneity specification in the complete dynamic simultaneous equation model. *Journal of Econometrics 7*, 163-185.

GOLDFELD, S.M. and BLINDER, A.S. (1972) Some implications of endogenous stabilization policy. *Brookings Papers on Economic Activity 3*, 585-640.

GRANGER, C.W.J. (1969) Investigating causal relations by econometric models and cross spectral methods. *Econometrica 37*, 424-438.

GRANGER, C.W.J. (1979) Testing for causality - a personal viewpoint. *Department of Economics Working Paper 79-15*, University of California, San Diego.

GRANGER, C.W.J. and NEWBOLD, P. (1974) Spurious regression in econometrics. *Journal of Econometrics*, 111-120.

GRANGER, C.W.J. and NEWBOLD, P. (1977) *Forecasting Economic Time Series.* Academic Press, New York.

HANNAN, E.J. (1970) *Multiple Time Series.* Wiley and Sons, New York.

HATANAKA, M. (1975) On the global identification of the dynamic simultaneous equation model with stationary disturbances. *International Economic Review 16*, 545-554.

HAUGH, L.D. (1976) Checking the independence of two covariance-stationary time series: a univariate residual cross correlation approach. *Journal of the American Statistical Association 71*, 378-385.

HAUGH, L.D. and BOX, G.E.P. (1977) Identification of dynamic regression (distributed lag) models connecting two time series. *Journal of the American Statistical Association 72*, 121-130.

KLEIN, L.R. (1950) *Economic Fluctuation in the United States 1921-41.* John Wiley and Sons, New York.

KOREISHA, S.G. (1977) Analysis of U. S. unemployment using Box-Jenkins techniques. (Unpublished manuscript), Harvard University.

KOREISHA, S.G. (1980) *The Integration of Transfer Functions with Econometric Modeling* (unpublished dissertation), Harvard University.

LIU, L. and HANSSENS, D. (1981) Identification of multiple-input transfer function models. Working Paper, Dept. of Biomathematics and School of Management, U.C.L.A.

MADDALA, G.S. (1977) *Econometrics*. McGraw-Hill, New York.

MCLEOD, A.I. (1979) Distribution of the residual cross correlation in univariate ARMA time series models. *Journal of the American Statistical Association 74*, 849-855.

MEHRA, Y.P. (1977) Is money exogenous in money-demand equations. *Journal of Political Economy 86*, 211-228.

MICHAEL, R.T. (1978) Causation among socioeconomic time series. *NBER Working Paper 246*.

NATIONAL BUREAU OF ECONOMIC RESEARCH (NBER) COMPUTER FILES, Tapes available through Harvard Business School Computer Center.

PALM, F. and ZELLNER, A. (1980) Large sample estimation and testing procedures for dynamic equations systems. *Journal of Econometrics 12*, 251-283.

PIERCE, D.A. (1977) Relationships-and the lack thereof-between economic time series, with special reference to money and interest rates. *Journal of the American Statistical Association 12*, 11-26.

PIERCE, D.A. and HAUGH, L.D. (1977) Causality in temporal systems. *Journal of Econometrics 5*, 265-293.

PROTHERO, D.L. and WALLIS, K.F. (1976) Modeling Macroeconomic time series. *Journal of the Royal Statistical Society A 139, Part 4*,468-500.

SARGAN, J. (1961) The maximum likelihood estimation of economic relationships with autoregressive residuals. *Econometrica 29*, 414-426.

SARGENT, T.J. (1976) A classical econometric model of the United States. *Journal of Political Economy 84*, 207-237.

SIMS, C. (1972) Money, income and causality. *The American Economy Review 62*, 540-552.

SIMS, C. (1980) Macroeconomics and Reality. *Econometrica 48*, 1-48.

VANDAELE, W. and KOREISHA, S. (1977) Formulation of Box-Jenkins models for U. S. residential construction. *Harvard Business School Working Paper No. 50*.

WALLIS, K. (1977) Multiple time series analysis and the final form of economic variables. *Econometrica 45*, 1481-1497.

WILLIAMS, D., GOODHART, C. and GOWLAND, D. (1976) Money income and causality: the U. K. experience. *American Economic Review 66*, 417-423.

ZELLNER, A. and PALM F. (1974) Time series analysis and simultaneous equation econometric models. *Journal of Econometrics 2*, 17-54.

ZELLNER, A. and PALM, F. (1975) Time series and structural analysis of monetary models of the U. S. economy. *Sankhya 37 Series C*, 12-56.

APPLIED TIME SERIES ANALYSIS
O.D. Anderson & M.R. Perryman (eds.)

SMALL SAMPLE METHODS FOR MAXIMUM LIKELIHOOD IDENTIFICATION OF DYNAMICAL PROCESSES

Wallace E. Larimore

The Analytic Sciences Corporation
One Jacob Way
Reading, Massachusetts
U.S.A.

Exact evaluation of the likelihood function using a Kalman filter is described for a finite order linear dynamical process with additive measurement noise. Parameter identifiability is exactly characterized by the null space of the Fisher information matrix. This gives sound theory for use of Fisher's method of "scores" even in the nonidentifiable case. Recent approximate small sample distribution theory for maximum likelihood estimates is described in terms of the Fisher information matrix and derivatives of the log likelihood function at the maximum likelihood estimate.

INTRODUCTION

Much of the current theory and practice of time series analysis is based upon relationships which hold only asymptotically for large samples. These relationships are usually assumed to hold approximately for moderate sample sizes. There is little advice for even checking the appropriateness of this assumption and much less discussion of methods for making appropriate corrections in small samples.

This paper addresses three areas of the small sample analysis of time series. First the theory of parameter identifiability is shown to be exactly characterized locally by the Fisher information matrix. Next, exact methods for the evaluation and maximization of the likelihood function including the case of nonidentifiable parameters are described. Finally we discuss the small sample distribution theory for maximum likelihood estimates which is necessary to determine the distribution of the estimation error and to test hypotheses about the parameters.

In the literature, identifiability of parameters has been largely treated asympotically. The Fisher information matrix arises naturally in describing the asymptotic normal distribution of maximum likelihood estimates. What seems to have been largely overlooked is that the Fisher information matrix exactly characterizes the identifiability of parameters for <u>any</u> sample size and <u>any</u> probability distribution. This was shown in Rothenberg (1971) for the full rank case and is extended to the singular case by Larimore (1981).

Exact, small-sample evaluation of the likelihood function using the Kalman filter has been done for some time in the engineering literature and has more recently been used in other fields. This approach is discussed, and practical maximization procedures using the method of scores based upon the Fisher information as an approximation to the Hessian are described. The identifiability theory gives a sound justification for use of such a procedure even when some parameters are not identifiable.

The attraction of recent developments in small sample maximum likelihood estimation is that the theory uses pivotal quantities directly computable from the log likelihood function. These pivotals can be used to determine reparameterizations making the log likelihood function more quadratic and the parameter estimates more nearly normally distributed. A more nearly quadratic log likelihood function can considerably improve the convergence rate of a quadratic hill-climbing maximization procedure.

IDENTIFIABILITY THEORY

The introduction of parametric statistical inference concepts and methods by R.A. Fisher (1922) was one of his major contributions to statistics. Unfortunately, a major difficulty in identifiability problems has been an overemphasis on inference about a parameter vector $\underline{\theta}$ rather than inference about a class $F = \{p(\underline{z},\underline{\theta})\}$ of probability densities indexed by the parameter $\underline{\theta}$. Most definitions of identifiability concern properties of the resulting parameter estimates rather than intrinsic properties of the class F indexed by the parameter $\underline{\theta}$. Indeed, some definitions require a hypothetical infinite sample and define identifiability in terms of asymptotic convergence of the estimates to the true parameter values.

This overemphasis on the parameter values rather than the class of probability densities indexed by the parameters has developed despite the early and fundamental contribution of Koopmans and Reiersøl (1950). Their paper explicitly defines identifiability of the class F if unique parameter values produce unique probability densities. This formulation of the problem of specifying probability models for statistical inference includes the identifiability problem - i.e., whether or not the specified models are "observationally" unique. Later developments in the literature seem to have largely overlooked this basic concept except for a few econometric papers (see Rothenberg (1971) and cited references).

In the approach taken in this paper, the properties of the parametrized class $\{p(\underline{z},\underline{\theta}), \underline{\theta} \in \Theta\}$ are the central issue. The above definition of identifiability (Koopmans and Reiersol (1950)) as formulated for the parametric case by Rothenberg (1971) is adopted:

> Two parameter points $\underline{\theta}_1$ and $\underline{\theta}_2$ are said to be observationally equivalent if $p(\underline{z},\underline{\theta}_1) = p(\underline{z},\underline{\theta}_2)$ with probability 1. A parameter point $\underline{\theta}_1$ is said to be globally identifiable if there is no other $\underline{\theta} \in \Theta$ which is observationally equivalent. A parameter point $\underline{\theta}_1$ is said to be locally identifiable if there exists an open neighborhood of $\underline{\theta}_1$ containing no other $\underline{\theta}$ in Θ which is observationally equivalent.

The approach of this paper exploits the equivalence between local identifiability and full rank of the Fisher information matrix as in Rothenberg (1971). To extend this connection much more generally, a powerful new result on existence of identifiable reparameterizations is used (Larimore 1981).

Reparameterization Theorem If the Fisher information matrix $F_{\underline{\theta}}$ of a parameterization $\underline{\theta}$ of the likelihood function has constant rank h in a neignborhood of a point $\underline{\theta}_o$ of parameter space, then there exists a reparameterization $\underline{\psi}(\underline{\theta})$ such that $(\psi_1,\ldots,\psi_h)$ is locally identifiable and the likelihood function is not a function of $\psi_{h+1},\ldots,\psi_k$. Furthermore the gradient vectors

$$(\partial\psi_j/\partial\theta_1,\ldots,\partial\psi_j/\partial\theta_k)^T \text{ for } j = h+1,\ldots,k \qquad (2\text{-}1)$$

span the null space of $F_{\underline{\theta}}$.

By using reparameterizations resulting in a nonsingular Fisher information matrix, a complete characterization of local identifiability by the Fisher information is possible in the singular case. Previous results have carefully avoided reparameterizations where existence is not trivially guaranteed by constraints, etc. The direct construction of a reparameterization with the null functions $\psi_{h+1},\ldots,\psi_k$ seems to be necessary to obtain these general results.

Another aspect of the identifiability approach is to exploit the special structure involving the Fisher information matrix to devise efficient and numerically well conditioned methods for maximizing the likelihood function. Using the general results on reparameterizations, it is possible to generalize and to make precise a procedure for using generalized inverses in the method of scoring when the Fisher information matrix is singular (Rao and Mitra (1971)). Specifically how the reparameterization result is useful in studying the special structure of the maximum likelihood optimization problem is summarized in Section 4 and discussed in detail in Larimore (1981).

LIKELIHOOD FUNCTION EVALUATIONS

For the purpose of discussion in this and following sections, a gaussian linear dynamical system is considered of the form

$$\frac{d\underline{x}(t)}{dt} = F(\underline{\theta},t)\underline{x}(t) + G(\underline{\theta},t)\underline{u}(t) + \underline{w}(t) \tag{3-1}$$

$$\underline{z}_i = \underline{z}(t_i) = H(\underline{\theta},t_i)\underline{x}(t_i) + J(\underline{\theta},t_i)\underline{u}(t_i) + \underline{v}(t_i) \tag{3-2}$$

where $\underline{x}$ is the state vector, $\underline{z}$ is the measurement vector, $\underline{u}$ is the vector of known deterministic inputs, $\underline{\theta}$ is a vector parameterizing the system, $\underline{w}$ and $\underline{v}$ are gaussian white noises with cross spectral density matrices respectively $Q(\underline{\theta},t)$ and $R(\theta,t)$. The state space matricies F, G, H and J are parameterized by $\underline{\theta}$ and may be known functions of time.

The initial condition $\underline{x}$ at time t_o also needs to be specified either as known constants or statistically as a mean $\underline{\mu}_o$ and a covariance matrix P_o with $\underline{x}(t_o)$ assumed to be gaussian. If F, H, Q and R are not time dependent, then the distribution of the observations $\underline{z}_i$ about the expected value of the deterministic input $\underline{u}(t)$ is a stationary process. The covariance P_o of the initial state $\underline{x}(t_o)$ at time t_o then satisfies the steady-state covariance equation (Gelb (1974))

$$FP(t_o) + P(t_o)F^T + Q = 0 \tag{3-3}$$

The solution P_o depends upon the parameters $\underline{\theta}$ since $Q(\theta)$ and $F(\underline{\theta})$ depends upon $\underline{\theta}$ so we write $P_o(\underline{\theta})$. Furthermore we will include the initial state mean $\underline{\mu}_o$ in the parameter vector $\underline{\theta}$.

A method is described for evaluating the joint likelihood function $p(\underline{z},\underline{\theta})$ of the observations $\underline{z}^T = (\underline{z}_1^T,\ldots,\underline{z}_N^T)$ for the model described in Eqs. 3-1 and 3-2 with a particular assumed value $\underline{\theta}$ for the parameters. The approach is to take advantage of the independence properties of the innovations of a Kalman filter.

Consider a value of $\underline{\theta}$ for which evaluation of the likilihood function is desired. If the observations $\underline{z}$ were produced by the model Eqs. 3-1 and 3-2 with true parameters equal to $\underline{\theta}_o$, then the Kalman filter based upon this model would produce an innovations sequence $\underline{\eta}_1,\ldots,\underline{\eta}_N$ with B_i the covariance matrix of $\underline{\eta}_i$ such that for i $\neq$ j, $\underline{\eta}_i$ is indepedent of $\underline{\lambda}_j$. Such a Kalman filter is based upon the initial state with mean $\underline{\mu}_o$ and covariance $P_o(\underline{\theta})$.

For a linear model the innovations have a Gaussian distribution, and the expression for the logarithm of the probability density is

$$\log p(\underline{z},\underline{\theta}_o) = -\frac{1}{2}\sum_{i=1}^{N}(\log|B_i| + \underline{\eta}_i^T B_i^{-1}\underline{\eta}_i) + \text{constant} \tag{3-4}$$

where the constant depends only on N and not $\underline{z}$ or θ_o. This expression was derived by taking $\underline{\theta}$ to be the true value $\underline{\theta}_o$ of the process generating the actual observations $\underline{z}$. However, this is not significant since a formula for evaluating the probability density $p(\underline{z},\underline{\theta}_o)$ as a function of both $\underline{z}$ and $\underline{\theta}_o$ must yield the correct result for any possible $\underline{z}$ whether or not it infact came from a model with $\underline{\theta}_o$ the true parameter value. Note that in general the innovations depend upon the value of $\underline{\theta}$ in the likelihood and we often write $\underline{\eta}_i(\underline{\theta})$ or $\hat{\underline{x}}_{i/i}(\underline{\theta})$ where $\hat{\underline{x}}_{i/i}$ is the Kalman filter state estimates at time i given the observations $\underline{z}_o,\ldots,\underline{z}_i$.

To maximize the likelihood function, gradients of the log likelihood function are needed. This is obtained by differentiating the log likelihood function with respect to each component θ_j of $\underline{\theta}$

$$\frac{\partial \log p(\underline{z},\underline{\theta})}{\partial\theta_j} = -\frac{1}{2}\sum_{i=1}^{N}[\mathrm{tr}(B_i^{-1}\frac{\mu B_i}{\partial\theta_j}) - \underline{\eta}_i^T B_i^{-1}\frac{\partial B_i}{\partial\varepsilon_j}B_i^{-1}\underline{\eta}_i$$

$$+ 2\,\underline{\eta}_i^T B_i^{-1}\frac{\partial\underline{\eta}_i}{\partial\theta_j}] \tag{3-5}$$

The sensitivities $\partial B_i/\partial\theta_j$ and $\partial\underline{\eta}_i/\partial\theta_j$ are expressed in Goldstein and Larimore (1980) for a given θ_j in terms of the sensitivities of the state estimation error covariance and state estimate which in turn are obtained by straightforward but lengthy differentiation of all the Kalman filter equations for each θ_j similar in form to the Kalman filter (and in order of computation) except that instead of propagating the state estimate $\hat{\underline{x}}_{i/i}$ and its covariance matrix $P_{i/i}$, the sensitivities $\partial\hat{\underline{x}}_{i/i}/\partial\theta_j$ and $\partial P_{i/i}/\partial\theta_j$ are propagated.

Finally, an approximation to the Fisher information matrix is obtained from the gradient calculations. The (i,j) element of the Fisher information is approximated as

$$F_{ij} = E\,\frac{\partial \log p(\underline{z},\underline{\theta})}{\partial\theta_i}\,\frac{\partial \log p(\underline{z},\underline{\theta})}{\partial\theta_j}$$

$$\cong \sum_k [\frac{\partial\underline{\eta}_k^T}{\partial\theta_i}B_k^{-1}\frac{\partial\underline{\eta}_k}{\partial\theta_j} + \frac{1}{2}\mathrm{tr}(B_k^{-1}\frac{\partial B_k}{\partial\theta_i}B_k^{-1}\frac{\partial B_k}{\partial\theta_j})] \tag{3-6}$$

MAXIMIZATION OF LIKELIHOOD FUNCTIONS

Lack of uniqueness, i.e., nonidentifiability, manifests itself as ill-conditioning in computation of least squares or maximum likelihood estimates. Even when the parameters are identifiable, ill-conditioning often arises because of the considerable difference in sensitivity of the likelihood or squared error functions to changes in different parameters. Some past work used the special structure of least squares problems to devise efficient optimization methods which recognize any nonuniqueness or ill-conditioning and solve for the parameters within the equivalence due to the nonuniqueness (Briggs (1977) and Fletcher (1967)). Very little progres along these lines has been made in the maximum likelihood problem although a very general methods has been proposed by Rao and Mitra (1971). The maximum liklihood method provides practical parameter estimation and test of hypothesis procedures for many complex random processes, and also provides the needed approximate distribution theory which is not generally available with alternative procedures. Most maximum likelihood methods require that the class of models be reparameterized uniquely so all parameters are identifiable. There are no conditions given for when such a reparameterization is possible. To answer questions of existence and to actually reparameterize involves solving a system of nonlinear partial differential equations.

The method proposed in Rao and Mitra (1971) presumes that there exists a reparameterization for which the Fisher information matrix is nonsingular and evidently equates such nonsingularity to identifiability although any rigorous discussion or even definition of identifiability in the nonlinear case is lacking. It is argued that the method of scoring (using the Fisher information matrix in place of the Hessian in a Newton type algorithm) can be implemented entirely in the original nonidentifiable parameterization using the pseudoinverse of the Fisher information matrix to restrict the maximization to a locally identifiabile subspace of parameter space. This would avoid any need to reparameterize in terms of an identifiable set of parameters. Presently the only alternative method to preclude identifiability difficulties is to actually carry out such a reparameterization which is often exceedingly difficult if at all feasible computationally.

One of the objectives of the present approach is to use the special structure of the maximum likelihood estimation problem to devise efficient, numerically accurate and stable maximization methods. In particular the method of Rao and Mitra (1971) can be shown to work very generally utilizing the new result described in Section 2 which guarantees the local existence of an identifiable reparameterization whenever the Fisher information matrix has constant rank locally. This also generalizes the equivalence of identifiability and full rank of the Fisher informatin matrix (Rothenberg (1971)) to the reduced rank case (a nonidentifiable parametric function of $\psi(\underline{\theta})$ has a gradient vector (2-1) in the null space of the Fisher information matrix).

The method of scoring is not only attractive in removing identifiability problems, but it has several other attractive computational features. In a number of problems the Fisher information can be computed from the graident computations with little additional work. Thus Hessian information is obtained from the gradients without additional computation. It has been found in practice that this Hessian approximation gives excellent approximation to the eigenvectors which dominate the numerical behavior of approximate Hessian methods. This suggests further special structure of the problem since Hessian approximations in illconditioned cases usually result in poor algorithm performance unless there is some special structure.

SMALL SAMPLE DISTRIBUTION THEORY

There have been a number of recent methods developed in improved small sample distribution theory for maximum likelihood estimates. These are available in a variety of forms involving a number of different approximations. One of the more precise appears to be that developed by Sprott (1973, 1980) which gives an expansion of the log liklihood function in terms of its derivatives evaluated at the maximum likelihood estimate. For simplicity we discuss only the single parameter case, the multiparameter case is given in Sprott (1980). The logarithm of the relative likelihood function is expanded in a Taylor series around the maximum and expressed as

$$\log \{p(\theta)/p(\hat{\theta})\} = -\frac{1}{2}(\theta-\hat{\theta})^2 I_\theta \{1 - \frac{\theta-\hat{\theta}}{3} \frac{\partial^3 \log p(\hat{\theta})/\partial\hat{\theta}^3}{I_\theta} - \frac{(\theta-\hat{\theta})^2}{12} \frac{\partial^4 \log p(\hat{\theta})/\partial\hat{\theta}^4}{I_\theta}\} \tag{5-1}$$

where I_θ is the observed information $\partial^2 \log p(\hat{\theta})/\partial\hat{\theta}^2$.

As a measure of departure from normality within ± 3 standard errors, set $\hat{\theta} - \theta = \pm 3F_\theta^{-1}$, where F_θ denotes the Fisher or expected information $E\{I_\theta\}$. The quantities

$$F_3 = \frac{\partial^3 \log p(\theta)/\partial\theta^3}{I_\theta^{3/2}} \tag{5-2}$$

$$F_4 = \frac{\partial^4 \log p(\theta)/\partial\theta^4}{I_\theta^{2}} \tag{5-3}$$

measure the departure due to skewness and Kurtosis respectively. Similarly, Welch and Peers (1963) show that the random variable

$$Z = (\hat{\theta} - \theta)\, I_\theta^{1/2} + \frac{1}{6}[(\hat{\theta} - \theta)^2 I_\theta + 2]\, F_3 + \frac{1}{2} F_\theta^{-3/2}\, \partial F_\theta/\partial\theta \tag{5-4}$$

is more nearly normally distributed than $\hat{\theta}$ in small samples. Sprott (1973, 1980) proceeds by choosing a reparameterization which will make F_3 as small as possible. But this involves determination of the third partials of the log likelihood function. The Fisher information is more readily available in maximum likelihood algorithms so that it is more feasible to determine if it is constant as a function of the parameters, i.e. is a <u>stable parameterization</u>. This would make $\partial F_\theta/\partial\theta$ zero and the third term of (5-4) constant and zero. This will also usually reduce and often make the second term quite small since the relationship

$$E \frac{\partial^3 \log p(\theta)}{\partial\theta^3} = \frac{\partial}{\partial\theta} E\left\{\frac{\partial^2 \log p(\theta)}{\partial\theta^2}\right\} - E\left\{\frac{\partial^2 \log p(\theta)}{\partial\theta^2} \frac{\partial \log p(\theta)}{\partial\theta}\right\}$$

$$= -\operatorname{Cov}\left\{\frac{\partial^2 \log p(\theta)}{\partial\theta^2}, \frac{\partial \log p(\theta)}{\partial\theta}\right\} \tag{5-5}$$

holds so that the entire departure from normality as expressed in (5-5) is a result of the covariance between the first and second partials of the log likelihood.

The rate of covergence of the quadratic algorithm depends upon the goodness of the quadratic approximation to the log likelihood function. An appropriate parameterization of the likelihood function which stabilizes the Fisher information matrix will usually improve the quadratic behavior of the likelihood. This is due to the reduction in the third partial derivative term of (5-1) as discussed following (5-5).

Returning to the multiparameter case, stabilizing reparameterizations are obtained from a given parameterization $\underline{\theta}$ and its Fisher information matrix F_θ by finding a reparameterization $\underline{\phi}(\underline{\theta})$ of $\underline{\theta}$ with a gradient $\nabla_\theta \phi$ such that

$$F_\phi = (\nabla_\theta \phi)^{-1T} F_\theta (\nabla_\theta \phi)^{-1} \tag{5-6}$$

is nearly constant independent of the point $\underline{\phi}$ of parameter space, where $\nabla_\theta \phi$ is the matrix of partial derivatives with (i,j) element $\partial\phi_i/\partial\theta_j$. This is in general difficult to do, but in many cases there are simple reparameterizations which will yield considerable improvement. Two examples are the variance parameter σ^2 and the correlation coefficient ρ which are improved by the reparameterizations

$$\phi(\sigma^2) = \ln \sigma^2 \tag{5-7}$$

$$\phi(\rho) = \frac{1}{2} \ln \frac{1+\rho}{1-\rho} = \text{arc tanh } (\rho) \tag{5-8}$$

In general for a single parameter θ the stabilizing reparameterization is given by integrating (5-6) as

$$\phi(\theta) = \int_{\theta_o}^{\theta} (F_\theta)^{1/2} \, d\theta \tag{5-9}$$

CONCLUSIONS

A variety of methods are available for the analysis of time series involving a small sample of data by the maximum likelihood method. Difficulties due to non-identifiable parameters can be circumvented by inspection of the Fisher information matrix which is usually available from maximum likelihood calculations. The exact likelihood function can be evaluated using a Kalman filter, and the method of scores has a sound basis even in the case of nonidentifiable parameters. Approximate small-sample distribution theory involves only derivatives of the log likelihood function.

REFERENCES:

1. Biggs, M.C. (1977). The estimation of the hessian in nonlinear least squares problems with non-zero residuals. Mathematical Programming 12, 67-80.

2. Fisher, R.A (1922). On the mathematical foundations of theoretical statistics. Phil. Trans. Roy. Soc. London, Ser. A, 222, 309-286.

3. Fletcher, R. (1967). Generalized inverse methods for the best least squares solution of nonlinear equations. Computer J., 10, 392-399.

4. Gelb, A., Ed. (1974). Applied Optimal Estimation, Cambridge, MIT Press.

5. Goldstein, J.D. and Larimore, W.E. (1980). Applications of Kalman filtering and maximum likelihood parameter identfication to hydorologic forecasting. The Analytic Sciences Corporation, Report No. TR-1480-1, March 1980.

6. Koopmans, T.C. and Reiersøl, O. (1950). The identification of structural characteristics. Ann. Math. Stat., 21, 165-181.

7. Larimore, W.E. (1981). Identifiability and Maximization of Likelihood. Draft in preparation.

8. Rao, C.R. and Mitra, S.K. (1971). Generalized inverse of matrices and its applications. New York, Wiley and Sons.

9. Rothenberg, T.J. (1971) Identification in parametric models. Econometrica, 39, 577-591.

10. Sprott, D.A. (1973) Normal likelihoods and their relation to large sample theory of estimation. Biometrika, 60, 457-65.

11. Sprott, D.A. (1980) Maximum likelihood in small samples: estimation in the presence of nuisance parameters. Biometrika, 67, 515-23.

APPLIED TIME SERIES ANALYSIS
O.D. Anderson & M.R. Perryman (eds.)

A STATISTICAL ANALYSIS OF THE SOCIAL CONSEQUENCES OF PLANT CLOSINGS

Yih-wu Liu
Richard H. Bee

Department of Economics
Youngstown State University
Youngstown, Ohio
U.S.A.

The objectives of the paper are to develop a model and to provide a statistical analysis of the impact of plant closings on a number of social variables. The Time Shared Reactive On-Line Laboratory (TROLL) computer package was used in estimating all equations. Local crime activities were found to be statistically related to a set of economic variables and, separately, to a set of social variables. Most social variables were found to be related to plant closings with varying time lags.

INTRODUCTION

Monday, September 19, 1977 will go down as a memorable day in the history of the steel industry in the city of Youngstown, Ohio. On this date, quickly labeled Black Monday, the announcement was made of the first of what were to be several steel mill closings. The dramatic announcement of the impending closing of the Campbell Works made Youngstown an apparent synonym for economic and social crises.

The weeks and months that followed Black Monday were filled with predictions, all dark, on various economic variables and how the local community would be adversely influenced. Over time, 15,000 or so steel related jobs were lost. It was feared that the shredding of the economic fiber of the community had begun. In addition, predictions pertaining to the social consequences resulting from Black Monday were also abundant. The social fabric of the community was certain to experience a dramatic increase in crime, a total collapse of the school system and an increase in domestic difficulties. The intangible social evils of individuals having to face the uncertainties and hardships associated with the task of searching for new employment, a possible revamping of family lifestyles and, in several cases, the necessity of leaving a place that had been home.

The economic impact of local plant shutdowns has been studied by many different researchers (Stocks and Buss, 1981). The focus of this paper will be on the impact of plant closings on a number of non-economic variables.

THE MODEL

Our model employs a number of economic and social variables to explore the consequences of plant shutdowns. Economic variables related to plant closings include the local unemployment rate, number of bankruptcies, number of real estate transactions, and a dummy variable representing the nationwide recession of 1973. Non-economic or social variables included in the model are the number of marriage applications, the number of divorces asked and granted, the number of court cases dealing with new complaints and the number of court cases dealing with domestic relations. Plant closings may have an adverse impact on these economic and non-economic variables which, in turn, may lead to an increase in the crime rate. The model will be useful in investigating these types of relationships.

For example, plant closings may lead to an increase in the local unemployment rate which in turn may contribute to an increase in the number of crimes committed, particularly those crimes against property. On the other hand, plant closings and the ensuing long-term layoff may cause an unemployed worker to lose hope and self-esteem and after a time lag tend to disrupt normal domestic relations. The end result of this particular chain of events may well be an increase in crimes committed against people.

Table 1 presents a list of variables used in the model, together with a brief description of each symbol used. Equations contained in the model are estimated using the ordinary least squares procedure. The Time Shared Reactive On Line Laboratory computer package maintained at the Massachusetts Institute of Technology was used to do the actual computing work.

TABLE 1

Glossary of Variables Used in the Model

VARIABLES	DESCRIPTION
CI	Quarterly Crime Index for Youngstown, Ohio
UR	Quarterly data for Mahoning County unemployment rate
PC	Plant closing dummy variable, 1977.4-1980.1 = 1, 0 otherwise
ADIV	Quarterly data for Mahoning County divorces and dissolutions asked
MLA	Quarterly data for Mahoning County marriage license applications
REAL	Quarterly data for Mahoning County real estate transfers
BKR	Quarterly data for Mahoning County bankruptcies
DR	Quarterly data for Mahoning County domestic relations
NC	Quarterly data for Mahoning County new complaints
GDIV	Quarterly data for Mahoning County divorces and dissolutions granted
RD	Recession dummy variable, 1973.4-1975.1 = 1, 0 otherwise
SD_1	Seasonal dummy variable, all first quarter = 1, 0 otherwise
SD_2	Seasonal dummy variable, all second quarter = 1, 0 otherwise
SD_3	Seasonal dummy variable, all third quarter = 1, 0 otherwise

The model developed in this paper centers around the analysis of three general equations. Initially, it is important to give the specifications for each of the equations in the model and a brief discussion of the rationale behind each specification. Estimation results and an analysis of the results obtained from the general equations are presented in the following section.

In Equation 1, the crime index for the city of Youngstown is expressed as a linear function of non-economic variables contained in the model as well as the number of real estate transfers and the number of local bankruptcies. The rationale is that if the crime index is found to be significantly related to these variables, there is justification to use the crime index as a proxy for various facets of stress, whether social or economic, brought about by the plant closings. Since crime data is readily available, a study of the relationship between the crime rate and plant closings may shed light on the adverse impact plant closings might have on the set of economic and non-economic social variables under consideration in the model. A lagged dependent variable is included as an explanatory variable intended to capture the influence of all other variables having an impact on the dependent variable which are not included here.

(1) $CI = A_0 + A_1*ADIV + A_2*MLA + A_3*GDIV + A_4*REAL + A_5*DR + A_6*NC + A_7*BKR + A_8*CI(-1)$

In Equation 2, the crime index is expressed as a linear function of plant closings, the local unemployment rate, a recession dummy variable for the 1973-1975 nationwide economic recession and three seasonal dummy variables. The rationale for this specification is that the crime index can be used to measure the economic stress induced by the local plant closings.

(2) $CI = B_0 + B_1*PC + B_2*UR + B_3*RD + B_4*CI(-1) + B_5*SD_1 + B_6*SD_2 + B_7*SD_3$

The specification of the regression equations for the remaining variables takes on the general form as presented in Equation 3.

(3) $\text{Dependent Variable} = A + \Sigma B_i * PC(-i) + C * (\text{Dependent})(-1) + \Sigma D_i * SD_i$

The rationale behind this specification is to analyze the possible adverse impact plant closings might have on a series of specific social variables. Regression analysis will be performed for each of the following variables: divorces and dissolutions asked and granted, marriage license applications, domestic relations cases, new complaints cases, number of bankruptcies and number of real estate transfers. As a result, seven equations will be estimated from the general specification given in Equation 3.

In estimating the set of equations that arise from specification of Equation 3, a second degree polynomial Almon distributed lag structure was applied to the plant closing variable. The rationale behind this is that the impact of plant closings on the local community might not be felt immediately due to the availability of unemployment compensation, trade adjustment and other transfer payments. These payments covered a maximum of twenty-four months. As these payments were exhausted, the impact of plant closings should influence more and more the behavior of those affected individuals. Also, it took sometime for plants to wind down operations following the announced closings.

DATA COLLECTION

As in any empirical study, difficulties were encountered in the securing of data. Decisions to use one data source over another were geared not so much to known or suspected inadequacies of the information, but rather to its availability. Time series data pertaining to various social and economic variables were collected by the Urban Studies Center at Youngstown State University. Crime statistics were obtained from quarterly Federal Bureau of Investigation Uniform Crime Reports published by the Department of Justice. The crime index is the aggregate number of certain offenses that occur in any quarter. These offenses include murder, rape, robbery, aggravated assault, burglary, larceny and motor vehicle theft.

REGRESSION RESULTS

Sample data used to estimate Equation 2 consisted of 45 quarterly observations from 1969.1 through 1980.1. Since data for most of the non-economic variables are not available prior to 1975, the sample period used to estimate the remaining equations in the model covers the 21 quarters from 1975.1 through 1980.1. Regression results are given in Table 2. Figures in parentheses are the computed "t" statistics. In addition, other relevant statistics such as the Durbin-Watson (DW) statistic and the estimated coefficient of determination are also provided (R^2).

The Durbin-Watson statistic measures possible serial correlation. A Durbin-Watson statistic around 2.00 indicates a lack of serial correlation in the error term. Most of the Durbin-Watson statistics reported in Table 2 indicate a lack of serial correlation. The coefficient of determination (R^2) measures the proportion of the total variation in the dependent variable explained by the regression analysis. R^2 values range from 0.41 for Equation 4 to 0.97 for Equation 5, indicating that while some of the dependent variables are highly related to the set of explanatory variables, other dependent variables are only marginally related to the independent variables.

An examination of Equation 1 in Table 2 reveals that 86% of the variation in the crime index can be accounted for by the set of explanatory variables. The process of aggregating all crime activities into one index has no doubt concealed some important relationships between crime activities and the individual explanatory variables. The seemingly inconsistent signs associated with a few of the estimated coefficients can largely be explained by the peculiar patterns of behavior of the data series during the sample period. For example, we would expect marriage license applications to be unrelated to local crime activities, yet the estimated coefficient indicates a highly significant positive relation between the two activities. One explanation for this peculiar result, after examining the original two time series, is that during the sample period both activities followed a very similar seasonal pattern with winter quarter being the slowest quarter for both activities.

The negative signs associated with the estimated coefficients of domestic relations, new court complaints and divorces/dissolutions granted is due largely to the fact that during the sample period, as crime increased the number of court cases related to domestic relations, new complaints and divorces or dissolutions granted either stayed the same or slightly trended downward.

Equation 2 in Table 2 indicates that the crime index is related to the level of local unemployment, plant closings and recession variables. As might be expected, the crime index is positively related to each of the economic variables. As the local unemployment rate increases, for instance, the crime rate also tends to increase. In addition, the crime index exhibits a strong seasonal pattern, with the first quarter having the lowest crime activities and the fourth quarter having the highest crime activities.

Regression results on individual social variables and plant closings are given in Equations 3 through 9 in Table 2. In general, the results tend to support the assumed adverse effects of plant closings on the social variables. However, the lag structures of the effects vary from equation to equation. For example, Equation 5 shows that the adverse impact of plant closings on marriage license applications becomes significant two to six quarters after plant closings. On the other hand, Equation 8 implies that real estate transactions were not significantly affected until five or more quarters after plant closings.

A study of the estimated coefficients of seasonal dummy variables, in general, suggests strong seasonal patterns. For example, regression results in Equation 5 suggests that the number of marriage license applications is the lowest in the first quarter and the highest in the third quarter.

CONCLUSION

The results of the study seem to suggest that most of the social variables under study in the model are statistically related in varying degrees to local plant closings. However, the impacts of plant closings on these social variables seem to have very different time lags. In some cases, the effect of plant closings is felt in the next quarter or so, while in other cases the lagged effect may occur in six or seven quarters.

TABLE 2

REGRESSION RESULTS

EQUATION — Estimated Coefficients

1

$$CI = \underset{(2.28)}{2072.17} + \underset{(1.51)}{2.26}\ ADIV + \underset{(3.06)}{1.96}\ MLA - \underset{(-.32)}{.08}\ REAL + \underset{(2.95)}{8.19}\ BKR - \underset{(-2.31)}{1.34}\ DR - \underset{(-1.82)}{2.49}\ NC - \underset{(-3.76)}{4.63}\ GDIV + \underset{(3.63)}{.60}\ CI(-1)$$

R^2=.86 DW=1.67

2

$$CI = \underset{(3.10)}{725.13} + \underset{(1.50)}{199.38}\ PC + \underset{(2.26)}{44.75}\ UR + \underset{(1.10)}{143.00}\ RD - \underset{(-5.89)}{706.36}\ SD_1 - \underset{(-.82)}{102.53}\ SD_2 - \underset{(-1.11)}{130.62}\ SD_3 + \underset{(3.67)}{.52}\ CI(-1)$$

R^2=.69 DW=2.20

3

$$ADIV = \underset{(2.75)}{469.80} + \Sigma B_i\ PC(-i) - \underset{(-.37)}{.12}\ ADIV(-1) + \underset{(2.75)}{93.26}\ SD_1 + \underset{(.49)}{21.29}\ SD_2 + \underset{(1.10)}{40.82}\ SD_3$$

LAG	B_i	t	LAG	B_i	t
0	-1.96	-.06	5	- 2.26	- .31
1	1.74	.20	6	- 7.97	-1.04
2	3.66	.48	7	-15.56	-1.05
3	3.51	.30	8	-25.03	.82
4	1.57	.12			

R^2=.59 DW=1.97

4

$$GDIV = \underset{(2.58)}{427.83} + \Sigma B_i\ PC(-i) + \underset{(.12)}{.04}\ GDIV(-1) + \underset{(.52)}{17.91}\ SD_1 + \underset{(1.87)}{65.62}\ SD_2 + \underset{(1.54)}{44.89}\ SD_3$$

LAG	B_i	t	LAG	B_i	t
0	-24.89	- .96	5	19.53	.76
1	-12.12	-1.07	6	6.58	.73
2	- 1.69	- .22	7	.29	.02
3	5.39	.43	8	-9.35	-.31
4	9.13	.63			

R^2=.41 DW=1.81

5

$$MLA = \underset{(1.09)}{356.14} + \Sigma B_i\ PC(-i) + \underset{(.83)}{.22}\ MLA(-1) - \underset{(-.87)}{65.34}\ SD_1 + \underset{(2.58)}{274.56}\ SD_2 + \underset{(9.26)}{293.23}\ SD_3$$

LAG	B_i	t	LAG	B_i	t
0	16.93	1.09	5	-12.27	-1.60
1	3.92	.60	6	-7.35	-1.22
2	-5.51	-1.05	7	1.15	.11
3	-11.35	-1.41	8	13.24	.65
4	-13.60	-1.51			

R^2=.97 DW=2.27

6

$$DR = \underset{(2.22)}{211.35} + \Sigma B_i\ PC(-i) + \underset{(.01)}{.01}\ DR(-1) + \underset{(1.69)}{75.05}\ SD_1 + \underset{(2.55)}{105.66}\ SD_2 + \underset{(1.97)}{82.75}\ SD_3$$

LAG	B_i	t	LAG	B_i	t
0	3.55	.12	5	15.65	1.05
1	-6.46	-.51	6	36.72	2.87
2	-10.26	-.94	7	64.01	2.63
3	-7.84	-.48	8	97.51	2.08
4	.79	.04			

R^2=.66 DW=1.56

TABLE 2 Continued

7 $NC = 627.47 + \Sigma B_i PC(-i) - .16 NC(-1) + 18.57 SD_1 - 52.02 SD_2 - 7.12 SD_3$
(4.77) (-.67) (.87) (-2.36) (-.27)

LAG	B_i	t	LAG	B_i	t
0	-20.02	-1.21	5	10.60	1.26
1	.53	.08	6	-4.92	-.83
2	13.87	2.19	7	-27.65	-2.40
3	19.99	2.08	8	-57.60	-2.41
4	18.90	1.81			

$R^2=.57$ DW=1.83

8 $REAL = 836.91 + \Sigma B_i PC(-i) + .55 REAL(-1) - 5.22 SD_1 + 834.28 SD_2 + 464.32 SD_3$
(1.32) (2.49) (-.04) (4.35) (5.13)

LAG	B_i	t	LAG	B_i	t
0	-10.37	-.16	5	-5.14	-.15
1	16.32	.58	6	-42.56	-1.72
2	30.19	1.21	7	-92.81	-2.13
3	31.24	.82	8	-155.89	-1.73
4	19.46	.47			

$R^2=.91$ DW=1.65

9 $BKR = 121.18 + \Sigma B_i PC(-i) -.39 BKR(-1) - 9.42 SD_1 - 2.57 SD_2 - 11.96 SD_3$
(6.02) (-1.58) (-1.09) (-.33) (-1.44)

LAG	B_i	t	LAG	B_i	t
0	-16.69	-2.75	5	10.92	3.53
1	-7.98	-2.88	6	11.66	4.17
2	-.86	-.43	7	10.81	2.55
3	4.66	1.59	8	8.36	1.09
4	8.59	2.55			

$R^2=.70$ DW=2.49

REFERENCES

Cohen, Lawrence E. (1981). Modeling Crime Trend: A Criminal Opportunity in Perspective. Journal of Research in Crime and Delinquency, 18, 138-159.

Johnson, J. Econometric Methods (McGraw-Hill, New York, 1972).

Stocks, A. and Buss, T.F. (1981). Socio-Economic Consequences of a Plant Shutdown. In Time Series Analysis (Proceedings of the International Conference held at Houston, Texas, August 1980). Eds: O.D. Anderson and M.R. Perryman, North-Holland, Amsterdam, 559-568.

TROLL User's Guide. MIT Information Processing Services, Massachusetts Institute of Technology, 1979.

Wellfor, Charles. Quantitative Studies in Criminology (Sage Publications, Beverly Hills, 1978).

APPLIED TIME SERIES ANALYSIS
O.D. Anderson & M.R. Perryman (eds.)

ESTIMATING THE SPECTRUM OF A SIGNAL RESTRICTED TO A SUBSPACE OF C^n

C. Roger Longbotham

Department of Statistics and Operations Research
University of Denver
Denver, Colorado 80208
U.S.A.

Many authors have considered the problem of estimating the spectrum of a multivariate signal that is observed with additive noise contamination. This paper considers the problem of estimating the spectrum of a contaminated signal when the signal is restricted to a subspace of C^n.

Let $\{Y(t),\ t = 0, 1, 2, \ldots\}$ be a complex-valued stationary nxl time series. For each t, assume the model

$$Y(t) = X\beta(t) + e(t),$$

where X is a constant nxp matrix and $\{X\beta(t)\}$ and $\{e(t)\}$ are stationary nxl time series.

Let $f_y(\lambda)$, $f_x(\lambda)$ and $f_e(\lambda)$ be the spectral density matrices of $\{Y(t)\}$, $\{X\beta(t)\}$ and $\{e(t)\}$, respectively, for each frequency λ. We will assume that $E\{X\beta(t)\ e(s)^*\} = 0$, an nxn matrix of zeroes, for all t and s, where * denotes conjugate transpose. Then $f_y(\lambda) = f_x(\lambda) + f_e(\lambda)$, for each λ. The objective is to estimate $f_x(\lambda)$ having observed Y(t) for t = 1, 2, ..., T.

Using the observations, an estimate for $f_y(\lambda)$, say $\hat{f}_y(\lambda)$, may be obtained. It may be desirable to estimate $f_x(\lambda)$ from $\hat{f}_y(\lambda)$. However, if we are able to determine an estimate for $f_e(\lambda)$, $\hat{f}_e(\lambda)$, we might want to use a function of $\hat{f}_y(\lambda) - \hat{f}_e(\lambda)$ to estimate $f_x(\lambda)$.

Let M(X) be the subspace of C^n spanned by the columns of X and let $M(\bar{X})$ be the subspace spanned by the columns of $\bar{X}$, where $\bar{X}$ is the complex conjugate of X. It is not difficult to show that each column of $f_x(\lambda)$ is in M(X) and each row of $f_x(\lambda)$ is in $M(\bar{X})$. One desirable property of an estimator of $f_x(\lambda)$ is that the estimator have columns in M(X) and rows in $M(\bar{X})$. Note: If X is real, the columns of $Re[f_x(\lambda)]$ and $Im[f_x(\lambda)]$ lie in M(X) and the rows of $Re[f_x(\lambda)]$ and $Im[f_x(\lambda)]$ lie in M(X). Results for the case of X real are in the second part of the appendix.

Let $C^n \times C^m$ be the space of all nxm matrices with complex components. Let M_n and M_m be subspaces of C^n and C^m, respectively. Denote by A_n the orthogonal projection operator onto M_n and by A_m the orthogonal projection operator onto M_m. The set of all nxm matrices with rows in M_m and columns in M_n is easily seen to be a closed linear subspace of $C^n \times C^m$. Denote this subspace by $M_n \boxtimes M_m$.

For an nxm matrix Q with columns $Q = (q_1, q_2, \ldots, q_m)$, denote by vec Q the vector of length n·m, $\text{vec } Q = (q_1', q_2', \ldots, q_m')'$. An inner product may be defined on $C^n \times C^m$ by

(1) $$(Q_1, Q_2) = (\text{vec } Q_1)^*(\text{vec } Q_2)$$

for any matrices Q_1 and Q_2 in $C^n \times C^m$.

<u>Theorem 1</u>

The projection of an nxm matrix Q onto $M_n \boxtimes M_m$ with respect to the inner product defined in equation (1) is given by $A_n Q A_m'$.
(All proofs are collected in the Appendix.)

<u>Lemma 1</u>

Let Q be any matrix in $C^n \times C^m$, then

$$Q = A_n Q A_m' + (I_n - A_n) Q A_m' + A_n Q (I_m - A_m') + (I_n - A_n) Q (I_m - A_m').$$

Also, each term in the sum is orthogonal to the remaining terms. (The four terms are, respectively, the projections of Q onto $M_n \boxtimes M_m$, $M_n^\perp \boxtimes M_m$, $M_n \boxtimes M_m^\perp$ and $M_n^\perp \boxtimes M_m^\perp$.)

This lemma implies that $C^n \times C^m$ may be decomposed into the four orthogonal subspaces listed. For the special case we are interested in, $m = n$, $M_n = M(X)$, $M_m = M(\bar{X})$ and the matrix $f_x(\lambda)$ is in $M_n \boxtimes M_m$.

Let A_1 be the orthogonal projection matrix onto M(X) and let A_2 be the orthogonal projection matrix onto $M(\bar{X})$. A natural choice for an estimator of $f_x(\lambda)$ derived from $\hat{f}_y(\lambda)$ is $A_1 \hat{f}_y(\lambda) A_2'$. If $\hat{f}_e(\lambda)$ is an estimator of $f_e(\lambda)$, an estimator of $f_x(\lambda)$ that should be considered is $A_1(\hat{f}_y(\lambda) - \hat{f}_e(\lambda))A_2'$.

The next lemma compares any estimator of $f_x(\lambda)$, say $f_x^+(\lambda)$ with the projected version, $\hat{f}_x(\lambda) = A_1 f_x^+(\lambda) A_2'$.

Lemma 2

For any estimator of $f_x(\lambda)$, $f_x^+(\lambda)$, the estimator $\hat{f}_x(\lambda) = A_1 f_x^+(\lambda) A_2'$ has strictly smaller mean square error (MSE) unless $f_x^+(\lambda) = \hat{f}_x(\lambda)$ with probability one. If $f_x^+(\lambda)$ is an unbiased estimator of $f_x(\lambda)$, then $\hat{f}_x(\lambda)$ is unbiased, also. (The MSE of an estimator $f_x^+(\lambda)$ of $f_x(\lambda)$ is defined by

$$MSE(f_x^+(\lambda)) = E\{vec(f_x^+(\lambda) - f_x(\lambda))^* \; vec(f_x^+(\lambda) - f_x(\lambda))\}.)$$

APPENDIX

I. Proofs of Results

Proof of Theorem

First, show that if $Q \in M_n \boxtimes M_m$, then $A_n Q A_m' = Q$. Obviously, if the columns of Q are in M_n, then $A_n Q = Q$. In the same manner, if the rows of Q are in M_m, then $(QA_m')' = A_m Q' = Q'$. Therefore, $A_n Q A_m' = Q$, if $Q \in M_n \boxtimes M_m$. In the same way, it is easy to show that, for any $Q \in C^n \times C^m$, $A_n Q A_m' \in M_n \boxtimes M_m$.

Now, we show that $Q - A_n Q A_m' \perp M_n \boxtimes M_m$. Let Q_1 and Q_2 be members of $C^n \times C^m$. Then Q_1' and Q_2' are in $C^m \times C^n$. On $C^m \times C^n$ define the inner product $(Q_1', Q_2')_{mxn} = (vec\ Q_1')^* (vec\ Q_2')$. Since A_n projects the columns of a matrix onto M_n, A_n is self-adjoint, i.e.

$$(Q_1, A_n Q_2) = (A_n Q_1, Q_2). \quad \text{Also}$$

$$(Q_1, Q_2 A_m') = (Q_1', A_m Q_2')_{mxn}$$

$$= (A_m Q_1', Q_2')_{mxn}$$

$$= (Q_1 A'_m, Q_2),$$

so A_m' is self-adjoint.

Let H be any member of $M_n \boxtimes M_m$. Then $(Q - A_nQA'_m, H) = (Q, H) - (A_nQA'_m, H)$. Now,

$$(A_nQA'_m, H) = (QA'_m, A_nH)$$

$$= (Q, A_nHA'_m)$$

$$= (Q, H), \text{ since } H \in M_n \boxtimes M_m.$$

Therefore, $Q - A_nQA'_m \perp M_n \boxtimes M_m$.

Proof of Lemma 1

Expanding the last three terms in the sum,

$$(I_n - A_n)QA'_m = QA'_m - A_nQA'_m$$

$$A_nQ(I_m - A'_m) = A_nQ - A_nQA'_m$$

$$(I_n - A_n)Q(I_m - A'_m) = Q - A_nQ - QA'_m + A_nQA'_m.$$

The sum of these three with $A_nQA'_m$ is Q.

To show orthogonality, observe that if

a. $Q \in M_n \boxtimes M_m$, then

$A_nQA'_m = Q$ and

$(I_n - A_n)Q = 0$ and $Q(I_m - A'_m) = 0$,

so the first term is orthogonal to the last three.

b. If $Q \in M_n^{\perp} \boxtimes M_m$, then

$(I_n - A_n)QA'_m = Q$ and

$Q(I_m - A'_m) = 0$, so the second term is orthogonal to the last two terms.

c. If $Q \in M_n \boxtimes M_m^{\perp}$, then

$A_nQ(I_m - A'_m) = Q$ and

$(I_n - A_n)Q = 0$, so the third and fourth terms are orthogonal.

Proof of Lemma 2

First, the unbiasedness question.

If $E(f_x^+(\lambda)) = f_x(\lambda)$, then

$$E(A_1 f_x^+(\lambda) A_2') = A_1 E(f_x^+(\lambda)) A_2'$$

$$= A_1 f_x(\lambda) A_2'$$

$$= f_x(\lambda), \text{ since}$$

$f_x(\lambda) \varepsilon M(X) \otimes M(\bar{X})$.

$$\text{Now, } MSE(f_x^+(\lambda)) = E\{vec(f_x^+(\lambda) - f_x(\lambda)) * vec(f_x^+(\lambda) - f_x(\lambda))\}$$

$$= E \| f_x^+(\lambda) - f_x(\lambda) \|^2$$

$$= E \| f_x^+(\lambda) \|^2 - 2E(f_x^+(\lambda), f_x(\lambda)) + \| f_x(\lambda) \|^2$$

Similarly,

$$MSE(\hat{f}_x(\lambda)) = E \| \hat{f}_x(\lambda) \|^2 - 2E(\hat{f}_x(\lambda), f_x(\lambda)) + \| f_x(\lambda) \|^2.$$

Using Lemma 1,

$$f_x^+(\lambda) = A_1 f_x^+(\lambda) A_2' + (I - A_1) f_x^+(\lambda) A_2' + A_1 f_x^+(\lambda)(I - A_2') +$$

$$(I - A_1) f_x^+(\lambda)(I - A_2'),$$

and all terms on the right are mutually orthogonal. So

$$\| f_x^+(\lambda) \|^2 = \| \hat{f}_x(\lambda) \|^2 + \| (I - A_1) f_x^+(\lambda) A_2' \|^2 + \| A_1 f_x^+(\lambda)(I - A_2') \|^2 +$$

$$\| (I - A_1) f_x^+(\lambda)(I - A_2') \|^2$$

Therefore, $E \| f_x^+(\lambda) \|^2 > E \| \hat{f}_x(\lambda) \|^2$ unless $\hat{f}_x(\lambda) = f_x^+(\lambda)$ with probability one.

Since

$$f_x(\lambda) = A_1 f_x(\lambda) A_2', \quad (f_x^+(\lambda),\ f_x(\lambda)) = (f_x^+(\lambda),\ A_1 f_x(\lambda) A_2')$$

$$= (A_1 f_x^+(\lambda) A_2',\ f_x(\lambda))$$

$$= (\hat{f}_x(\lambda),\ f_x(\lambda)).$$

Therefore, $MSE(\hat{f}_x(\lambda)) < MSE(f_x^+(\lambda))$ unless $\hat{f}_x(\lambda) = f_x^+(\lambda)$ with probability one.

II. Results for a real time series $\{Y(\varepsilon),\ t = 0, \pm 1, \pm 2, \ldots\}$.

Let M(X) be the subspace of R^n spanned by the columns of X. Then the columns of $Re[f_x(\lambda)]$ and $Im[f_x(\lambda)]$ are in M(X) and the rows of $Re[f_x(\lambda)]$ and $Im[f_x(\lambda)]$ are in M(X).

Let M_n be a subspace of R^n and M_m a subspace of R^m. Let $A_n(A_m)$ be the orthogonal projection matrix onto $M_n(M_m)$ in R^n (R^m).

Let $C^n \times C^m$ be the space of all nxm complex matrices. It is easy to see that the set of all matrices (e.g., Q) that are in $C^n \times C^m$ such that the columns of Re[Q] and Im[Q] lie in M_n and the rows of Re[Q] and Im[Q] lie in M_m is a subspace of $C^n \times C^m$. Denote this subspace by $M_n \boxtimes M_m$.

<u>Theorem 1'</u>

The projection of an nxm complex matrix, Q onto $M_n \boxtimes M_m$ with respect to the inner product $(Q_1, Q_2) = (\text{vec } Q_1) * (\text{vec } Q_2)$ is given by $A_n Q A_m$.

<u>Lemma 1'</u>

Let Q be any nxm complex matrix, then

$$Q = A_n Q A_m + (I_n - A_n) Q A_m + A_n Q (I_m - A_m) + (I_n - A_n) Q (I_m - A_m),$$

where each term in the sum is orthogonal to the remaining terms.

Lemma 2'

Let A be the real projection matrix onto the subspace of R^n spanned by X. For any estimator of $f_x(\lambda)$, $f_x^+(\lambda)$, the estimator $\hat{f}_x(\lambda) = Af_x^+(\lambda)A$ has strictly smaller MSE unless $f_x^+(\lambda) = \hat{f}_x(\lambda)$ with probability one. If $f_x^+(\lambda)$ is an unbiased estimator of $f_x(\lambda)$, then $\hat{f}_x(\lambda)$ is unbiased, as well.

The proofs for the real case follow along the same lines as in the complex case.

REFERENCES:

[1] Brillinger, David R. (1975), Time Series: Data Analysis and Theory, (Holt, Rinehart, and Winston, New York).

[2] Clendenen, G., Gallucci, V.F. and Gara, R.I. (1978), "On the Spectral Analysis of Cyclical Tussock Moth Epidemics and Corresponding Climate Indices with a Critical Discussion of the Underlying Hypotheses," Time Series and Ecological Processes, 279-293 (Ed: H.H. Shugart, SIAM-SIMS Conference Series).

[3] Hannan, E.J. (1970), Multiple Time Series (Wiley, New York).

[4] Longbotham, C. Roger (1978), Estimation of Time Dependent Parameters in the Gauss-Markov Model, Ph.D. dissertation, Department of Statistics, Florida State University.

APPLIED TIME SERIES ANALYSIS
O.D. Anderson & M.R. Perryman (eds.)

FASHION IN WOMEN'S DRESS: MODELLING STYLISTIC CHANGE AS A STOCHASTIC PROCESS

John W. G. Lowe

Department of Anthropology
University of Illinois
Urbana-Champaign, Illinois
U.S.A.

A mathematical analysis of six dimensions of women's evening dress suggests that stylistic change is an inherently stochastic process, neither deterministic with an overlay of white noise, nor random movement within constraints set by propriety and physical possibility. The six dimensions interact with one another to create a causal network exhibiting marginal stability; while events elsewhere in the sociocultural system--economic and political unrest in particular--are reflected indirectly in stylistic change, periods of unrest operating to "heat up" the process of fashion change.

This paper looks at stylistic change in women's evening dress from the perspective of time series analysis and mathematical systems theory. Styles may be defined most broadly as a particular way of doing things (Kroeber 1957), a delineator of possibilities, and more specifically as a syntax of forms admitting varying degrees of intensity and delicacy of statement. A quantitative, metrical analysis of stylistic change necessarily fails to capture nuances of form and expression. But this loss is in part offset by a greater rigor and clarity of analysis and exposition.

Evening dress was chosen because the emphasis of the study is on style and stylistic change, and formal evening wear is "about as far removed from utilitarian motivations as dress could be" (Richardson and Kroeber 1940:111). A set of six persistant features--skirt length, skirt width, waist length, waist width, decolletage length, decolletage width--may be discerned, which, taken together, define a basic pattern or configuration in women's evening wear. Fashion then, in western civilization, can be characterized by the movement up and down and in and out of a decolletage or neckline, a waistline, and a line defined by the bottom of the skirt.

The data employed in this paper derive from Richardson's and Kroeber's (1940) examination of almost three and a half centuries of women's evening dress. These data may be divided into two parts, an almost continuous record of fashion plates from 1789 to 1936, a span of just under 150 years, and before that time, fragmentary data from portraiture extending back to 1605. Raw measurements of the six basic dimensions were converted to ratios by dividing each dimension by the total length of the figure. To avoid difficulties associated with changing hairstyles, the total length was measured from the center of the mouth downward. A number of measures were taken for each year and a mean calculated for each dress dimension.

Looking first at the means of each of the six measures, Richardson and Kroeber found cycles of about a century in length, by measuring from peak to peak and trough to trough and averaging. Next, employing within year variability, they defined an ideal configuration on the basis of a minimum coefficient of variation, that is, when the ratio of the within year standard deviation to the mean was at a minimum.

To Kroeber, the ideal functioned as an equilibrium state, but there was no convergence. The necessity in fashion for incessant change resulted in a "slow pendulum like swing between extremes" (Kroeber 1957:9), but always a return to the same ideal configuration. Long cycles with periods of 70 to 150 years in length could hardly have been generated by a single individual. This confirmed Kroeber's basic thesis that people were embedded in a cultural matrix which changed through time according to its own inner logic, carrying the individual along willy-nilly.

This, Richardson and Kroeber considered the primary mechanism of stylistic change; however, high within year variability was also confined to certain periods, specifically, the times around the French Revolution/Napoleonic Wars and their immediate aftermath and the interval around W.W.I. Thus Richardson and Kroeber postulated that socio-political tension tended to impart a kind of stylistic tension, expressed as a departure from the ideal.

Though Richardson and Kroeber engaged in a quantitative analysis of stylistic change, they employed a minimum of quantitative/statistical analysis. One might ask, are these century-long cycles real, or are they merely an "artifact of a more complex, perhaps chaotic time series process" (Lumsden and Wilson 1981:176)? The fact is, the visual regularity in these cycles is somewhat less than overwhelming (far less, say, than that of Canadian lynxes or Wolfer's sunspots).

Time series analysis, then, can be employed to distinguish and test two prevailing hypotheses concerning fashion in women's dress. First, a notion proposed by Kroeber (1919, 1948, 1957) and Richardson and Kroeber (1940) postulates the existence of long period oscillations determining the course of stylistic change. The other, rather more pervasive notion, is that fashion is an inherently chaotic process, a random jumble of innovation, acceptance, and discard with little or no direction or order.

Since the data before 1789 contain multiple gaps, it seems most reasonable to apply time series analysis to the period between 1789 and 1936 (148 points in all) and extend any model estimated with these data backwards as a kind of independent test.

The first question that has to be addressed is that of stationarity since the assumption of stationarity is a precondition for almost all time series analyses. The simplest test for stationarity is inspection of the correlogram which in all six cases does not fall off with inordinate rapidity. Under these conditions Box and Jenkins (1976) propose differencing. Inspection of the autocorrelation and partial autocorrelation suggest an MA(1) or AR(1) process on the first difference for each measure. The time series is fairly short and the fall-off in both the autocorrelation and partial autocorrelation is so rapid that it is nearly impossible to distinguish between these two possibilities. While it is true that a large number of other processes can be modelled by a simple moving average of order one (0,1,1) (Granger and Newbold 1977), casting the process into AR(1) form seems to have greater cognitive salience and produces an equation of the form

$$\nabla x_t = a\nabla x_{t-1} + e_t$$

The coefficient, a, is invariably negative and, for example, takes a value of approximately -.25 in the case of skirt length.

Certainly we are not dealing here with a purely random process, such as white noise or a simple random walk, containing a complete absence of structure, order, and predictability. Changes in women's fashions are not entirely chaotic. And, in fact, there is probably more structure than the above equations would suggest.

Simulation of these equations does not work well; for instance, when a value of -.25 is used for (a), e_t is estimated from the residual variance, and the above equation is simulated, skirt length tends to rise somewhat too high--up to the chin and beyond. By taking the first difference, the low frequencies are being filtered out and this is where most of the real structure seems to lie.

There are tests for stationarity; e.g. Brillinger (1974) describes one based on the cumulative spectral density function. Unfortunately, these are indecisive for much the same reason--the probable existence of waves with periods approximating that of the data set.

An alternate approach is to extract trends; however these typically produce impossible values, either within the 148 year span or slightly outside of it. For example, a linear trend on skirt length suggests that skirts go below the floor in the mid-18th century and continue to become ever longer as you go backwards in time.

In essence, movement within a bounded domain suggests a process in stable equilibrium, as does Kroeber's thesis of fashion as movement toward and away from an equilibrium configuration; and, of course, all stationary processes are in stable equilibrium. Therefore, it seemed best to assume stationarity and see if the resulting analysis proved consistant.

Both the correlogram and spectrum in all six measures point to the existence of long period oscillation. A wave pattern exists in the autocorrelation and spectral analysis demonstrating a peak near the extreme low frequency end. In both respects this is completely consistent with the conclusions of Richardson and Kroeber. Pursuing their ideas to a logical conclusion leads to a deterministic model of fashion as a superposition of one or more sinusoidal trends, that is

$$x_i = \Sigma A_{ij}\sin(\omega_{ij}t) + \beta_{ij}\cos(\omega_{ij}t) + e_i(t) \qquad i = 1,2,3, . . . 6$$

where $e_i(t)$ represents an error term superimposed on an essentially deterministic process. The coefficients A_{ij}, β_{ij}, ω_{ij} were estimated by fitting the equation to the six time series using a least squares algorithm. In all cases several waves had to be extracted before the residual spectrum and correlogram approximated white noise. Though only a fairly rough estimation of parameters was possible given the shortness of the time series, there was a definite tendency to sweep out the same set of frequencies in all measures, implying that the measures were not fluctuating independently of one another, but rather formed some sort of interdependent system.

Such deterministic models have the property of producing forecasts whose exactitude does not decay in time. Consequently, it is possible to construct a model based on the 1789-1936 data and project it forwards and backwards in time. In general, the fit in both instances is rather poor; for example, the model predicts a large peak in skirt width in the mid-20th century. But even without the aid of metrical data, it is apparent that skirts failed to expand out to Civil War proportions when skirt width equalled or exceeded the height of the figure. Likewise, the extrapolation backwards in time fails to reproduce known results--even with limits set by the residual error on the one hand and sampling error on the other.

Thus, statistical analysis confirms neither of the two initial hypotheses. Stylistic change in women's dress appears to be not completely random and unpredictable, nor is it essentially deterministic with an overlay of noise due to measurement error, sampling error, and so forth. If these two hypotheses are considered according to Hegelian logic--thesis and antithesis--then viewing stylistic change as a stochastic system, at once partly random and partly structured, forms the synthesis. Of course, deterministic systems also have structural and random parts, but the manner in which these two portions interact is greatly different.

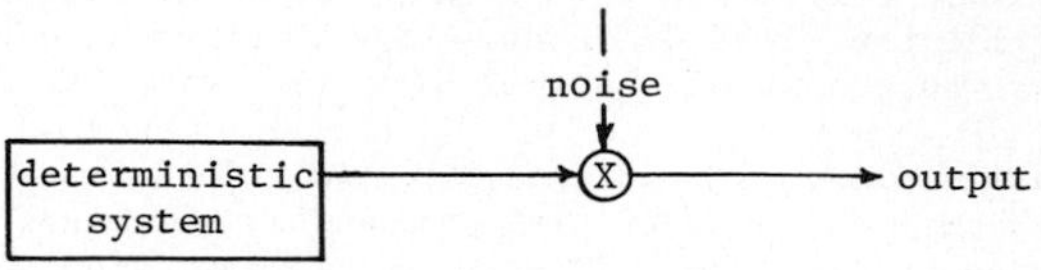

versus

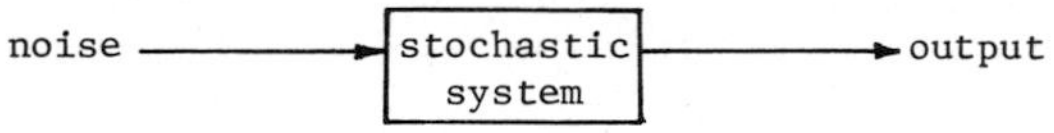

A stochastic model of stylistic change can be constructed from three basic assumptions. First, the fact that many of the same frequencies tend to show up in all of the measures strongly suggests a single interactive system. Therefore, rather than assume an ideal skirt length, skirt width, waist length, etc. existed, it seems more reasonable to suppose ideal proportions exist between each of these measures, a kind of aesthetic grammar or syntax. It is not that skirts cannot get short, but rather that if they are short, they should not become too wide, or else the result tends to appear absurd or grotesque.

Like grammar in language, these rules may be expected to vary in form from one cultural group to another, not diffusing easily across societal boundaries (language borrows words far more readily than organizational patterns). Unlike rules in grammar, it is assumed these aesthetic principles, or ideal proportions, may be bent as well as broken; but if so, an elastic restoring force develops tending to return the system to an ideal balance between dimensions.

The existence of a set of ideal proportions may be inferred from an inspection of the scatterplots.

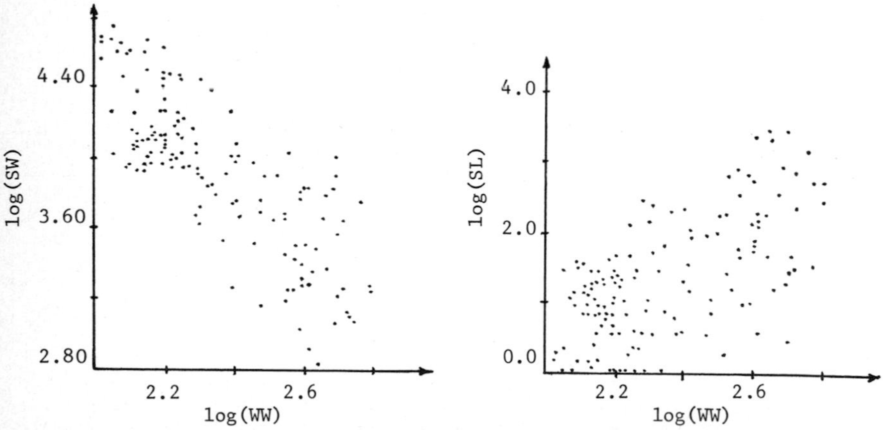

Figure 1
An example of the interrelations between dimensions: the logarithm of skirt width versus the logarithm of waist width, and the logarithm of skirt length versus the logarithm of waist width.

If log-log plots are made, most pairs demonstrate a linear relationship, suggesting that these dimensions are not fluctuating in an independent manner and that the ideal proportions tend to take one of two forms: an allometric relation, $y^b = kx^a$ (e.g. SW - DW, WW - WL, SL - WW, DL - WL, SL - DW, SW - WL, DL - DW, WL - DW, DL - WW) or be expressed as a kind of multiplicative conservation rule, $y^b x^a = k$ (e.g. SW - WW, SL - SW, DW - WW, SL - WL, SL - DL, SW - DL).

The second basic assumption derives from indications of oscillatory behavior in the correlogram and spectrum, suggesting that inertia is operating. Here inertia is used in the same sense as in Newton's Second Law of Motion. Inertia represents the tendency of an entity at rest to remain at rest and once set in motion to remain in motion. Because of inertia a variable will tend to overshoot its equilibrium point, giving rise to oscillatory behavior. The simplest way to express inertia is to make the dependent variable an accelleration $\{\nabla^2 x_i(t)\}$ rather than a velocity $\{\nabla x_t\}$, that is, to create a second order system.

The third assumption is that damping operates.

$$\nabla^2 x_t = - k \nabla x_t$$

or

$$\nabla x_t = (1 - k) \nabla x_{t-1}$$

Damping expresses the principle of cultural continuity, a resistance to change inherent within all cultural systems. If a change occurs, it engenders forces favoring movement back the other way so that the greater the rate of change, the stronger the opposing force.

As with so many economic time series, error proved to be multiplicative and a logarithmic transformation of the variables was required. By choosing reference points with some care, heteroscedasticity was alleviated, the effect of physical constraints such as the floor on skirt length minimized, and of course the interaction between dimensions expressed as linear relations.

$x_1 = \log(100 - SL + 1)$	
$x_2 = \pm \log(\lvert WL - 25.5\rvert + 1)$	If WL > 25.5, $x_2 < 0$.
$x_3 = \pm \log(\lvert DL - 13.3\rvert + 1)$	If DL > 13.3, $x_3 < 0$.
$x_4 = \log(SW)$	And,
$x_5 = \log(WW)$	25.5 = mean WL
$x_6 = \log(DW)$	13.3 = mean DL

A model based on these three assumptions is described below.

$$\nabla^2 x_i(t) = \gamma_i \nabla x_i(t-1) + \Sigma k_{ij}\{x_{eqij}(t-1) - x_i(t-1)\} + e_i(t) \qquad (1)$$

$$i = 1,2, \ldots 6$$

Where:

e_i = random white noise

$\gamma_i \nabla x_i$ = a damping term

γ_i and k_{ij} are constants (parameters)

γ_i determine the magnitude of the damping force on x_i

k_{ij} determine the magnitude of the elastic restoring force on x_i

x_{eqij} = the equilibrium value of x_i for a particular value of x_j

Assuming x_i and x_j are linked by a linear equation expressing an aesthetic ideal, then:

$$x_{eqij} = x_i = \alpha_{ij} x_j + \beta_{ij} \quad (2)$$

Substituting equation (2) into equation (1) and combining terms, we obtain:

$$\nabla^2 x_i(t) = \gamma_i \nabla x_i(t-1) - c_i x_i(t-1) + \sum_{j=1}^{5} a_{ij} x_j(t-1) + b_i + e_i(t) \quad (3)$$

where:

$$c_i = \sum_{j=1}^{5} k_{ij}$$

$$a_{ij} = k_{ij} \alpha_{ij}$$

$$b_i = \sum_{j=1}^{5} k_{ij} \beta_{ij}$$

where the parameters may be estimated using ordinary least squares.

RESULTS

Not too surprising, the term representing cultural continuity and a resistance to change--the analog of a viscous damping force in mechanics--dominates all the others. This is just the expression that is captured by a simple-minded use of the Box and Jenkins method.

Though smaller in magnitude, interactive effects also operate. The interactive terms prove to be not only statistically significant but dynamically decisive. If only damping operates, the system is not stable; its equilibrium point exhibits random drift, but with the interactive terms added, such is no longer the case and stable equilibrium prevails.

A methodological problem ensues because not all the interactive terms are statistically significant--which terms should be left in and which discarded? There is every reason to expect that interactive effects between variables will range from relatively strong to very weak. How then can one distinguish between interactive coefficients representing essentially random fluctuations about zero, and very real, but weak, interrelationships? To include relations that do not really exist is to commit one type of error, to exclude those that do really exist is to commit another.

One solution is to investigate the qualitative dynamic properties of the interactive network of these six dimensions. Because the system is linear, its dynamic behavior is relatively easy to ascertain. Starting with only those relations that are statistically significant at the .05 level, all eigenvalues are less than one, so the system is stable, but only marginally so. The largest eigenvalue is approximately .95; consequently equilibration is rather weak.

As more relations are added, the network remains marginally stable, but a new mode of behavior appears--oscillations. For small values of connectance the network is so heavily damped that the eigenvalues are invariably real, but as the number of interactions included increases, the number of feedback loops proliferate and complex eigenvalues (oscillations) begin to appear.

If the number of interactions included continues to be increased, multiple oscillations occur and the period of the longest and largest wave drops to about 100-150 years in length. Interestingly, this long period wave tends to dominate the others, being the largest or next to the largest eigenvalue. Therefore, a quantitative model was constructed assuming *a priori* maximum connectance. Six out of 30 possible interactions were excluded because the signs of the regression coefficients were inconsistent with the scatterplot relations described previously. From the equations (see Table I), it is possible to compute the equilibrium point. The results appear reasonable--it lies very near the mean values for each of the six measures. A large, complex eigenvalue is compatible with the autocorrelation and spectral results; the period is approximately 150 years.

The residuals are indistinguishable from white noise. Since a nonrandom process may appear random using only one set of tests (Shannon 1975), a series of tests for randomness was employed: the Durbin-Watson test, the first 30 autocorrelations and cross-correlations were checked and the cumulative periodogram was compared to that of white noise using the Kolmogorov-Smirnov statistic (Fuller 1976). In no case could the null hypothesis of simple white noise be rejected (see Figure 2).

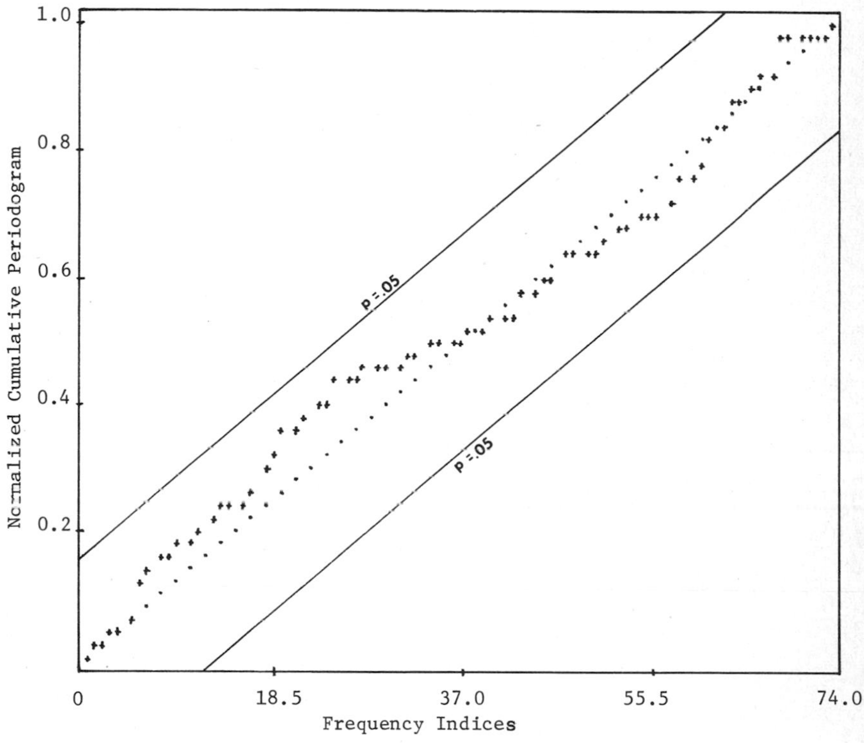

Figure 2
Cumulative Periodogram for the Residual Error in Skirt Length

The residuals do vary through time. The time period 1789-1936 was divided into four equal segments; the means were not statistically distinguishable, but the variances were ($p < .05$), being higher in the periods, 1789-1827 and 1901-1936, than in the intervening periods.[1]

Apparently stylistic change in women's evening dress is only partially an autonomous subsystem. Herein perhaps lies an explication of how style changes. As impinging noise from the remainder of the socio-cultural system increases in magnitude, extremes in fashion are produced. And, if the intensity and duration

TABLE I. FINAL REGRESSION RELATIONSHIPS FOR EACH DIMENSION.

	Initial Variance	Error Variance
$\nabla^2 x_1 = -1.066\nabla x_1 - .202x_1 - .227x_4 + .165x_5 + .233x_6 + .155$	.320	.124
$\nabla^2 x_2 = -1.195\nabla x_2 - .107x_2 - .090x_1 + .056x_3 + .223x_4 + .639x_5 - 2.267$	.704	.256
$\nabla^2 x_3 = -1.083\nabla x_3 - .636x_3 - .129x_1 + .109x_2 - .399x_4 + .488x_6 + .362$	1.932	.439
$\nabla^2 x_4 = -1.358\nabla x_4 - .231x_4 - .015x_1 + .009x_2 - .013x_3 - .273x_5 + .194x_6 + 1.029$	.096	.026
$\nabla^2 x_5 = -1.162\nabla x_5 - .278x_5 + .014x_1 + .009x_2 + .015x_3 - .073x_4 - .035x_6 + 1.015$	.028	.007
$\nabla^2 x_6 = -1.411\nabla x_6 - .104x_6 + .015x_1 + .008x_2 + .024x_4 + .172$	.053	.012

$$x_p = \text{the particular solution} = \text{the equilibrium point} = \begin{cases} x_1 = 1.024 & SL = 98.2 \\ x_2 = -.068 & WL = 27.0 \\ x_3 = -.032 & DL = 13.6 \\ x_4 = 4.056 & SW = 57.5 \\ x_5 = 2.288 & WW = 9.9 \\ x_6 = 2.822 & DW = 16.8 \end{cases}$$

The eigenvalues are: $\lambda_1 = .98$, $\lambda_2 = .94 + .033i$, $\lambda_3 = .94 - .033i$, $\lambda_4 = .77$, $\lambda_5 = .67$, $\lambda_6 = -.47$, $\lambda_7 = -.45$, $\lambda_8 = -.42$, $\lambda_9 = -.22 + .009i$, $\lambda_{10} = -.22 - .009i$, $\lambda_{11} = -.16$, $\lambda_{12} = -.085$, where the four complex numbers produce two oscillations with periods between 100-200 years.

of the random input is great enough, eventually structural change will occur. The aesthetic grammar will be rewritten and a new set of ideal relations created.

CONCLUSION

In regions where time series analysis is only marginally applicable (where, for example, there is difficulty with series length, questions of stationarity, and weak, but decisive interaction effects), it may sometime be possible to construct fairly elaborate models by concentrating on qualitative results, e.g. stability, oscillatory behavior, etc. and by using arguments of internal consistancy as opposed to simple statistical significance. Systems analysis and stochastic simulation are invaluable tools in both of these endeavors.

In particular, stylistic change in women's evening dress appears to be neither entirely determined nor completely random,[2] but rather a stochastic process whose non-random portion is composed of three parts: (1) inertia, the tendency for stylistic change, once set in motion to remain in motion, (2) cultural continuity, an inherent resistance to rapid change, and (3) a rule system of aesthetic proportions.

How general this finding is for stylistic change in general awaits metrical investigation of other stylistic systems in which chronological control is adequate and a basic configuration of persistent elements can be specified.

NOTES

[1]Using Levine's test for equality of the variance (Brown and Forsythe 1974a) and a Brown-Forsythe test for the means (Brown and Forsythe 1974b).

[2]In the course of discussion following the presentation of this paper, one point was raised that perhaps ought to be addressed here, viz, can the hypothesis of randomness really be rejected? Given the fact that gyrations in the dimensions of dress are limited in their compass, then a random walk model is unlikely to be adequate, but perhaps these measures undergo random movement within reflecting boundaries. Such a proposition is difficult to test directly by time series analysis, but a number of indirect arguments indicate that a simple random model of this sort is inadequate.

First, such boundaries are not generally that well defined. Decolletage length and waist length have clear limits, but what is the upper boundary for skirt length--just above the ankle, mid-calf, knee length? All of these seem to have functioned as limits at one time or another, and if there is no reflection off an upper boundary, what keeps skirts from going too high? It seems that a random walk between reflecting boundaries may be a feasible model for decolletage length and waist length, but not for the other measures. Even with these two measures, however, such a model fails to account for the apparent interaction between dimensions.

REFERENCES

BOX, G.E.P. and JENKINS, G.M. (1976). *Time Series Analysis: Forecasting and Control*, Revised edition. Holden-Day, Inc., San Francisco.

BRILLINGER, D.R. (1974). *Time Series: Data Analysis and Theory.* Holt, Rinehart, and Winston, New York.

BROWN, M.B. and FORSYTHE, A.G. (1974a). Robust tests of equality of variances. *Journal of the American Statistical Association 69*, 364-367.

BROWN, M.B. and FORSYTHE, A.G. (1974b). The small sample behavior of some statistics which test the equality of several means. *Technometrics 16*, 129-132.

FULLER, W.A. (1976). *Introduction to Statistical Time Series.* John Wiley & Sons, Inc., New York.

GRANGER, C.W.J. and NEWBOLD, P. (1977). *Forecasting Economic Time Series.* Academic Press, Inc., New York.

KROEBER, A.L. (1919). On the principle of order in civilization as exemplified by changes of fashion. *American Anthropologist 21*, 235-263.

KROEBER, A.L. (1948). *Anthropology: Culture Patterns and Processes.* Harcourt, Brace & World, Inc., New York.

KROEBER, A.L. (1957). *Style and Civilizations.* Cornell University Press, Ithaca, N.Y.

LUMSDEN, C.J. and WILSON, E.O. (1981). *Genes, Mind, and Culture: The Coevolutionary Process.* Harvard University Press, Cambridge, Mass.

RICHARDSON, J. and KROEBER, A.L. (1940). Three centuries of women's dress fashions: a quantitative analysis. *Anthropological Records 5*(2), 111-153.

SHANNON, R.E. (1975). *Systems Simulation: The Art and Science.* Prentice-Hall, Inc., Englewood Cliffs, N.J.

APPLIED TIME SERIES ANALYSIS
O.D. Anderson & M.R. Perryman (eds.)

DIRECT QUADRATIC SPECTRUM ESTIMATION FROM UNEQUALLY SPACED DATA

Donald W. Marquardt and Sherry K. Acuff

E. I. du Pont de Nemours and Company, Inc.
Engineering Department
Wilmington, DE 19898
USA

One objective of this paper is to emphasize the importance of unequally spaced time series and to provide contacts with the literature. The principal objective is to describe a versatile, general-purpose method of direct quadratic spectrum estimation, and to explore various properties of the method. Numerical examples from several applications are presented to demonstrate the practical results that can be obtained. The examples include synthetic data, chronobiology data (oral temperature, blood pressure), and environmental data (stratospheric ozone).

1. CAUSES OF UNEQUALLY SPACED TIME SERIES DATA

It is useful to recognize five qualitatively different causes by which data may be unequally spaced.

(i) Some data points may be missing in otherwise equally spaced data.

(ii) The data point locations may vary unavoidably about equally spaced target locations.

(iii) The series may have one or more extended gaps.

(iv) The data points may inherently be obtained at random locations, eg, Poisson sampling.

(v) The data points may deliberately be Poisson-sampled to provide "alias"-free data.

It will be assumed that the pattern of unequal spacing is "unrelated" to the stochastic properties of the process y(t) being sampled.

More than one of these causes may simultaneously be operative. For example, blood pressure and body temperature may be scheduled to be taken from a patient at approximately equal intervals during waking hours (cf, cause ii) but less frequently or not at all during periods of sleep (cf, cause iii).

In this time series process the independent variable t may represent time units or distance units, depending on the application. It will be assumed throughout that the process y(t) is at least weakly stationary. Thus, the covariance between two points at times $t_i < t_j$ depends only on the time difference $(t_j - t_i)$. It is also assumed that the mean of the process is zero, whence the covariance for time difference $(t_j - t_i)$ is $E[y(t_j)y(t_i)]$.

2. SOURCES OF UNEQUALLY SPACED TIME SERIES

Data obtained manually or data obtained over a long period of time are likely to suffer from one or more of these causes of unequal spacing. This is exemplified by medical data such as the Halberg data and the stratospheric ozone data examples discussed in this paper, where one encounters over a period of years both missing points and gaps. Another example is found in the insect population cycle data discussed by Thrall (1979) where counts of births and deaths were taken "frequently, but not every day" during the 600-day experiment. Dunsmuir (1981) discusses analysis of the 663 observations of daily average carbon monoxide taken over 798 days at Boston, MA. Both isolated missing days and long missing gaps are encountered. The measurement of fluid velocity by means of a laser anemometer prompted the work of Gaster and Roberts (1975). Such data inherently follow Poisson sampling, an important special case of unequal spacing.

Some of the examples cited could potentially be treated by methods that can deal with missing data but not unequally spaced data. The Bibliography lists a number of such references. Most such work is based on the use of an "amplitude modulating sequence" to describe the missingness pattern, as originated by Parzen (1963). The goal of this paper is to develop an omnibus approach that will be suitable for arbitrary unequal spacing due to any one or more of the five causes.

3. TYPES OF TIME SERIES ANALYSIS

Table 1 lists four types of time series analysis. Three of these are familiar to large numbers of practitioners. Autocovariance analysis computes and displays the autocovariance versus lag. From this display the cyclic content of the data often can be diagnosed. For example, the autocovariance decay pattern provides information. Data containing two nearby sharp cycles show a beat pattern in the auto-covariance display. Spectral analysis displays the relative variance, or power spectral density, of the process y(t) along the frequency scale. Spectral analysis is extremely useful as a tool to display and diagnose the stochastic structure of the data. It will reveal whether the spectrum has sharp features (peaks of variance in narrow frequency ranges) or whether the spectrum is featureless (shows only gradual changes in the relative variance over a wide frequency range).

TABLE 1

TYPES OF TIME SERIES ANALYSIS

Type	Domain	Principal Use
Autocovariance	Time	"Display and Diagnose" cyclic content of data and possible properties that cause beat frequencies or autocovariance decay patterns.
Spectrum	Frequency	"Display and Diagnose" the frequency content of the data; sharp peaks and/or featureless.
Variance-Length Curve	Time	"Display and Diagnose", especially for sampling guidance.
Box-Jenkins (ARMA)	Time	"Model and Predict", perhaps "Control".

Both the theory and the practice of spectral analysis make heavy use of the fact that the autocovariance function and the power spectral density are an integral transform pair. They each display fully the stochastic content of a stationary time series. For many practical purposes the spectrum is the most informative display, while for other purposes the autocovariance function is more informative. Knowledgeable practitioners regularly compute and use both displays when interpreting time series data.

A third type of time series analysis is the variance-length curve. This approach is known to very few practitioners, although it is an old concept. The variance-length curve is a "display and diagnose" type of method. For many applications it is a preferred way to display the relevant structure of the time series. It is especially useful to characterize a time series for sampling guidance, as for quality control purposes. An estimate of the variance-length curve can be computed from the estimated autocovariance, or from the power spectral density, or it can be computed directly from the data, even when the data points are unequally spaced. Thus the variance-length curve should claim an important place in both equally and unequally spaced time series analysis (Marquardt and Acuff 1979, 1980, Meketon 1980). Some early references are Cox and Townsend (1951), Townsend and Cox (1951), Yule (1945).

The autoregressive and moving average type analysis (ARMA models) operate in the time domain. This fourth group of techniques were formalized and systematized by Box and Jenkins (1970) and a vast literature now exists. Some key references are included in the Bibliography. The ARMA methods do not merely display and diagnose. Instead they "model" the process y(t), from which it often becomes possible to "predict" and sometimes to "control" the process. However, autocovariance methods are vital to identifying the appropriate form of model to use, Box and Jenkins (1970), Gray, Kelley and McIntire (1978).

4. OBJECTIVES OF THIS PAPER

One objective of this paper is to draw attention to the importance of unequally spaced time series in practice. Unfortunately, the textbook literature to date has almost completely ignored the unequally spaced data problem. Consequently, very few practitioners are prepared to handle such problems. There is now a growing body of research papers. The Bibliography of this paper provides contacts with that literature.

Principally, this paper develops a versatile, general-purpose method of direct quadratic spectrum estimation. With this new estimator excellent results can be obtained for many practical problems. This paper gives direct quadratic spectrum estimation results for several numerical examples.

Although ARMA models are not discussed here, the Bibliography lists a number of key references to important current work, including work on unequally spaced data.

5. PRELIMINARIES ON NOTATION AND ESTIMABILITY

The spectral estimate is based upon a time series sample from the process y(t). The observations are made at known times t_i within a single continuous total period of observation [0,T]. The data are denoted

$$(t_i,\ Y_i) \qquad i = 1, 2, \ldots, n \tag{1}$$

with $0 \leq t_i \leq T$.

For equally spaced data with interval Δt, the estimated spectrum will cover the frequency range from zero cycles per unit time, to the Nyquist cutoff frequency of $1/(2(\Delta t))$ cycles per unit time. Any cyclic structure in y(t) that has higher frequency (shorter cycles) than the Nyquist frequency will be "aliased with" (ie, added to) the computed power somewhere in the estimable frequency range 0 to $1/(2(\Delta t))$. One does know where the aliased power will appear because the aliasing causes the frequency scale to be folded back upon itself in an accordion-pleated fashion, with the successive folds alternately at the ends of the estimable scale. Good design practice for equally spaced data is to select Δt small enough so that no cycle with significant energy will have length less than $2(\Delta t)$. When the data are unequally spaced and there is a wide range of spacings throughout the total period of observation, the Nyquist effect is spread out, and does not create a sharp cutoff between the nominally estimable frequency range and the nominally aliased range. The more variable the interpoint spacing in the data, the more gradual the cutoff. Hence, the sensitivity for exhibiting high frequency cycles on the spectrum is diminished at frequencies below the Nyquist frequency associated with the average Δt, but spectrum estimates can be obtained at frequencies above that nominal Nyquist frequency. The resulting spectral window provides a quantitative measure of the sensitivity achieved.

As shown in Shapiro and Silverman (1960), only certain kinds of unequal spacing, for example, Poisson sampling, create (asymptotically) "alias free" data. Nevertheless, unequally spaced data provide inherently more "robust" spectral estimates than equally spaced data. This is illustrated in the examples.

6. HISTORICAL PERSPECTIVE

As noted by Jones (1962a), several historical perspectives should be recognized. Daniell (1946) showed that the inconsistency of the periodogram can be eliminated by smoothing over neighboring frequencies. Bartlett (1948) and Tukey (1949) showed that the periodogram can be smoothed by weighting the estimated autocovariances. Grenander (1951) showed that smoothing the periodogram with positive weighting is equivalent to a nonnegative quadratic form of the sample.

From this evolved the work of Grenander and Rosenblatt (1953, 1957). They influenced the comprehensive paper of Jones (1962a) which provides a theoretical basis for all succeeding work on direct quadratic spectrum estimation from unequally spaced data. The symmetric quadratic form of the sample provides a unifying framework for all such methods. The succeeding papers of Jones (1970, 1971, 1972, 1975, 1977) are particularly important.

The term "direct quadratic spectrum estimation" (DQSE) is introduced here to describe estimation of the spectrum by direct numerical evaluation of the quadratic form, without requiring factorability of the form. By contrast, the classical procedure for equally spaced data, which first computes the estimated autocovariance function, then estimates the smoothed spectrum as an integral transform (numerical weighted integration) of the autocovariance, is a consequence of the factorability of the quadratic form when the data are equally spaced. Recently, the Fast Fourier Transform methods have become widely used for computing spectral estimates. They take extremely elegant advantage of the factorability of the quadratic form for equally spaced data.

Gaster and Roberts (1975) developed independently the direct quadratic spectrum estimation procedure appropriate to Poisson sampled data, which had also been one of the results in Jones (1962a). Gaster and Roberts (see also 1977) apply these methods to an important class of problems in physics. See also Masry and Lui (1975).

7. DIRECT SPECTRUM ESTIMATION AS A QUADRATIC FORM IN THE DATA

Throughout this paper spectral estimates of the form

$$Z(\omega) = \frac{1}{T'} \sum_{i<j}^{n} \sum^{n} W_{ij} Y(t_i) Y(t_j) \tag{2}$$

are considered. In this quadratic form

- T' is the effective record length, nominally T but precisely defined by Equation 6,
- W_{ij} is the weight represented by the i,j element in the symmetric quadratic form ($W_{ij} = W_{ji}$).

Note that evaluation of the quadratic form is a numerical double integration.

Tailoring the weights to the circumstances of a particular problem is a central issue in spectral estimation. The general form of such weights is

$$W_{ij} = D(t_j - t_i) F(t_i, t_j) \cos 2\pi\, \omega(t_j - t_i), \tag{3}$$

where

- $D(\tau)$ is the lag window for lag τ, $\tau = t_j - t_i$,
- $F(t_i, t_j)$ is the "data spacing factor", precisely defined by Equation 5' or 5",
- ω is expressed as cycles per unit time.

Previous authors have not explicitly defined the data spacing factor, but doing so helps to clarify some critical issues.

8. LAG WINDOW

The literature of equally spaced time series analysis investigates extensively the relationship between the shapes of proposed lag windows and their effect on the estimated spectrum. All such windows have the general shape shown in Figure 1. Such results apply also to unequally spaced data. For all the numerical examples in this paper, we have used one specific window, the Hanning window

$$\begin{aligned} D(\tau) &= \frac{1}{2}\left[1 + \cos \pi \frac{(t_j - t_i)}{\tau_m}\right], & t_j - t_i &\leq \tau_m \\ &= 0 \qquad , & t_j - t_i &> \tau_m . \end{aligned} \tag{4}$$

This window is equivalent to smoothing the raw spectrum with Hanning weights (0.25, 0.50, 0.25) when analyzing equally spaced data.

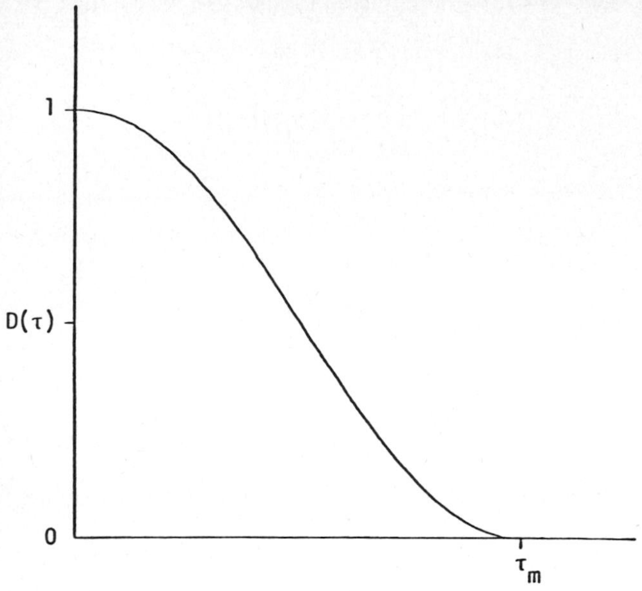

Figure 1
Typical lag window (Hanning)

9. DATA SPACING FACTOR

The data spacing factor defines, in the t_i,t_j-plane, the area which the covariance ordinate $Y(t_i)Y(t_j)$ represents in the numerical double integration.

It is instructive to examine the results of Jones (1962a) who derived the W_{ij} required for asymptotically unbiased estimates of $Z(\omega)$ for several special cases. The Data Spacing Factors implied by those results are:

Data Sampling	$F(t_i, t_j)$, Jones	$F(t_i,t_j)$ in terms of area in t_i, t_j plane
Equally spaced		$(\Delta t)^2$
Poisson Sampling With Sampling Intensity, ρ	$1/\rho^2$	$(E(\Delta t))^2$
Poisson Sampling With Time Varying Intensity, $\rho(t)$	$1/\rho(t_i)\rho(t_j)$	$E(\Delta t)_{t=t_i}E(\Delta t)_{t=t_j}$

Jones (1962a) derived the variance of these asymptotically unbiased estimators and found that the variance of the constant intensity Poisson sampling case is somewhat larger than the variance of the equally spaced case. See also Gaster and Roberts (1975). Jones found the variance of the time varying intensity Poisson sampling case is still larger, but there is no evidence that these variances are so large as to become a serious problem.

The key question is how to define the data spacing factor for data with arbitrary unequal spacing. Such a definition is suggested by the last column of the preceeding table. First, however, it is helpful to examine the plane of integration and to consider Jones (1970) Methods 2 and 3 from the geometric viewpoint. (Jones' Method 1 is the constant intensity Poisson sampling case already considered.)

Figure 2 shows the plane of integration. Each of t_i and t_j has range [0,T], but since $t_j > t_i$ only the upper triangle is to be considered. Also since $D(\tau) = 0$ for $\tau > \tau_m$ the domain of integration includes only the indicated strip from the principal diagonal up to τ_m. Jones' Method 2 (Discretizing) is to choose a small interval width and to construct successive diagonal strips of that width. Then $F(t_i,t_j) = n_\ell/(\text{width})$ for all lags $(t - t_i)$ that fall in diagonal strip ℓ. Jones' Method 3 (Ordering) involves ordering all time differences (t_j-t_i). Then a diagonal strip is defined for each distinct lag and $F(t_i-t_j)$ is defined similarly in terms of the width of the strip and the number of points therein.

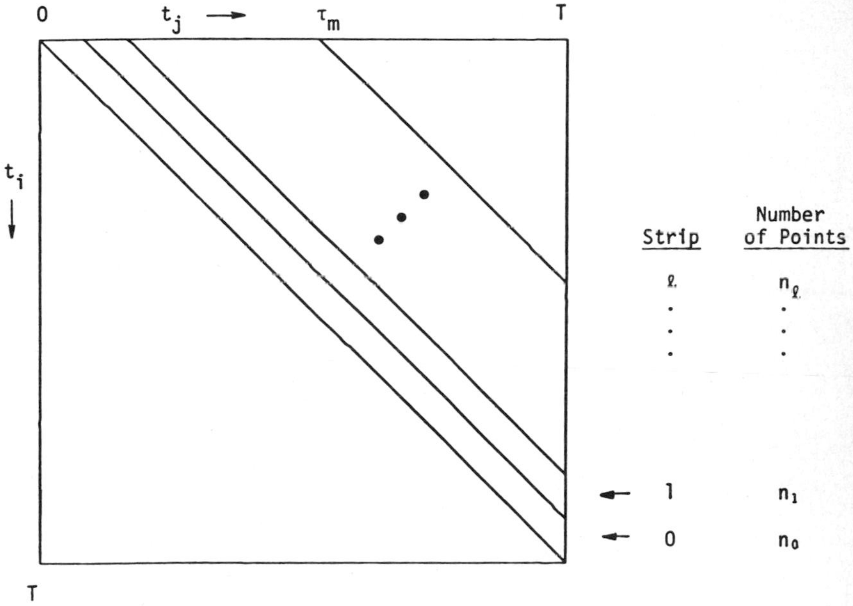

Figure 2
The plane of integration, illustrating geometrically
Jones' (1970) Method 2 (Discretizing) and Method 3 (Ordering)

These definitions of $F(t_i-t_j)$ are constant throughout a diagonal strip based on the average density of covariance ordinate locations falling in that strip. Large local biases can occur for finite data sets if the unequal data spacing creates wide variations in the density of the covariance ordinate locations within a strip.

Recalling that $F(t_i,t_j)$ defines the area in the t_i,t_j-plane which the covariance ordinate $Y(t_i)Y(t_j)$ represents in the numerical double integration, it is possible to define a data spacing factor for each i,j combination. In its simplest form

$$F(t_i,t_j) = \left(\frac{t_{i+1}+t_i}{2} - \frac{t_i+t_{i-1}}{2}\right)\left(\frac{t_{j+1}+t_j}{2} - \frac{t_j+t_{j-1}}{2}\right) \quad (5)$$

$$= (t_{i(+)} - t_{i(-)})(t_{j(+)} - t_{j(-)})$$

$$= \delta_i \delta_j$$

with endpoint conditions $t_{i-1} \geq 0$, $t_{j+1} \leq T$.

Equation 5 has the characteristic that the domain of integration is represented by a sum of rectangular contiguous nonoverlapping areas, with a covariance ordinate contained within each such area. $F(t_i,t_j)$ is large where the data are sparse, and conversely. This illustrates how the numerical double integration can be adapted to any pattern of unequal data spacing.

The asymptotic properties of the Poisson sampling result, $F(t_i,t_j) = 1/\rho^2$, rest on the fact that the area associated with each covariance ordinate is appropriate <u>on the average</u>. By using a data spacing factor such as Equation 5, the area can be made appropriate for <u>every ordinate</u>, for arbitrary unequal spacing. This suggests that the estimator can have good small sample properties, in addition to being asymptotically unbiased.

The acronym DQSE can be understood equally well from two perspectives:

- "Direct <u>quadratic</u> spectrum estimation", emphasizing that the statistical properties are those deriving from the estimator being a quadratic form in the data, and
- "Direct <u>quadrature</u> spectrum estimation", emphasizing that the properties of the estimates are enhanced by tailoring the weights so that the quadratic form constitutes a proper algorithm for the numerical quadrature of the double integral.

Viewing the DQSE method as a numerical quadrature scheme, Equation 5 implies that every covariance ordinate is contained within a rectangular area in the domain of integration in the (t_i,t_j)-plane. The nomenclature is defined in Figure 3, for the general case of unequal spacing.

When the data are equally spaced, with no gaps or missing data, the covariance ordinates are at the centers of a square grid of dimension Δt in the domain of integration (ignoring the triangular local areas along each of the diagonal boundaries). Thus the dimension $\delta_i = \delta_j$ of each local area represents less than one-half cycle for any value of ω below the Nyquist frequency. Correspondingly, the magnitude of the covariance ordinate is a suitable representation of the value of the integrand throughout the local area. Moreover, for each value of ω the collection of local areas represents a balanced sampling over the domain of integration.

For unequally spaced data, or data with gaps, the sampling is not uniform. The data spacing factor of Equation 5 seeks to tailor the integration weights so that the nonuniform sampling is nevertheless representative. We have obtained good practical results on a variety of unequally spaced data sets, using Equation 5 in its unmodified form (Marquardt and Acuff 1980). However, the estimated spectra sometimes have anomolous "positive peaks" or "negative valleys" at the upper portion of the frequency range. The cause of such anomolous behavior is that the simple data spacing factor defined by Equation 5 will not satisfy the Nyquist criterion for all values of δ_i and δ_j at each value of ω for which an estimated spectral ordinate can appropriately be computed. The covariance ordinate contained in a local area cannot be representative of the integrand throughout the local area if the boundary of the local area is more distant than one-fourth of the cycle length corresponding to the Nyquist frequency for the specified value of ω. To satisfy this requirement, it is desirable to place an upper bound on the size of the local area (ie, the weight) wherever the data are sparse, or contain large gaps compared to the Nyquist cycle length.

The modified data spacing factor then becomes

$$F(t_i,t_j) = \delta_i\delta_j$$

$$= (t_{i(+)} - t_{i(-)})(t_{j(+)} - t_{j(-)}) \qquad (5')$$

where

$$t_{i(+)} = \min[(\frac{t_{i+1} + t_i}{2}), (t_i + \frac{1}{4\omega})],$$

$$t_{i(-)} = \max[(\frac{t_i + t_{i-1}}{2}), (t_i - \frac{1}{4\omega})]$$

The factor δ_j is defined similarly, substituting j for i in the last two expressions, and ω is expressed as cycles per unit time.

End conditions at $t = T$ and $t = 0$ are imposed by defining $t_{n+1} = 2T-t_n$ and $t_0 = -t_1$ respectively, in Equation 5'. Moreover, when $\omega = 0$ or is very small, it is desirable to protect against extraordinarily large relative weighting of data points at the beginning and the end of long gaps (including long gaps at the beginning or end of the period of observation). For this purpose the value of ω used in Equation 5' can be constrained by

$$\omega(\text{Equation 5'}) = \max[\omega, M\omega_M]$$

where M is an arbitrary multiplier (say, $M = 0.1$) and $\omega_M = (n-1)/2T$ is the Nyquist frequency associated with the average inter-point interval in the period of observation.

Equation 5' implies a decrease in the effective length of the time series whenever the data are unequally spaced or contain gaps. The effective length $T'(\omega) \leq T$ is defined by

$$T'(\omega) = \sum_{i=1}^{n} \delta_i \qquad (6)$$

where the δ_i are as defined in Equation 5'.

At $\omega = 0$, Equations 5' and 6 have minimal or no numerical effect on the results from Equation 2. As ω increases, the modifications become progressively more necessary and the effects become numerically larger.

Figure 3 illustrates the geometry with unequally spaced data when the rectangular areas intersect the boundaries of the domain of integration.

One characteristic that is immediately obvious from Figure 3 is that a covariance ordinate is not at the center of its associated data-spacing-factor rectangle unless the intervals immediately surrounding each of t_i and t_j are equal. However, we shall take the value of the covariance function throughout each associated local area of integration to be equal to the value at the ordinate location contained in that local area.

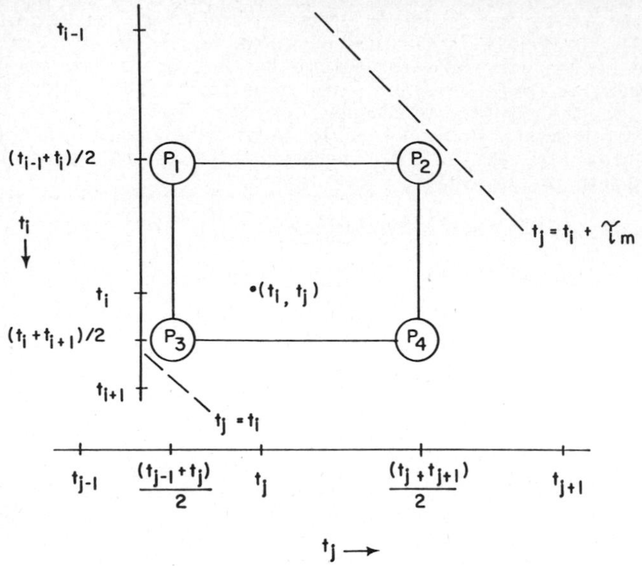

Figure 3a
Nomenclature for a rectangular local area associated with a covariance ordinate

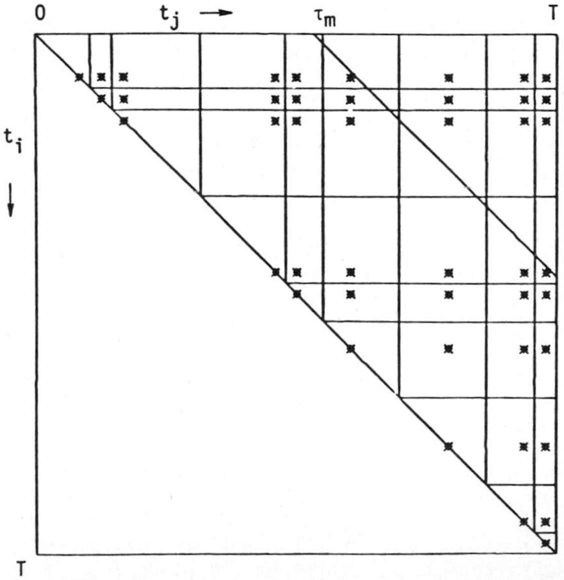

Figure 3b
Example of partial rectangular local areas at domain boundaries

Figure 3 shows that portions of some local-area rectangles are outside the domain of integration. For geometrical simplicity Figure 3 uses only nine values of t_i, but this suffices to illustrate all cases, including the case where t_1 and t_n are strictly interior to the period of observation [0,T]. Figure 3 defines nomenclature needed to diagnose the cases that can occur. The corners of the rectangle are denoted P_1, P_2, P_3, P_4, respectively, with coordinates:

Point	t_i-dimension	t_j-dimension
P_1	$t_{i(-)}$	$t_{j(-)}$
P_2	$t_{i(-)}$	$t_{j(+)}$
P_3	$t_{i(+)}$	$t_{j(-)}$
P_4	$t_{i(+)}$	$t_{j(+)}$

A further refinement of the data spacing factor can be considered. The strategy is to derive a subtractive adjustment to the local area defined by Equation 5'. The adjustments will make the local areas associated with the individual covariance ordinates precisely combine to form the overall domain of integration, except for long gaps per Equation 5'.

Thus, Equation 5' can be modified to

$$F(t_i,t_j) = \delta_i\delta_j - A_{ij}. \tag{5"}$$

The A_{ij} represent areas to be (in most cases) subtracted from the rectangular local area described by Equation 5', and Equation 6 applies.

The following diagnostic criteria are required:

P_2: If $t_{j(+)} \leq t_{i(-)} + \tau_m$, then P_2 is inside the domain of integration; otherwise P_2 is outside.

P_1: If $t_{j(-)} \leq t_{i(-)} + \tau_m$, then P_1 is inside the domain of integration; otherwise both P_1 and P_2 are outside.

P_4: If $t_{j(+)} \leq t_{i(+)} + \tau_m$, then P_4 is inside the domain of integration; otherwise both P_2 and P_4 are outside.

P_3: If $t_{j(-)} \leq t_{i(+)} + \tau_m$, then P_3 is inside the domain of integration; otherwise P_1, P_2, P_3 and P_4 are outside.

These criteria define the area cut off the local rectangle (Equation 5') by the diagonal line $t_j = t_i + \tau_m$. Note that a point on the boundary is defined to be inside the domain. It will be convenient to adopt the following notation:

d_{1j} = length cut off the rectangle in the i-direction along the side measured from point P_1,

d_{2j} = length cut off in j-direction, measured from point P_2,

etc.

Note that when $t_j - t_i > \tau_m$, the areas of these partial rectangles need not be computed. The areas of such rectangles do not contribute to the DQSE because their covariance ordinate location (t_i,t_j) is outside the domain of integration, where the lag window factor $D(\tau)$ is zero.

Case 1: P_2 is inside.

$A_{ij} = 0.$

Case 2: P_2 is outside. P_1, P_3, P_4 are inside.

Then $d_{2j} = t_j(+) - t_i(-) - \tau_m > 0$

$$A_{ij} = d_{2j}^2/2.$$

Case 3: P_1 and P_2 are outside. P_3 and P_4 are inside.

Then
$$d_{1i} = t_j(-) - t_i(-) - \tau_m > 0$$
$$d_{2j} = t_j(+) - t_j(-) > 0$$
$$d_{2i} = t_j(+) - t_i(-) - \tau_m > 0$$

$$A_{ij} = (d_{1i} + d_{2i})d_{2j}/2$$

Case 4: P_2 and P_4 are outside. P_1 and P_3 are inside.

Then
$$d_{2i} = t_i(+) - t_i(-) > 0$$
$$d_{2j} = t_j(+) - t_i(-) - \tau_m > 0$$
$$d_{4j} = t_j(+) - t_i(+) - \tau_m > 0$$

$$A_{ij} = (d_{2j} + d_{4j})d_{2i}/2$$

Case 5: P_1, P_2 and P_4 are outside. P_3 is inside.

Then
$$d_{2i} = t_i(+) - t_i(-) > 0$$
$$d_{2j} = t_j(+) - t_j(-) > 0$$
$$d_{1i} = t_j(-) - t_i(-) - \tau_m > 0$$
$$d_{4j} = t_j(+) - t_i(+) - \tau_m > 0$$

$$A_{ij} = d_{2i}d_{4j} + (d_{1i}+d_{2i})(d_{2j} - d_{4j})/2$$

Case 6: The period of observation is from $t = 0$ to $t = T$, by definition. If the period of observation begins before the first time point (ie, $t_1 > 0$), or if $t_n < T$, the local areas at the ends of the domain of integration are extended by Equation 5' to fill out the domain, but the magnitude of such end-corrections is never large and is bounded by Equation 5'.

Case 7: Bordering the main diagonal are triangular areas containing no covariance ordinate. The variance ordinates on the main diagonal itself, where $t_i=t_j$, are not used in the DQSE. However, the areas of these right isosceles triangles can be associated with the areas of the "above" and "right" adjacent local rectangular areas. For example, subject to the constraints of Equation 5', each such triangle can be represented by two right isosceles triangles by dropping a perpendicular from point P_4 of the rectangle "above", to the main diagonal of the domain of integration and a perpendicular from point P_1 of the "right" rectangle. In the absence of Equation 5' constraints these points P_4 and P_1 are coincident.

It is beyond the scope of this paper to optimize the strategy in using the refinements of Equations 5' and 5", and the geometrical Cases 1 through 7. Our practical experience to date confirms the high importance of the Equation 5' boundary. However, in using Equation 5", we believe that Cases 1 to 5 have secondary importance, presumably due to the small value of $D(\tau)$ near the τ_m boundary. In the examples reported in Section 12 we have used Equation 5", incorporating Cases 1 through 6. In computing the examples, we have not explicitly included the small-ω constraints described immediately following Equation 5', but these constraints would have little effect in the examples. With equally spaced data Equations 5, 5' and 5", are identical.

10. SPECTRAL WINDOW AND ROBUSTNESS PROPERTIES

We recommend that the spectral window be evaluated in parallel with estimation of the spectrum from Equation 2, Jones (1970, 1972). The actual spectral window has the shape

$$Z^*(\omega) = \frac{1}{T'} \sum_{i<j}^{n} \sum^{n} W_{ij}$$

$$= \frac{1}{T'} \sum_{i<j}^{n} \sum^{n} D(t_j - t_i) F(t_i, t_j) \cos 2\pi\, \omega(t_j - t_i). \tag{7}$$

For equally spaced data $Z^*(\omega)$ is a periodic function of ω. The sharp center-lobe peak is repeated at all even multiples of the Nyquist frequency. This is the property that gives rise to 100 percent aliasing of all frequencies beyond the Nyquist frequency. For unequally spaced data $Z^*(\omega)$ is an almost periodic function. Hence for unequally spaced data the actual spectral window may have important side lobes far from the center frequency but within the range of frequencies over which it is meaningful to compute the spectrum estimate.

The spectral window for unequally-spaced data has two moderate deficiencies compared to the window for equally-spaced data.

- The center-lobe is less sharp. This leads to somewhat higher variances of the spectral estimates, as mentioned in the introductory paragraphs of Section 9.
- The interpretation of estimated spectra must be done with a little extra care due to possible partial aliasing if important side lobes are present in the actual spectral window.

As illustrated by the examples to follow, neither of these often becomes a serious practical deficiency. In fact, these moderate deficiencies are more than compensated by the remarkable robustness properties of spectra estimated from unequally spaced data. Although only certain types of unequal spacing (cf, Section 5) are truly alias-free, spectra computed by the DQSE method from data with typical unequal spacing encountered in practice are extremely robust to isolated outlier data points. Undoubtedly this is due to the fact that aliasing is only partial (much less than 100 percent) even at side lobes that may occur.

In the important special case of equally spaced data with gaps, the DQSE spectra may be markedly affected by isolated outlier data points, due to near 100 percent aliasing. However, since the DQSE method readily accommodates gaps, it is easy to compute spectra after trimming the outlier data point(s). In some cases two or more degrees of trimming may be appropriate. Usually the estimated spectra will achieve stability after deletion of only the most extreme outlier observation(s), and will be stable (robust) to further modest trimming.

11. PROPERTIES OF DIRECT QUADRATIC SPECTRUM ESTIMATION

The DQSE method has the following advantages:

- It is versatile in handling unequally spaced or missing data.
- It makes maximum use of all available data.
- It is computationally stable.
- Many theoretical properties are known.

- It is easy to program on a digital computer.
- The estimates can be robust to isolated outlier observations.
- The DQSE is capable of much finer frequency resolution than classical methods (Blackman and Tukey 1959), and because the DQSE can be evaluated at arbitrary small intervals of ω, it sometimes can be demonstrated that a region of high spectral density really resolves into two or more separate peaks.

There is one disadvantage:

- It requires much more computing time than conventional methods that apply only to equally spaced data.

With today's fast computers (the examples described in this paper were computed on a Univac 1100/83) analysis of unequally spaced time series with n ~4000 is routinely practical using direct quadratic spectrum estimation.

There are steps that can be taken to reduce the computing time. A feature of this method is that the computation for any value of ω is independent of all other values of ω.

- The computing time can be reduced by making τ_m a decreasing function of ω. For example, adequate precision may be obtained if τ_m covers, say, 100 cycles for a given frequency ω.

  This is particularly important when the spectrum must be computed over several decades of ω. Multiple decades of ω are much more feasible with unequally spaced data than equally spaced data.

- Computing time can be reduced by computing only at values of ω that are of known or potential interest.

12. EXAMPLES

Example 1. (Jitter Spaced Data)

The data consist of a sine wave plus independent random normal error. Characteristics of the data are:

Sine Wave: Amplitude = 1
Period - 12

Noise: Standard Deviation = 0.20 (Variance = 16 percent of signal)

Number of Observations = 2000

Data Spacing: Target times equally spaced, $\Delta t = 1$, but actual times vary about targets according to a uniform distribution on $[-\frac{\Delta t}{2}, \frac{\Delta t}{2}]$.

The direct quadratic spectrum estimate was computed from Equation 5" using $\tau_m = 120$, and $\Delta\omega = 1/96$ from $\omega = 0$ to 0.5. The upper end of the range corresponds to the nominal Nyquist frequency for the equally spaced target locations. The power density spectrum estimate is shown in Figure 4. The period of length 12, corresponding to a frequency of 1/12 = 0.083 is revealed by the sharp peak.

The spectral window is also shown in Figure 4. The spectral window is normalized to unit height at $\omega = 0$. The window is amply sharp, having negligible values beyond one multiple of the selected $\Delta\omega$.

DQSE FOR EXAMPLE 1 (JITTER SPACED DATA)

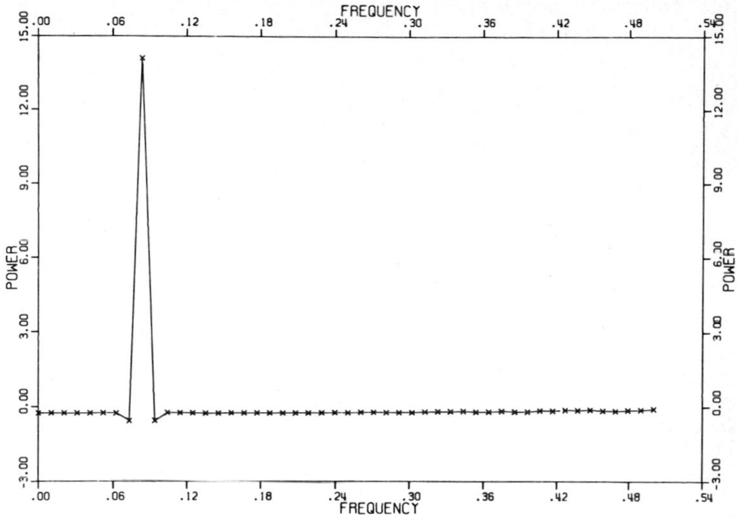

SPECTRAL WINDOW FOR EXAMPLE 1

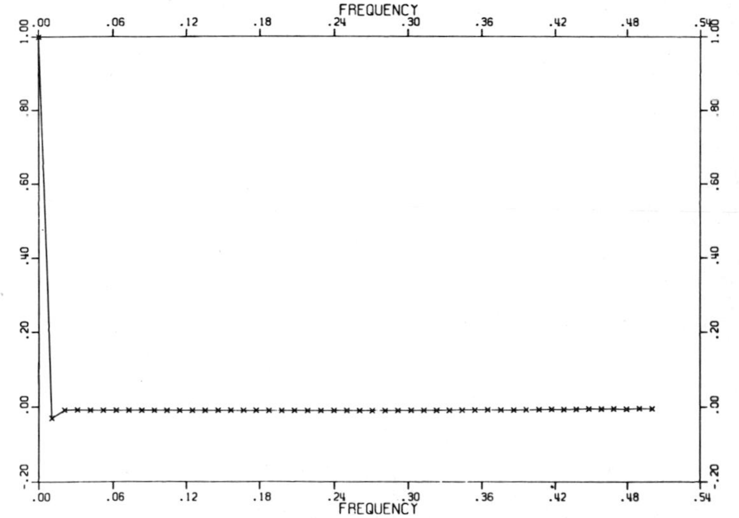

Figure 4
Example 1 (Jitter Spaced Data)

Further numerical checks on the computing procedure, and the functions of Equations 5' and 6, were accomplished by adding a second sine wave near the upper end of the spectrum (Period = 2.2 and Amplitude = 1), and by computing DQSE results, both with and without long gaps in the data. The results, not shown here, confirm the ability to sharply display both cycles, with comparable peak heights/areas.

Example 2. (Halberg Oral Temperature Data)

The data are 2813 self-observations of oral temperature taken by an adult woman about 4 times daily on most days during the period February, 1965 to April, 1969. The raw data are plotted in Figure 5. As seen in the Figure, there are some long gaps in the data; there is one brief period of elevated temperature in February, 1966, and one brief period of depressed temperature in July, 1967. The original data tape covers a six year period, but the present data were selected as a convenient length of data for this example. The direct quadratic spectrum estimate is shown in Figure 6. The spectrum is computed over the frequency range $\omega = 0$ to 1.5 cycles per day, at frequency intervals of 1/60 cycles per day. The upper end of this frequency range lies beyond the Nyquist frequency associated with the average interval, but the average interval is much larger than the typical interval, due to the gaps in the data. There is, in fact, an ample proportion of intervals shorter than 1/3 day, so that reliable spectrum estimates can be computed up to 3/2 cycles per day. For this calculation $\tau_m = 30$ was specified. This rather small value of τ_m resulted in relatively short computing times.

The power density spectrum in Figure 6 shows a peak centered at a cycle length just below 30 days. A second peak is centered at a cycle length of one day. These features are consistent with medical expectations. Cycles of approximately once per day are found in almost all chronobiological responses; these are called circadian cycles, cf Halberg, et. al. (1978, 1979). We conclude that the selected value of τ_m is adequate. Interestingly, the brief episodes of elevated and depressed temperature have not impaired the ability of the direct quadratic spectrum estimate to give good results. In fact, this spectrum was recomputed omitting these deviant observations, and the numerical results (not shown here) differed only in the second or third decimal place.

Figure 6 shows the spectral window for this example. Of course, a larger or smaller value of τ_m would affect the shape of the spectral window. The spectral window has a side lobe whose height is two percent of the center lobe, and located 60 multiples of the selected $\Delta\omega$ away from the center. This illustrates the comment made in Section 10. In this spectrum there is essentially no power at $\omega = 0$, hence the peak at $\omega = 1$ cannot be interpreted as leakage from a peak at $\omega = 0$ due to the remote side lobe in the spectral window.

Example 3. (Halberg Systolic Blood Pressure)

The data are 2696 observations of systolic blood pressure. The data for examples 2 and 3 comprised 2840 observation times total, but not all responses were measured at every observation time. The raw systolic data are shown in Figure 7. The direct quadratic spectrum estimate is shown in Figure 8. In this spectrum there is a large amount of variance in the vicinity of $\omega = 0$, suggesting that a long trend, or a cycle longer than 60 days may be present. Indeed, examination of Figure 7 shows an overall decrease in systolic pressure.

The spectrum in Figure 8 also shows a strong peak at one cycle per day, which is consistent with the widespread occurrence of circadian cycles. However, the spectral window, Figure 8, shows a side lobe whose height is two percent of the main lobe that will affect estimated power at one cycle per day for power actually present at $\omega = 0$. Consequently, the peak at $\omega = 1$ may be due in part to power from $\omega = 0$. However, the $\omega = 1$ peak is so strong that it must be due

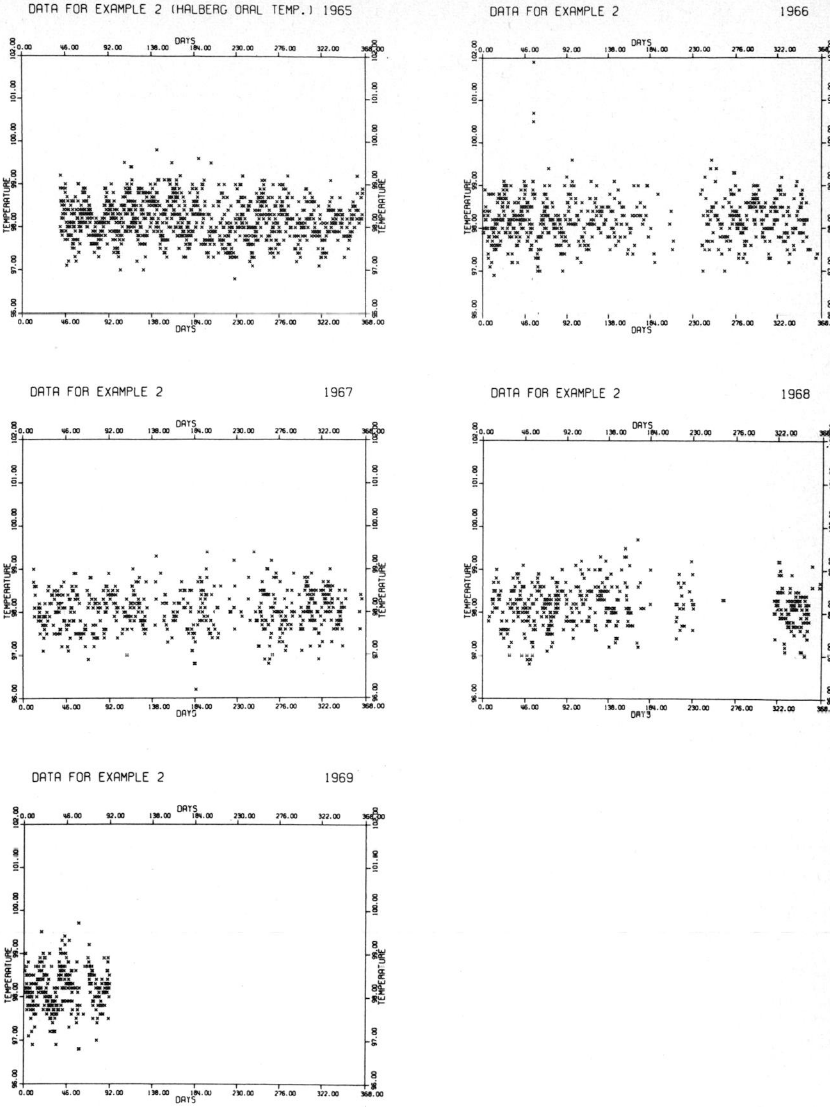

Figure 5
Example 2 (Halberg Oral Temperature)

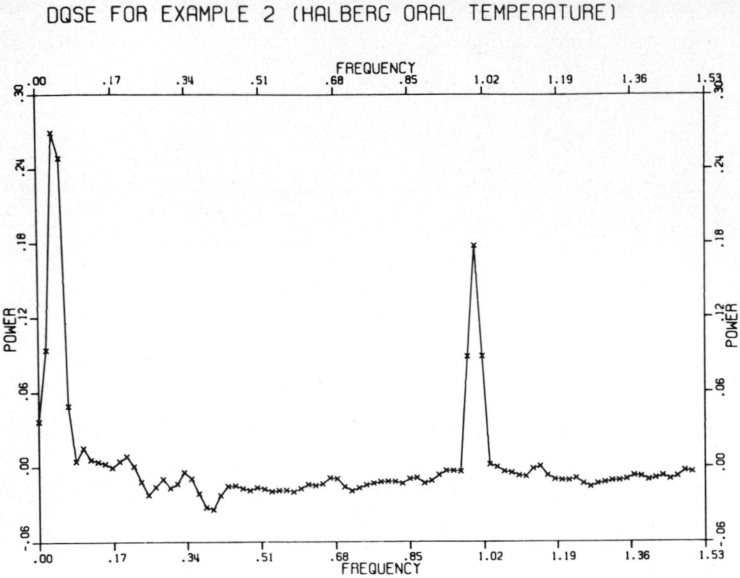

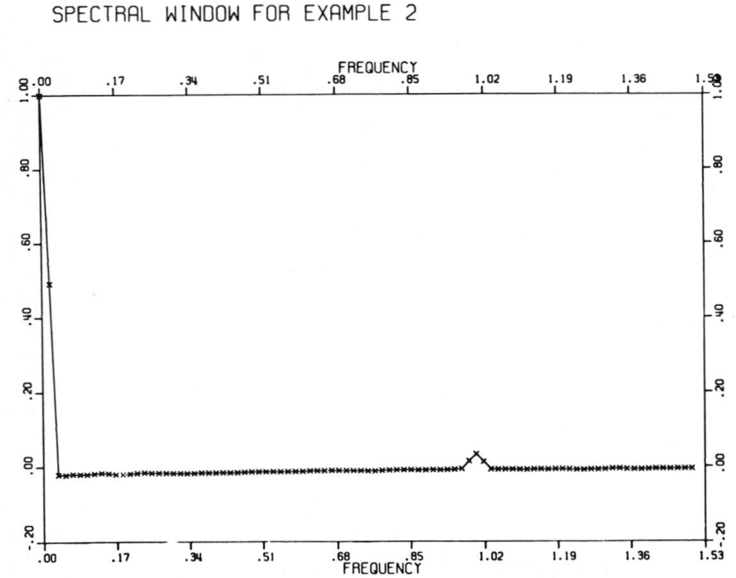

Figure 6
Example 2 (Halberg Oral Temperature)

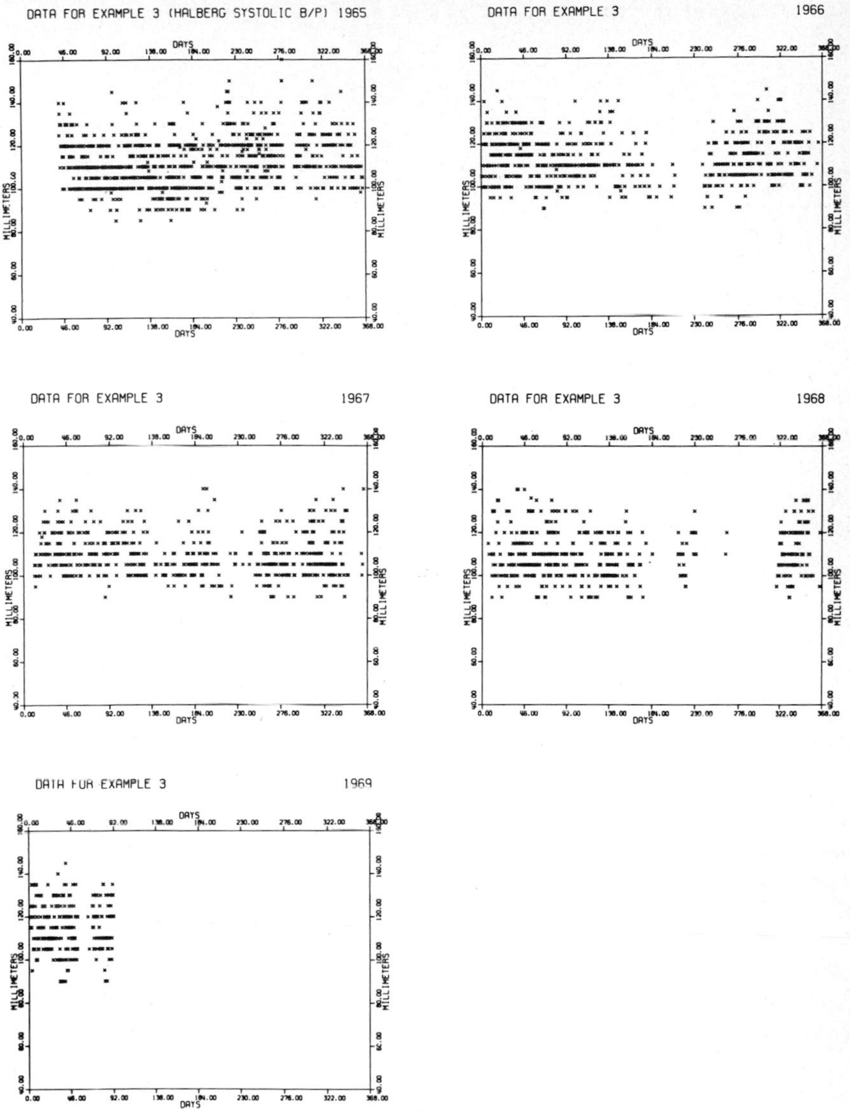

Figure 7
Example 3 (Halberg Systolic Blood Pressure)

DQSE FOR EXAMPLE 3 (HALBERG SYSTOLIC BLOOD PRESSURE)

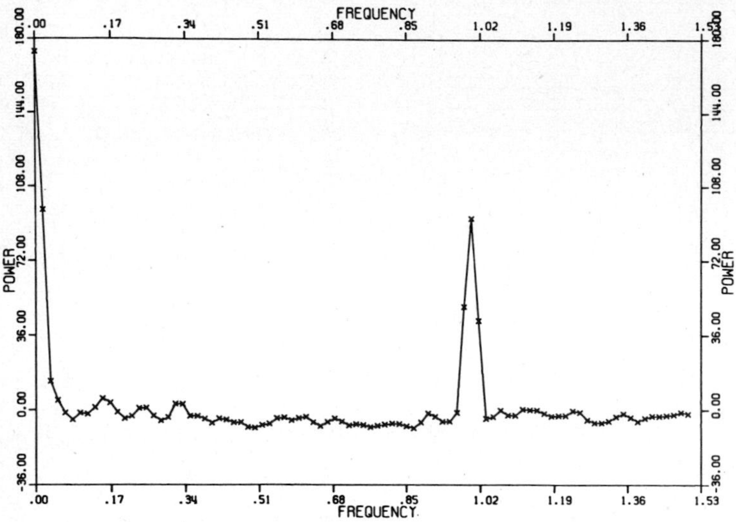

SPECTRAL WINDOW FOR EXAMPLE 3

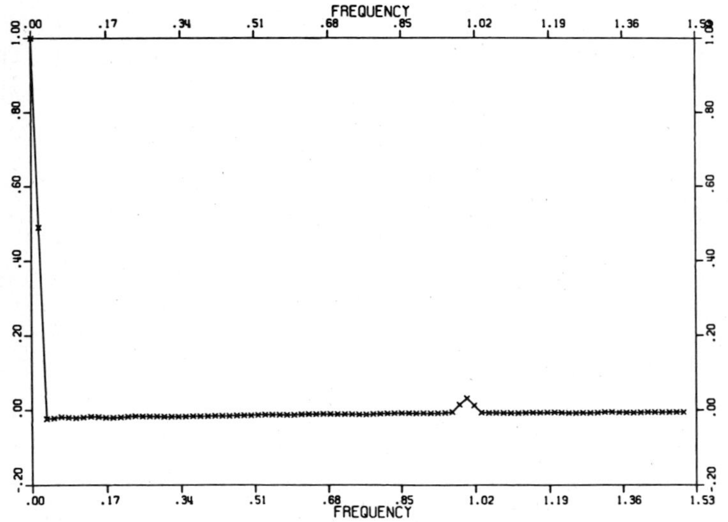

Figure 8
Example 3 (Halberg Systolic Blood Pressure)

primarily to a circadian cycle, even on the evidence of this spectrum. This aliasing question can be investigated further by detrending the raw data to remove actual power near $\omega = 0$, or by changing the value of τ_m to change the shape of the spectral window, or both.

Example 4. (Stratospheric Ozone - Resolute Data - Monthly Averages)

The data for this example are 249 monthly averages of stratospheric ozone during the 270 months from July, 1957 through December, 1979, measured by a Dobson spectrophotometer at Resolute, Canada (74.41°N, 94.54°W). This station is one of about 40 world-wide making regular measurements. Because of the high northern latitude of this station, meteorological conditions do not permit measurements every day, and as indicated 21 months have no data. Conditions are particularly unfavorable during some months, eg, October. Even so, ozone measurements at extreme northern latitudes are of interest because the cyclic behavior is more pronounced, and such data are important contributing information for assessing global trends and their causes.

The raw data are shown in Figure 9. An annual cycle is visually evident. The direct quadratic spectrum estimate is shown in Figure 10. Due to the small number of data points in this series, we set $\tau_m = 120$, about half the total number of data points. The annual cycle is well exhibited, but little additional structure can be diagnosed. The spectral window, Figure 10, is amply sharp and is negligible outside the main lobe.

These data have significant limitations. The meteorological conditions that cause missing days result in wide variations of the numbers of daily observations included in the monthly averages. This limitation could be handled easily within the direct quadratic spectrum estimation method by placing another factor in the quadratic form weights, Equation 3, to account for the unequal variances of the monthly average data. We do not pursue this further in this paper. A second limitation is the possibility of significant cycles shorter than one month that might not be uniformly averaged out in the equally spaced monthly averages obtained from the unequally spaced daily data. A third limitation is the loss of time structure resolution due to the small number of monthly averages available. All of these limitations suggest that analysis of the unequally spaced daily observations might be worthwhile.

Example 5. (Stratospheric Ozone - Resolute Data - Daily Observations)

The data are the 4154 daily observations made during most of the years, 1960 through 1979, for which monthly averages were available, as used in Example 4. The data are shown in Figure 11. The direct quadratic spectrum estimate is shown in Figure 12, for the frequency range $\omega = 0$ through $\omega = 1/6$ at intervals of 1/720. The annual cycle at $\omega = 1/365.25 = 0.003$ is clearly displayed. We used $\tau_m = 400$.

The spectral window is shown in Figure 12. The window has a negligible side lobe (about one percent of main lobe amplitude as read from the tabular computer output) at 0.036 frequency units distance from the center lobe. The effect of this small side lobe can be seen in the spectrum, as a negligible peak about one percent of the annual peak amplitude, located at $\omega = 0.003 + 0.036 = 0.039$.

The spectrum also shows two other negligible peaks, at $\omega = 0.007$ and $\omega = 0.030$. These have amplitudes about four percent and one percent of the annual peak, respectively. The practical result of this analysis is confirmation of the validity of using monthly averages for detecting stratospheric ozone trends, because there is negligible cyclic structure at cycle lengths shorter than one year. Not shown here are DQSE computations on these same data that show negligible cyclic structure from $\omega = 1/6$ through $\omega = 1/2$.

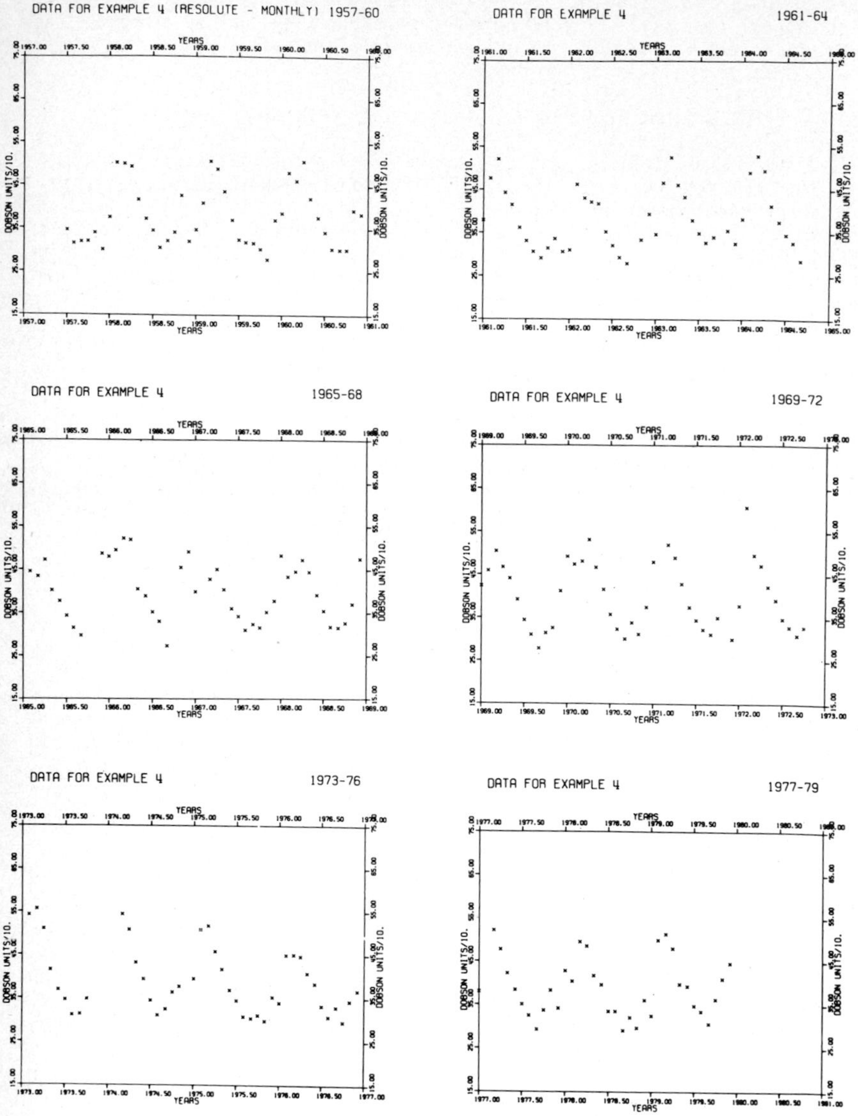

Figure 9
Example 4 (Stratospheric Ozone - Resolute Data - Monthly Averages)

DQSE FOR EXAMPLE 4 (OZONE - RESOLUTE - MONTHLY)

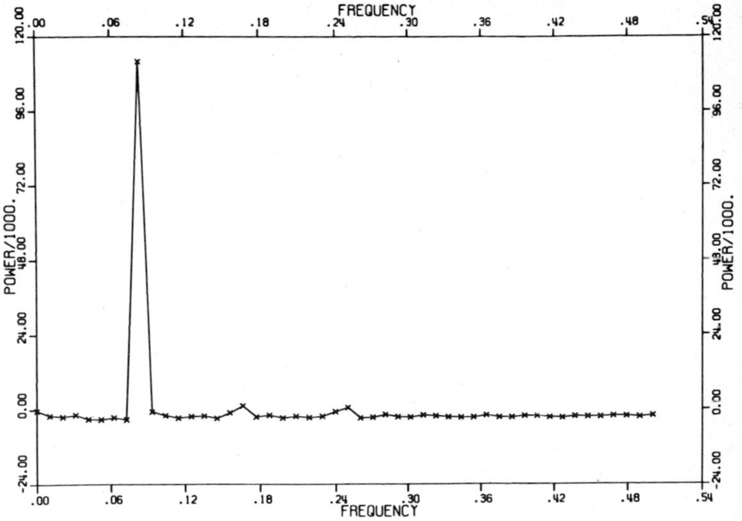

SPECTRAL WINDOW FOR EXAMPLE 4

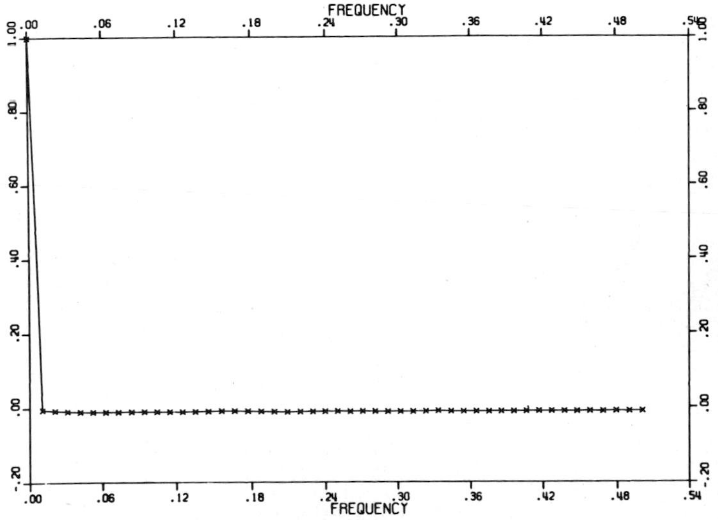

Figure 10
Example 4 (Stratospheric Ozone - Resolute Data - Monthly Averages)

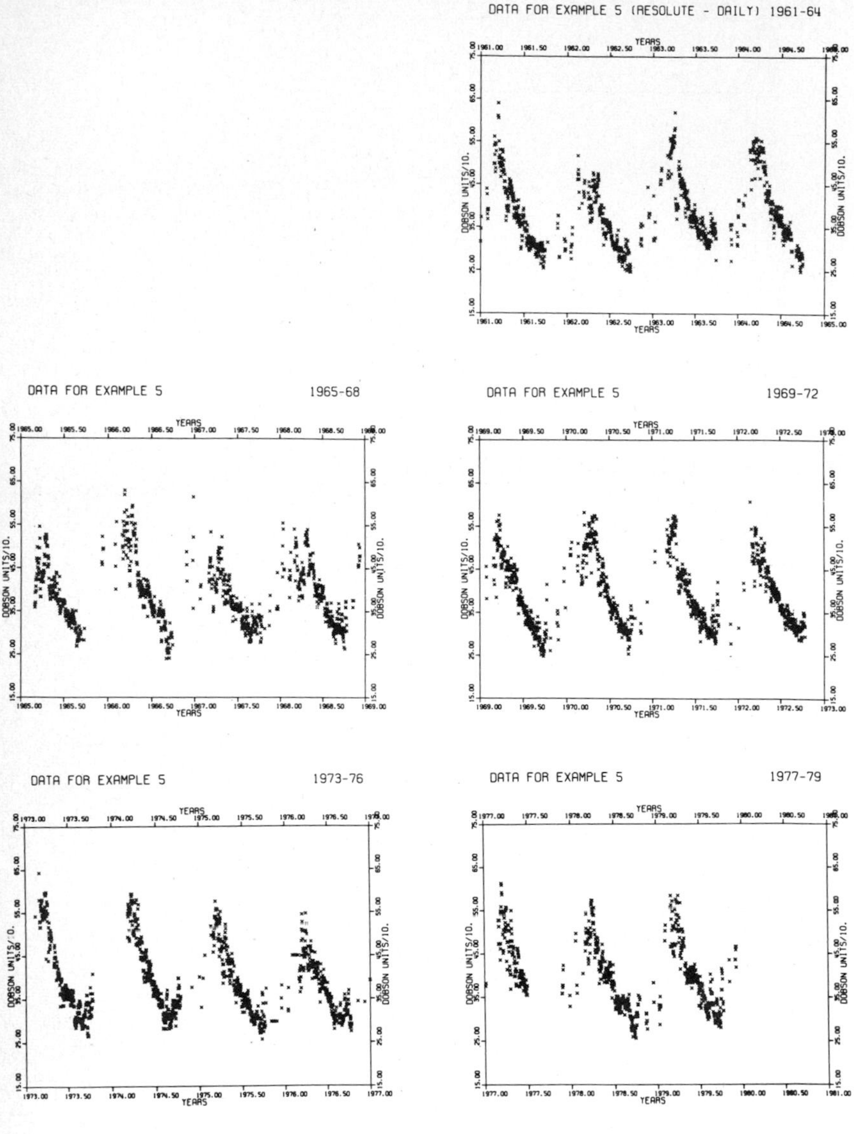

Figure 11
Example 5 (Stratospheric Ozone – Resolute Data – Daily Averages)

DQSE FOR EXAMPLE 5 (OZONE - RESOLUTE - DAILY)

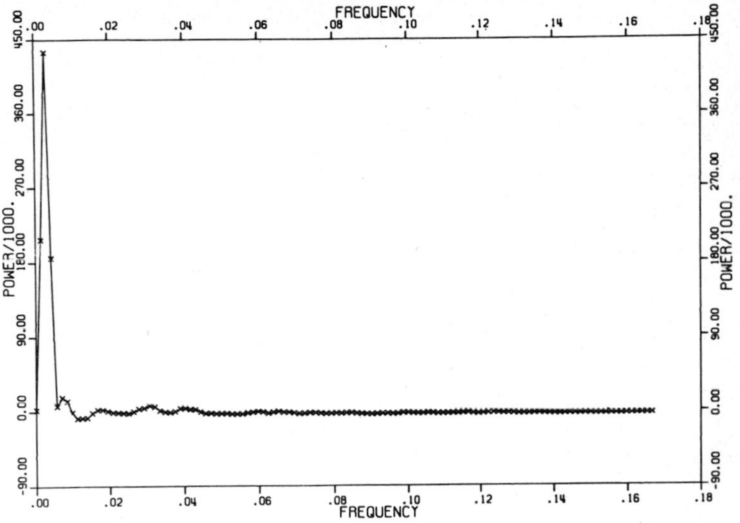

SPECTRAL WINDOW FOR EXAMPLE 5

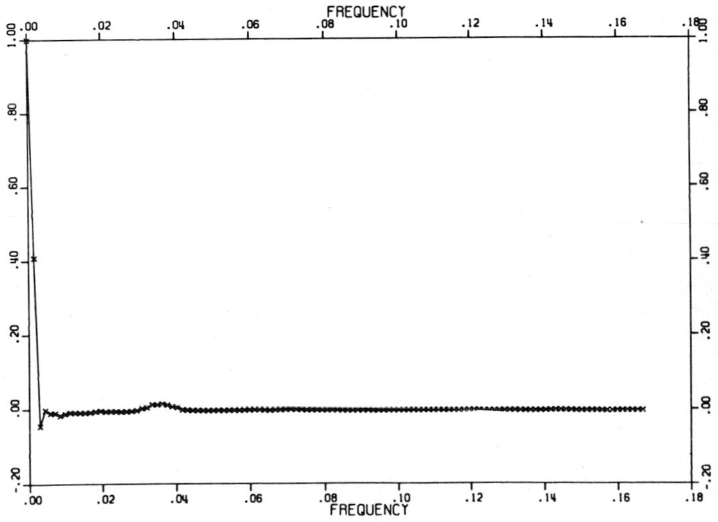

Figure 12
Example 5 (Stratospheric Ozone - Resolute Data - Daily Averages)

13. PRESENT SUMMARY AND FUTURE POSSIBILITIES

The DQSE method, in the form presented here, can be used with confidence to display and diagnose the cyclic content of unequally spaced data.

The remainder of this section summarizes some prospective future enhancements.

- Data having nonconstant variance can be handled as described in connection with Example 4.

- Confidence limits for the spectrum can be computed by utilizing the quadratic form structure of the DQSE.

- The spectral windows computed for the examples in this paper are slightly negative except at the central lobe. This slight negative bias, almost constant at all values of ω, is reflected in the spectrum estimates. For the practical purposes of these examples, the bias is negligible, being typically less than one percent of the height of the center-lobe of the window. We conjecture that the negative bias arises from not including correction for the Case 7 geometry discussed earlier. Recall that:

  1. the local areas involved in Case 7 border the main diagonal of the domain of integration,

  2. the variance associated with points precisely on the diagonal corresponds to a positive bias, constant at all frequencies on the spectrum, and

  3. the lag window has large values near the main diagonal.

  Development of a suitable correction for the Case 7 geometry would be a useful enhancement.

- The DQSE could be used on an on-line basis by developing suitable procedures for adding new data and dropping old data.

- The DQSE formalism can be generalized to cross-spectral analysis by associating the subscript j with a second data series, and by considering both positive and negative lags.

ACKNOWLEDGEMENTS

We are grateful to Professor Franz Halberg, Director of Chronobiology Laboratories, University of Minnesota, who emphasized to D. W. Marquardt the critical need for practical methods for analyzing time series data with unequal spacing, and who subsequently supplied to us the unique chronobiological data. We are grateful to Dr. William Hill, Allied Chemical Corporation, and Dr. J. M. Minor, Du Pont Company, for providing the stratospheric ozone data. D. W. Marquardt is pleased to acknowledge helpful correspondence from a number of authors shown in the Bibliography, especially those who supplied preprints of papers not yet published. Finally, we are grateful to Michael S. Saccucci for assistance in preparing the examples and other aspects of this paper.

BIBLIOGRAPHY

Box, G.E.P. and G. M. Jenkins (1970), "Time-Series Analysis, Forecasting and Control", Holden-Day, San Francisco, CA, 2nd Ed. (1976).

Blackman, R. B. and J. W. Tukey (1959), "The Measurement of Power Spectra", Dover Publications, New York.

Bloomfield, P. (1976), "Fourier Analysis of Time Series: An Introduction", John Wiley and Sons, New York.

Cox, D. R. and M. W. Townsend (1951), "The Use of Correlograms for Measuring Yarn Irregularity", J. Textile Institute, 42, pp. 145-151.

Dunsmuir, W. (1981), "Estimation for Stationary Time Series when Data are Irregularly Spaced or Missing", Applied Time Series Analysis II, Academic Press, New York and London, pp. 609-649.

Dunsmuir, W. and P. M. Robinson (1979), "Asymptotic Theory for Time Series Containing Missing and Amplitude Modulated Observations", Technical Report No. 4, Dept. of Math., Mass. Instit. Technol., Cambridge, MA.

Dunsmuir, W. and P. M. Robinson (1980), "Estimation of Time Series Models in the Presence of Missing Data", Technical Report No. 9, Dept. of Math., Mass. Instit. Technol., Cambridge, MA.

Dunsmuir, W. and P. M. Robinson (1981), "Parametric Estimators for Stationary Time Series with Missing Observations", Advanced Applied Probability, 13. To appear.

Gaster, M. and J. B. Roberts (1975), "Spectral Analysis of Randomly Sampled Signals", J. Inst. Maths. Applics., 15, pp. 195-216.

Gaster, M. and J. B. Roberts (1977), "The Spectral Analysis of Randomly Sampled Records by a Direct Transform", Proc. Roy. Soc. London A, 354, pp. 27-58.

Glasbey, C. A. (1980), "Modeling Time Series Observed at Irregular Intervals", (Personal Communication), ARC Unit of Statistics, University of Edinburgh, Scotland.

Gray, H. L., G. D. Kelley, and D. D. McIntire (1978), "A New Approach to ARMA Modeling", Commun. Statist. - Simula. Computa., B7, pp. 1-96 (with discussion).

Halberg, E., F. Halberg, E. Haus, G. Cornelissen, L.A. Wallach, M. Smolensby, M. Garcia-Sainz, H. W. Simpson, and A. R. Shons (1978), "Toward a Chronopsy: Part I. A Chronobiologic Case Report and a Thermopsy Complementing the Biopsy", Chronobiologia, 5, pp. 241-249.

Halberg, E., et al. (1979)(ibid) "Part II. A Thermopsy Revealing Asymmetrical Circadian Variation in Surface Temperature of Human Female Breasts and Related Studies", Chronobiologia, 6, pp. 231-257.

Jenkins, G. M. and D. G. Watts (1968), "Spectral Analysis and Its Applications", Holden-Day, San Francisco, CA.

Jones, R. H. (1962a), "Spectral Estimates and Their Distributions", Skandinavisk Aktuarietidskrift, Part I and Part II, pp. 39-153.

Jones, R. H. (1962b), "Spectral Analysis with Regularly Missed Observations", Ann. Math. Statist., 33, pp. 455-461.

Jones, R. H. (1970), "Spectrum Estimation with Unequally Spaced Observations", Proc. Kyoto Int'l Conf. on Circuit and Syst. Theory, Instit. Electron. and Commun. Engrs. of Japan, pp. 253-254.

Jones, R. H. (1971), "Aliasing with Unequally Spaced Observations", J. Appl. Meteorol., 11, pp. 245-254.

Jones, R. H. (1971), "Spectrum Estimation with Missing Observations", Ann. Instit. Statist. Math., 23, pp. 387-398.

Jones, R. H. (1975), "Estimation of Spatial Wavenumber Spectra and Falloff Rate with Unequally Spaced Observations", J. Atmos. Sci., 32, pp. 260-268.

Jones, R. H. (1977), "Spectrum Estimation from Unequally Spaced Data", Fifth Conference on Probability and Statistics, Amer. Meterol. Soc., Boston, Preprint Volume, pp. 277-282.

Jones, R. H. (1979), "Fitting Rational Spectra with Unequally Spaced Data", Sixth Conf. on Probability and Statistics in Atmospheric Sciences, Amer. Meteorol. Soc., Boston, Preprint Volume, pp. 231-234.

Jones, R. H. (1980), "Maximum Likelihood Fitting of ARMA Models to Time Series with Missing Observations", Technometrics, 22, pp. 389-395.

Jones, R. H. (1981), "Fitting a Continuous Time Autoregression to Discrete Data", Applied Time Series Analysis II, Academic Press, New York and London, pp. 651-682.

Marquardt, D. W. and S. K. Acuff (1979), "Analysis of Unequally Spaced Time Series Using the Variance-Length Curve", Presented at American Statistical Association Annual Meeting, August, Washington, DC.

Marquardt, D. W. and S. K. Acuff (1980), "Time Series Analysis of Unequally Spaced Data", Unpublished Manuscript, July 21, 1980. (Basis of presentation at Gordon Conference on Statistics in Chemistry and Chemical Engineering.)

Masry, E. and M. C. Lui (1975), "A Consistent Estimate of the Spectrum by Random Sampling of the Time Series", SIAM J. Appl. Math., 28, pp. 793-810.

Masry, E. and M. C. Lui (1976), "Discrete-Time Spectral Estimation of Continuous Parameter Processes - A New Consistent Estimate", IEEE Trans. Information Theory, IT-22, pp. 298-312.

Meketon, M. S. (1980), "The Variance Time Curve: Theory, Estimation and Applications", PhD Dissertation, Cornell University, University Microfilms, Order No. 8102964.

Parzen, E. (1963), "On Spectral Analysis with Missing Observations and Amplitude Modulation", Sankhya, Series A, 25, pp. 383-392.

Owens, A. J. (1977), "The Nested Variance Power Spectrum", Journal of Geophysical Research, 82, pp. 3315-3318.

Robinson, P. M. (1977), "Estimation of a Time Series Model from Unequally Spaced Data", Stochastic Processes and Their Applications, 6, pp. 9-24.

Shapiro, H. S. and R. A. Silverman (1960), "Alias-Free Sampling of Random Noise", J. Soc. Indust. Appl. Math., 8, pp. 225-248.

Thrall, A. D. (1979), "Spectral Estimation for a Randomly Sampled Time Series", PhD Dissertation, University of California, Berkeley, University Microfilms, Order No. 8000544.

Townsend, M.W.H. and D. R. Cox (1951), "The Analysis of Yarn Irregularity", J. Textile Institute, 42, pp. 107-113.

Yuen, C. K. (1978), "Quadratic Windowing in the Segment Averaging Method for Power Spectrum Computation", Technometrics, 20, pp. 195-200.

Yule, G. U. (1945), "On a Method of Studying Time-Series Based on Their Internal Correlations", J. Roy. Statist. Soc., 108, pp. 208-225. With appended note by M. G. Kendall, pp. 226-230.

APPLIED TIME SERIES ANALYSIS
O.D. Anderson & M.R. Perryman (eds.)

THE LIKELIHOOD FUNCTION OF A TIME-DEPENDENT ARMA MODEL

Guy Mélard

Institut de Statistique et Centre d'Economie
mathématique et d'Econométrie
Campus Plaine U.L.B. - C.P. 210
Boulevard du Triomphe
B-1050 Bruxelles
BELGIUM

An algorithm is given for the evaluation of the exact likelihood function of a mixed autoregressive-moving average process of order (p,q), with time-dependent coefficients, including a time-dependent innovation variance. The algorithm, a faster version of a method briefly described in a previous paper (Mélard and Kiehm (1981)), is convenient for computers which can perform short loops of instructions very quickly. Some results on computing time performance are given when the coefficients are either constant, or linear or exponential functions of time. The time is, at most, proportional to n, the series length. The amount of storage necessary for a computation, seemingly equivalent to the inversion of an nxn matrix, is reduced to an R x (p+q+2R+4) matrix, where $R = \max(p,q) + 1$.

1. INTRODUCTION

A few years ago, computing the exact likelihood function of an autoregressive-moving average (ARMA) model seemed to be a very hard numerical problem. The likelihood function is

$$L(\alpha, \sigma^2; w) = (2\pi)^{-n/2} (\det \Gamma_w)^{-1/2} \exp\left\{-\frac{1}{2} w^T \Gamma_w^{-1} w\right\}, \qquad (1)$$

where $w^T = (w_1, \ldots, w_n)$ is the vector of n observations of the ARMA(p,q) model

$$w_t = \phi_1 w_{t-1} + \ldots + \phi_p w_{t-p} + a_t - \theta_1 a_{t-1} - \ldots - \theta_q a_{t-q}, \qquad (2)$$

Γ_w is the covariance matrix of w, σ^2 is the variance of the innovations a_t, assumed to be normally and independently distributed with mean zero, and $\alpha^T = (\alpha_1, \ldots, \alpha_h)$ is the set of parameters included in the autoregressive coefficients $\phi_1, \ldots, \phi_p$, and in the moving average coefficients $\theta_1, \ldots, \theta_q$. The direct evaluation of (1) requires a number of operations which is $O(n^3)$ and about n^2 memory cells for storing the symmetric matrix Γ_w and its inverse. By operation we will always mean multiplications and divisions. Several algorithms have been proposed in order to solve this problem in a more efficient way, i.e. with O(n) operations.

Our main purpose is to extend one of these algorithms to ARMA models with time-dependent coefficients. Recursive methods have been proposed, e.g. Harrison and Stevens (1976), Ledolter (1979), if the coefficients of the model are assumed to change stochastically. On the contrary, if the coefficients are deterministic functions of time, the estimation approach of Box and Jenkins (1976) used in the constant case can be extended, see Mélard and Kiehm (1981). The model is written

$$w_t = \phi_{t,1} w_{t-1} + \ldots + \phi_{t,p} w_{t-p} + g_t a_t - \theta_{t,1}(g_{t-1} a_{t-1}) - \ldots - \theta_{t,q}(g_{t-q} a_{t-q}), \qquad (3)$$

where $g_t a_t$ is the innovation at time t, and where $\phi_{t,i}(i = 1, \ldots, p)$, $\theta_{t,j}$ $(j = 1, \ldots, q)$ and g_t are deterministic functions of t, which may depend on parameters.
The exact likelihood function is still given by (1). The conditional likelihood function, given $a_*^T = (a_{1-q}, \ldots, a_o)$ and $w_*^T = (w_{1-p}, \ldots, w_o)$ is

$$L(\alpha, \sigma^2; w|a_*, w_*) = (2\pi)^{-n/2} \sigma^{-n} (\prod_{t=1}^{n} g_t)^{-1} \exp\left\{-\frac{1}{2\sigma^2} a^T a\right\}, \tag{4}$$

where the elements of $a^T = (a_1, \ldots, a_n)$ are obtained recursively from (3). Normalizing g_t so that their geometric mean for t = 1, ..., n is equal to 1 seems natural. Then, maximizing the conditional likelihood is equivalent to minimizing the sum of squares $a^T a$.

The unconditional method of estimation, Box and Jenkins (1976), is not so easily generalized. Let us paraphrase Box and Jenkins's derivation for the MA(q) model, i.e. (3) with p = 0. It can be written $a = Lw + Pa_*$ where L is a nxn lower triangular matrix and P is a nxq matrix. Then (1) can be put under the form

$$L(\alpha, \sigma^2; w) = (2\pi)^{-n/2} \sigma^{-n} (\det B^T B)^{-1/2} \exp\left\{-\frac{1}{2\sigma^2} (\hat{a}_*^T \hat{a}_* + \hat{a}^T \hat{a})\right\} \tag{5}$$

where $B^T = (I_q, P^T)$, $\hat{a} = Lw + P\hat{a}_*$ and

$$\hat{a}_* = -(B^T B)^{-1} B^T Aw. \tag{6}$$

Similarly, the results of Galbraith and Galbraith (1974) and Newbold (1974) can be partly extended to ARMA models with time-dependent coefficients. The number of operations is $O(n^2)$. In the approximate method proposed by Box and Jenkins, which is O(n), the factor $(\det B^T B)^{-1/2}$ is ignored and $\hat{a}_*$ is computed by a backforecasting procedure. This procedure can be iterated, if necessary, in order to approach the true value. For time-dependent processes, the problem has already been discussed by Mélard and Kiehm (1981). Kiehm, Lefèvre and Mélard (1980) have shown, in the case of the MA(1) model with constant θ_{t1} but variable g_t, that the indefinitely iterated backforecasting procedure leads to a value for $a_* = (a_o)$ equal to

$$-\frac{1}{g_o} \sum_{t=1}^{n} \frac{v_t(\theta_1)}{v_o(\theta_1)} \theta_1^t w_t \tag{7}$$

with

$$v_t(\theta_1) = 1 - \theta_1^{2n+2-2t},$$

instead of the result deduced from (6), which is (7) together with

$$v_t(\theta_1) = \sum_{i=0}^{n-t} \theta^{2i} g_{i+t}^{-2}$$

This negative property gives more importance to the calculation of the exact likelihood function.

After a discussion about a possible extension to time-dependent models of existing methods for constant ARMA models (Section 2), we state the basic recursions for the covariance function of ARMA models with time-dependent coefficients (Section 3). We give a time-series approach of the algorithm of Ansley (Section 4). Then we are able to describe the basic algorithm and the complete algorithms for constant and time-dependent ARMA models (Sections 5, 6 and 7). We conclude this study by practical measures of performance of our algorithms.

2. SOME OTHER METHODS

Among the fastest methods proposed for constant ARMA models, the method published by Ljung and Box (1979) makes use of a direct generalization of (5) since the first step is to compute $\hat{a}*$ and $\hat{w}*$, the starting values required by (2). However, all the arguments are not applicable if the coefficients vary with time.

The algorithm of Gardner, Harvey and Phillips (1980) is based on a state space representation of the ARMA process. Recursions are given by the Kalman filter. Of course, they are still valid in the case of a time-dependent state space representation defined as following. Let $r = \max(p,q+1)$ and the rx1 state vector x_t, at time t, such that

$$(x_t)_k = \phi_{t+k-1,k}\, w_{t-1} - \theta_{t+k-1,k-1}(g_t a_t) + (x_{t-1})_{k+1}\,, \tag{8}$$

for $k = 1, \ldots, r$, where $\phi_{t,k} = 0\ (k > p)$, $\theta_{t,0} = -1$, $\theta_{t,k-1} = 0\ (k > q)$ and $(x_t)_{r+1}$ is set equal to zero. The state space representation is

$$\begin{cases} w_t = U^T x_t \\ x_t = F_t\, x_{t-1} + G_t a_t \end{cases} \tag{9}$$

where $U^T = (1, 0, \ldots, 0)$ and

$$F_t = \left[\begin{array}{c|c} \begin{matrix}\phi_{t,1} \\ \phi_{t+1,2}\end{matrix} & I_{r-1} \\ \hline \begin{matrix}\vdots \\ \phi_{t+r-1,r}\end{matrix} & 0^T_{r-1} \end{array}\right], \quad G_t = \begin{bmatrix} 1 \\ -\theta_{t+1,1} \\ \vdots \\ -\theta_{t+r-1,r-1} \end{bmatrix}. \tag{10}$$

Two more efficient algorithms have been proposed by Pearlman (1980). The Kalman filter recursions are replaced by a set of recursions which make use of three 1xr vectors instead of two 1xm vectors for the algorithm of Gardner, Harvey and Phillips, where $m = r(r+1)/2$. Except for small values of p and q, it is much faster than any other algorithm. However, it is argued in the Appendix that the alternate set of recursions cannot be extended to the case where F_t and G_t depend on time.

The algorithm published by Ansley (1979) is not so efficient but it offers the degree of generality we are asking for. The fundamental device discovered by Ansley is a transformation from w to the nx1 vector x* defined as following

$$x^*_t = \begin{cases} w_t & (t = 1, \ldots, r^*) \\ y_t & (t = r^* + 1, \ldots, n), \end{cases}$$

where $r^* = \max(p,q)$ and

$$y_t = w_t - \phi_{t,1}\, w_{t-1} - \cdots - \phi_{t,p}\, w_{t-p}. \tag{11}$$

We have preferred to use a variant in which the nx1 vector x is defined by

$$x_t = \begin{cases} w_t & (t = 1, \ldots, p) \\ y_t & (t = p+1, \ldots, n). \end{cases} \tag{12}$$

3. BASIC RECURSIONS FOR TIME-DEPENDENT ARMA MODELS

The transformation from w to x (or x*) has unit Jacobian so that (1) can be expressed as

$$L(\alpha, \sigma^2; w) = (2\pi)^{-1/2} (\det \Gamma_x)^{-1/2} \exp\left\{-\tfrac{1}{2} x^T \Gamma_x^{-1} x\right\},$$

where the (t,s)th element of the nxn matrix Γ_x is $\lambda_{t,t-s} = \mathrm{cov}(x_t, x_s)$. This matrix is a symmetric positive definite band matrix with a maximum number of 2r-1 elements different from zero on each row. The elements of Γ_x can be computed by means of three relations given by Mélard and Kiehm (1981).

First, for $t > p$ and $s > p$, we have $\lambda_{t,t-s} = \mathrm{cov}\,(y_t, y_s) = \beta_{t,t-s}$, say, where

$$\beta_{t,k} = \sum_{j=k}^{q} \theta_{t-k,j-k}\, \theta_{tj} g_{t-j}^2 \qquad (0 \leqslant k \leqslant q), \tag{13}$$

and $\beta_{t,k} = 0$ for $k > q$, letting $\theta_{t,0} = -1$.

For $p+1 \leqslant t \leqslant t+q$ and $1 \leqslant s \leqslant p$, we have $\lambda_{t,t-s} = \mathrm{cov}(y_t, w_s) = \mu_{t,t-s}$, say, where

$$\mu_{t,k} = \mathrm{cov}(y_t,\ y_{t-k} + \sum_{i=1}^{p} \phi_{t-k,i}\, w_{t-k,i})$$

$$= \beta_{t,k} + \sum_{i=1}^{\min(p,q-k)} \phi_{t-k,i}\, \mu_{t,k+i} \quad (0 \leqslant k \leqslant q), \tag{14}$$

and $\mu_{t,k} = 0$ for $k > q$. (Equation (15) of the above-mentioned paper is incorrect).

Finally, for $1 \leqslant t \leqslant p$ and $1 \leqslant s \leqslant p$, we have $\lambda_{t,t-s} = \mathrm{cov}(w_t, w_s) = \gamma_{t,t-s}$, say, where

$$\gamma_{t,k} = \mathrm{cov}(y_t + \sum_{i=1}^{p} \phi_{t,i}\, w_{t-i},\ w_{t-k}) = \mu_{t,k} + \sum_{i=1}^{p} \phi_{t,i}\, \gamma_{t-j,k-i}. \tag{15}$$

This recurrence relation requires initial conditions which are obtained in the constant case by solving a system of p+1 equations in the p+1 unknowns, $\gamma_k (0 \leqslant k \leqslant p)$ (McLeod (1975)). Here we should have to use the Wold-Cramér decomposition (Cramér (1961)) of the stochastic process $\{w_t;\ t \in \mathbb{Z}\}$

$$w_t = g_t a_t + \sum_{j=1}^{\infty} \psi_{t,j}\, (g_{t-j} a_{t-j}),$$

where the $\psi_{t,j}$'s are obtained by repeated substitution of $w_{t-1}, w_{t-2}, \ldots$ in (3). Then, $\gamma_{t,k}$ can be derived in the same way as $\beta_{t,k}$ in (13) but with an infinite series instead of a finite sum. From the practical point of view, this series must be approximated by a finite sum by letting $\psi_{t,j} = 0$ for $j > Q$, say, and for all t :

$$\gamma_{t,k} = \sum_{j=k}^{Q} \psi_{t-k,j-k}\, (\psi_{t,j} g_{t-j}^2) \quad (0 \leqslant k \leqslant p).$$

The number of operations is proportional to Q; but Q can sometimes be very large and its choice depends on the remote past of the process. We have preferred another approach in which the stochastic process is assumed to be stationary for $t \leqslant 0$.

It means that g_t, $\phi_{t,i}$ and $\theta_{t,j}$ are supposed to be constant with respect to time, for $t \leqslant 0$, and to be so that the stationary and invertibility conditions are fulfilled. Hence, the covariances $\gamma_{0,k}$ $(1 \leqslant k \leqslant p)$ are obtained as in the constant case whereas all the requested $\gamma_{t,k}$ are computed by (13-14-15). We need $q(q+1) + p(p+1)$ operations at each t such that $t \leqslant p$, if $p > q$, and fewer otherwise. When $p+1 \leqslant t \leqslant p+q$, this number is reduced to $q(q+1)$ or less. When $t \geqslant p + q$, there remains only $q(q+1)/2$ operations. It is worth noting that the direct evaluation of line t of Γ_w would require $q(q+1) + tp$ operations, hence $O(n^2)$ for the complete evaluation of Γ_w.

4. A TIME-SERIES APPROACH OF ANSLEY'S METHOD

Once Γ_x is calculated, Ansley's proposal consists in a Cholesky decomposition of Γ_x using an algorithm for a band matrix with maximum bandwidth r*, or r if the variant (12) is used. It is equivalent to find the Wold-Cramér decomposition of the process $\{x_t;\ t \geqslant 1\}$. Let $h_t e_t$ be the innovation at time t, where $\text{var}(e_t) = \sigma^2$. We can write

$$x_t = h_t e_t - \sum_{j=1}^{q(t)} \varphi_{t,j}\ (h_{t-j} e_{t-j}) \tag{16}$$

where $q(t) = t - 1$ for $1 \leqslant t \leqslant p$, and $q(t) = q$ for $p+1 \leqslant t \leqslant n$, since y_t is defined by (11) as a time-dependent MA(q) process. If we introduce the 1xn vector $e^T = (e_1,\dots,e_n)$, the nxn diagonal matrix H such that $H_{t,t} = h_t$ and the lower triangular matrix Φ such that $\Phi_{t,t} = 1$, $\Phi_{t,s} = -\varphi_{t,s}$ for $1 \leqslant s \leqslant q(t)$ and $\Phi_{t,s} = 0$ for $s > q(t)$. Then (16) can be written as $x = \Phi He$. Consequently

$$\Gamma_x = E(xx^T) = E(\Phi Hee^T H\Phi^T) = \Phi HH\Phi^T\sigma^2,$$

by definition of $h_t e_t$ as the innovation at time t. Hence

$$\det \Gamma_x = \sigma^2 (\det H)^2 = \sigma^2 (\prod_{t=1}^{n} h_t)^2,$$

$$x^T \Gamma_x^{-1} x = \sigma^{-2} e^T H\Phi^T (\Phi^T)^{-1}\ H^{-1} H^{-1} \Phi^{-1} \Phi He$$
$$= \sigma^{-2} e^T e.$$

Thus, the likelihood function can be computed by

$$L(\alpha, \sigma^2; w) = (2\pi)^{-n/2} \sigma^{-n}\ (\prod_{t=1}^{n} h_t)^{-1} \exp\left\{- \frac{1}{2\sigma^2}\ e^T e\right\}. \tag{17}$$

This presentation shows that $h_t e_t$ can be interpreted as the one-step-ahead forecast error. Moreover, there is no problem in handling an ARMA model with time-dependent coefficients since $\{x_t;\ t \leqslant 1\}$ is already a nonstationary process if the coefficients are constant. In the algorithm, it will be necessary to combine the computation of the covariances with the derivation of the $\varphi_{t,j}$'s, h_t^2 and $h_t e_t$ in order to reduce the amount of storage. We recall now the basic algorithm, which is equivalent to the computation of one row of the Cholesky decomposition of Γ_x. After that we will describe the two algorithms for constant and time-dependent ARMA models.

5. THE BASIC ALGORITHM

We want to obtain the coefficients $\varphi_{t,j}$ of a decomposition of x_t in terms of the present and past innovations $h_{t-j}e_{t-j}$ $(0 \leqslant j \leqslant q(t))$. We have the q(t)+1 equations

$$\lambda_{t,k} = -\varphi_{t,k}h^2_{t-k} + \sum_{j=k+1}^{q(t)} \varphi_{t-j,j-k}\,\varphi_{t,j}\,h^2_{t-j}\,, \tag{18}$$

$$\lambda_{t,0} = h^2_t + \sum_{j=1}^{q(t)} \varphi^2_{t,j}\,h^2_{t-j}. \tag{19}$$

The computation of the $\varphi_{t,j}$'s, h^2_t and h_te_t is made through a recursion over t :

1° we determine $\varphi_{t,j}h^2_{t-j}$ for $j = q(t), q(t) - 1, \ldots, 1$ by using respectively $\lambda_{t,q(t)}, \lambda_{t,q(t)-1}, \ldots, \lambda_{t,1}$; the determination of $\varphi_{t,k}$ requires q(t)-k multiplications and one division;

2° we obtain h^2_t by means of (19) and q(t) multiplications;

3° h_te_t is obtained from (16).

Notice also that det H can be obtained recursively. The number of operations at time t is (q(t)+1)(q(t)+4)/2.

6. THE ALGORITHM FOR CONSTANT ARMA PROCESSES

The complete algorithm for the likelihood function of an ARMA (p,q) model with constant coefficients is composed of q+2 steps as shown in Table 1. Each step makes use of the basic algorithm of Section 5.

TABLE 1 - The algorithm for constant ARMA models

Step	t	$\lambda_{t,k}$	q(t)
1	$1 \leqslant t \leqslant p$	$\gamma_0, \gamma_1, \ldots, \gamma_{p-1}$	from 0 to p-1
2	p + 1	$\beta_0, \mu_1, \ldots, \mu_q$	min(p,q)
3	p + 2	$\beta_0, \beta_1, \mu_2, \ldots, \mu_q$	min(p+1,q)
...	...	...	...
q + 1	p + q	$\beta_0, \beta_1, \ldots, \beta_{q-1}, \mu_q$	q
q + 2	$p+q+1 \leqslant t \leqslant n$	$\beta_0, \beta_1, \ldots, \beta_q$	q

The covariances γ_k, μ_k and β_k are the analogs of $\gamma_{t,k}$, $\mu_{t,k}$ and $\beta_{t,k}$ in the constant case. When $p > q$, this algorithm is better than the original algorithm of Ansley because the bandwidth of the band matrix Γ_X is reduced after t = p. The number of operations at time t, $t > p + q$, is indeed p+(q+1)(q+4)/2 instead of p+(r*+1)(r*+4)/2 for the algorithm proposed by Ansley.

7. THE ALGORITHM FOR TIME-DEPENDENT ARMA PROCESSES

The algorithm is composed of 5 stages :

Stage 1 : 1° The coefficients $\phi_{t,i}$(i = 1,...,p) and $\theta_{t,j}$(j = 1,...,q) are stored for t = 1-p, 2-p,...,0;

2° The values of g_t are stored for t = 1-q, 2-q,...,0;

3° The covariances $\gamma_{t,k}(k = 0,\ldots,p)$ are determined for $t = 1-p, 2-p,\ldots,0$ as if the stochastic process were stationary, with constant coefficients equal to $\phi_{0,i}$, $\theta_{0,j}$ and g_0^2.

Stage 2 : For each t, $t = 1, \ldots, p$

1° The $\beta_{t,k}$ $(k = 0,\ldots,q)$ are computed using (13) followed by $\mu_{t,k}$ $(k = q, q-1,\ldots,0)$ in that order, using (14), and then the $\gamma_{t,k}$ $(k = 1, \ldots, q)$ using (15);

2° The latter are used in (15) for $k = 0$, to give us $\gamma_{t,0}$;

3° The basic algorithm of Section 5 is applied at time t, with the autocovariance sequence $(\gamma_{t,k}; k\text{-}0,\ldots,t-1)$

Stage 3 : y_t is computed by (11) for $t = p+1,\ldots,n$

Stage 4 : For each t, $t = q+1,\ldots,p+q$:

1° The $\beta_{t,k}(k = 0,\ldots,q)$ are computed, then $\mu_{t,k}(k = q,q-1,\ldots,t-1)$ in that order;

2° The basic algorithm is applied at time t with the autocovariance sequence $(\beta_{t,0},\ldots,\beta_{t,t-p-1}, \mu_{t,t-p},\ldots, \mu_{t,\min(t-1,q)})$

Stage 5 : The basic algorithm, suitably modified with built-in calculation of the covariances $\beta_{t,k}$, is applied for $t = p+q+1$ to n. It means that (18-19) are replaced by

$$\varphi_{t,k}h_{t-k}^2 = \theta_{t,k}g_{t-k}^2 + \sum_{j=k+1}^{q} \left[\varphi_{t-k,j-k}(\varphi_{t,j}h_{t-j}^2) - \theta_{t-k,j-k}(\theta_{t,j}g_{t-j}^2)\right] \quad (k = 1,\ldots,q) \tag{20}$$

$$h_t^2 = g_t^2 - \sum_{j=1}^{q} \left[\varphi_{t,j}^2\, h_{t-j}^2 - \theta_{t,j}^2\, g_{t-j}^2\right]. \tag{21}$$

If r is small enough with respect to n and p is small compared with q, the 5th stage, like the (q+2)nd step in the constant case, will take most of the computation time. The number of operations at time t, $p + q^2 + 4q$, is about less than twice what is reached in the constant case. The total number of operations to compute $x^T\Gamma_x^{-1}x$ is $O(n)$, which is to be compared to $O(n^2)$ required merely for the computation of the elements of Γ_w. The amount of storage is also kept at a very low level : $R(p+q+2R+4)$ memory cells, where $R = \max(p,q) + 1$. This is a dramatic improvement over the algorithm of Gardner et al, even in the constant case.

8. PERFORMANCE OF THE ALGORITHMS

The algorithms of Sections 6 and 7 have been programmed in Fortran. Two implementations of the algorithm for constant models were written. In implementation I, the $\varphi_{t,j}$'s for a given t are stored as a column in a 2-dimensional array with circular allocation.In implementation II, they are stored in a vector as a column of the lower triangular matrix Φ. The algorithm for time-dependent models was programmed like in implementation I, but with three supplementary arrays for the covariances, the $\phi_{t,i}$'s and the $\theta'_{t,j}$'s. As a matter of comparison, the conditional sum of squares for constant coefficients was also computed, starting at $t = R$: a loop over t contains two loops, one over i and the other over j.

The programs were run on two computers from Control Data : the CDC 6500 and Cyber 170-750. The central processor of the latter has several functional units and an instruction stack which allows for faster execution of small loops. The compiler FTN 4.8 can produce three levels of optimization : level 0 code is easy for debugging, level 1 code is intermediate and level 2 code is highly optimized in order to use the 24 registers whenever possible. The approximate CPU execution time in milliseconds was averaged over N series, where N = 100, most of the time. The ratio of the amount of computer time required for the computation of the exact likelihood over that of the conditional sum of squares is also reported.

The models are the same as those already used by Ansley (1979) and Ljung and Box (1979), and also a sequence of MA(q) models with q ranging between 2 and 20. Four cases were considered for the algorithm for time-dependent models :

(i) the algorithm is applied with constant coefficients;
(ii) g_t varies and the coefficients vary linearly with time;
(iii) g_t varies and the coefficients vary exponentially;
(iv) only g_t varies and the coefficients are constant, so that their calculation is skipped over.

Most of the time, the series length was equal to 100, except in Table 3.

Table 2 shows selected results for the algorithm of Section 6 run on the two machines, under implementations I and II, and three compiler optimization levels. The ratio exact likelihood/simple sum of squares of the CPU times is not always affected in a simple way by code optimization. The most impressive changes are obtained for large values of q, where level 2 code obtains the best from the central processor capabilities in performing the loops contained in equations (18) (19). Implementation II needs instructions for index handling which take more time than the manipulation of the 2-dimensional array. The unsophisticated architecture of the CDC 6500 leads to worse results. The discrepancies between our different columns indicate that it is difficult to compare the results of Table 2 with those given by Ansley or Ljung and Box.

TABLE 2 - Ratio of the average CPU times for computation of the exact and conditional likelihood functions, for multiplicative seasonal models with constant coefficients. Series length n = 100

Computer	Cyber 170-750(N=100)				CDC 6500 (N = 10)
Implementation	I	I	I	II	I
Optimization level	0	1	2	2	2
Models :					
(1,0)	1.3	1.2	1.0	1.0	1.0
(0,1)	5.1	3.7	4.6	3.7	5.9
(2,0)	1.4	1.4	1.2	1.1	1.2
(0,2)	6.2	4.7	5.5	5.1	5.9
(1,1)	4.6	3.5	4.3	3.5	4.4
$(1,0) \times (1,0)_{12}$	5.4	5.3	5.7	5.9	7.7
$(0,1) \times (1,0)_{12}$	5.6	5.3	6.3	6.1	8.4
$(1,0) \times (0,1)_{12}$	16.7	11.4	9.9	12.9	17.4
$(0,1) \times (0,1)_{12}$	18.1	12.3	10.9	14.9	18.6

Table 3 shows how the computation time depends on the series length n for the algorithm of Section 7. Only in the last two columns are the computation times proportional to n. Otherwise, especially when p is large, the preliminary calculations take too much time. This is due to the algorithm of McLeod (1975) used in the computation at stage 1 - 3°.

In Table 4, the performance of the algorithms for MA(q) models is studied in function of q. The quadratic function stated in Sections 6 and 7 is only slightly apparent with implementation I. The computation time with the algorithm for time-dependent MA models converge to twice the one for the constant model. Implementation II is obviously disqualified on our computer.

TABLE 3 - Average (N=100) CPU times (in ms) for the exact likelihood of ARMA(p,q) models with time dependent coefficients (i). Various series length. Cyber 170-750. Optimization level 2

n	(1,0)	(0,1)	(2,0)	(0,2)	(1,1)	(13,0)	(12,1)	(1,12)	(0,13)
40	1.10	2.45	1.65	3.31	3.24	30.6	27.2	17.4	16.9
60	1.28	3.36	1.94	4.74	4.23	31.5	29.3	23.8	24.0
80	1.46	4.33	2.25	6.07	5.41	32.9	31.1	30.8	31.0
100	1.62	5.32	2.49	7.48	6.38	34.2	33.3	37.3	38.6

TABLE 4 - Average (N=100) CPU times (in ms) for the three algorithms for series with n = 100, represented by MA(q) models. Various q. Cyber 170-750. Optimization level 2.

Coefficient		constant		variable(i)
Implementation		I	II	I
q	conditional			
2	0.8	4.7	4.5	7.5
4	1.1	7.5	7.4	12
6	1.3	10	11	17
8	1.5	13	15	23
10	1.7	16	20	29
12	2.1	21	28	38
14	2.3	25	35	47
16	2.5	29	42	55
18	2.7	33	50	62
20	2.8	37	59	73

TABLE 5 - Average (N=100) CPU times (in ms) for the two algorithms in implementation I, where the coefficients of the time-dependent ARMA models are linear, exponential or constant functions of time. Cyber 170-750, optimization level 2. Series length n = 100

algorithm	constant	time-dependent		
coefficients	constant	linear	exponential	constant
model	(i)	(ii)	(iii)	(iv)
(1,0)	0.7	1.5	1.7	1.5
(0,1)	3.6	5.3	5.3	5.2
(2,0)	1.0	2.5	2.4	2.2
(0,2)	4.7	7.7	7.5	7.2
(1,1)	4.1	6.3	6.6	6.1
$(1,0) \times (1,0)_{12}$	13	35	34	33
$(0,1) \times (1,0)_{12}$	14	35	33	31
$(1,0) \times (0,1)_{12}$	20	39	36	36
$(0,1) \times (0,1)_{12}$	21	40	37	37

Finally, Table 5 gives the computation time for the algorithm for time-dependent ARMA models applied in cases (ii),(iii),(iv). The computation of the coefficients only affects marginally the CPU time.

APPENDIX

In this appendix, we indicate why the recursions proposed by Pearlman (1980) cannot be extended to the time-dependent case. Suppose that the following relations are valid up to time t

$$P_t = F_t P_{t-1} F_{t-1}^T + G_t G_t^T - K_{t-1} K_{t-1}^T/h_{t-1}^2$$
$$= P_{t-1} - L_{t-1} L_{t-1}^T/h_{t-1}^2$$
$$K_t = F_{t+1} P_t U^T = K_{t-1} - \alpha_{t-1} F_t L_{t-1}$$
$$L_t = F_t L_{t-1} - \alpha_{t-1} K_{t-1}$$
$$h_t^2 = U^T P_t U = h_{t-1}^2 (1 - \alpha_{t-1}^2),$$

where α_{t-1} is written for $U^T L_{t-1}/h_{t-1}^2$. We have to show, if possible, that

$$P_{t+1} =_{\text{def}} F_{t+1} P_t F_{t+1}^T + G_{t+1} G_{t+1}^T - K_t K_t^T/h_T^2 = P_t - L_t L_t^T/h_t^2$$
$$K_{t+1} =_{\text{def}} F_{t+2} P_{t+1} U^T = K_t - \alpha_t F_{t+1} L_t$$
$$h_{t+1}^2 =_{\text{def}} U^T P_{t+1} U = h_t^2 (1 - \alpha_t^2),$$

where $\alpha_t = U^T L_t/h_t^2$.

Let us first take the case where F_t and G_t are constants. We start from

$$h_t^2 P_{t+1} = h_t^2 F P_t F^T + h_t^2 G G^T - K_t K_t^T , \qquad (A.1)$$

where

$$h_t^2 F P_t F^T + h_t^2 GG^T = h_t^2 F (P_{t-1} - L_{t-1} L_{t-1}^T/h_{t-1}^2) F^T + h_t^2 GG^T$$
$$= h_t^2 (F P_{t-1} F^T + GG^T) - \frac{h_t^2}{h_{t-1}^2} F L_{t-1} L_{t-1}^T F^T \qquad (A.2)$$
$$= h_t^2 P_t + \frac{h_t^2}{h_{t-1}^2} (K_{t-1} K_{t-1}^T - F L_{t-1} L_{t-1}^T F^T). \qquad (A.3)$$

But $h_t^2/h_{t-1}^2 = 1 - \alpha_{t-1}^2$. Also

$$K_t K_t^T = (K_{t-1} - \alpha_{t-1} F L_{t-1})(K_{t-1}^T - \alpha_{t-1} L_{t-1}^T F^T). \qquad (A.4)$$

Putting together (A.3) and (A.4), we obtain for (A.1)

$$h_t^2 P_{t+1} = h_t^2 P_t - (FL_{t-1} - \alpha_{t-1} K_{t-1})(L_{t-1}^T F^T - \alpha_{t-1} K_{t-1}^T)$$
$$= h_t^2 P_t - L_t L_t^T.$$

Then

$$K_{t+1} = F(P_t - L_t L_t^T/h_t^2) U^T = K_t - \alpha_t F L_t$$

and $h_{t+1}^2 = U^T(P_t - L_t L_t^T/h_t^2) U = h_t^2 - (U^T L_t)^2/h_t^2 = h_t^2(1 - \alpha_t^2)$.

A close inspection of these relations show that nothing can be said if F and/or G depend on time. For instance the analog of (A.2) :

$$h_t^2(F_{t+1} P_{t-1} F_{t+1} + G_{t+1} G_{t+1}^T) - \frac{h_t^2}{h_{t-1}^2} F_{t+1} L_{t-1} L_{t-1}^T F_{t+1}^T.$$

can no longer be related to $h_t^2 P_t$ plus a term like the one in (A.3).

REFERENCES

ANSLEY, C.F. (1979). An algorithm for the exact likelihood of a mixed autoregressive-moving average process. *Biometrika 66*, 59-65.

BOX, G.E.P. and JENKINS, G.M. (1976). *Time Series Analysis, Forecasting and Control*. Holden-Day, San Francisco (revised edition).

CRAMER, H. (1961). On some classes of nonstationary stochastic processes. In *Proceedings of the Fourth Berkeley Symposium on Mathematical Statistics and Probability*. Vol. 2, 57-78, University of California Press, Berkeley and Los Angeles.

GALBRAITH, R.F. and GALBRAITH, J.I. (1974). On the inverses of some patterned matrices arising in the theory of stationary time series. *J. Appl. Prob. 11*, 63-71.

GARDNER, G., HARVEY, A.C. and PHILLIPS, G.D.A. (1980). An algorithm for exact maximum likelihood estimation of autoregressive-moving average models by means of Kalman filtering. *J. Roy. Statist. Soc. Ser. C Appl. Statist. 29*, 311-322.

HARRISON, P.J. and STEVENS, C.F. (1976). Bayesian forecasting. *J. Roy. Statist. Soc. Ser. B 38*, 205-228.

LEDOLTER, J. (1979). A recursive approach to parameter estimation in regression and time series models. *Comm. Statist. - Theor. Meth. A8(12)*, 1227-1245.

LJUNG, G.M. and BOX, G.E.P. (1979). The likelihood function of stationary autoregressive-moving average models. *Biometrika 66*, 265-270.

KIEHM, J.-L., LEFEVRE, C. and MELARD, G. (1980). Forme rétrospective d'un modèle FARIMAG. *Cahiers du CERO 22*, 375-383.

McLEOD, I. (1975). Derivation of the theoretical autocovariance function of autoregressive-moving average time series. *J. Roy. Statist. Soc. Ser. C Appl. Statist. 24*, 255-256.

MELARD, G. and KIEHM, J.-L. (1981). ARIMA models with time-dependent coefficients for economic time series. In *Time Series Analysis*, (Proceedings of the International Conference held at Houston, Texas, August 1980). Eds : O.D. Anderson and M.R. Perryman. North-Holland, Amsterdam, 355-363.

NEWBOLD, P. (1974). The exact likelihood function for a mixed autoregressive-moving average process. *Biometrika 61*, 423-426.

PEARLMAN, J.G. (1980). An algorithm for the exact likelihood of a high-order autoregressive-moving average process. *Biometrika 67*, 232-233.

APPLIED TIME SERIES ANALYSIS
O.D. Anderson & M.R. Perryman (eds.)

A NONPARAMETRIC APPROACH TO TIME SERIES DISCRIMINATION

Kris Moore, Fred Hulme, and Dean Young

Hankamer School of Business
Baylor University
Waco, Texas
U.S.A.

Linda Jennings

The University of Texas at Dallas
Dallas, Texas
U.S.A.

Time series discrimination is an increasingly important topic because of its many possible applications. This paper introduces a nonparametric method for time series discrimination based on the rank transformation. A feature selection technique is suggested and a Monte Carlo simulation is performed to evaluate the performance of the suggested feature selection approach and the proposed nonparametric discrimination method.

INTRODUCTION

Let Π_1 and Π_2 be two distinct homogeneous populations of individuals with some quantifiable characteristic, C. Also, let $X_N(I) = \{X(1), X(2), \ldots, X(N)\}$ be a univariate time series of measurements on the characteristic C taken from an individual $I \varepsilon U\Pi_i$, i=1,2, such that $X_N(I)$ has mean and covariance matrix μ_1 and Σ_1 if the series is obtained from an individual belonging to Π_1, and μ_2 and Σ_2 if obtained from an individual belong to Π_2. The time series discrimination problem may be described as attempting to correctly classify an empirical time series $x_N(I)$ of observations on the characteristic C as being taken from an individual belonging to population Π_i, i=1,2. This type of discrimination problem is important in several different disciplines, such as speech recognition as discussed in Kashyap and Mittal (1976), and Jelinek (1976), the recognition of a speaker from a spoken wave form as analyzed by Bricker (1970), Atal (1974), Kashyap (1976), Hoehne and Hoefker (1974), and the classification of geological shock waves which has been investigated by Sandvin and Tjøstheim (1979).

Several approaches to time series discrimination are currently employed. One of the most popular methods is the linear discriminant function (LDF), which is well documented in Anderson (1958) and Shumway and Unger (1974) and is given by $(\mu_1 - \mu_2)\, \Sigma^{-1} X_N$. An observation x_N is classified into Π_1 if:

$$(\mu_1 - \mu_2)\, \Sigma^{-1} x_N \geq c_1$$

and into Π_2 if: (1.1)

$$(\mu_1 - \mu_2)\, \Sigma^{-1} x_N < c_1$$

for an appropriate constant c_1. Note that it is assumed that $\Sigma_1 = \Sigma_2 = \Sigma$ in the LDF. If, in fact $\Sigma_1 = \Sigma_2 = \Sigma$, the classical linear discriminant function minimizes the expected probability of misclassification.

Another popular method is the quadratic discrimnation function (QDF), which may be found in Andrews (1972). The discrimination rule given by the QDF is that an observation x_N is classified into Π_1 if:

$$(x_N - \mu_1)^T\, \Sigma_1^{-1}(x_N - \mu_1) - (x_N - \mu_2)^T\, \Sigma_2^{-1}(x_N - \mu_2) \geq c_2$$

and is classified into Π_2 if: (1.2)

$$(x_N - \mu_1)^T \Sigma_1^{-1}(x_N - \mu_1) - (x_N - \mu_2)^T \Sigma_2^{-1}(x_N - \mu_2) < c_2$$

where c_2 is an appropriate constant. It should be noted that both of these techniques assume normality of the series X_N and that the parameters μ_i and Σ_i, i=1,2, must be estimated in the actual data case using training data. If, however, these two techniques are applied to the data in the time domain, problems may occur due to the small number of training series usually available when contrasted with the dimensionality of the time series $X_N^{(I)}$, as noted in Sandvin and Tjøstheim (1979).

Raudys (1979) and Chandrasekaren and Jain (1975) have shown that the expected probability of misclassification (EPM) increases significantly if the training sample size n_i, i=1,2, is small relative to the series dimensionality, N. This effect is caused by the inadequate information in the training data needed to estimate the large number of parameters, especially in the QDF since Σ_1 and Σ_2 must be independently estimated. To combat this dimensionality problem, some type of feature selection is often performed prior to the classification process. Feature selection may be described as the process of performing a transformation $B=(b_1, b_2, \ldots, b_M)^T$, $1<M<N$, such that $Y_M = B(X_N)$ with $Y_i = b_i(X_N)$, $1<i<M$, so that the transformed series is of smaller dimension than N with a minimal increase in the expected probability of misclassification.

A DIFFERENT APPROACH

The feature selection approach proposed in this paper is to transform the data from the original time domain into the periodogram in the spectral domain and then use the coefficients of the periodogram $I_k(w_j)$ as the new features $Y_M = \{Y(1), Y(2), \ldots, Y(M)\}$ for performing the classification. For an odd number of time periods, N, the dimensionality of the new feature set is $M = N/2 + 1$, which represents a considerable reduction in the dimensionality. However, it should be noted that if $X_N(I) = \{X(1), X(2), \ldots, X(N)\}$ is normally distributed, then the periodogram ordinates $I_k(w_j)$ are approximately independently distributed as $2\pi f(w_j)\chi^2$, that is, as a multiple of a chi-square random variable with two degrees of freedom where $f(w_j)$ is the spectral density function evaluated at the frequency w_j. Since the new feature set is not normally distributed, the assumptions underlying the classical LDF and QDF are violated. We propose to apply a nonparametric discrimination procedure based on the rank transformation.

RANK TRANSFORMATION DISCRIMINANT FUNCTIONS

Moore and Smith (1975) introduced a rank transformation method of discrimination in which the multivariate sample is replaced by its rank. Conover and Iman (1980) illustrated that a rank transformation method performs quite well when compared to other methods. The use of rank transformations allows application of the method on data that is only ordinal in level of measurement and is unrestricted by the assumptions required in parametric methodologies, such as the assumption of normality of the data. Consequently, the rank methods are inherently robust.

The rank transformation quadratic discriminant function method (RQDF) consists of ranking each of the N components of all the training observations from the smallest, with rank 1, to the largest, with rank $n_1 + n_2$. Average ranks are used whenever ties occur. Each time component is ranked separately across the training observations. Then the sample means $\bar{x}_i(R)$ and sample covariance matrices $S_i(R)$ are computed using the transformed values from each of the respective populations. The new time series, which is to be classified, $x'_N(R)$, is then transformed by replacing each component among the corresponding values of the same time period for the combined training data from the two populations. Then $x'_N(R)$ is classified as being taken from an individual belonging to Π_i, i=1,2, if the QDF of (1.2) is performed with $x'_N(R)$, $S_i(R)$, and $\overline{x}_i(R)$, respectively, in (1.2).

The RLDF method is similar to the RQDF method, except that a pooled estimate S(R) of the common covariance matrix is used in all estimated densities instead of the individual estimates $S_i(R)$. The rank transformation procedure (RLDF) corresponding to the LDF method involves computing a common covariance matrix on all of the training ranks. The rest of the RLDF procedure is as described above for the LDF with S(R) replacing S and $\overline{x}_i(R)$ replacing $\overline{x}_i$ in (1.1).

A criticism of rank transformation methods is that a loss of information occurs. It is well known that when the correct assumptions are made, many of the classical parametric tests are designed to be and are most powerful. Indeed, as Rao (1973) has noted, a transformation of the original data can result only in a loss of information. However, in those cases where comparison studies have been made, as in Gibbons (1971), nonparametric tests are frequently almost as powerful, especially for small samples. Thus, they may be considered more desirable whenever there is doubt concerning the validity of the assumptions. Even when the assumptions of the usual parametric test, such as the normality assumption, is satisfied, the loss of efficiency is surprisingly small. Conover (1971) has shown that the efficiency of tests using only the ranks of the observations is approximately .95.

SIMULATION RESULTS

A Monte Carlo simulation was performed to explore the efficacy of the rank-discrimination methodology. Five different time series models, each with two sets of distinguishable parameters, were simulated with a length of N=50 time periods (see Table 1). Four discrimination methodologies LDF, QDF, RLDF, and RQDF were applied to each of the time series models with 25 repetitions of the classification process per model. If a covariance matrix used in a discrimination algorithm was less than full rank, then the pseudoinverse Σ^+, as defined by Moore (1920), was used in the algorithm replacing Σ^{-1}. This technique is discussed in Rao and Mitra (1975). Also, for each of the ten time series discrimination models, the separability measure known as the divergence was calculated in order to choose parameters that would promote a reasonable probability of correct classification (i.e., $.5<PCC<1.0$). The results of the simulation experiment are shown in Table 2.

The Friedman test, which is discussed in Conover (1971), was applied to test for differences among the four discrimination methodologies and was found to be significant at the .001 level. A nonparametric multiple comparisons procedure found in Miller (1966) was applied to test for individual differences in methodologies. The results are given in Table 4.

COMMENTS AND CONCLUSIONS

The results of the Monte Carlo simulation do not indicate that either of the rank discrimination methods, the RLDF or the RQDF, is superior to the LDF or the QDF methods of discrimination. However, this simulation experiment is not intended to be an exhaustive study. The experiment does raise many questions, such as: How does the rank procedure perform as a function of the length N? Does the methodology work for some types of models better than others? Is there a relationship between the rank of the covariance matrices $S_i(R)$, which is a function of the training sample size and the series length, and the loss of information due to the rank transformation? Would the nonparametric rank methodology be more competitive relative to the LDF and the QDF if the data in the time domain was not normally distributed?

Several comments may also be in order. It seems plausible that additional feature selection techniques, such as the M method of dimension reduction developed by Odell (1978), might be applied to the data following the transformation to the periodogram ordinates. Also, the less-than-full-rank inversion problem might be improved since the periodogram estimates are asymptotically independent and therefore a better approach might be to estimate only the variances. Finally, if the LDF, QDF, and RLDF are not significantly different in their classification performance, the RLDF is slightly more computationally complex and thus not as efficient as the LDF or QDF methods. Further study employing a good experimental design to explore the rank transformation methodology may reveal that the RLDF is superior to the LDF and the QDF for certain combinations of population models, sample sizes, and series lengths which have not been considered in this paper.

SIMULATION MODELS

Table 1

1. AR(1) vs AR(1): $X_t = \phi_i X_{t-1} + e_t \sim N(0,\sigma^2)$, i = 1,2

   a) Group 1: $\sigma^2=1, \phi_1=-.6$ — Divergence = 110.2500
      Group 2: $\sigma^2=1, \phi_2=.6$

   b) Group 1: $\sigma^2=1, \phi_1=.2$ — Divergence = 10.2917
      Group 2: $\sigma^2=1, \phi_2=.6$

2. AR(1) vs AR(2): $X_t = \phi_{1i} X_{t-1} + \phi_{2i} X_{t-2} + e_t$, $e_t \sim N(0,\sigma^2)$, i = 1,2

   a) Group 1: $\sigma^2=1, \phi_{11}=-.6, \phi_{21}=0$ — Divergence = 14.7741
      Group 2: $\sigma^2=1, \phi_{12}=-1, \phi_{22}=0$

   b) Group 1: $\sigma^2=1, \phi_{11}=-.6, \phi_{21}=0$ — Divergence = 3.4294
      Group 2: $\sigma^2=1, \phi_{12}=-.5, \phi_{22}=.25$

3. Freq 1 vs Freq 2 with WN: $X_t = A\cos(f_i t) + e_t$, $e_t \sim N(0,\sigma^2)$, i = 1,2

   a) Group 1: A=.25, $f_1=1.0$, $\sigma^2=1$ — Divergence = 3.1832
      Group 2: A=.25, $f_2=2.3562$, $\sigma^2=1$

   b) Group 1: A=.5, $f_1=1.0$, $\sigma^2=1$ — Divergence = 12.7328
      Group 2: A=.5, $f_2=2.3562$, $\sigma^2=1$

4. Amp1 vs Amp2 with WN: $X_t = A_i\cos(ft) + e_t \sim N(0,\sigma^2)$, i = 1,2

   a) Group 1: $A_1=1.0$, f=1.0, $\sigma^2=1$ — Divergence = 99.5371
      Group 2: $A_2=3.0$, f=1.0, $\sigma^2=1$

   Group 1: $A_1=.25$, f=1.0, $\sigma^2=1$ — Divergence = 3.4993

Table 1 (continued)

b)
Group 2:A_2=.625, f=1.0, σ^2=1

5. Linear trend vs Linear trend: $X_t = C_i + D_i t/N + e_t$, $e_t \sim N(0, \sigma^2)$, i = 1,2

a)
Group 1:C_1=0, D_1=1, σ^2=1 Divergence = 9.0956
Group 2:C_2=.75, D_2=.25, σ^2=1

b)
Group 1:C_1=0, D_1=1, σ^2=1 Divergence = 3.0331
Group 2:C_2=.5, D_2=.25, σ^2=1

Table 2

Estimated Probability of Correct Classification

	LDF	QDF	RLDF	RQDF
AR1 vs AR1 (a)	.9284	.9832	.9616	.8876
AR1 vs AR1 (b)	.7388	.7496	.7424	.6012
AR1 vs AR2 (a)	.6132	.6104	.6104	.5366
AR1 vs AR2 (b)	.8012	.7748	.7788	.6352
Freq 1 vs Freq 2 (a)	.6904	.6412	.7344	.5636
Freq 1 vs Freq 2 (b)	.5484	.5028	.5372	.5000
Amp1 vs Amp2 (a)	.9828	.9384	.8048	.5704
Amp1 vs Amp2 (b)	.6368	.6108	.6812	.5380
Linear trend vs Linear trend (a)	.5936	.5648	.6088	.5428
Linear trend vs Linear trend (b)	.7580	.6944	.7848	.5604

Table 3

Rankings of Estimated Probability of Correct Classification

	Linear	Quadratic	Ranked Linear	Ranked Quadratic
AR1 vs AR1	3	1	2	4
AR1 vs AR1	3	1	2	4
AR1 vs AR2	1	3	2	4
AR1 vs AR2	1	2.5	2.5	4

Table 3 (continued)

Freq 1 vs Freq 2	2	3	1	4
Freq 1 vs Freq 2	1	3	2	4
Ampl vs Amp2	2	3	1	4
Linear trend vs Linear trend	2	3	1	4
Linear trend vs Linear trend	2	3	1	4
Total Ranks	18	24.5	17.5	40
	Friedman Test		Significant at less than the .001 level	

Table 4

RLDF, LDF, QDF, RQDF

REFERENCES

ANDERSON, T. W. (1958). An Introduction to Multivariate Statistical Analysis. John Wiley and Sons, New York.

ANDREWS, H. C. (1972). Introduction to Mathematical Techniques in Pattern Recognition. Wiley-Interscience, New York.

ATAL, B. S. (1974). Effectiveness of linear prediction characteristics of the speech wave for automatic speaker verification. J. Acoust. Soc. Amer. 55, 1304.

BRICKER, P. D., et al (1970). Statistical techniques for talker identification. Bell Syst. Tech. J. 50, 1427-1454.

CHANDRASEKAREN, B. and JAIN, A. K. (1975). Independence, measurement complexity, and classification performance. IEEE Trans. Systems, Man, and Cybernetics. SMC-5, No. 2.

CONOVER, W. J. (1971). Practical Nonparametric Statistics. John Wiley and Sons, New York.

CONOVER, W. J. and IMAN, R.T. (1980). The rank transformation as a method of discrimination with some examples. Commun. Stat.--Theor. Meth. A9(5), 465-486.

GIBBONS, J. D. (1971). Nonparametric Statistical Inference. McGraw-Hill, New York.

HOEHNE, H. D. and HOEFKER, U. (1976). Pattern recognition in speaker verification and identification tasks. Proc.III Joint Int. Conf. Pattern Recognition, San Diego.

JELINEK, F. (1976). Continuous speech recognition by statistical methods. Proc. IEEE 64, 532-556.

KASHYAP, R. L. (1976). Speaker recognition from an unknown utterance and speaker speech interaction. IEEE Trans. Acoust., Speech, and Signal Processing AASP-24, No. 6, 481-487.

KASHYAP, R. L. and MITTAL, M. C. (1976). Word recognition in a multitalker environment using syntactic methods. Proc. III Joint Int. Conf. Pattern Recognition, San Diego. See also Tech. Rep. TREE-76-28, School of Electrical Engineering, Purdue University, Lafayette, IN.

MILLER, R. G. (1966). Simultaneous Statistical Inference. McGraw-Hill, New York.

MOORE, E. H. (1920). Abstract. Bulletin Am. Math Soc. 26, 394-395.

MOORE, K. K. and SMITH, W. B. (1975). A rank order approach to discriminant analysis. Proc. Bus. and Eco. Stat. Section Am. Stat. Assn., 451-455.

ODELL, P. L. (1978). A model for dimension reduction in pattern recognition using continuous data. Pattern Recognition 11, 51-54.

RAO, C. R. (1973). Linear Statistical Inference and Its Applications, 2nd ed. John Wiley and Sons, New York.

RAO, C. R. and MITRA (1975). Generalized Inverses of Matrices and Its Applications. John Wiley and Sons, New York.

RAUDYS, S. J. (1979). Determination of optimal dimensionality in statistical pattern classification. Pattern Recognition 2, 263-270.

SANDVIN, O. and TJϕSTHEIM, D. (1979). Multivariate autoregressive feature extraction and the recognition of multichannel waveforms. IEEE Trans. Pattern Analysis and Machine Intelligence Pami-1, No. 1.

SHUMWAY, B. H. and UNGER, A. N. (1974). Linear discriminant functions for stationary time series. J. Am. Stat. Assn. 69, 948-956.

YOUNG, T. Y. and CALVERT, T. W. (1974). Classification, Estimation and Pattern Recognition. American Elsevier, New York.

APPLIED TIME SERIES ANALYSIS
O.D. Anderson & M.R. Perryman (eds.)

A TIME SERIES EXTENSION OF A SPECIFICATION TEST DUE TO RAMSEY

Mathew J. Morey
Department of Economics
Indiana University
Bloomington, Indiana, USA

David E. Spencer
Department of Economics
Washington State University
Pullman, Washington, USA

In recent years, increased attention has been paid to tests of specification in econometric models. This is especially true of tests which are intended to assess the validity of the "exogeneity" (or "orthogonality") assumption in the standard linear regression model. Several tests have been developed but none is satisfactory for time series applications. This paper extends a test due to J.B. Ramsey in such a way as to accommodate the special problems associated with time series models.

INTRODUCTION

In recent years, increased attention has been paid to tests of specification in econometric models. This is especially true of tests which are intended to assess the validity of the "exogeneity" (or "orthogonality") assumption in the standard linear model. The several tests which have been developed have come from two different perspectives. One group of tests has been generated from a time series perspective as a consequence of the concept of Weiner-Granger causality, Granger (1969), and its relationship to exogeneity; see, e.g., Sims (1972), Sargent (1976), Pierce and Haugh (1977), and Ashley, Granger, and Schmalensee (1980). Tests from a more traditional econometric perspective have been developed by Ramsey (1969), Wu (1973), Revankar (1978), and Hausman (1978).

Several problems arise in the practical application of these two classes of tests, however. It has been shown theoretically that the condition Y does not cause X in Granger's sense is necessary <u>but not sufficient</u> for exogeneity of X. This clearly limits the usefulness of the time series tests. On the other hand, the more traditional tests are generally very sensitive to serial correlation in the errors and thus are not appropriate in time series applications. The purpose of this paper is to extend one of the tests in this latter group, Ramsey's regression specification error test (RESET), to a time series regression model with serially correlated errors.

Ramsey's specification error test is based on the distribution of Theil's (1965) BLUS residuals under the null and alternative hypotheses. Given a bivariate time series model, RESET is applied to a final form equation for Y.[1] The regressors X, X_{-1}, ... of this final form equation will be uncorrelated with the error in that equation only under the null hypothesis that X is exogenous. It should be noted that this corresponds to a complete lack of causality from Y to X in Granger's sense. Unfortunately, Ramsey's test is applicable only with some modification since the final form equation will have serially correlated errors and it is well-known that Theil's residuals do not retain the scalar covariance property in such a case. This paper demonstrates two ways to overcome this problem. The first of these derives a new vector of residuals by defining an appropriate residual transformation which does in fact obtain BLUS residuals under serial correlation. The second involves a direct transformation of the model itself so that the Theil residuals can be obtained in the usual manner.

The next section of the paper briefly reviews the nature of Ramsey's test in the context of the standard linear regression model. Section III introduces a bivariate time series model and examines the testable implications of the model under the alternative hypothesis. Section IV actually extends the test, RESET, in such a way that it is applicable in the time series model of Section III. The fifth and final section contains some concluding remarks.

RAMSEY'S TEST

The "exogeneity" test which we extend in this paper is the regression specification error test (RESET) developed by Ramsey (1969). This test is appropriate when the null hypothesis is characterized by the standard linear regression model; i.e.,

(2.1) $$y = X\beta + u$$

where $E(u|X) = 0$ and $E(uu') = \sigma^2 I$. Under the alternative hypothesis $E(u|X) \neq 0$. Suppose that we have T observations and K regressors (including the constant term).

Ramsey's RESET is based on the distribution of the OLS residuals (or, alternatively, Theil's BLUS residuals) under the null hypothesis of no specification error and under the alternative hypothesis that specification error exists. RESET exploits the fact that under the null hypothesis, the vector of BLUS residuals has zero mean while under the alternative hypothesis that the orthogonali ty assumption is violated, the vector of BLUS residuals no longer has zero mean. Let the mean of the BLUS residual vector, $\tilde{u}$, under the alternative hypothesis be given by $A'\xi$ where ξ is a nonstochastic vector, the definition of which is determined by the specification error involved, and A' is the $(T-K) \times T$ matrix defined by the procedure to obtain Theil's BLUS residuals. A is chosen such that

i) $A'X = \phi$

ii) $A'A = I_{T-K}$, and

iii) $E[u'(A-J)'u]$ is a minimum

where $(A-J)'u$ is the "error in predicting" u and J is the matrix obtained by dropping K columns from the identity matrix I_T.[2] As shown by Ramsey, this mean vector can be approximated by

(2.2) $$A'\xi \doteq \alpha_o + \alpha_1 q_1 + \alpha_2 q_2 + \ldots$$

where the $(T-K) \times 1$ vectors q_j, $j = 1, 2, \ldots$ are defined by

(2.3) $$q_j = A'\hat{y}^{(j+1)} \quad j = 1, 2, \ldots$$

and $\hat{y}^{(j+1)}$ is the estimated conditional mean vector of the dependent variable raised to the j+1 power.

Since ξ is unknown, RESET considers the regression equation

(2.4) $$\tilde{u} = \alpha_0 + \alpha_1 q_1 + \alpha_2 q_2 + \ldots + \alpha_s q_s + e.$$

Under the assumption that $e_t \sim$ i.i.d. $N(0,\sigma_e^2)$, RESET becomes a test of the joint significance of the s+1 regression coefficients in equation (2.4) using the appropriate F statistic. To obtain reasonable power under the alternative hypothesis, Ramsey has suggested that s should be chosen such that $s \geq 3$.

THE MODEL

Consider a bivariate covariance-stationary time series $\{(X_t,Y_t)'\}$ which can be expressed in an autoregressive form

$$(3.1) \qquad \alpha(L)X_t + \beta(L)Y_t = a_t$$

$$(3.2) \qquad \gamma(L)X_t + \delta(L)Y_t = b_t$$

or

$$(3.3) \qquad \begin{bmatrix} \alpha(L) & \beta(L) \\ \gamma(L) & \delta(L) \end{bmatrix} \begin{pmatrix} X_t \\ Y_t \end{pmatrix} = \begin{pmatrix} a_t \\ b_t \end{pmatrix}$$

where $\{(a_t,b_t)'\}$ is a vector white noise process statisfying

$$(3.4) \qquad E\begin{pmatrix} a_t \\ b_t \end{pmatrix} = 0 \qquad \text{for all } t$$

$$(3.5) \qquad E\left[\begin{pmatrix} a_t \\ b_t \end{pmatrix}(a_s b_s)\right] = \begin{cases} \Sigma & t = s \\ 0 & t \neq s \end{cases}$$

and $\alpha(L)$, $\beta(L)$, $\gamma(L)$, $\delta(L)$ are polynomials in nonnegative powers of the lag operator L such that

$$(3.6) \qquad \alpha(L)X_t = \sum_{i=0}^{P_x} \alpha_i L^i X_t = \sum_{i=0}^{P_x} \alpha_i X_{t-i}.$$

P_x is the order of the operator on variable X. It is assumed that all operators are square summable; i.e., $\sum_{j=0}^{\infty} \alpha_j^2 < \infty$, $\sum_{j=0}^{\infty} \beta_j^2 < \infty$, $\sum_{j=0}^{\infty} \gamma_j^2 < \infty$, $\sum_{j=0}^{\infty} \delta_j^2 < \infty$.[3]

We want to test the hypothesis that $E(a_t|x) = 0$ in equation (3.1) where x is a vector of current and lagged values of X included in that equation. This is a test of the hypothesis that X is exogenous in equation (3.1). It is well-known that there is a close relationship between econometric exogeneity and the notion of Granger noncausality. See Sims (1972), Theorem 2, pp. 543-544. It is therefore useful to note that in the above model, Y does not cause X only if $\delta(L) \equiv 0$ in equation (3.2). See Pierce and Haugh (1977) for additional characterizations.

We also note that if we assume a and x are jointly normally distributed, then

$$(3.7) \qquad E(a|x) = \mu_a + \Sigma_{ax}\Sigma_{xx}^{-1}(x-\mu_x)$$

where μ_a is the unconditional mean of a, Σ_{ax} is the matrix of covariance terms between a and x in their joint density function, Σ_{xx} is the variance-covariance matrix in the marginal density for x, and μ_x is the unconditional mean vector of x. Since $\mu_a = 0$ by assumption, (3.6) becomes

(3.8) $E(a|x) = \Sigma_{ax}\Sigma_{xx}^{-1}(x-\mu_x)$.

In general, $E(a|x) = 0$ only if $\Sigma_{ax} = 0$, but Σ_{ax} can be written

(3.9) $\Sigma_{ax} = E\{a[-\gamma(L)^{-1}\delta(L)y + \gamma(L)^{-1}b]\}$.

It therefore follows that $\Sigma_{ax} = 0$ only if $\delta(L) \equiv 0$ and $\sigma_{ab} = 0$ where σ_{ab} is the covariance between a_t and b_t.[4]

The problem with testing the hypothesis that $\delta(L) = 0$ in the structural model given by (3.1) and (3.2) is that the model is not identifiable. Thus, without appropriate a priori information, hypotheses involving structural parameters cannot be tested. It is the case, however, that the reduced form and final form equations are always identified under the null hypothesis. The following question therefore arises: Can one find a testable restriction on the reduced form or final form which holds only under the null hypothesis that X is exogenous in (3.1)? As we demonstrate below, the answer to this question is yes.

Assuming invertibility, we write (3.1) as a final form equation for Y_t under the assumption that X is exogenous

(3.10) $Y_t = -\beta(L)^{-1}\alpha(L)X_t + \beta(L)^{-1}a_t$

or

(3.11) $Y_t = \theta(L)X_t + \pi(L)a_t$

where $\theta(L) = -\beta(L)^{-1}\,\alpha(L)$ and $\pi(L) = -\beta(L)^{-1}$. The solution for the θ_i $(i = 0, 1, 2, \ldots)$ in terms of the $\beta_j (j = 0, 1, \ldots, P_y)$ and the α_ℓ $(\ell = 0, 1, \ldots, P_x)$ is obtained by equating terms in powers of L in the expression

(3.12) $-\beta(L)\theta(L) = \alpha(L)$.

Since, in practice, the sample will be finite, it will be necessary to truncate the operator on X_t in (3.11) so that

(3.13) $$Y_t = \sum_{i=0}^{M} \theta_i X_{t-i} + v_t$$

where

(3.14) $$v_t = \sum_{i=M+1}^{\infty} \theta_i X_{t-i} + \sum_{j=0}^{\infty} \pi_j a_{t-j} \qquad M < T$$

$$= \eta_t + \varepsilon_t$$

If we allow M to increase with T, following Jorgenson (1966), any square summable distributed lag may be approximated to any degree of accuracy by a mixed autoregressive-moving-average (or a rational distributed lag) process of finite order.

Since $v_t \doteq \sum_{i=0}^{\infty} \pi_i a_{t-i}$, it is the case that X is independent of the error in equation (3.13) only if X is independent of a in equation (3.1). Thus, in theory, using Ramsey's RESET to test that X is exogenous in (3.13) can be viewed as a test that X is exogenous in (3.1). It should be noted that a failure to reject the null hypothesis of exogeneity is a sufficient condition for Granger noncasuality.

THE EXTENSION OF RESET

The results of the previous section suggest that application of Ramsey's RESET to the final form equation (3.13) will yield a test of the hypothesis that X is exogenous in (3.1). As indicated in Section II, RESET requires estimation of the regression equation

$$\tilde{v} = \alpha_o + \alpha_1 q_1 + \ldots + \alpha_s q_s + e \tag{4.1}$$

where q_i i = 1, ..., s are defined in equation (2.3) and $\tilde{v}$ is an appropriate (perhaps BLUS) residual vector of length T-K. A test of the null hypothesis is given by a test of the hypothesis that the regression coefficients in (4.1) are jointly zero.

There are two important problems that arise in this context which must be resolved before RESET is applicable, however. The first arises because of the possibility of "spurious" exogeneity which would result if the elements of $E(\tilde{v}|X)$ are identical to those of the orthogonal projector, say $M= I-W(W'W)^{-1}W'$. In this case, the conditional mean of $\tilde{v}$ will be zero regardless of the specification error involved. Therefore, the model given by equations (3.1) and (3.2) is modified to include at least one other variable (say Δ) which is known (assumed) to be exogenous in both equations. We note that this does not alleviate the identification problem for that model, nor does it complicate the essential features of the model for the purposes of our test.

The second problem concerns the choice of an appropriate residual vector, $\tilde{v}$, for use in equation (4.1). Since OLS residuals are neither homoscedastic nor independent, Ramsey has recommended the use of Theil's BLUS residuals. Unfortunately, BLUS residuals have a scalar covariance matrix only when the regression error is serially uncorrelated, a condition which is clearly not true in equation (3.13). We will discuss two ways to overcome this problem. The first approach actually derives new residuals which are BLUS even when the regression error is serially correlated. The second involves a transformation of (3.13) so that Theil's procedure does in fact give residuals with a scalar covariance matrix.

BLUS residuals under serially correlated errors

In order to understand clearly why new residuals are necessary it will be useful to review the general idea of Theil's BLUS residuals in the standard linear model (i.e., the model in which the ideal assumptions hold). Recall from section II that the problem is to choose a matrix A such that conditions i) - iii) hold. Condition i)$A'X = 0$ is required in order to yield linearity and unbiasedness while condition ii) $A'A = I_{T-K}$ is required to yield a scalar covariance matrix for the residuals and follows from the assumption that the errors of the original model have a scalar covariance matrix. In our case, however, this assumption cannot be maintained and therefore condition ii) will not yield a scalar covariance matrix. It will therefore be necessary to alter condition ii) and derive the appropriate vector of BLUS residuals.

For notational convenience write (3.13) as

$$y = Z\theta + v \tag{4.2}$$

where y is the $T \times 1$ vector of observations on Y_t, Z is the $T \times K$ matrix of observations on current and lagged X as well as any additional exogenous variables,[5] θ is the $K(=M+1+K_\Delta) \times 1$ vector of regression coefficients and v is the vector of serially correlated errors; see equation (3.14). We assume that $E(vv') = \sigma_v^2\,\Omega$ where Ω is known or consistently estimated.[6] Our objective is to obtain estimates of the errors which are BLUS.

The condition of linearity requires that our residual vector be of the form H'y where H is a $T \times (T-K)$ matrix of nonstochastic terms. Since the residuals should be conditionally unbiased we obtain

$$(4.3) \qquad E(H'y|Z) = H'Z\theta = 0$$

which implies

i') $H'Z = 0$ or, in large samples, $\operatorname{plim}_{T \to \infty} (1/T\ H'Z) = 0$.

Furthermore H should yield residuals with a scalar covariance matrix. Since, conditional on Z (and neglecting η),

$$(4.4) \qquad E[H'yy'H] = E[H'(Z\theta + v)(Z\theta + v)'H] = E[H'vv'H] = \sigma_v^2 H'\Omega H$$

we require

ii') $H'\Omega H = I_{T-K}$

or at least $\operatorname{plim}_{T \to \infty} (1/T\ H'v\ v'H) = \sigma_v^2\ H'\Omega H$. For the "best" estimator we choose that estimator which minimizes the conditional expected length of the error vector $(H-J)'v$ where J is the $T \times (T-K)$ matrix formed by deleting K columns from the identity matrix I_T (i.e., J is a selection matrix). We therefore want to minimize

$$(4.5) \qquad E[(H'v - J'v)'(H'v - J'v)] = E[v'(H-J)(H-J)'v]$$

(or the corresponding probability limit) subject to i') $H'Z = 0$ and ii') $H'\Omega H = I_{T-K}$. This implies

$$(4.6) \qquad \min_H \ \sigma^2[\operatorname{tr} H'\Omega H + \operatorname{tr} J'\Omega J - 2 \operatorname{tr} J'\Omega H]$$
$$= \sigma^2[(T-K) + \operatorname{tr} J'\Omega J - 2 \operatorname{tr} J'\Omega H]$$

subject to the above constraints. The minimization problem in (4.6) is the same as choosing H to maximize tr $J'\Omega H$. Therefore, we form the following Lagrangian function

$$(4.7) \qquad L = \operatorname{tr} J'\Omega H - \operatorname{tr} G_1(H'Z - 0) - \operatorname{tr} G_2(H'\Omega H - I)$$

where G_1 and G_2 are matrices of Lagrange multipliers. Differentiating (4.7) with respect to H and setting equal to zero gives

$$(4.8) \qquad \Omega J - ZG_1 - \Omega H(G_2 + G_2') = 0.$$

Premultiplying (4.8) by $Z'\Omega^{-1}$ we obtain

$$(4.9) \qquad Z'J - Z'\Omega^{-1}ZG_1 - Z'H(G_2 + G_2') = 0$$

which, under the restriction that $Z'H = 0$, implies

$$(4.10) \qquad G_1 = (Z'\Omega^{-1}Z)^{-1}Z'J.$$

Substituting (4.10) into (4.8) and premultiplying by H' leads to

$$(4.11) \qquad (G_2 + G_2') = H'\Omega J.$$

Substituting (4.11) into (4.8) we get

$$(4.12) \qquad \Omega J - Z(Z'\Omega^{-1}Z)^{-1}Z'J - \Omega HH'\Omega J = 0$$

$$(\Omega - Z(Z'\Omega^{-1}Z)^{-1}Z' - \Omega HH'\Omega)J = 0.$$

Since J is nonzero (4.12) implies that

$$(4.13) \qquad (\Omega - Z(Z'\Omega^{-1}Z)^{-1}Z' - \Omega HH'\Omega) = 0$$

Premultiplying and postmultiplying (4.13) by Ω^{-1} results in

$$(4.14) \qquad (\Omega^{-1} - \Omega^{-1}Z(Z'\Omega^{-1}Z)^{-1} Z'\Omega^{-1} - H H') = 0$$

To find H (or H') we partition Z, H, and Ω^{-1} according to

$$(4.15) \qquad Z' = \underset{\;\;K \quad T-K}{[Z_0' \; Z_1']}K \quad , \quad H = \underset{\;\;K \quad T-K}{[H_0 \; H_1]}K$$

$$\Omega^{-1} = \begin{bmatrix} \Omega^{00} & \Omega^{01} \\ \Omega^{10} & \Omega^{11} \end{bmatrix} \begin{matrix} K \\ T-K \end{matrix}$$
$$\qquad\qquad K \qquad T-K \qquad .$$

Given these partitions $(Z'\Omega^{-1}Z)^{-1}$ can be expressed as

$$(4.16) \qquad C = [Z_0'\Omega^{00}Z_0 + Z_1'\Omega^{10}Z_0 + Z_0'\Omega^{01}Z_1 + Z_1'\Omega^{11}Z_1]^{-1}$$

and

$$(4.17) \qquad \Omega^{-1} - \Omega^{-1}Z(Z'\Omega^{-1}Z)^{-1}Z'\Omega^{-1}$$
$$= \begin{bmatrix} \Omega^{00}-(\Omega^{00}Z_0+\Omega^{01}Z_1)C(Z_0'\Omega^{00}+Z_1'\Omega^{10}) & \Omega^{01}-(\Omega^{00}Z_0+\Omega^{01}Z_1)C(Z_0'\Omega^{01}+Z_1'\Omega^{11}) \\ \Omega^{10}-(\Omega^{10}Z_0+\Omega^{11}Z_1)C(Z_0'\Omega^{00}+Z_1'\Omega^{10}) & \Omega^{11}-(\Omega^{10}Z_0+\Omega^{11}Z_1)C(Z_0'\Omega^{01}+Z_1'\Omega^{11}) \end{bmatrix}$$
$$= \begin{bmatrix} W_{00} & W_{01} \\ W_{10} & W_{11} \end{bmatrix} = W^{7} \quad .$$

Now from (4.14) we have

$$(4.18) \qquad HH' = W$$

and

$$(4.19) \qquad H_1H_1' = W_{11} = \Omega^{11} - (\Omega^{10}Z_0 + \Omega^{11}Z_1)C(Z_0'\Omega^{01} + Z_1'\Omega^{11}).$$

Since W_{11} is symmetric it can be diagonalized as

$$(4.20) \qquad W_{11} = P \, \Delta^2 \, P'$$

where $\Delta^2 = \text{diag}(\lambda_1 \ldots \lambda_{T-K})$ is a diagonal matrix with the eigenvalues of W_{11} on the main diagonal and P is a nonsingular matrix of the T - K eigenvectors associated with the eigenvalues of W_{11}. It therefore follows that

(4.21) $$H_1 = P \Delta P'.$$

It also follows from

(4.22) $$H_0 H_1' = W_{01} = \Omega^{01} - (\Omega^{00} Z_0 + \Omega^{01} Z_1) C (Z_0' \Omega^{01} + Z_1' \Omega^{11})$$

that

(4.23) $$H_0 = W_{01} W_{11}^{-1} = [\Omega^{01} (\Omega^{00} Z_0 + \Omega^{01} Z_1) C (Z_0' \Omega^{01} + Z_1' \Omega^{11})] P \Delta^{-1} P'.$$

Thus, we have shown that a BLUS residual vector can still be derived when the original disturbances are serially correlated. Such residuals are defined by

(4.24) $$\tilde{v} = H'y$$

where

(4.25) $$H' = [H_0' \; H_1']$$

$$H_1 = P \Delta P'$$

$$H_0 = [\Omega^{01} - (\Omega^{00} Z_0 + \Omega^{01} Z_1) C (Z_0' \Omega^{01} + Z_1' \Omega^{11})] \cdot P \Delta^{-1} P.$$

This particular BLUS residual transformation should guarantee that the joint F-test of the signifiance of the α_i i = 1,...,s in equation (4.1) is a valid test; i.e., $\hat{e}'\hat{e}$ will have a central chi square distribution with T-K-s-1 degrees of freedom under the null hypothesis.

Transforming the model

As was seen above, Theil's procedure requires the original disturbance to have a scalar covariance matrix in order for the Theil residuals to have a scalar covariance matrix. The response of the previous subsection to this problem was to find a new transformation which yields residuals with the desirable properties. Another approach, discussed below, is to transform the original model so as to obtain disturbances which do in fact have a scalar covariance matrix and then apply the usual transformation. This approach is probably easier to use with existing computer programs.

An efficient estimate of θ in (4.2) can be obtained by the application of generalized least squares (GLS)[8] to obtain

(4.26) $$\hat{\theta} = (Z'\Omega^{-1}Z)^{-1} Z'\Omega^{-1} y.$$

This estimator can be obtained by applying OLS to the following transformation of (4.1),

(4.27) $$y^* = Z^*\theta + v^*$$

where $y^* = \Omega^{-\frac{1}{2}} y$, $Z^* = \Omega^{-\frac{1}{2}} Z$, and $v^* = \Omega^{-\frac{1}{2}} v$. The (transformed) disturbance in equation (4.27), v^*, has a scalar covariance matrix and therefore the application of the usual Theil procedure will yield true BLUS residuals. Note, however, that the BLUS residuals are given by $\tilde{v} = A'y^*$ and not $A'y$. It can be shown that the two sets of BLUS residuals derived in this section are identical.

Practical application of the test

The essence of the specification test described here is to obtain BLUS residuals and apply OLS to (4.1). Both methods of obtaining BLUS residuals discussed above have assumed that Ω is known. In general this is not the case. If, however, a consistent estimate of Ω is available, $\hat{\Omega}$ say, then the test will hold asymptotically if $\hat{\Omega}$ is substituted for Ω.[9] We conclude this section by describing how such a consistent estimate can be obtained in the practical application of the test.

The first step in applying the test procedure is to estimate (3.13) by OLS. The length of the lag distribution should be generous so that the truncation contributes explanatory power of trivial variance.[10]

The next step requires modelling the infinite moving average process in (3.13) by a finite-order autoregressive process (or a finite-order mixed ARMA process).[11] This parallels a suggestion by Durbin (1960) for approximating finite MA processes with finite AR processes. A consistent estimate of Ω, $\hat{\Omega}$, can be obtained from the estimates of the autocorrelation coefficients; see Anderson (1971), Theorem 8.4.6, p. 489.

The consistent estimate of Ω can be used to obtain (asymptotic) BLUS residuals in either of the ways suggested above. One can substitute $\hat{\Omega}$ directly into expression (4.24) or use $\hat{\Omega}$ to transform the model as in (4.27) and use Theil's procedure to obtain BLUS residuals.

Finally, the test is completed by estimating equation (4.1) using OLS and testing the joint significance of the coefficients.

CONCLUSION

The quality of empirical economic analysis depends critically on the "correctness" of the empirical economic models used. Thus, the use of tests of specification to improve construction of empirical models is very important. The purpose of this paper has been to present an extension of an exogeneity test due to Ramsey to the case of an important class of time series models. This extended test should be powerful against a wide range of specification errors; e.g., omitted variables, incorrect functional form, as well as endogeneity (of which "instantaneous Granger causality" is a special case) and the more commonly considered case of Granger feedback mechanisms.

FOOTNOTES

[1]The precise definition of "final form" will be given in the third section and should not be confused with Jorgenson's (1966) use of that term.

[2]See Theil (1965) or Ramsey (1969) for further details.

[3]Even if the assumption of finiteness of the polynomial operators in (3.3) is relaxed, following Haugh (1976), we assume they are rational; e.g., $\alpha(L) = \alpha_1(L)/\alpha_2(L)$ where $\alpha_1(L)$ and $\alpha_2(L)$ are finite order polynomials. We maintain the assumption of square summability.

[4]Thus the conclusion, noted in the first section, that Granger noncausality is necessary (but not sufficient) for exogeneity.

[5]Recall that at least one additional exogenous variable, denoted Δ above, is required to eliminate the possibility of "spurious" exogeneity.

[6]We note that the mixed finite ARMA approximation will lead to consistent estimates of the elements of Ω. The contribution of η to v vanishes with increasing M and T so that plim (1/T vv') = plim (1/T $\varepsilon\varepsilon'$). T.W. Anderson (1971) has shown (theorems 5.5.2 (p. 195), 5.5.3 (p. 196), 8.4.2 (p. 478), and 8.4.6 (p 489)) that sample covariances and correlations are consistent and asymptotically normally distributed.

[7]Note that W is orthogonal to Z.

[8]Recall the assumption that Ω is known.

[9]Two further points should be noted. 1) Since the v_t are likely to follow a moving average, inversion of $\hat{\Omega}$ to obtain $\hat{\Omega}^{-1}$ is not as straightforward as for autoregressive processes; see Pesaran (1973), Balestra (1980), or Judge, et al (1980). 2) The asymptotic distribution theory which allows assessment of the properties of the test depend, as for time series tests of this kind, on the sample size. Therefore the dimension of the parameter space must be permitted to increase with the sample size, see Hannan (1970), Amemiya (1973), or Geweke (1978).

[10]Sims (1975, p. 5) suggests this as an expedient way of estimating final form equations when, a priori, the orders of $\alpha(L)$ and $\beta(L)$ are unknown.

[11]We note that the correspondence between MA (∞) and finite ARMA processes is not one-one so any representation will not be unique. Uniqueness, however, is not of importance here; only consistency of the estimator of the variance-covariance matrix.

REFERENCES

AMEMIYA, T., "Generalized Least Squares with an Estimated Autocovariance Matrix," Econometrica (1973), 723-732.

ANDERSON, T.W., The Statistical Analysis of Time Series (New York: Wiley, 1971).

ASHLEY, R., C.W.J. GRANGER, and R. SCHMALENSEE, "Advertising and Aggregate Consumption: An Analysis of Causality," Econometrica (1980), 1149-1168.

BALESTRA, P., "A Note on the Exact Transformation Associated with the First-Order Moving Average Process," Journal of Econometrics, 14 (1980), 381-394.

DURBIN, J., "Estimation of Parameters in Time-Series Regression Models," Journal of the Royal Statistical Society, Series B (1960), 139-153.

GEWEKE, J., "Testing the Exogeneity Specification in the Complete Dynamic Simultaneous Equation Model," Journal of Econometrics (1978), 163-185.

GRANGER, C.W.J. "Investigating Causal Relations by Econometric Models and Cross Spectral Methods," Econometrica (1969), 424-438.

HANNAN, E.J., Multiple Time Series (New York: Wiley, 1970).

HAUSMAN, J.A., "Specification Tests in Econometrics," Econometrica (1978), 1251-1271.

JORGENSON, D.W., "Rational Distributed Lag Functions," Econometrica, (1966), 135-149.

JUDGE, G.G., W.E. GRIFFITHS, R.C. HILL, and T.C. LEE, The Theory and Practice of Econometrics (New York: Wiley, 1980).

PESARAN, M.H., "Exact Maximum Likelihood Estimation of a Regression Equation with a First-Order Moving-Average Error," Review of Economic Studies (1973), 529-536.

PIERCE, D.A., and L.D. HAUGH, "Causality in Temporal Systems," Journal of Econometrics (1977), 265-293.

RAMSEY, J.B., "Tests for Specification Errors in Classical Linear Least Squares Regression Analysis," Journal of the Royal Statistical Society, Series B (1969), 350-371.

REVANKAR, N.S., "Asymptotic Relative Efficiency Analysis of Certain Tests of Independence in Structural Systems," International Economic Review (1978), 165-179.

SARGENT, T.J., "A Classical Macroeconometric Model for the United States," Journal of Political Economy (1976), 207-237.

SIMS, C.A., "Money, Income, and Causality," American Economic Review (1972), 540-552.

SIMS, C.A., "Exogeneity Tests and Multivariate Time Series: Part 1," Discussion Paper No. 75-54, University of Minnesota (1975).

THEIL, H., "The Analysis of Disturbances in Regression Analysis," Journal of the American Statistical Association (1965), 1067-1079.

WU, D.M., "Alternative Tests of Independence Between Stochastic Regressors and Disturbances," Econometrica (1973), 733-750.

APPLIED TIME SERIES ANALYSIS
O.D. Anderson & M.R. Perryman (eds.)

A MONTE CARLO STUDY OF BAYESIAN PREDICTIONS FOR JOIN LOCATIONS IN SEGMENTED LINEAR MODELS

J. A. Norton

Statistics Department
California State University, Hayward
Hayward, California
U.S.A.

A monte carlo study is described which investigates the effects of various assumptions and model properties of two line segments that join at an unknown location. Both bayesian and maximum likelihood estimates are considered. The emphasis of the study is to forecast the join location with as few observations as possible in order to administer treatment coincident with or prior to the join location.

For the conditions studied it was apparent that the prior assumptions for bayesian estimates were not important under many conditions. The noise present in the data in relation to the underlying model is the major determinant of accuracy in forecasting. The use of the mode of the posterior distribution of the join location as the basis for action resulted in forecasts from fewer observations than decisions based on the posterior mean.

1. INTRODUCTION

Observations which appear to abruptly change direction over time are often modelled as intersecting line segments. For example, Cook and Smith (1980) used this model when attempting to infer the time of rejection for transplanted kidneys. Norton (1977) applied this model to estrogen and pregnandiol secretions in urine over a menstrual cycle. This work attempted to forecast mature ova for artifically inducing ovulation. In both of these examples there is a need to forecast a change over time in conditions as near to the change location as possible, preferably before the change occurs. Both papers developed bayesian analysis methods to incorporate existing information into their prediction schemes.

Inference for joined line segment models and large sample distributions are studied by Hinkley (1971). Maximum likelihood estimates for joined line segment models are discussed by Hudson

(1966). Methods which produce approximate maximum likelihood estimates are described by Hawkins (1976). Both maximum likelihood estimates and their approximate counterparts require data sequences known to contain at least two data points past the join location. Thus these estimates and large sample inference cannot be used in early forecasting.

The aim of this paper is to investigate a bayesian method for making forecasts. Some relationships among the prior information, the underlying distribution of the data and decisions for stopping based on incomplete data sequences are studied. Preliminary investigations of values used in this monte carlo study are described in Norton and Lovell (1980).

2. THE MODEL AND BAYESIAN ESTIMATES

Suppose that observations $y_1, y_2, \ldots, y_T$ arrive at times $t = 1,2, \ldots, T$. For an unknown integer r between 2 and $T-1$, the sequences $y_1, y_2, \ldots, y_r$, and $y_{r+1}, y_{r+2}, \ldots, y_T$ are simple linear functions of the arrival times of the observations. Two line segments join at an unknown location τ between r and $r+1$. Thus a two segment model is defined by the equations

$$y_t = a_1 + b_1 t + e_t \qquad (t = 1, \ldots, r)$$

$$y_t = a_2 + b_2 t + e_t \qquad (t = r+1, \ldots, T)$$

$$a_1 + b_1 \tau = a_2 + b_2 \tau \, .$$

The parameters a_1, b_1, a_2, b_2 are the customary regression parameters constrained to intersect at τ. The e_t are independent, identically distributed random variables with mean 0 and variance σ^2 (white noise).

In a bayesian formulation the e_t are usually assumed to follow a normal distribution and that is the case here. This makes the conjugate prior distribution of the linear coefficients $\beta = (a_1, b_1, b_2)$ multivariate normal with mean vector γ and covariance matrix $\sigma^2 \Xi$. (The second intercept a_2 is determined by the constraint equation.) The variance element σ^2 follows an inverse gamma distribution with parameter n.

Assuming that the marginal prior for τ is $p(\tau)$, the marginal posterior density of τ given the first j observations is proportional to

$$p(\tau|y_1,...,y_j) \propto \frac{p(\tau)\ |\Gamma|^{1/2}}{(y'y+\gamma'\Xi^{-1}\gamma-\Delta'\Gamma^{-1}\Delta)^{j+n-2}} .$$

In this form Γ is the posterior covariance matrix for β and Δ is the posterior mean vector for β, both are conditional on τ . with

$$\Gamma = (X'X + \Xi^{-1})^{-1}$$

$$\Delta = \Gamma(X'Y + \Xi^{-1}\gamma) .$$

The matrix X is a function of r and t and is the regression design matrix for estimating a_1, b_1, and b_2 conditional on r and t.

This form of the posterior density is that given by Norton (1977) and is equivalent to that reported by Smith and Cook (1980) with $\nu = n - 2$.

3. THE EXPERIMENTAL DESIGN

The monte carlo study was planned to study two stopping rules for sampling data, two prior covariance matrices Ξ, two error scales for e_t, a flat versus a normal prior for the line parameters β, a flat versus a normal prior for the join point τ and three levels of prior means for the line slopes b_1 and b_2. With 50 replications per experimental unit, 96 experimental units were examined. The experiment was carried out on an Apple II plus microcomputer using available uniform random number generator and normal deviates generated using standard polar techniques.

For each of the 50 replications, the join point was sampled according to its prior distribution and was the same over 48 experimental units, differing only for the two sampling plans, normal prior versus flat prior. The error variance σ^2 was sampled according to its prior inverse gamma distribution and either scaled by 10 or not, depending upon the experimental unit.

The line parameters were sampled from a multivariate normal distribution with marginal mean vector $\gamma = (0,\gamma_1,-\gamma_1)'$. The two covariance matrices were chosen to show near independence between β and the join point τ at one level and high correlation between b_1 and τ at the second level. The conditional covariance matrix for β given τ was obtained from the joint distribution covariance matrix and the conditional mean vector was obtained from conditional properties of the multivariate normal,

$$\gamma_{\beta|\tau} = \gamma + \Xi_{\beta}^{-1}\Xi_{\beta\cdot\tau}(\tau-E(\tau)) .$$

In order to sample the line parameters, the conditional covariance matrix was decomposed so that

$$\Xi_{\beta|\tau} = \Xi_{\beta|\tau}^{1/2}\Xi_{\beta|\tau}^{1/2} .$$

For a particular replication the line parameters were computed from a vector Z of three normal deviates as

$$\beta = \beta|\tau + \Xi_{\beta|\tau}^{1/2} Z .$$

After all the parameters were chosen for a replication, error terms (noise) were generated and added to the chosen model to obtain observations. The posterior density for τ was obtained as each observation $y_1,\ldots,y_j$ $(j \le 12 = T)$ arrived. The sampling of observations ceased if the stopping rule criterion was met. The two stopping rules were

mean stopping:

Sampling should be discontinued and appropriate action taken if the posterior mean of τ after j observations was greater than or equal to j.

mode stopping rule:

Sampling should be discontinued and appropriate action taken if the posterior mode of τ after j observations was greater than or equal to j.

The experimental units, then, consisted of the 96 conditions comprised by all combinations of the following variables.

Distribution of τ

1) $E(\tau) = 6$ $V(\tau) = 4$ normal
2) $E(\tau) = 6$ uniform

Scale of σ^2

1) $V(e_t) = \sigma^2$
2) $V(e_t) = 10\ \sigma^2$

Marginal prior mean for b_1 and $-b_2$

1) $\gamma_2 = 1 = -\gamma_3$
2) $\gamma_2 = 4 = -\gamma_3$
3) $\gamma_2 = 16 = -\gamma_3$

Joint correlation matrix for β and τ

1)
$$\begin{bmatrix} 1 & .375 & 0 & -.2 \\ .375 & 1 & .1 & -.9 \\ 0 & .1 & 1 & 0 \\ -.2 & -.9 & 0 & 1 \end{bmatrix}$$

2)
$$\begin{bmatrix} 1 & .1 & .05 & 0 \\ .1 & 1 & .1 & 0 \\ .05 & .1 & 1 & 0 \\ 0 & 0 & 0 & 1 \end{bmatrix}$$

Prior marginal variance for a_1, b_1, b_2

1) 1
2) 100

Stopping rule

1) mean stopping rule.
2) mode stopping rule.

4. MEASUREMENTS FOR COMPARISONS

For each experimental unit, the means, standard deviations and correlations for the 50 replications of the following statistics were kept.

1) stopping point. The observation number j where the stopping rule criterion was met.
2) stop point error. The difference between the integer part of τ and the stopping point. The optimal condition is to have this error be zero.
3) slope error. The difference between the model slope b_1 and the conditional posterior mean for b_1 evaluated at the marginal posterior mean for τ.
4) posterior mode. The mode of the marginal posterior distribution for τ when the distribution is discretized over the intervals of observation $t = 1,2,...,12$. This value is taken at the time the stopping point is reached.
5) posterior mean. The mean of the marginal posterior distribution for τ when the distribution is discretized over the intervals of observation $t = 1,2,...,12$. This value is taken at the time the stopping point is reached.
6) posterior variance. The variance of the marginal posterior for τ as above. This value is taken at the time the stopping point is reached.
7) tau. The sampled value of τ.
8) b_1. The sampled slope of the first line segment.
9) final estimate. Bayes estimate for τ based on all 12 observations.
10) m.l.e. estimate. Maximum likelihood estimate for τ based on all 12 observations.

From these quantities the statistics for the error in forecasting τ by using the posterior mean, posterior mode, final estimate and m.l.e. estimate can be computed. The average reduction in sampling can also be computed from the stopping point statistics.

Two additional statistics were computed to estimate the proportional reduction in error achieved by using the stopping rule, bayesian estimation procedure as opposed to using the prior mean for τ. These two measurements are

$$\text{PRE} = 1 - \frac{\sum_{i=1}^{50} |\text{posterior mean}_i - \tau_i|}{\sum_{i=1}^{50} |\text{prior mean} - \tau_i|}$$

and

$$\text{PRE 2} = 1 - \frac{\Sigma\,(\text{posterior mean}_i - \tau_i)^2}{\Sigma\,(\text{prior mean} - \tau_i)^2} .$$

These measurements represent the proportion of error (from using the prior mean of τ as the estimate) which is reduced by using the posterior mean as the estimate instead.

Counts were kept for the 50 trials which categorized how accurate the stopping criterion was. If the difference between the number of observations required to reach a decision to make a forecast and the integer part of τ was zero, then the method was forecasting join points appropriately. A difference of $-i$ indicated a decision to make a forecast which was i observations early. A difference of $+i$ indicated a decision to stop which was i observations late.

5. RESULTS OF THE EXPERIMENT

The analysis of variance was used to examine main effects of the design for some of the measurements. All interactions were assumed negligible and combined with the error term. Examining the available data supported the assumption of no interactions except as noted.

Using the PRE measure, the level of the slope mean contributed the most variation in PRE. The stopping rule was next in influence, followed by the error level. The prior scale was just significant at the .05 level while the distribution of τ and the correlation structure were not significant at the .05 level.

When the second measure, PRE 2, was used for the analysis of variance, the order of decreasing significance was the slope mean, the error level and the stopping rule. Following these highly significant (< .01) effects, the prior scale was significant at the .01 level and the correlation structure was just significant at the .05 level.

Using the stopping point error as the analysis of variance measurement, the most influential effect on variation was the stopping rule. The mode stopping rule stopped with fewer observations than the mean stopping rule so that there were more opportunities for early predictions and lower errors. The error level, slope level and the distribution of τ were all highly significant (< .01). The normal distribution of τ resulted in fewer early predictions and larger errors than the uniform distribution of τ. The prior distribution of the linear parameters was significant at .05 with larger errors for the normal prior than the diffuse prior. The prior correlation structure was not significant.

Table 1 summarizes the treatment means for all levels of the experimental variables and all four measurements. The two measures PRE and PRE 2 which reflect the proportional reduction in error in estimating the join point show a larger reduction for the mean stopping rule. This fact is reasonable since the mean stopping rule in general requires more observations than the mode stopping rule. Both the stop point error measure and slope error favor the mode as having smaller errors. The distribution of τ does not significantly effect the PRE measures, but a uniform sampling rule lowers stop point errors and increases slope errors.

As would be expected, larger data errors decrease PRE measures, but, surprisingly, they also decrease both error measurements. The most reasonable explanation is that the lack of variability in the data forces the likelihood to overpower the prior information because there is no indication of a down trend. This explanation is supported by the fact that low slope means have the lowest average stop point error. The low, medium and high slope means indicate that increasing the slope means increases slope error but decreases the PRE measures. More accurate τ forecasts are obtained at the expense of β estimates and early

Table 1

Averages for experimental conditions of four measurements

Experimental Conditions	Measurement Averages			
	PRE	PRE 2	Stop Point Error	Slope Error
Mean stopping rule	.586	.708	-.569	1.063
Mode stopping rule	.388	.523	.180	.714
Uniform τ distribution	.488	.613	-.124	.998
Normal τ distribution	.485	.619	-.291	.779
Normal β prior	.508	.651	-.259	.775
Diffuse β prior	.465	.581	-.156	1.001
Low data errors	.553	.712	-.349	.972
High data errors	.420	.520	-.066	.805
Low slope mean	.228	.335	-.075	.181
Medium slope mean	.510	.652	-.158	.520
High slope mean	.722	.860	-.390	1.964
Low correlation	.472	.593	-.188	.928
High correlation	.502	.639	-.227	.849

forecasts. Prior correlation is only significant in the PRE measures. High correlation increases PRE.

Applying the analysis of variance to the counts of early and late categorization reveals that the only important experimental variable in the category of zero counts is the distribution of the join point. The analyses with τ selected from a normal distribution averaged significantly more counts in this category.

The variation in one early counts is attributed in decreasing order to stopping rule, slope mean level, error level, and the sampling distribution of the join point. The prior distribution for β and the correlation structure are not significant contributors. The mode stopping rule selects 3 more early intervals in 50 on the average than does the mean stopping rule. If an early forecast is preferred to a late forecast, the mode stopping rule is better. The mode stopping rule has significantly fewer one late counts as well.

The category of one late counts identifies cases where the join interval is correctly identified as $(j-1, j)$ with j observations or one time unit later than 0. The variation in these counts is due in decreasing order to the level of β, stopping rule and data error. High levels of the slope mean cause sampling to continue past the join point because of strong likelihood influence. High data errors lower the number of late counts as well since larger errors will have a tendency to show "false" changes in slope. An interaction between error level and slope mean seems likely.

Table 2 summarizes the average counts for each of the experimental conditions. The averages favor the mode stopping rule for early forecasting. A low slope mean leads to more early decisions but more late decisions also. A high slope mean resulted in the one time unit late category fifty-four percent of the time. This was the most extreme effect in the category.

6. SUMMARY

The effects of applying bayesian methods to a sequence of observations arriving at equally spaced intervals and modeled by

Table 2

Averages for experimental conditions of count categories

Experimental Conditions	Average Counts for Fifty Trials				
	<-1	-1	0	+1	>+1
Mean stopping rule	4.06	5.38	6.79	23.71	10.04
Mode stopping rule	11.77	8.04	6.90	18.44	4.90
Uniform τ distribution	9.67	6.23	5.04	21.44	7.62
Normal τ distribution	6.17	7.19	8.65	20.71	7.31
Normal β prior	7.46	6.62	6.54	21.73	7.69
Diffuse β prior	8.38	6.79	7.15	20.42	7.25
Low data error	6.44	6.21	7.02	22.42	7.96
High data error	9.40	7.21	6.67	19.73	6.98
Low slope mean	9.94	8.19	7.25	14.19	10.50
Medium slope mean	8.44	6.47	6.66	21.84	6.56
High slope mean	5.38	5.47	6.62	27.19	5.34
Low correlation	7.90	6.79	7.04	21.10	7.17
High correlation	7.94	6.62	6.65	21.04	7.79

two intersecting line segments were studied in an attempt to improve forecasting of trends by finding the location of a join point. The method which leads to the earliest forecasts uses a rule based on the posterior mode of the join point. While the particular choice of prior distribution has some effects in the forecasting process, vague priors work suitably.

REFERENCES

HAWKINS, D.M. (1976). Point estimation of the parameters of piecewise regression models. *Applied Statistics* 25, 51-57.

HINKLEY, D.V. (December, 1971). Inference in two phase regression. *Journal of the American Statistical Association* 66, 736-743.

HUDSON, D.J. (December, 1966). Fitting segmented curves whose join-points have to be estimated. *Journal of the American Statistical Association* 61, 1097-1129.

NORTON, J.A. (January, 1977). Estimation and sequential prediction for unknown join points in broken line segment models. Ph.D. Thesis, Harvard University, Cambridge, Massachusetts.

NORTON, J.A. and LOVELL, J.D. (January, 1981). Statistical modeling at the University of the South Pacific Extension Services. Extension Services Report, University of the South Pacific, Suva, Fiji.

SMITH, A.F.M. and COOK, D.G. (1980). Straight lines with a change-point: A Bayesian analysis of some renal transplant data. *Applied Statistics* 29, 180-189.

APPLIED TIME SERIES ANALYSIS
O.D. Anderson & M.R. Perryman (eds.)

EXPECTED INCOME, PERMANENT INCOME, AND OPTIMAL FORECASTING

Douglas K. Pearce
Department of Economics
University of Missouri-Columbia
Columbia, Missouri
U.S.A.

James R. Schmidt
Department of Economics
University of Nebraska-Lincoln
Lincoln, Nebraska
U.S.A.

The terms "expected income" and "permanent income" have sometimes been used synonomously in macroeconomic contexts, consumption and money demand theories being prime examples. Recently, Meyer and Neri [7] and Sargent [8] have made potentially useful distinctions between expected and permanent income, the former authors by incorporating permanent income into an expected income construct and the latter author by doing the reverse. This paper re-examines the relationship between Friedman's [5] permanent income concept and short-run expected income. In particular, we approach this issue from the viewpoint of optimal forecasting and rational expectations. We show that from this perspective, there are serious problems with the definitions proposed by Meyer and Neri and Sargent.

Since Meyer and Neri argue that their measure of expected income corresponds to the optimal forecast, section II evaluates this assertion and presents evidence contradicting it. Section III outlines our methodology for constructing an expected income series. Section IV relates our measure of expected income to the model of permanent income suggested by Sargent. Section V summarizes our results.

I. Expected Income and Permanent Income

Is there a difference between short-run expected income and Friedman's notion of permanent income? While we do not wish to indulge in exegesis, it appears that the two terms have often been used interchangeably. Friedman [5] is not explicit on this point. He does suggest that one should "tentatively regard y_t^p (permanent income) as the 'expected' or predicted value of current measured income" (p. 143), while earlier in the same work he states that one possible interpretation of permanent income is the "mean of the probability distribution anticipated for future years" (p. 25). However, Friedman does argue that it is "neither necessary nor desireable to decide in advance the precise meaning to be attached to permanent'" (p. 23). Feige [2] uses as his measure of expected income the standard construct of permanent income involving exponentially declining weights on current and past income. Similarly, Laidler [6], in his review of the money demand literature, equates expected and permanent income but does not make the horizon of the expectation explicit.

At least two general distinctions, one conceptual and one empirical, can be made between permanent and expected income in the context of the time series behavior of aggregate income. Suppose, for example, that actual income fluctuated randomly over time about a sine curve. Short-run expected income should roughly trace out the sine curve while permanent income, when viewed as an average income from this experience, should flatten out the peaks and troughs. The second distinction concerns the conventional measure of permanent income used in macroeconomic studies:

$$(1) \quad y_t^p = \lambda y_t + (1-\lambda)y_{t-1}^p \qquad 0<\lambda<1,$$

or

$$y_t^p = \sum_{i=0}^{\infty} \lambda(1-\lambda)^i y_{t-i}.$$

When defined by this classic exponential weighting scheme on current and past incomes, permanent income includes the actual income of the current period and it receives the maximum weight. The inclusion of current income is at odds with the notion of expected income which, being an expectation, should incorporate information only through the time period prior to the one for which the expectation is being made.

We limit the information set to past income and write the general form of expected income in time t as a conditional expectation

$$(2) \quad y_t^e = \underset{t-1}{E}(y_t) = E[y_t | y_{t-1}, y_{t-2}, \ldots].$$

In forming the expectation, agents are assumed to use the optimal linear predictor of y_t obtained from a suitable model of the stochastic process which characterizes the income series. Suppose that the first difference of income follows a first-order moving average process in the error terms, a_t,

$$(3) \quad y_t = y_{t-1} + a_t - \Theta\, a_{t-1},$$

or,

$$y_t = \sum_{i=1}^{\infty} \lambda(1-\lambda)^{i-1} y_{t-i} + a_t$$

where $\lambda = 1-\Theta$. The optimal linear predictor of y_t is then

$$y_t^e = \underset{t-1}{E}(y_t) = \sum_{i=1}^{\infty} \lambda(1-\lambda)^{i-1} y_{t-i}.$$

Actual income from the current period is correctly excluded from the expectation of income for the current period.

One opportunity for reconciling the above measures of permanent and expected income does exist but the required conditions are overly restrictive. Specifically, permanent income of the current period is equal to expected income of the first period in the future given the definition in (1) and assuming that (3) happens by chance to be the correct process for income with $\lambda = 1-\Theta$. This forced equivalence would also hold between permanent income in the current period and expected income in any future period since forecasts of income from (3) are identical for all future periods.

II. The Meyer and Neri Model of Expected Income

Meyer and Neri [7] present a distinction between expected and permanent income in the context of an argument that the finance motive for holding money balances, emphasized by Keynes in his post-*General Theory* writings, requires the scale variable in the demand for money function to be the one-period ahead expectation of income, y^e_{t+1} or $E_t(y_{t+1})$ in our notation. It is assumed that this expectation is formed as a regression towards "normal income", y^n_t. The model is

$$(4) \quad y^e_{t+1} = y_t + \lambda(y^n_t - y_t)$$

$$(5) \quad y^n_t = \sum_{i=0}^{\infty} (1-\beta)\beta^i y_{t-1}$$

and so

$$(6) \quad y^e_{t+1} = (1-\lambda\beta)y_t + \lambda\beta(1-\beta) \sum_{i=0}^{\infty} \beta^i y_{t-1-i}.$$

Normal income is identical to the classic empirical definition of permanent income given in (1) above. Meyer and Neri go on to assert that equation (6) yields optimal one-period ahead forecasts of income for their data set: "the optimal forecast for next period, a modified exponentially weighted average of past income, corresponds with our short-run expected income construct based on doctrinal history" (p. 610). Equation (6), however, gives optimal one-period ahead forecasts only if income happens by chance to be generated by a particular stochastic process, namely the ARIMA (1,1,1) process:

$$(7) \quad y_t = (1+\phi)y_{t-1} - \phi y_{t-2} + a_t - \Theta a_{t-1}$$

with the constraints $\Theta = \beta$ and $(\Theta - \phi) = \lambda\beta$. Since equation (6) is chosen for the expected income model on an <u>a priori</u> basis, Meyer and Neri's approach appears contrary to the notion of rational expectation formation according to which agents utilize available information to make the best forecast. If the information set is limited to past observations of the series, agents are assumed to learn the stochastic process generating the data and employ this estimate of that process in making their predictions. As time passes they must learn if the process has changed and make the appropriate adjustments. The data known at the time the expectation is made should, however, dictate the model chosen. In the next section, such an approach for quantifying expected income is developed.

We now turn to an empirical evaluation of the expected income model of Meyer and Neri. As noted, the process generating income must be of the form given in equation (7) for their model to yield optimal forecasts of future income within the income sample.[1] Analyzing their data set which consists of the log of real income over 1897-1960, we find that the income series is adequately modeled as

$$(8) \quad y_t = y_{t-1} + .031 + a_t,$$

the first difference of income with a constant. Thus, the ARIMA(1,1,1) process of (7) represents a case of overspecification. To substantiate this claim, the forecast errors from their estimated models will be compared with the errors from our time series model in equation (8).

Meyer and Neri obtained values for λ and β by substituting their expected income construct, (6), into a linear model of the demand for money and then estimating the resulting specification.[2] Two money supply definitions, M1 and M2, and two interst rate definitions were used. Their estimates of λ and β can be used to compute the ϕ and Θ parameters of the implied ARIMA(1,1,1) forecasting model in (7). Table 1 contains the mean absolute error and mean square error of the one-period ahead forecasts from the Meyer and Neri models and the time series model in equation (8). Also presented is a χ^2 statistic, Q, which tests the hypothesis that forecast errors are serially independent and thus devoid of predictive information. The table clearly shows that our equation (8) yields dramatically better forecasts than the Meyer and Neri model in (6) and exhibits serially uncorrelated errors while (6) does not. We conclude that the expected income construct in (6) is not an optimal forecasting model. The construct is certainly at odds with even the crudest notion of rational or optimal expectation formation.

TABLE 1

Forecast Error Statistics for Models of Expected Income, 1897-1960

Model	Mean Absolute Error	Mean Square Error	Q
Meyer-Neri (M1 and r_s)	.1537	.0286	65.0
Meyer-Neri (M2 and r_s)	.2480	.0687	68.6
Meyer-Neri (M1 and r_L)	.1302	.0224	115.3
Meyer-Neri (M2 - r_L)	.2040	.0483	77.2
Equation (8)	.0584	.0057	25.0

Notes: M1 - currency plus demand deposits, M2 - M1 plus time deposits, r_s - 4-6 month commercial paper rate, r_L - 20 year corporate bond rate. The Q statistic is distributed χ^2 with degrees of freedom equal to the number of lagged autocorrelations of the errors. Forty autocorrelations were used. Thus, the null hypothesis of randomness is rejected for the forecast errors from the Meyer-Neri models.

III. Modeling Expected Income

Rather than positing an expectation formation process, the rational expectations theory first requires an assumption of the information set that agents are employing and then asserts that expectations will be identical to the best forecasts conditional on that information set. Initially, we shall limit the set to past income so expectations generated by the scheme below might be considered only narrowly or partly rational.[3] An immediate issue is the length of the historical record which agents believe relevant. In the context of annual observations on income, we have arbitrarily set the historical record at fifty years. Our income data consists of observations on real net national product over 1900-1974 and the methodology for contructing expected income is as follows.[4] Using an initial period of 1900-1949, we estimate the underlying stochastic process. This estimated model may then be used to form the conditional expectation or forecast of income for 1950, the forecast representing expected income for 1950. The sample then changes to 1901-1950 and a new model is estimated which may be used to forecast income for 1951. This method is repeated until the terminal sample, 1925-1974, is reached. Thus, agents are allowed to adjust their expectation models over time. Note that only the past history of income is used to determine the expectation model at any given point in time. This feature contrasts sharply with the common practice of estimating an expectations model from a sample of time series data and then inferring the period by period expectations within the sample from the single model of the entire sample. Our strategy of sliding the modeling process through time and allowing each expectation to be made from a unique model is a more realistic approach.[5]

Models of the sequential set of income series were estimated using the raw data on real income as well as natural logarithms of the data. A relatively simple process characterizes each of the sliding samples of income for both the raw data and the logarithmic data. The general model is

$$(9) \qquad y_t = y_{t-1} + \mu_i + a_t - \Theta_i a_{t-1} \qquad i=1949,\ldots,1974.$$

Thus, the first difference of income follows a first-order moving average process, an ARIMA(0,1,1), with a trend.

Agents could have utilized the sequential set of models above in forming their expectations through time but exceptional events occurring during the time spans of the income series may have had a significant effect upon expectation formation. For example, consider the forecast of income in, say, 1965 from the model estimated over 1915-1964. Income observations from the depression years are being given normal weight in the determination of the forecasting model when, in reality, the depression experience may have been discounted or given special treatment by agents. The income experiences during the two world wars are perhaps similar cases. Allowance for these potential effects can be made in the model of (9) by including parameters which quantify the events. Following Box and Tiao [1], the time series models are respecified for the income samples

$$(10) \qquad y_t = \sum_{j=1}^{k} \delta_{ij} I_{ijt} + \frac{\mu_i + (1-\Theta_i B)a_t}{1-B} \qquad i=1949,\ldots,1974$$

where B is the lag operator. I_{i1t} is a 0-1 indicator variable for the first world war having the value of one during 1917 and 1918, I_{i2t} is the indicator variable for the depression years having the value of one during 1930-1936, and I_{i3t} is the indicator variable for the second world war having the value of one during 1942-1945. The parameter δ_{ij} quantifies the effect of the respective events and k is the number of events occuring in the time span of the particular income sample. If the δ_{ij} equal zero, then the model in (10) collapses to (9).

Estimates of (10) using the raw and logarithmic data are given in Tables 2 and 3, respectively. The parameter representing the effect of the first world war was not significant in any of the models and was dropped but the parameter representing the depression was significant in the models containing it. Finally, the parameter representing the second world war was significant only in the models which employed the raw data on real income.

TABLE 2

Parameters of the Sequential Time Series
Models of Real Income, 1949-1974

Terminal Year	μ_i	Θ_i	δ_{i2}	δ_{i3}
1949	2.1162	-.5663	-11.3209	8.7011
1950	2.3778	-.5743	-11.3838	8.2085
1951	2.5821	-.6366	-12.0162	7.6170
1952	2.5941	-.6396	-12.0452	7.5499
1953	2.8065	-.7072	-12.7716	6.8796
1954	1.9357	-.6761	-14.6408	5.5301
1955	2.9156	-.4858	-10.5911	9.2289
1956	2.7249	-.4167	-10.0594	9.9345
1957	3.0429	-.4341	-10.1856	9.7572
1958	2.8216	-.4139	-10.0394	9.9634
1959	3.1576	-.3644	-9.7106	10.4684
1960	3.1458	-.3419	-9.5753	10.6981
1961	3.2449	-.3341	-9.5306	10.7776
1962	3.4357	-.3647	-9.7127	10.4650
1963	3.7044	-.3390	-9.5587	10.7274
1964	4.0485	-.3668	-9.7255	10.4441
1965	4.0703	-.4253	-10.1211	9.8471
1966	4.4973	-.4424	-10.2474	9.6733
1967	4.4249	-.4371	-10.2075	9.7275
1968	4.6537	-.4216	-10.0946	9.8846
1969	4.7754	-.4059	-9.9834	10.0447
1970	4.7574	-.4085	-10.0017	10.0190
1971	5.0210	-.3986	-9.9333	10.1191
1972	5.1807	-.4407	-10.2276	9.6901
1973	5.4459	-.4580	-10.3530	9.5135
1974	5.0803	-.4258	-10.1368	9.8425

Notes: δ_{i2} and δ_{i3} are the parameters for the depression and second world war, respectively.

TABLE 3

Parameters of the Sequential Time Series
Models of Real Income (Log data), 1949-1974

Terminal Year	μ_i	Θ_i	δ_{i2}
1949	.0295	-.4108	-.1151
1950	.0286	-.4447	-.1180
1951	.0306	-.4684	-.1202
1952	.0293	-.4833	-.1216
1953	.0313	-.4941	-.1226

1954	.0292	-.4969	-.1229
1955	.0289	-.4090	-.1149
1956	.0258	-.5039	-.1236
1957	.0328	-.7353	-.1499
1958	.0274	-.6763	-.1428
1959	.0298	-.6326	-.1376
1960	.0285	-.6029	-.1340
1961	.0295	-.5916	-.1329
1962	.0282	-.5796	-.1315
1963	.0309	-.4875	-.1220
1964	.0347	-.6239	-.1366
1965	.0292	-.6881	-.1445
1966	.0326	-.6873	-.1446
1967	.0308	-.6796	-.1437
1968	.0303	-.6832	-.1440
1969	.0309	-.6779	-.1430
1970	.0321	-.6665	-.1420
1971	.0334	-.6717	-.1423
1972	.0307	-.6529	-.1385
1973	.0305	-.6506	-.1384
1974	.0295	-.6436	-.1381

The one-period ahead expectation of income for any particular year in the span of 1950-1975 can be determined from the forecasting function of the model whose terminal sample point was the previous year. Note that forecasts of future income from the models will not utilize the estimated parameters that quantify the impacts of the depression and second world war since the values of the indicator variables are zero. For example, the forecasting function from the 1900-1949 sample of raw data is

$$\hat{y}_{1949}(\ell) = y_{1949} + 2.1162\,\ell + .5663a_{1949},$$

where ℓ is the number of periods ahead for which the forecast or expectation of income is being made.[6]

IV. Sargent's Definition of Permanent Income

Sargent [8] has employed the following definition of permanent income which can incorporate our construct of expected income

$$y_t^p = (1-\alpha)\sum_{\ell=0}^{\infty} \alpha^{\ell} E[y_{t+\ell}].$$

Permanent income is the weighted sum of expected incomes for future time periods, the sequence of expectations all being formed at time t. The parameter α functions as a discount factor and is constrained by $0<\alpha<1$. It is apparent from the above definition that permanent income will be determined from the stochastic process that is generating observed income, given that the information set for expectation formation is restricted to the past history of income. Using the above definition along with our scheme of modeling expected income would require placing the forecast function, $E_t[y_{t+\ell}]$ or $\hat{y}_t(\ell)$, for the model having the relevant terminal year t into the definition and then determining the convergence properties of y^p.[7] The forecast function for each of our models in the sequential set is

$$y_t(\ell) = y_t + \ell\mu - \Theta a_t.$$

Substituting this expression into y_t^p above yields an infinite series which converges to

$$y_t^P = y_t + \frac{\mu\alpha}{1-\alpha} - \alpha\Theta a_t.$$

Transitory income is therefore

$$y_t^T = y_t - y_t^p = \alpha\Theta a_t - \frac{\mu\alpha}{1-\alpha}.$$

The rightmost term is always positive and Θ is always negative in our empirical results. Thus, transitory income will be positive only if a_t is a sufficiently large negative number.

A set of permanent income series were generated for the years 1950 to 1975 using values of α, the discount parameter, which ranged from .1 to .9 in increments of .1. Values of μ, Θ, and a_t were taken from the estimates of the sequential set of models given in Table 5. Transitory income was negative for all but two years in the span using α values of .1 and .2, negative for all but one year in the span using α values of .3 through .7, and negative for all years in the span using α values of .8 and .9. Thus, the transitory income measure based upon Sargent's definition certainly does not satisfy a randomness criterion. In addition, the mean transitory income over the span for each α value was significantly below zero.

The difficulties are even more apparent if income should happen to follow a first-difference with a trend. In that case, the forecast function at time t is

$$\hat{y}_t(\ell) = y_t + \ell\mu$$

and the y_t^p definition converges to

$$y_t^p = y_t + \frac{\mu\alpha}{1-\alpha}.$$

Transitory income is always positive or negative depending upon the sign of μ. If income is consistently trending upward or downward over time, the randomness property usually attributed to transitory income is clearly violated as is the property of a zero mean or expectation.

V. Summary

This paper has presented a strategy for modeling the income expectaions process when the information set is limited to past incomes. The procedure involves estimation of time series models for income samples having terminal periods prior to the period or periods for which the income expectation is being formed. The approach has several desirable features. Expectations for any particular time period are not imputed from a model which has been estimated (either by a time series method or within the context of a regression equation) using information that includes income from subsequent time periods. No mechanisms are forced upon the expectations process whereas doing so is almost sure to jeopardize the feature of optimal forecasting. In addition, economic events or situations which are hypothesized to be atypical of income behavior during any sample span may be given special consideration by our modeling strategy.

Finally, when the expected income models are used to construct a permanent income series from Sargent's definition, the implied transitory income series violated the randomness assumption, casting doubt on the appropriateness of Sargent's approach.

Footnotes

1. Optimal is used in the sense of minimum mean square forecast error.

2. Meyer and Neri used two money demand models in their investigation, one which assumed equality between actual and desired money balances and one which employed the partial adjustment mechanism. Much of their discussion was based upon the former model so it is the one selected for evaluation here.

3. See Feige and Pearce [3] for a discussion of these issues.

4. The unpublished income data was obtained from the National Bureau of Economic Research.

5. Feldstein and Summers [4] have used the technique of sliding the modeling process through time for the purpose of computing inflation rate forecasts. Their rationale for doing so is that "forecasts made at anytime are to be based only on the information available at that time." (pg. 84).

6. We regret that computed values of expected income cannot be reported. To do so might jeopardize the unpublished nature of the income data.

7. A detailed discussion of convergence properties is available in Sargent [8].

References

1. Box, G. E. P. and G. C. Tiao, "Intervention Analysis with Applications to Economic and Environmental Problems." Journal of the American Statistical Association, March 1975, 70-79.

2. Feige, E. L., "Expectations and Adjustments in the Monetary Sector." American Economic Review, May 1967.

3. Feige, E. L. and D. K. Pearce, "Economically Rational Expectations: Are Innovations in the Rate of Inflation Independent of Innovations in Measures of Monetary and Fiscal Policy?" Journal of Political Economy, June 1976, 499-522.

4. Feldstein, M. and L. Summers, "Inflation, Tax Rules, and the Long-Term Interest Rate." Brookings Papers on Economic Activity, 1:1978, 61-99.

5. Friedman, M. A Theory of the Consumption Function. Princeton, N.J.: Princeton University Press, 1957.

6. Laidler, D. E. W. The Demand for Money: Theories and Evidence. New York: Dun-Donnelley, 1977.

7. Meyer, P. A. and J. A. Neri. "A Keynes-Friedman Money Demand Function." American Economic Review, September 1975, 610-623.

8. Sargent, T. J., "Rational Expectations, Econometric Exogeneity, and Consumption." Journal of Political Economy, August 1978, 673-700.

APPLIED TIME SERIES ANALYSIS
O.D. Anderson & M.R. Perryman (eds.)

CAUSALITY AND THE TEMPORAL CHARACTERIZATION OF MONETARY RESPONSES

M. Ray Perryman

Center for the Advancement of Economic Analysis
Baylor University
Waco, Texas
U.S.A.

Recent advances in the detection of temporal causal patterns are employed to examine the nature of "effective" monetary responses. (This approach captures the impact of monetary variables as a result of endogenous behavior rather than the simple policy response of control theory. The paper offers evidence on the selection of an optimal monetary indicator and finds that (1) there is a problem of reverse causation in developing reaction functions from traditional monetary aggregates and (2) policy responds most predictably to changes in income and international conditions.

INTRODUCTION

In recent years, the orthodox Keynesianism which dominated macroeconomic analysis for several decades has been confronted with numerous theoretical and empirical dilemmas and Monetarism has surfaced as a significant subdiscipline of economic thought. Consequently, a substantial literature has developed regarding both the nature and conduct of monetary policy in the United States and other advanced societies. This analysis has been pursued along a variety of diverse strains, including the targets and indicators issue,[1] the measurement and characterization of the lag in monetary policy,[2] and the assessment of the relative importance of monetary and fiscal policy, i.e., the St. Louis equation controversy.[3] An additional area of inquiry concerns the degree to which Federal Reserve actions respond to various aspects of contemporaneous economic activity, i.e., the estimation of "reaction functions."[4] In previous work within this realm, the approach employed to measure the stance of monetary policy has been to utilize a "target" variable through which authorities implement their objectives (e.g., the federal funds rate, the Federal Reserve security portfolio, or the monetary base). This specification is generally derived from a basic optimal control model of an economy and is designed to reflect "the overall thrust of monetary policy."[5]

From a pragmatic standpoint, the existing literature on monetary reaction functions suffers from two major shortcomings:

1. The variables utilized in the estimation of these relationships do not attempt to distinguish between "dynamic" and "defensive" monetary actions. Consequently, it is impossible to discern whether a particular perceived policy response is the result of endogenous forces in the economy, random external events, or a combination of the two. For money applications, however, the value of a reaction function for analytical purposes would be greatly enhanced by an ability to ascertain the nature of <u>dynamic</u> actions.
2. The typical reaction function contains explanatory variables which seek to reflect fundamental arguments in a policy goal function, i.e., output, prices, employment, and the international position of the economy. The resultant specification implicitly assumes a unidirectional causal process which, in reality, is not well grounded

in either standard theory or empirical evidence. The consequence of this simplification is, of course, simultaneous equation bias and a lack of confidence in the analytical conclusions. This problem, which has also been a source of considerable concern in the St. Louis equation literature, is generally referred to as "reverse causation".

Both of these difficulties serve to minimize the validity and usefulness of existing functions.

The present analysis seeks to overcome the difficulties outlined above through the development and implementation of alternative approaches to the reaction function problem. Initially, the first of the two factors noted above is addressed through the use of monetary indicators which are purged of cyclical influences and, thus, may be used to distinguish between dynamic and defensive policy. These measures of the stance of policy include the neutralized money stock (NMS) and two noncyclical indicators (M_1* and M_2*) derived from a simultaneous equation model.[6] A brief synposis of these measures is given at the outset of the paper. The second contingency noted above, the likelihood of mutual causation, should also be overcome through the utilization of variables which are constructed to be independent of endogenous economic activity.[7] Moreover, these relationships may be pre-tested for bi-directional causal processes using recent advances in time series and econometric procedures. The techniques employed in this aspect of present study are also summarized within the study. The results of the application of causal methodologies as a basis for reaction function development are then presented. Following the preliminary examination of the bivariate relationships, this information is utilized in order to formulate and estimate policy response measures with standard distributed lag models. The estimates are generated for the period from 1953 through 1975. Finally, a concluding section summarizes the findings of the study and assesses their impact for policy analysis.

THE NEUTRALIZED MONEY STOCK

The present section describes the procedure utilized by Hendershott in the derivation and construction of the neutralized money stock.[8] This indicator of monetary policy was the first systematic attempt to delineate dynamic and defensive monetary actions. In developing this variable, a basic conception of exogeneity is embodied which maintains that a policy indicator should only be independent of endogenous cyclical forces. The NMS thus reflects the systematic influence of monetary policy and the random impact of exogenous shocks on the economy, i.e., it is purged of business cycle phenomena. The rationale for utilizing this framework is provided by Hendershott (1968, pp. 98-99) as follows:

> The rationale for removing the impact of changes in endogenous variables, but not that of noncontrolled exogenous variables and random shocks, is best illustrated by an example. Say the endogenous forces (e.g., a domestic downswing accompanied by falling interest rates) and noncontrolled exogenous forces (e.g., military expenditures abroad) simultaneously generate a U.S. gold outflow. Also assume that the Federal Reserve offsets the entire gold flow by purchasing securities in the open market. This offsetting action can be interpreted in at least three ways. If one takes the (unchanged) observed money stock [or the (unchanged) extended adjusted monetary base] as the indicator of cyclical policy actions, one interprets the action as no action; the Federal Reserve is considered as pursuing policies that are neutral with respect to the business cycle. Alternatively, if one takes the impact of explicit Federal Reserve actions (open-market operations, etc.) on the money stock as the measure of Federal Reserve cyclical policies--if one contends, a la Brunner and Meltzer,

> that the policy indicator should be independent of both endogenous and noncontrolled exogenous variables--the Federal Reserve is interpreted as taking easing (contracyclical) actions equal to the size of the total purchase. Finally, if one takes the neutralized money stock as the measure of contracyclical actions--if one contends, as we do, that the money stock after the removal of the influence of the business cycle is the appropriate indicator--then only that part of the purchase that offsets the recession-induced component of the gold loss is viewed as contracyclical actions. The latter would seem to be the most reasonable interpretation of the Federal Reserve's policy stance vis-a-vis the business cycle. The Federal Reserve is clearly taking contracyclical actions; it is offsetting the impact of the downswing on bank reserves and thus on the money stock. However, the entire purchase is not contracyclical actions; only that part of the purchase that offsets the influence of the cycle can be properly classified as contracyclical.

It should thus be apparent that the NMS and other variables with comparable theoretical properties may be used to delineate dynamic and defensive policies. Given this characteristic, such measures become extremely useful in the development of effective monetary reaction functions and in the resulting measurement of Federal Reserve responses.

In the actual derivation of the neutralized money stock indicator, Hendershott makes use of the following procedure:

1. The basic Federal Reserve definition of the money supply (M1), the definition of the total level of required reserves in the monetary system, and the sources-and-uses-of-reserves identity are arithmetically manipulated in order to develop the following complex money stock identity:

$$M1 = S/rr_d + C_t/rr_d - V/rr_d - T_c/rr_d - A_o/rr_d + [(1-rr_d)/rr_d]DD_f$$

$$- DD_{gm} + DD_n + B/rr_d - R_e/rr_d + [(1-rr_d)rr_d]C_o + [(1-rr_d/rr_d]F$$

$$- (rr_t/rr_d)TD_m + G/rr_d,$$

where:

$M1$ = Basic Federal Reserve definition of the money stock
S = Federal Reserve holdings of U.S. Government securities
C_t = Treasury currency outstanding
V = Vault cash <u>not</u> countable as legal reserves
T_c = Treasury cash holdings
A_o = Accounts other than foreign or member bank deposits at the Federal Reserve
DD_f = Foreign demand deposits at member banks
DD_{gm} = U.S. government demand deposits at member banks
DD_n = Demand deposits at nonmember banks
B = Borrowed reserves
R_e = Excess reserves
F = Federal Reserve float
C_o = Currency in circulation outside banks
TD_m = Time deposits at member banks
G = Gold Stock
rr_d = Average legal reserve requirements against time deposits at member banks
rr_t = Average legal reserve requirements against time deposits at member banks.

2. After the construction of this identity, six of the included

variables (B, R_e, F, C_o, TD_m, and G) are classified as endogenous and thus influenced by the business cycle. Regression equations to explain movements in each of these variables are then specified and individually estimated using the simple ordinary least squares (OLS) technique. It should be noted that these equations do not contain variables reflecting Federal Reserve policy and contain few exogenous variables other than seasonal dummies. They represent primarily a statement of the endogenous influences on the relevant components of the money stock identity.

3. Following the estimation of these equations, each of the six endogenous components of the money supply is neutralized with respect to the business cycle in the following manner:
   a. Trend values are calculated for all explanatory variables in the six equations which are cyclically influenced, i.e., endogenous to the economic system.
   b. These computed values, which are purged of movements generated by business fluctuation, are then substituted for the actual variables in each of the six equations in order to obtain neutralized values of B, R_e, F, C_o, TD_m, and G.
4. Finally, these neutralized components are substituted for the actual endogenous values in the money stock identity, thus providing an acyclical measure of the money stock to be employed as a money indicator.

As this summary indicates, the methodology employed in the derivation process seeks to implement empirically the concept of exogeneity which was described at the outset of this section and which underlies the effective monetary reaction function.

In order to utilize the neutralized money stock in the present analysis, it is initially necessary to reconstruct the series for the period 1953 through 1975.[9] In his original derivation of this "unbiased" monetary aggregate, Hendershott specified and estimated equations explaining the variables from the money stock identity which are viewed as endogenous to the money sector. His estimation technique may be generally described as ordinary least squares regression applied to first-differenced equations in which the intercepts have been suppressed. In addition, Hendershott tests several alternative specifications of each equation and subsequently employs those which provide the highest coefficient of determination (R^2) in his derivation of the neutralized money stock. Although less than ideal, this procedure of selecting specifications is also incorporated into the current estimation process in order to retain consistency with the original exposition.

One extension of Hendershott's empirical analysis which is incorporated into the present reconstruction is the correction for the existence of first-order serial correlation in the estimated equations. Autoregressive disturbances are a perennial source of difficulty in statistical processes which make extensive use of cyclically influenced data. The correction procedure involves application of the maximum likelihood technique of Beach and MacKinnon (1978) to obtain an estimate of the autocorrelation coefficient, traditionally denoted by ρ. Once this estimate of ρ is generated, it is employed in a generalized differencing operation to produce a regression equation for which the error terms are serially independent. Aside from the obvious need to approximate the conditions of the Gauss-Markov theorem, the rationale for adding this correction procedure to Hendershott's estimation techniques lies in the fact that the use of first-differencing tends to systematically introduce autocorrelation into models which otherwise would exhibit serially uncorrelated errors. Specifically, denoting the current period disturbance by ε_t, there is a definitional correspondence between say, $(\varepsilon_{t+1} - \varepsilon_t)$ and $(\varepsilon_t - \varepsilon_{t-1})$ even if ε_{t+1} and ε_{t-1} are completely independent.[10]

NONCYCLICAL INDICATORS DERIVED FROM A SIMULTANEOUS EQUATION MODEL

As previously noted, the concept of exogeneity embodied in the neutralized money stock and its resulting theoretical basis as an indicator of dynamic monetary policy is quite viable for the purpose of analyzing Federal Reserve reactions to movements in major elements of the policy goal function, i.e., it conceptually delineates dynamic and defensive monetary actions. Despite the allowances made for autocorrelation, however, the actual computation of the neutralized money stock contains at least three significant shortcomings which cast doubt on its reliability. These empirical difficulties are outlined below:

1. The six equations do not contain explanatory variables reflecting either the influence of policy actions or exogenous shocks. While the author is only interested in neutralizing the endogenous explanatory variables, the inclusion of all relevant variables in the specification and estimation of the regression equations would in no way impair or complicate the calculation of a noncyclical measure of the money stock. The result of this "omitted variables" specification is, of course, that the resulting parameter estimates are both biased and inconsistent. These coefficients are directly employed in the computation of the neutralized money stock and, consequently, the reported values for this policy indicator may be unreliable.
2. The entire empirical analysis which underlies the construction of this policy gauge is void of temporal dynamics and, hence, is based on the implicit assumption that the endogenous variables adjust entirely to cyclical movements and exogenous shocks within the current quarter. Consequently, the statistical relationships fail to reveal the time paths of movement in the six endogenous components of the money stock.
3. Finally, the use of single equation estimation techniques such as OLS is inappropriate for the task of removing cyclical influence from a group of variables. By definition, a set of endogenous economic variables is determined within the context of a simultaneous equation model. As is well known, when variables which are endogenous to the economic structure are present in a particular equation, the disturbance term is not independent of the explanatory variables and, consequently, the parameter estimates are inconsistent. Furthermore, even if the estimation of a group of equations explaining movements of endogenous variables is conducted via a limited information method which accounts for mutual interactions, the possibility that the disturbance terms are correlated across equations is ignored. If such relationships are indeed present, then any non-system method will produce estimates which are asymptotically inefficient.

In an effort to overcome the problems outlined above, Perryman suggested a method for deriving more appropriate noncyclical indicators within the context of a simultaneous equation econometric model.[11] These "dynamic" policy gauges, which are denoted by M1* and M2*, are generated through an empirical methodology which is outlined as follows:

1. The construction begins with a simple definitional statement of the money stock identity, i.e.,

$$M1 = DD_c + C$$
$$M2 = DD_c + C + TD_c \tag{2}$$

where M1 and M2 = traditional Federal Reserve definitions of the money stock
DD_c = Adjusted demand deposits at commercial banks
C = Currency outside banks
TD_c = Time deposits at commercial banks (excluding negotiable certificates of deposit).[12]

2. An econometric model of the U.S. monetary sector with a simple linkage to real sector activity is then specified and estimated. This structure is a dynamic simultaneous equation system and is empirically determined via system methods. The model employs the basic framework of the system which was developed and estimated by Teigen (1970) but contains several significant modifications.[13] The present model is also influenced by the major financial sector models of the de Leeuw (1965) and Goldfeld (1966) and by the seminal theoretical analysis of Patinkin (1965). The primary rationale for the adoption and adaptation of the Teigen model lies in the fact that it contains explicit relationships for each of the major individual components of the money supply. These equations, once estimated, may be easily substituted into identities (2) and (3).
3. Following the estimation of the complete structural system, the equations explaining DD_c, C, and TD_c are employed as a basis from which to purge M1 and M2 of the influence of the business cycle. The coefficients of the model are determined through an application of the iterative three-stage least squares (I3SLS) technique. In all cases, it is assumed that the true underlying functional forms may be adequately approximated by linear formulations. In addition, the equations are tested for the presence of serial correlation among the disturbances via the h statistic developed by Durbin (1970). This measure provides an approximately valid normal distribution test in simultaneous equations which contain lagged dependent variables. In all relevant equations, the hypothesis of serially independent errors is not rejected at the five percent level of significance. The removal of cyclical forces from the money supply components is completely analogous to the procedure embodied in the neutralized money stock and described above, i.e., trend values of the endogenous explanatory variables are substituted into the three relevant equations in order to obtain acyclical measures for DD_c, C, and TD_c. These noncyclical values of the money stock components are then employed to replace the actual values in (2) and (3), thus yielding M1* and M2*.

The approach summarized in this section for the development of noncyclical indicators of monetary policy overcomes the major objections to Hendershott's derivation of the NMS which were discussed above. Specifically, the model permits both exogenous forces and Federal Reserve policy to affect the system. Consequently, this approach results in estimates which are more accurately reflective of cyclical patterns than are the simple equations of the neutralized money stock procedure. Additionally, complete reliance on the more direct formulation of the money stock identity avoids the questionable and somewhat treacherous process of classifying the various components as either exogenous or endogenous on an <u>a priori</u> basis. Within the modified Teigen model utilized in the derivation of M1* and M2*, all of the variables which appear in the identities (DD_c, C, and TD_c) are explicitly treated as endogenous.

Beyond the advantages enumerated above, the methodology utilized in developing M1* and M2* also overcomes an admitted tendency of the neutralized money stock to overstate the degree of tightness or ease in monetary policy. The implicit assumption that all of the endogenous variables in the Hendershott equations, particularly interest rates, are affected in the current period only by business fluctuations and not by contemporaneous monetary policy results in a neutralization which exaggerates the magnitude of policy actions. The simultaneous equation system employed herein contains current period policy variables which exert reduced-form effects on interest rates and, hence, resolves this difficulty.[14]

CAUSAL METHODOLOGIES AND THEIR VIABILITY FOR PRE-TESTING REACTION FUNCTIONS

Beyond the utilization of a dependent variable which exhibits superior theoretical properties to the traditional monetary aggregates, interest rates, and bank reserve and credit measures, the specification of a reaction function may be further examined for the presence of reverse causation through a pre-testing of the causal structure of the relevant empirical relationships. In the bivariate case, i.e., when the temporal correspondence between two variables is being investigated, all potential analytical procedures are derived from the seminal conception of statistical causality which was originally suggested by Granger (1969). One such technique, which was initially developed by Sims (1972) offers a direct test for the existence of identifiable causal patterns between variables and has been widely applied in a variety of contexts. In particular, the demonstration by Sims that the notion of statistical exogeneity is essentially equivalent to that of a unidirectional causal relationship as defined by Granger is of critical relevance and significance to the present study. Given this result, the dominant theme in the literature on monetary indicators is readily integrated with efforts to describe the causal process which underlies the interaction of biased policy gauges and various aspects of economic activity. Hence, this conception yields compatibility between optimal policy measurement and the specification of effective reaction functions.[15]

In terms of specific approaches, the present analysis utilizes a generalized adaptation of the Sims procedure, by far the most widely employed of the causality techniques. This approach is, in the case of a bivariate model, implemented through the estimation of a pair of two-sided distributed lag equations. One of these expressions relates the current value for one of the relevant variables to future, present and past values in the second series while the other simply reverses the respective roles of the two measures. As a simple example, the specification of the equations for the temporal pattern between M1* and income (Y) is given by

$$Y_t = \sum_{i=-d}^{c} \alpha_i M1^*_{t-1} + \varepsilon_{1t} \qquad (4)$$

$$M1^*_t = \sum_{i=-d}^{c} \beta_i Y_{t-1} + \varepsilon_{2t}, \qquad (5)$$

where the ε_{jt} are stochastic, though not necessarily random, disturbance terms, the α_i and β_i are system parameters to be estimated, and t is a time subscript. Within the context of expression (4), the failure to reject the null hypothesis that the coefficients on future values of M1*, and α_i, where i=-1, -2, ..., -d, are simultaneously zero supports the notion that M1* is exogenous to, or "causes", Y. Similarly, the corresponding null hypothesis for (5) must also be examined in an effort to ascertain comparable evidence regarding the presence (or absence) of reverse or bidirectional causality. These procedures may readily be applied to the reaction function problem. This approach would merely involve the testing of relationships between the indicators described above and the major endogenous variables which enter the policy goal function, i.e., income, employment, prices, and the balance of payments. Because of the temporal nature of the empirical process and the associated lag structures, estimation of the relevant regression models must inevitably correct for the presence of serial correlation among the residuals.

Standard hypothesis tests for the causal process are, of course, not valid unless the existence of autoregressive disturbances within the model under examination is explicitly accounted for. Several methods have been employed as a basis for producing "white noise" errors, the most popular being a simple prefiltering application suggested by Sims (1972). Specifically, his seminal paper utilized the prefilter $(1-.75B)^2$, where B is the lag or backshift operator, to transform or "filter" relevant data series and further maintained that the implementation of this formula would yield purely random error terms in the relevant regression models. This procedure, in essence, assumes that the true underlying disturbances which characterize the regression model follow a second-order autoregressive process with known parameters. In an excellent overview of the causality issue, Feige and Pearce (1979) have correctly observed that this prefiltering process is "arbitrary". Furthermore, they recommend the consideration of both simple first differencing and a filter determined from a Box-Jenkins Autoregressive Integrated Moving Average (ARIMA) model of the dependent variable in the particular expression being examined.[16] Moreover, Mehra and Spencer (1979) recently developed a somewhat different approach which also begins with the general proposition that the stochastic error sequence follows a second-order autoregressive process. Their filter may be expressed as $(1-B) \cdot (1-kB)$, where B is again the lag operator and k is ascertained through an iterative search procedure with a convergence criterion which stipulates that the residuals in the final form of the regression do not exhibit significant evidence of serial correlation. Hence, the Mehra and Spencer technique embodies the basic first difference filter as a special case. Additionally, Williams, Goodhart, and Gowland (1976) advocate a two-stage estimation approach, the first step of which basically involves a simple first-differencing of the data. Following this initial transformation, the optional filter is determined via a maximum likelihood procedure which explicitly identifies the most appropriate autoregressive coefficients. The implementation of this second stage is greatly facilitated by estimators which have been developed by Beach and MacKinnon (1978) for the case of first-order serial correlation and by Schmidt (1972) in the second-order case. Because of the greater flexibility which is afforded by these procedures, they are employed within the present analysis. The results of the application of the modified Sims procedure to the reaction function problem are now described.

A CAUSAL ANALYSIS OF EFFECTIVE REACTION FUNCTION SPECIFICATIONS

As noted previously, the application of the relevant causal methodology involves both (1) the empirical determination of a pair of two-sided distributed lag models and (2) the subsequent testing of the joint significance of the coefficients on the future values of the explanatory variables. Within the context of the effective reaction function problem, the requisite relationships include expressions relating each of the unbiased monetary indicators (M1*, M2*, and NMS) to each of the variables assumed to comprise the arguments of the ultimate goal function of the monetary authorities, i.e., gross national product (Y), the unemployment rate (U), the rate of inflation as measured by the producer price index (P), and the international economic situation as represented by the balance on the current account (BP). Hence, twelve pairs of regressions are needed in order to implement the procedure. For comparative purposes, a traditional monetary aggregate (M1) is also examined, thus necessitating an additional set of four equation estimates.

The computed F-test values which are utilized in the causal analysis are presented in Table I. Recall that the failure to reject the null hypothesis that the parameter estimates on the future values are zero, i.e., "acceptance" of the notion that the future does not cause the presence, provides evidence of unidirectional causality from the explanatory to the dependent variable.[17] To insure that the reaction function is properly specified and, thus, not subject to significant simultaneous equation bias, the following two patterns are possible:

1. Unidirectional causality from the policy goal to the monetary indicator.
2. No causal relationship between the two variables under examination.

The first of these cases is, of course, the "ideal" relationship. The remaining two situations not listed above, i.e., bidirectional causality or causality from the indicator to an element of the goal function, provide evidence which suggests that the reaction function is improperly developed.

TABLE I

CAUSALITY ANALYSIS OF COMPONENTS OF EFFECTIVE MONETARY REACTION FUNCTIONS

Dependent Variable	Independent Variable	First-Order F-Value	Second-Order F-Value
M1*	Y	3.42*	5.35*
Y	M1*	2.16	1.08
M1*	U	7.46*	4.64*
U	M1*	2.84*	5.23*
M1*	P	2.54*	4.87*
P	M1*	2.69	2.25
M1*	BP	3.97*	5.044*
BP	M1*	0.253	0.289
M2*	Y	4.02*	3.31*
Y	M2*	5.04	12.78*
M2*	U	8.58*	4.84*
U	M2*	1.01	0.875
M2*	P	3.96*	3.77*
P	M2*	0.388	0.473
M2*	BP	3.83*	4.86*
BP	M2*	0.841	1.14
NMS	Y	4.14*	4.55*
Y	NMS	7.05*	4.02*
NMS	U	2.82*	0.570
U	NMS	5.69*	4.67*
NMS	P	4.45*	5.64*
P	NMS	3.53*	0.634
NMS	BP	2.81*	6.63*
BP	NMS	1.19	0.888
M1	Y	2.74*	15.80*
Y	M1	0.741	0.328
M1	U	9.01*	0.575
U	M1	4.31*	7.13*
M1	P	7.02*	0.938
P	M1	2.20	3.66*
M1	BP	4.22*	4.90*
BP	M1	0.970	1.02

*Significance at the .05 level, i.e., absence of a causal relationship.

As the table illustrates, the noncyclical indicators (M1* and M2*) seem to be superior to the neutralized money stock as dependent variables in effective reaction functions. Moreover, this conclusion is robust with respect to the two autoregressive filters employed in the analysis. Specifically, two of the goal variables appear to "cause" M1* and three tend to "cause" M2* when the first-order correction is applied. With respect to the NMS, however, only the balance of payments exhibits the ideal causal pattern. There are no reverse causation problems in this case. When the second-order autoregressive filter is utilized, both M1* and M2* have the "ideal" relationships with three of the four arguments in the policy goal function. The neutralized money stock, however, only has two such results and, additionally, has a reverse causation problem in relation to the unemployment rate. Finally, it should be noted that the measured money stock variable seems appropriate under the first-order autoregressive scheme. In the second-order case, however, there is evidence of reverse causation with respect to both employment and prices. Thus, the more general procedure strongly suggests the existence of simultaneous equation bias in the traditional development of policy reaction functions. The evidence summarized in this section is now employed as a basis for the derivation of alternative monetary response models.

ESTIMATION OF EFFECTIVE POLICY REACTION FUNCTIONS

The causal analysis of the preceding section reveals that, for purpose of describing effective policy reactions, M1* and M2* seem superior to both the neutralized money stock and M1 as appropriate monetary indicators. Consequently these variables are now employed as dependent variables in the specification and estimation of reaction functions. Both the theoretical properties of these gauges and the statistical results given above strongly suggest the absence of significant reverse causation in the resulting relationships. The basic technique which is employed to empirically determine these equations is the polynomial distributed lag model which was originally developed by Almon.[18] This procedure is based upon the assumption that, given a basic regression model in which the total effect of the independent variables on the dependent variable occurs over several periods, the coefficients for each period in the lag structure may be adequately approximated by a polynomial of known degree. The total impact of an independent variable is then ascertained by simply summing the coefficients for the individual periods. This structure seems highly plausible in the case of policy reactions, as it may be reasonably assumed that the impact reaches a peak and then declines. This pattern is, of course, readily approximated by a polynomial. For the present models, the Almon approach is applied to expressions containing quarterly observations of each of the relevant monetary indicators (M1* and M2*) as dependent variables and the critical elements of the policy goal functions which were noted in the preceding section, i.e., income (Y), unemployment (U), the price level (P), and the balance of payments (BP) as explanatory variables. For comparative purposes, an equation is also estimated using the traditional M1 aggregate as the dependent variable. In all cases, the current and four lagged values are utilized. Thus, in terms of, say, M1*, the relevant reaction function is given as

$$M1^* = \alpha + \sum_{i=0}^{4} \beta_i Y_{t-i} + \sum_{i=0}^{4} \gamma_i U_{t-i} + \sum_{i=0}^{4} \delta_i P_{t-i} + \sum_{i=0}^{4} \phi_i BP_{t-i} + \varepsilon_t, \qquad (5)$$

where α, β_i, γ_i, δ_i, ϕ_i = parameters to be estimated
ε_t = a random disturbance term
t = index of current time period.

The equations for M2* and M1 are, of course, completely analogous in form. The expressions are tested for the presence of serial correlation and, when necessary, are corrected via the application of a maximum likelihood procedure recently developed by Beach and MacKinnon (1978). This technique searches a grid of values in order to obtain an optimal estimate of the autocorrelation coefficient, traditionally denoted by ρ. Once the estimate of ρ is generated, it is employed in a generalized differencing operation in order to produce a regression equation for which the error terms are serially independent.

The parameter estimates derived from the empirical implementation of the procedures noted above are exhibited in Table II, along with t-values and adjusted coefficients of determination ($\overline{R}^2$'s). Each of these equations utilizes a fourth-degree Almon polynomial and, as the Table indicates, each has a very high percentage of explained variance. With respect to the traditional M1 measure, the most significant responses tend to occur as (1) a lagged reaction to price movements and (2) an immediate response to international economic conditions. It is interesting to note, however, that the relationship between the money stock and inflation is <u>positive</u>. Thus, the results seem to suggest that the phenomenon being observed is the contribution of money to inflation rather than a Federal Reserve reaction to changes in the price level. This evidence tends to indicate that "reverse causation" is indeed a problem when traditional monetary aggregates are employed to estimate effective reaction functions and, of course, tends to confirm the results presented in Section V.

The equations for M_1* and M_2* both indicate that (1) monetary actions responded both rapidly and countercyclically to changes in income[19] and (2) Federal Reserve policy reacted significantly to international problems. The consistency and proper signs observed in these findings tend to lend credence to their characterization of policy. It must be noted, however, that despite the Almon polynomial transformations, the simultaneous existence of (1) relatively few significant coefficients, (2) t-values which are generally quite low in magnitude, and (3) high values for $\overline{R}^2$ often gives evidence of the presence of multicollinearity. In such cases, it is often difficult to reject the null hypothesis that a particular parameter estimate is zero.

CONCLUSION

The analysis which has been provided in the present paper has proceeded from the following basic premises:

1. There is a need to measure not only the reaction of monetary policy to economic conditions within the immediate operational sense of an optimal control problem, but also within the context of the "effective" policy reaction.
2. The estimation of the relevant relationships is subject to problems of reverse causation when traditional monetary variables are employed. This difficulty arises by the very nature of the economic process in that, in addition to its policy element, money is jointly determined within the framework of a simultaneous equation system which also includes the arguments of the goal function as endogenous variables.
3. The reverse causation or simultaneous equation bias found in policy reaction functions may be ameliorated by the use of noncyclical monetary indicators, i.e., variables from the financial sector which are purged of the influence of fluctuations in economic activity.

The first of these propositions is merely a pragmatic statement which is applicable to any case in which discretionary policy is pursued. The second two propositions may be examined empirically through the use of time series and econometric causality techniques. Initially, the nature and derivation of the relevant policy measures (M1*, M2*, and NMS) is briefly described. These indicators are shown to be consistent with a conception of exogeneity which separates dynamic and defensive monetary actions and, thus, they should offer an ideal instrument for determining effective dynamic Federal Reserve responses. Following this description, the indicators are tested for the presence of reverse causation. The results of this aspect of the investigation strongly support the conclusion that M1* and M2* are appropriate variables for the estimation of effective reaction functions, while potential problems arise with respect to both the neutralized money stock and traditional monetary measures. Consequently, these two gauges are utilized in the formulation and estimation of the policy relationships. The equations employed for this purpose are empirically determined through the utilization of the Almon polynomial distributed lag procedure. The

TABLE II

EFFECTIVE FEDERAL RESERVE REACTION FUNCTIONS FOR ALTERNATIVE MONETARY VARIABLES, 1953-1975

Dependent Variable	Coefficient: Y_t	Y_{t-1}	Y_{t-2}	Y_{t-3}
M1	.03981 (.0872999)	.005177 (.4582585)	.043497 (.0241593)	-.016083 (-.0009333)
M1*	-.14928* (-2.3671416)	.0606206 (.7364879)	.0106576 (.1271558)	-.0309174 (-.00175736)
M2*	-.29181 (-2.5171007)	.13573 (.9065605)	.19591 (1.37235)	-.063051 (-.28171927)

Dependent Variable	Coefficient: Y_{t-4}	P_t	P_{t-1}	P_{t-2}
M1	.024317 (.2099807)	.095828 (.44870655)	-.028359 (0.13780607)	.337456 (1.5377528)
M1*	-.0523744 (-.1166398)	-.22408 (-.87501882)	.5134 (.11604919)	.3438 (.0275731)
M2*	-.169326 (-.38607821)	-.51724 (1.0507889)	.12927) (.19595813)	.31452 (.36013816)

Dependent Variable	Coefficient: P_{t-3}	P_{t-4}	U_t	U_{t-1}
M1	.488081* (6.0353778)	.067836 (.24917689)	-.46787 (-.53576804)	.0615457 (.067840221)
M1*	.4862 (.02045524)	-1.2848 (-.03648283)	.84572 (.706475)	1.48307 (.9590375)
M2*	1.38647 (.87688552)	-1.11948 (-.34977257)	-.6335 (-.2923984)	3.7772 (1.4128694)

Dependent Variable	Coefficient: U_{t-2}	U_{t-3}	U_{t-4}	BP_t
M1	.1481212 (.19556787)	.3573217 (.7835549)	1.0802692 (.655863)	-.0010975* (-2.2375127)
M1*	-.43688 (-.407676)	.69647 (.350834685)	-1.16908 (-.255910269)	-.0024497 (-3.6801662)
M2*	1.41506 (.45801778)	-2.25332 (-.490289)	-.65206 (-.094518)	-.0033216 (-2.5202777)*

Dependent Variable	Coefficient: BP_{t-1}	BP_{t-2}	BP_{t-3}	BP_{t-4}	$\bar{R}^2$
M1	-.008341714 (-1.2861482)	-.0004514 (-.07507705)	-.0002997 (-.09075905)	-.0007269 (-.02814421)	.9023
M1*	.0002396004 (.30705352)	.00099038 (.03147031)	.00086167 (.07645482)	-.002206116 (-.21247102)	.9962
M2*	-.000209627 (-.11193165)	-.000518992 (-.02031883)	-.000848367 (-.3460914)	.001212448 (.009823253)	.9976

evidence obtained from this analysis suggests that (1) there is a problem of reverse causation in the development of reaction functions using traditional monetary aggregates and (2) policy responds most rapidly and predictably to changes in income and international conditions. Additional applications which might be undertaken within a context comparable to that of the present study include the utilization of temporal subsamples to determine shifts in the nature of policy responses under alternative economic climates and the examination of the causal relationships which are relevant to the reaction function problem utilizing alternative empirical methodologies. Investigations of this nature should provide a much more thorough understanding of the actual impact of monetary policy on the course of economic activity.

ACKNOWLEDGEMENTS

The author is Herman Brown Professor of Economics and Director of the Center for the Advancement of Economic Analysis, Baylor University. Appreciation is expressed to the staff of the Center for their computational assistance and to Leigh Humphrey for manuscript preparation. The author also wishes to acknowledge the Brown Foundation and the Hankamer School of Business for released time and other financial support.

FOOTNOTES

1. See Blinder and Goldfeld (1976), Brunner (1970), Brunner and Meltzer (1970), Dewald (1970), Friedman (1975), Froyen (1976), Guttentag (1966), Hamburger (1970), Hendershott (1968), Hendershott (1971), Hendershott and Horwich (1969),Kaufman (1967), Keran (1970), Pankcratz (1977), Perryman (1978b), Perryman (1980a), Perryman (1982), Savings (1967), Starleaf and Stephenson (1969), Tanner (1972), Tanner (1975), Zecher (1970).

2. See, for example, Kareken and Solow (1963), Mayer (1967), Perryman (1978a), Perryman (1980c), Willes (1970).

3. See, for example, Andersen and Jordan (1968), Anderson and Jordan (1969), Blinder and Goldfeld (1976), Davis (1969), de Leeuw and Kalchbrenner (1969), Elliott (1971), Hamburger (1969), Kareken (1967), Keran (1969), Perryman (1980b), Perryman (1978c), Waud (1974).

4. See, for example, Froyen (1974), Froyen (1976), Havrilesky, Sapp, and Schweitzer (1975), Teigen (1970).

5. Froyen (1974, p. 178).

6. See Hendershott (1968), Hendershott and Horwich (1969), Perryman (1978b), Perryman (1980a). An indicator with similar properties is described in Starleaf and Stephenson (1969). This measure is not described herein,

however, because of a number of difficulties in its construction which were enumerated in Hendershott (1971).

7. This problem also appears in the analysis of the St. Louis equation. The situation within the context is addressed in Perryman (1978c) and Perryman (1980b).

8. For more detail, see Hendershott (1968).

9. Actually, it is the modified neutralized money stock which is reconstructed. This variable, which is employed by Hendershott in his domestic policy analysis, treats the gold stock as exogenous.

10. A complete discussion of this reconstruction is given by Perryman (1979a).

11. See Perryman (1978b), Perryman (1980a), and Perryman (1982).

12. During the period subsequent to 1975, the rapid development of highly liquid financial instruments has led to a number of alternative money supply definitions and related constraints. It is useful to note, however, that <u>any</u> variable from the monetary sector which is properly purged of cyclical components should provide an accurate reflection of the stance of dynamic policy (see Hendershott (1968), Hendershott (1971)). Consequently, M1* and M2*, if extended to the present, should continue to accurately assess the effective impact of Federal Reserve actions.

13. A full description of the relevant model is given in Perryman (1979b).

14. A more thorough discussion of this problem in relation to the construction of the neutralized money stock is given in Hendershott (1968).

15. The exogeneity issue is formally addressed in Perryman (1979c).

16. These models were originated in Box and Jenkins (1970).

17. It should be noted at this point that the references to "causality" in an empirical setting apply only to the usage of the term in contemporary econometric and time series literature and in no way seek to embrace its entire philosophical conception. See, for example, Jacobs, Leamer, and Ward (1979).

18. See Almon (1965). Extensions of the basic procedures which were utilized in the present analysis are given in Almon (1968) and Tinsley (1967).

19. Evidence supporting this conclusion in an entirely different context is given in Perryman (1981).

REFERENCES

ALMON, S. (1965). The distributed lag between capital appropriations and and expenditures. *Econometrica 33*, 178-196.

ALMON, S. (1968). Lags between investment decisions and their causes. *Review of Economics and Statistics 50*, 193-206.

ANDERSEN, L.C. and JORDAN, J.L. (1968). Monetary and fiscal actions: a test of their relative importance in economic stabilization. Federal Reserve Bank of St. Louis, *Review 50*, 11-23.

ANDERSEN, L.C. and JORDAN, J.L. (1969). Monetary and fiscal actions: a test of their relative importance in economic stabilization--reply. Federal Reserve

Bank of St. Louis, *Review 51*, 12-16.

BEACH, C. and MACKINNON, J. (1978). A maximum likelihood procedure for regression with autocorrelated errors. *Econometrica 46*, 51-59.

BLINDER, A.S. and GOLDFELD, S.M. (1976). New measures of fiscal and monetary, 1958-73. *American Economic Review 66*, 180-196.

BOX, G.E.P. and JENKINS, G.M. (1970). *Time Series Analysis: Forecasting and Control*. Holden Day, San Francisco.

BRUNNER, K. (1970). *Targets and Indicators of Monetary Policy*. Chandler Publishing, San Francisco.

BRUNNER, K. and MELTZER, A.H. (1967). The meaning of monetary indicators. In *Monetary Process and Policy: A Symposium*. Ed: G. Horwich, Irwin, Homewood, Illinois, 3-27.

DAVIS, R.G. (1969). How much does money matter: a look at some recent evidence. Federal Reserve Bank of New York, *Monthly Review 51*, 119-131.

DE LEEUW, F. (1965). A model of financial behavior. In *The Brookings Quarterly Econometric Model of the United States*. Eds: J. Duesenberry, G. Fromm, L.R. Klein, and E. Kue, Rand McNally, Chicago, 465-532.

DE LEEUW, F. and KALCHBRENNER, J. (1969). Monetary and fiscal actions: a test of their relative importance in economic stabilization--comment. Federal Reserve Bank of St. Louis, *Review 51*, 6-11.

DEWALD, W.G. (1970). A review of the conference on targets and indicators of monetary policy. In *Targets and Indicators of Monetary Policy*. Ed: K. Brunner, Chandler Publishing, San Francisco, 313-330.

DURBIN, J. (1970). An alternative to the bounds test for testing for serial correlation in least-squares regression. *Econometrica 38*, 422-429.

ELLIOT, J.W. (1971). The influence of monetary and fiscal actions on total spending. *Journal of Money, Credit, and Banking 3*, 181-192.

FEIGE, E.L. and PEARCE, D.K. (1979). The casual causal relationship between money and income: some caveats for time series analysis. *Review of Economics and Statistics 61*, 521-533.

FRIEDMAN, B.M. (1975). Targets, instruments, and indicators of monetary policy. *Journal of Monetary Economics 1*, 443-473.

FROYEN, R.T. (1974). A test of the endogeneity of monetary policy. *Journal of Econometrics 2*, 175-188.

FROYEN, R.T. (1976). An alternative neutralized monetary policy variable and the implications for the reverse causation controversy. *Journal of Economics and Business 29*, 16-21.

GOLDFELD, S. (1966). *Commercial Bank Behavior and Economic Activity: A Structural Study of Monetary Policy in the United States*. North-Holland, Amsterdam.

GRANGER, C.W.J. (1969). Investigation of causal relations by econometric models and cross-spectral methods. *Econometrica 37*, 424-438.

GUTTENTAG, J.M. (1966). The strategy of open market operations. *Quarterly*

Journal of Economics 80, 1-30.

HAMBURGER, M.J. (1969). The impact of monetary variables: a survey of recent econometric literature. In *Essays in Domestic and International Finance*. Federal Reserve Bank of New York, New York, 33-39.

HAMBURGER, M.J. (1970). Indicators of monetary policy: the arguments and the evidence. In *American Economics Review 60*, 33-39.

HAVRILESKY, T.M., SAPP, R.H. and SCHWEITZER, R.L. (1975). Tests of the federal reserve's reaction to the state of the economy: 1964-74. *Social Science Quarterly 55*, 835-52.

HENDERSHOTT, P.H. (1968). *The Neutralized Money Stock: An Unbiased Measure of Federal Reserve Policy Actions*. Irwin, Homewood, Illinois.

HENDERSHOTT, P.H. (1971). The full-employment interest rate and the neutralized money stock: comment. *Journal of Finance 26*, 127-136.

HENDERSHOTT, P.H. and HORWICH, G. (1969). Money, interest and policy. In *Savings and Residential Financing: 1969 Conference Proceedings*. United States Savings and Loan League, 1969, Chicago, 32-52.

JACOBS, R.L., LEAMER, EE. and WARD, M.P. (1979). Difficulties with testing for causation. *Economic Inquiry 17*, 401-413.

KAREKEN, J.H. (1967). The mix of monetary and fiscal policies. *Journal of Finance 22*, 241-246.

KAREKEN, J.H. and SOLOW, R.M. (1963). Lags in monetary policy. In *Stabilization Policies*. Prentice-Hall, New York, 110-130.

KAUFMAN, C.C. (1967). Indicators of monetary policy: theory and evidence. *National Banking Review*, 481-491.

KERAN, M.W. (1969). Monetary and fiscal influence on economic activity--the historical evidence. Federal Reserve Bank of St. Louis, *Review 51*, 5-23.

KERAN, M.W. (1970). Selecting a monetary indicator--evidence from the United States and other developed countries. Federal Reserve Bank of St. Louis, *Review 52*, 8-19.

MAYER, T. (1967). The lag in the effect of monetary policy: some criticisms. *Western Economic Journal 5*, 324-342.

MEHRA, U.P. and SPENCER, D.E. (1979). The St. Louis equation and reverse causation: the evidence reexamined. *Southern Economic Journal 45*, 1104-1120.

PANKRATZ, A. (1977). Bank portfolio composition: indicator or instrument? *Southern Economic Journal 44*, 99-107.

PATINKIN, D. (1965). *Money, Interest and Prices*. Harper and Row, New York.

PERRYMAN, M.R. (1978a). An alternative empirical strategy for the measurement of the lag in the countercyclical response of monetary policy. *Proceedings of the American Statistical Association*, 220-225.

PERRYMAN, M.R. (1978b). Noncyclical indicators of dynamic monetary policy derived from a simultaneous equation econometric model. (Paper delivered to the 1978 *Confernece of the Western Economic Association*, Honolulu, Hawaii.)

PERRYMAN, M.R. (1978c). The relative impacts of monetary and fiscal policy: a consistent assessment. (Paper delivered to the *Southern Economic Association*, Washington, D.C.

PERRYMAN, M.R. (1979a). An alternative characterization of monetary policy, 1953-1975. *Papers and Proceedings of the Southwestern Society of Economists*, 214-223.

PERRYMAN, M.R. (1979b). An amended simultaneous equation model of the U.S. monetary sector and its viability for indicator construction. *Baylor Business Studies 121*, 7-28.

PERRYMAN, M.R. (1979c). Bias and exogeneity among alternative monetary indicators: a theoretical and empirical comparison. (Paper delivered to the 1979 Conference of the *Atlantic Economic Society*, Washington, D.C.)

PERRYMAN, M.R. (1980a). An alternative indicator of dynamic monetary policy: construction and application. *Journal of Economics 6*, 33-37.

PERRYMAN, M.R. (1980b). The economic impact of economic policy: some consistent estimates of the St. Louis equation. *Review of Business and Economic Research 15*, 57-72.

PERRYMAN, M.R. (1980c). Some evidence regarding the lag in the initial countercyclical impact of monetary policy. *Nebraska Journal of Economics and Business 19*, 65-71.

PERRYMAN, M.R. (1981). Policy intent, policy formulation, and policy effectiveness: an appraisal of federal reserve actions. *Applied Economics 13*, 91-105.

PERRYMAN, M.R. (1982). The optimal indicator of dynamic monetary policy: a theoretical, empirical, and historical perspective. *Journal of Economics 8* (forthcoming).

SAVING, T.R. (1967). Monetary-policy targets and indicators. *Journal of Political Economy 75*, 446-465.

SCHMIDT, P. (1972). Estimation of a distributed lag model with second order autoregressive disturbances: a Monte Carlo experiment. *International Economic Review 12*, 372-380.

SIMS, C.A. (1972). Money, income, and causality. *American Economic Review 62*, 540-552.

STARLEAF, D. and STEPHENSON, J. (1969). A suggested solution to the monetary indicator problem: the monetary full-employment interest rate. *Journal of Finance 24*, 623-631.

TANNER, J.E. (1972). Indicators of monetary policy: an evaluation of five. *Banco Nazionale del Lavoro Quarterly Review*, 3-19.

TANNER, J.E. (1975). A Wicksellian indicator of monetary policy. *Journal of Monetary Economics 1*, 171-185.

TEIGEN, R.L. (1970). An aggregated quarterly model of the U.S. monetary sector, 1953-1964. In *Targets and Indicators of Monetary Policy*. Ed: K. Brunner, Chandler Publishing, San Francisco, 175-218.

TINSLEY, P.A. (1967). An application of variable weight distributed lags. *Journal of the American Statistical Association 62*, 1277-1289.

WAUD, R.M. (1974). Monetary and fiscal effects on economic activity: a reduced form examination of their relative importance. *Review of Economics and Statistics 56*, 177-187.

WILLES, M.H. (1971). Lags, fine tuning, and rules of monetary policy. Federal Reserve Bank of Philadelphia, *Business Review*, 630-638.

WILLIAMS, D., GOODHART, C.A.E., and GOWLAND, D.J. (1976). Money, income, and causality: the U.K. experience. *American Economic Review 66*, 417-423.

ZECHER, R. (1970). Implications of four econometric models for the indicators issue. *American Economic Review 60*, 47-54.

APPLIED TIME SERIES ANALYSIS
O.D. Anderson & M.R. Perryman (eds.)

TIME SERIES ANALYSIS, ARIMA MODELS, AND COMPREHENSIVE REGIONAL EVALUATIVE PROCESSES

M. Ray Perryman

Center for the Advancement of Economic Analysis
Baylor University
Waco, Texas
U.S.A.

This paper examines a regional information system which was conceived by the author and is presently being implemented for the State of Texas. Included in the analysis are (1) a synopsis of the requisite characteristics and components of the structure and (2) a summary exposition of the Texas Econometric Model, which represents the core of the framework. The emphasis of the paper is on (1) the critical role of ARIMA models and other forms of time series analysis in the system and (2) the rationale for regional analysis within a time series context.

INTRODUCTION

During the past few years, there has been a tremendous increase in the number of states, regions, and metropolitan areas which have devoted considerable effort and expense to the development of econometric models.[1] A number of factors have contributed to this phenomenon, including (1) the widespread acceptance of major national models in the corporate and government sectors, (2) increased uncertainty regarding the future course of economic activity, (3) the trend toward more systematic approaches to governmental planning at the state and local levels, (4) major advances in computer technology with respect to mass information storage, manipulation, and retrieval, (5) significant progress in the derivation of estimators for large dynamic temporal systems,[2] (6) gradual improvement in the data base for regional analysis, and (7) the contemporaneous emergence of urban and regional economics as an independent and distinct area of academic inquiry. Despite the specification and estimation of numerous sub-national econometric models, a substantial number of them (approximately 75%) are not maintained and utilized on a continuing basis. Moreover, even in the limited cases in which the models are frequently and effectively employed, they tend to be primarily an adjunct to the overall analytical process which accompanies policy formulation. A more viable and realistic approach would seem to be the integration of an econometric model into a comprehensive information network for regional planning at all relevant governmental levels and corporate enterprises.

The present paper describes several significant aspects of an ongoing effort to create a comprehensive and efficient regional information structure for the state of Texas. The overall design of this system represents an attempt to provide maximal benefit to all potential users. Initially, a brief analysis of the requisite characteristics and components of a complete regional system is provided. Following this discussion, a summary exposition of the State of Texas Econometric Model is presented in both graphical and descriptive form. This large scale dynamic simulation structure, which was recently completed under the direction of the author, is the central focus of the entire information-based project. The investigation then proceeds to an enumeration of the various applications of time series analysis which are essential to the successful implementation of the suggested approach. Finally, a concluding section provides a synopsis of the study and assesses its implications for regional analysis within the time series context.

THE COMPLETE REGIONAL INFORMATION SYSTEM

Several properties are crucial to the development of an effective regional information system. In particular, it must be characterized by comprehensiveness, flexibility, continuity, and consistency. With respect to comprehensiveness, for example, it is necessary for the network to include a complete set of models, an accompanying data base and a support program which is conducive to a large number of applications. Specifically, the empirical structure should include, at a minimum, a large scale econometric model for the state and a detailed set of appropriate sectoral and geographical submodels.[3] Similarly, the data base should include all major areas which are considered useful by a significant group of actual or potential subscribers. These data would generally include information on the region as a whole, as well as extensive categorical and geographical disaggregations. Moreover, the base should reflect information which is obtained from a wide variety of sources, including a large number of federal, state, and local government agencies, private corporations, trade organizations, and standard publications. Finally, the support system should encompass sufficient computational and storage capability to permit frequent simulation under alternative scenarios, accessibility to all parts of the data base, and the generation of timely forecasts of relevant economic aggregates.

As previously indicated, the complete regional information system must be characterized by great flexibility. In the area of simulation, for example, the ideal framework would permit economic analysis under alternative national policy situations, alternative state and local policy structures, alternative general levels of national and regional economic activity, and a large number of major exogenous shocks (e.g., an oil embargo). Additionally, it is highly beneficial if the model can be simulated at varying degrees of disaggregation within and among major sectors. In some applications, it would be necessary to have a high degree of disaggregation in, say, the agricultural sector, while in others the manufacturing sector might be more significant. The data base and support systems must also be highly flexible. It should be possible to call, transform, retrieve, implement, and store any individual series or any combination of series from the information bank without jeopardizing the permanent storage capability.

Regarding continuity, it is necessary to provide a rigorous and efficient maintenance program. This program will involve such items as (1) constant update of the data base as more information becomes available, (2) periodic revision of the model as new information becomes available, (3) constant adaptation of the model and/or simulation capability to changes in the economic climate, and (4) an ability to promptly incorporate advances in computer software, model structure, and estimation procedures into the information system. In this manner, users can be assured of a reliable system which can be safely incorporated as a critical part of the decision making process.

The final requisite property of a complete regional information system is consistency. The data base should be fully coordinated with the model structure in order to facilitate applications by primary and secondary users. Moreover, the models within the system should evolve from a basic structural design which assures the internal consistency of individual submodels as well as an overall mathematical linkage of the entire system. In achieving this goal, it may at times be necessary to construct key data series which are not generally published. Within the state of Texas Economic Model project, for example, a major series on gross state product by industry and a manufacturing capital stock series have been generated.[4] It is also essential that data definitions, seasonal adjustment factors, and related maintenance activity strictly adhere to the basic tenents of consistency. The software support system must, of course, also be developed in a manner which permits consistent interaction between the model and the related data.

The essential components of a complete regional information system are probably quite obvious at this point. The dominant aspect of the system is, of course, the aggregate regional econometric model. As noted earlier, the model should be flexible, comprehensive, and amenable to a variety of submodels. Within the State of Texas model, there are detailed submodels of retail sales, agriculture, financial institutions, manufacturing, and government revenues. There are also geographic submodels of both the Dallas and Houston metropolitan areas. Additionally, the mining and energy sectors are currently under development at a reasonable level of disaggregation. All of these submodels are designed in a manner which permits them to be simulated both as "stand alone" structures or as functioning parts of the general system. The second major component of the complete regional information system is the supporting data base. At a minimum, the data base must include all major national, regional, and subregional variables which are employed in the various aspects of the model. To achieve maximum effectiveness, however, the information set should also include other data series which are relevant to corporate and governmental decision processes. It must also encompass an active maintenance and seasonal adjustment strategy. The final component of the overall regional system is the support base. This computer oriented framework must include extensive data base management software, extensive access to major fiscal and econometric estimation routines, access to other related data bases, access to major simulation algorithms, and a rigorous maintenance program. Given these items, a highly pragmatic and effective system may be generated. A brief exposition of the basic econometric model is presently given.

AN OVERVIEW OF THE TEXAS ECONOMETRIC MODEL[5]

As noted previously, the hub of an effective regional information network is a comprehensive econometric model of the area under investigation. Major limitations in the existing data base for states, however, preclude the development of an "ideal" structural system. In such a framework, which is characteristic of most of the major national models constructed in recent years, the expenditure sector, i.e., consumption, investment, government spending, and net exports, are explored in great detail. Sufficient information is available to explore a viable "second best" model structure, the basic outline of which is adopted for Texas. Specifically, it is plausible to develop a reasonable and highly flexible equation system around a basic "income-output" identity of the form

$$GSP \equiv TO \equiv TI, \tag{1}$$

where:

GSP = Gross State Product
TO = Total Output
TI = Total Income.

A reasonable output series may be generated on a sectoral basis. The methodology for developing such a series on an annual basis was originated by Kendrick and Jaycox (1965). Modifications of this approach have been suggested both for more accurate measurement in selected sectors and for computation in quarterly form.[7] The basic identity for this procedure is given by

$$GSP = TO = GPO_{Mi} + GPO_C + GPO_{Ma0} + GPO_{MaN} + GPO_W + GPO_R + GPO_F + GPO_{Tr} + GPO_S + GPO_G + GPO_{Ag}, \tag{2}$$

where:

GPO_{Mi} = gross product originating in mining

GPO_C = gross product originating in construction

GPO_{Ma0} = gross product originating in durable manufacturing

GPO_{MaN} = gross product originating in non-durable manufacturing

GPO_W = gross product originating in wholesale trade

GPO_R = gross product originating in retail trade

GPO_F = gross product originating in finance, insurance, and real estate

GPO_{Tr} = gross product originating in transportation, communication, and public utilities

GPO_S = gross product originating in government

GPO_G = gross product originating in government

GPO_{Ag} = gross product originating in agriculture, forestry, and fisheries.

With respect to income, it is only possible to develop an identity of the form

$$GSP = TI = PI + NPI \tag{3}$$

where:

PI = personal income
NPI = non-personal income.

Personal income may be further disaggregated into various wage categories and other types of income received (e.g., other labor income and proprietors' income). Unfortunately, there is no direct data on non-personal income and, hence, it may only be approximated as

$$NPI = GSP - PI. \tag{4}$$

Despite the computational difficulties involved, expression (1) provides a reasonable framework for the consistent development of a regional econometric model.

The employment relationships within the model are related to sectoral outputs in a mutually casual, i.e., simultaneous, manner. Specifically, supply-side gross product expressions are essentially production functions, while labor demand is related to output by the corresponding cost minimization processes.[8] In an ideal setting, data on other inputs would also be utilized. Unfortunately, such information is not generally available. Data on materials are published for manufacturing, although with a considerable lag. Moreover, a viable capital stock series for manufacturing may be developed from the published series on new capital expenditures and the capital decay factor recently compiled by the Labor Department. Hence, relatively effective production relationships are feasible in manufacturing. Given that several other sectors are highly labor intensive, however, the loss of information is probably not extremely critical. The remaining structure of the model is primarily based on a detailed examination of the key individual sectors noted above and several demographic factors. Depending upon the degree of disaggregation in various parts of the model, any individual simulation may be conducted with as few as fifty equations or as many as several thousand.

To illustrate the sectoral flexibility of the Texas model, the agricultural sector is presently examined in explicit detail. This sector derives basically from the GPO_{Ag} component of equation (2). Specifically, the relevant gross product originating term may be decomposed into

$$GPO_{Ag} = GFP + GPO_{AS} + GPO_{Fo} + GPO_{FT} \tag{5}$$

where:

GFP = Gross Farm Product

GPO_{AS} = Gross product originating in agricultural services

GPO_{Fo} = Gross product originating in forestry

GPO_{FT} = Gross product originating in fishing and trapping.

Unlike other sectors at the regional level, there is sufficient information with respect to farming to permit direct calculation, at least on an annual basis, of gross farm product without the Kendrick-Jaycox adjustments. At its most aggregated level, the farm product identity may be given by

$$GFP = VFP - TIE, \tag{6}$$

where:

GFP = value of total farm product
TIE = total current intermediate product expenses.

Relations explaining VFP and TIE could, thus, comprise a highly simplistic farming sector of a regional model. The initial disaggregation of this expression may be decomposed into the following expressions:

$$VFP = CRM + VHC + GVD + OI + \Delta Inv - NR \tag{7}$$

and

$$TIE = LSE + SE + FE + ROE + ME, \tag{8}$$

where:

CRM = cash receipts from marketings
VHC = value of home consumption
GVD = gross rental value of dwellings
OI = other income

ΔInv = net change in inventories
NR = net rent to non-farm landlords
LSE = livestock expenses
SE = seed expenses
FE = feed expenses
ROE = repairs and operation of equipment
ME = miscellaneous expenses.

With respect to receipts from marketings, several levels of breakdown in the data are available. As a point of departure, sales may be dichotomized into two basic categories, i.e.,

$$CRM = LS + CS, \tag{9}$$

where:

LS = livestock sales
CS = crop sales.

Moreover, livestock receipts can be broken into meat animal sales, dairy product sales, and poultry and egg sales. Similarly, crop marketings consist of food crops, feed crops, cotton crops, oil crops, vegetables, and fruit and nuts. This level of aggregation seems appropriate for the agricultural sector of a general regional model. Each of the livestock and crop categories enumerated above may be further delineated into specific animals, grains, vegetables, etc. This very detailed system is perhaps feasible to serve as a separate agricultural submodel to be simulated apart from the entire structure. Given this design procedure, it is possible to specify the interactions of production, sales, prices, income and other relevant variables within the farm sector of a regional economy at virtually any level of detail. Moreover, the sectoral system is directly linked to the overall model via the set of identities regarding agricultural and total product which were previously described. Finally, the consistency of the inter-relationship between the sector and the model as a whole is preserved irrespective of the level of aggregation.

The example given immediately above illustrates the precise development of a sectoral submodel. Utilizing comparable frameworks, it is possible to systematically disaggregate other sectors. Consider, for example, the manufacturing sector. Gross product originating in this area may be divided into outputs for various sectors in accordance with the Standard Industrial Classification (SIC) system. Data which are available from the Annual Survey of Manufacturers and the Census of Manufacturers permit the both demand and supply side specifications of manufacturing output (value added) at varying levels of detail. It is probably desirable, for example, to include both all major two-digit industries and the dominant three-digit industries within the state in the structure of a general model. Smaller systems might, of course, include only a summary manufacturing relationship. Moreover, a detailed subsector model may be formulated by the use of the more disaggregated three and four-digit SIC classification. In some cases, however, the classification of individual firms within four-digit categories is highly unstable from year to year. In terms of linkage, the four-digit industries may be linked to the three-digit industries by a set of definitional identities. Similarly, the three-digit classifications are merely disaggregations of the two-digit industries. Finally, the two-digits may be summed into the one-digit (durable and non-durable) categories and then into total manufacturing output as described in the proceeding section. The submodel is thus readily embodied into the general model in a mathematically consistent manner.

As a further example, it is also possible to design a simple submodel structure for various categories of retail sales. Initially, a functional relationship between retail sales and gross product originating in wholesale and retail trade is established. Given this expression, it is then plausible to disaggregate retail sales into the various categories of retail sales for which data are available (e.g., food, apparel, household furnishings). Individual equations for each category may then be developed for each type of sales. These disaggregated relationships are thus linked into the general framework via (1) their definitional status as components of total retail sales and (2) the functional correspondence between retail sales and gross product originating in wholesale and retail trade. As with the agriculture and manufacturing sectors, mathematical consistency has been preserved.

As a final illustration, a sectoral design from a subregional model is briefly described. As previously noted, equation systems for the Dallas and Houston areas are being formulated as adjuncts to the Texas structure. These models revolve around geographic identities which are completely analogous to the sectoral specifications which were utilized above. For example, a fundamental intraregional expression may be given by

$$RS_T = RS_H + RS_D + RS_O \qquad (10)$$

where:

RS_T = Total retail sales in Texas

RS_H = Total retail sales in the Houston area

RS_D = Total retail sales in the Dallas area

RS_O = Total retail sales in other areas of the state.

It is also true, of course, that total retail sales equals the sum of its various product components. Hence, the geographic models are easily linked to their sectoral counterparts and subsequently to the overall structure. Data which are available by standard metropolitan statistical area (SMSA) permit the estimation of behavioral relationships for retail sales, banking, manufacturing output, employment and numerous other aggregates at the substate level. Models of this nature provide a valuable resource for the development of an overall regional information system.

TIME SERIES ANALYSIS AND THE REGIONAL INFORMATION SYSTEM

As noted at the outset of this analysis, one of its principal objectives is to demonstrate the viability and importance of time series analysis within the context of a complete regional information network. There is a popular misconception, particularly among applied practitioners, that econometrics and time series analysis are generally competitive approaches to effective empirical analysis. In reality, however, the two methodologies are complementary in many significant cases.[9] In the situation being examined herein, for example, both procedural techniques are essential to a successful system. Several major roles for time series analysis and related approaches are briefly described as follows:

1. Traditional time series methods are employed to ascertain both preliminary and subsequent information regarding the underlying structure of all of the economic series employed in the model and

maintained in the data base. The results of an extensive set of initial empirical investigations are highly useful for exploratory purposes prior to estimation, for obtaining a more complete understanding of the nature of the various series, and for assisting in the development of alternative applications.

2. Recently developed time series methodologies, e.g., the X-11 ARIMA procedure which was established at Statistics Canada,[10] are utilized for seasonal adjustment of many quarterly series from the data base. One of the major requirements for consistency is, of course, a basic means of accounting for non-endogenous seasonal factors which is universally applicable and permits evolving weighting patterns.

3. As is the case with virtually all econometric models based on a large set of temporal data, the existence of autoregressive disturbances poses a significant problem in constructing regional systems. In addition to the statistical inefficiency engendered by serial correlation, its presence in a forecasting procedure inherently increases error variance. Correction for autocorrelation in large scale dynamic simultaneous equation models is, thus, an extremely important area of interaction between the time series and econometric conceptual frameworks which is vital to the development of a regional information network.[11] The relevant estimation processes are employed extensively in the Texas system.

4. Autoregressive Integrated Moving Average (ARIMA) models of the type suggested by Box and Jenkins (1970) are invaluable as tools for short-term forecasting within a comprehensive and systematic regional structure. As indicated previously, flexibility is a fundamental requirement for the program outlined in Section II. One feature of this versatility is the ability to adapt to the cost and time constraints of individual projects. When only a limited number of predictions are required and when the underlying interactions are of no significant consequences, then simple single equation reduced-form expressions and ARIMA models provide substantial advantages in terms of expense with respect to the simulation of a large regional model. Moreover, it should be noted that while extrinsic, i.e., econometric, systems are unique in their ability to emulate alternative economic scenarios, they are not able to consistently provide optimal short-term forecasts.[13] Hence, the simple ARIMA forecast of important aggregates provide substantial direct benefits to the overall information process.

5. Time series processes also play a significant indirect role in the performance evaluation system being implemented for Texas. Specifically, the use of these techniques for forecasting purposes permits a constant comparison of the forecasting and simulation capacities of the state and substate models with respect to the inherent autoregressive processes of the series being analyzed and, thus, provides a benchmark by which the performance of the model may be continually evaluated. The use of virtually any standard measure of forecast accuracy is predicated upon the existence of an alternative set of estimates for comparative purposes, i.e., for providing a basis for relative assessment. Among the measures utilized for testing purposes within the project are the root mean square error, the Theil inequality coefficient, the Janus quotient, and the Granger and Newbold "conditional efficiency" measure.[14]

6. ARIMA and econometric approaches to temporal causality testing in a multivariate context are utilized in order to examine the underlying

causal processes of the various models.[15] Individual expressions are examined for evidence of unidirectional and bidirectional temporal interrelationships among the major variables within the Texas system and its various submodels. In many cases, these structures are, by nature, extremely complex and a much greater degree of understanding is gained by examining their underlying causality properties. As a result of this analysis, the model of the Texas economy is the first major regional model to be developed in accordance with the inherent temporal causal processes which exist within the economy.

7. ARIMA forecasts of relevant exogenous variables may be generated as a basis for simulation of the various models. The simulation of any large scale econometric model requires, of course, the preliminary generation of a set of predicted values for all of the exogenous variables in the system. Ideally, such estimates are obtained from the output of national models or, perhaps, predictions generated by state and national governmental entities based on current policy plans. It is often the case, however, that information of this nature is not readily available for all requisite variables or that the values generated by other methods are inadequate. In such instances, an effective and inexpensive means of acquiring the necessary forecasts is, of course, to generate them through ARIMA processes. This approach, which has been termed "hybrid" by Spivey and Wecker (1971), is designed to utilize econometric and time series models as complementary tools in the prediction and simulation process.

As the above examples clearly indicate, time series analysis is a vitally important procedural method for use in the development of large scale regional information systems.

CONCLUSION

This paper has sought to (1) describe the essential features which characterize a comprehensive regional information network, (2) outline some of the efforts in this direction which are currently being implemented within the State of Texas, (3) illustrate the value of ARIMA and other time series methods in such a context. Frameworks of this nature, if developed and maintained, could serve to greatly enhance the capacity for effective and systematic analysis as a basis for decision making in both the governmental and private sectors. With the specification and estimation of a large number of regional econometric models around the country, the potential for major advancement toward the goals set out in the exposition on a national basis is indeed substantial. Should this type of system evolve on a broad scale, then the emergence of regional econometric models as a primary statistical tool for prediction would not necessarily bring an accompanying decline in the application of time series procedures. In fact, if such empirical systems are embedded in the broader context suggested herein, then all types of temporal analysis may be expected to play a greatly expanded role in the process of regional economic planning and investigation.

ACKNOWLEDGEMENTS

The author is Herman Brown Professor of Economics and Director of the Center for the Advancement of Economic Analysis, Baylor University. Appreciation is expressed to Leigh Humphrey for manuscript preparation and to the Brown Foundation and the Hankamer School of Business for released time and other financial support.

FOOTNOTES

1. A comprehensive summary of several existing models is given in Knapp, Fields, and Jerome (1976). The structure of selected models is discussed in Glickman (1977).
2. See Brundy and Jorgenson (1971); Dhrymes (1971); Fair (1971); Kloek and Mennes (1960); Perryman (1980); Perryman (1981a); Theil (1973).
3. A detailed discussion of this strategy for model development is given in Perryman (1981b) and Perryman (1981c).
4. The gross state product series is subsequently discussed. The capital stock is derived from capital expenditure data and industrial discard and decay functions.
5. For a more detailed discussion, see Perryman (1981b) and Perryman (1981c).
6. The major modifications are described in Niemi (1972) and L'Esperance, Nestel, and Fromm (1969). The method for generating a quarterly series is given in Weber (1979).
7. A detailed derivation and application of these interactions is given in Perryman and Green (1980).
8. The basic elements of this argument are outlined in Perryman (1981d).
9. See, for example, Dagum (1981).
10. See note 2 for a sample of the relevant literature.
11. For a theoretical demonstration, see Reid (1969) and Granger and Newbold (1974).
12. The inequality coefficient is described in Theil (1966). The Janus quotient is given in Gadd and Wold (1964). The "conditional efficiency" measure is described in Granger and Newbold (1973).
13. See especially Granger (1969); Haugh (1976); Mehra and Spencer (1979); Pierce and Haugh (1977); Sims (1972).

REFERENCES

BOX, G.E.P. and JENKINS, G.M. (1970). *Time Series Analysis: Forecasting and Control.* 2nd ed., Holden-Day, San Francisco.

BRUNDY, J.M. and JORGENSON, D.W. (1971). Efficient estimation of simultaneous equations by instrumental variables. *Review of Economics and Statistics 53*, 207-224.

DAGUM, E.B. (1981). Diagnostic checks for the ARIMA models of the X-11-ARIMA seasonal adjustment method. In *Time Series Analysis*. Eds: O.D. Anderson and M.R. Perryman, North-Holland, Amsterdam, 133-145.

DHRYMES, P.J. (1971). A simplified estimator for large scale econometric models. *Australian Journal of Statistics 23*, 27-46.

FAIR, R.C. (1971). Efficient estimation of simultaneous equations with autoregressive errors by instrumental variables. *Review of Economics and Statistics 54*, 444-449.

GADD, A. and WOLD, H. (1964). The Janus quotient: a measure for the accuracy of prediction. *Econometric Model Building.* Ed: H. Wold, North-Holland, Amsterdam, 229-234.

GLICKMAN, N.J. (1977). *Econometric Analysis of Regional Systems: Explorations in Model Building and Policy Analysis.* Academic Press, New York.

GRANGER, C.W.J. (1969). Investigating causal relations by econometric models and cross-spectral methods. *Econometrica 37*, 424-438.

GRANGER, C.W.J. and NEWBOLD, P. (1973). Some comments on the evaluation of economic forecasts. *Applied Economics 5*, 35-47.

GRANGER, C.W.J. and NEWBOLD, P. (1974). Experience with forecasting univariate time series and the combination of forecasts. *Journal of the Royal Statistical Society - Series A 137*, 131-164.

HAUGH, L.D. (1976). Checking the independence of two covariance-stationary time series: a univariate residual cross-correlation approach. *Journal of the American Statistical Association 71*, 378-386.

KENDRICK, J.W. and JAYCOX, C.M. (1965). The concept and estimation of gross state product. *Southern Economic Journal 31*, 153-168.

KLOEK, T. and MENNES, L.B.M. (1960). Simultaneous equations estimation based on principal components of predetermined variables. *Econometrica 28*, 45-61.

KNAPP, J.L.; FIELDS, T.W. and JEROME, R.T. (1976). *A Survey of State and Regional Econometric Models*. Tayloe Murphy Institute, Charlottesville.

L'ESPERANCE, W.L.; NESTEL, G.; and FROMM, D. (1969). Gross state product and an econometric model of a state. *Journal of the American Statistical Association 64*, 787-807.

MEHRA, Y.P. and SPENCER, D.E. (1979). The St. Louis equation and reverse causation: the evidence reexamined. *Southern Economic Journal 45*, 1104-1120.

NIEMI, A.W. (1972). A reexamination of the Kendrick-Jaycox method of estimating gross state product. *Review of Regional Studies 12*, 123-131.

PERRYMAN, M.R. (1980). A simple multi-stage consistent estimator for dynamic equations from large simultaneous systems which exhibit first-order autoregressive disturbances. Paper delivered to the *Conference of the Econometric Society,* Denver, Colorado.

PERRYMAN, M.R. (1981a). Estimation of large dynamic temporal systems with autoregressive disturbances. In *Time Series Analysis*. Eds: O.D. Anderson and M.R. Perryman, North-Holland, Amsterdam, 425-435.

PERRYMAN, M.R. (1981b). A mathematically consistent structural specification for regional econometric models. In *Modeling and Simulation 12*, 1067-1074.

PERRYMAN, M.R. (1981c). On the mathematical properties of a complete regional model. Paper delivered to the *Third International Conference on Mathematical Modeling*, Los Angeles, California.

PERRYMAN, M.R. (1981d). Time series analysis as an ex poste evaluation criterion for evaluating econometric models. In *Time Series Analysis*. Eds: O.D. Anderson and M.R. Perryman, North-Holland, Amsterdam, 437-445.

PERRYMAN, M.R. and GREEN, S.L. (1980). The supply side interactions of employment and output within a large regional econometric simulation model. Paper delivered to the *Conference of the Atlantic Economic Society*, Boston, Massachusetts.

PIERCE, D.A. and HAUGH, L.D. (1977). Causality in temporal systems: characterizations and a survey. *Journal of Econometrics 5*, 265-293.

REID, D.J. (1969). *A Comparative Study of Time Series Prediction Techniques on Economic Data*. Ph.D. thesis, Nottingham University.

SIMS, C.A. (1972). Money, income, and causality. *American Economic Review 62*, 540-552.

SPIVEY, W.A. and WECKER, W.E. (1971). Regional economic forecasting: concepts and methodology. *Papers of the Regional Science Association 28*, 257-276.

THEIL, H. (1966). *Applied Economic Forecasting*. Rand McNally, Chicago.

THEIL, H. (1973). A simple modification of thw two-stage least squares procedure for undersized samples. In *Structural Equation Models in the Social Sciences*. Eds: A.S. Goldberger and O.D. Duncan, Seminar Press, New York, 113-129.

WEBER, R.E. (1979). Estimating quarterly gross state product. *Business Economics 14*, 38-43.

APPLIED TIME SERIES ANALYSIS
O.D. Anderson & M.R. Perryman (eds.)

EVIDENCE ON THE IDENTIFICATION AND CAUSALITY DISPUTE ABOUT THE DEATH PENALTY

Llad Phillips and Subhash C. Ray†
Department of Economics
University of California
Santa Barbara, California
U.S.A.

This paper reports the application of multivariate time series techniques to resolve questions of identification and Granger-Causality in the controversy about the deterrent effect of the death penalty. In this paper we identify a causal dynamic model of the homicide rate in California using data for the years 1945-1978. The theory suggests three possible relations between the offense rate and the probability of imprisonment: control (negative), definitional (negative) and response (positive). We find that the direction of Granger-Causality runs from the probability of imprisonment to the homicide rate.

INTRODUCTION

The finding by Ehrlich (1975) that the probability of execution given conviction had a significant effect in reducing homicides in the United States by 8 for every execution has been widely challenged. Ehrlich recognized that the supply of offenses function was one of a system of equations, and reported estimates obtained from a three round procedure. Nonetheless, critics have contended that there is scant knowledge of the exogenous forces determining homicide. The *a priori* specification of the supply of offenses function with exclusion of certain exogenous variables and inclusion of such restrictions as necessary for identification rests on a weak basis.

Hoenack and Weiler (1980) argue that Ehrlich's failure to completely specify a system of equations describing the behavior of enforcement agencies, as well as the supply of homicides, led to an improper identification of the latter. They suggest that

> "the causality underlying this equation is the response of the punishments meted out by the criminal justice system to murder behavior. - - - - - Ehrlich's murder supply function, if not identified, could represent criminal justice system behavior and not murder behavior.[1]"

Klein, Forst, and Filatov (1978) contend that Ehrlich's significant finding for the probability of execution given conviction could be vitiated by omitted causal variables and by measurement error.

---

† This research was supported under grant #79-NI-AX-0069 from the National Institute of Law Enforcement and Criminal Justice. Points of view are those of the authors and do not necessarily reflect the position of the U.S. Department of Justice.

> "The use of Q [homicides] both as the numerator of the homicide rate and as part of the denominator of the execution rate could have biased Ehrlich's estimate of the regression coefficient for $P_{e/c}$ [Probability of execution given conviction] toward the appearance of a deterrent effect.[2]"

Fisher and Nagin (1978) analyze whether it is feasible to identify a supply of offenses function in a simultaneous model of crime rates and sanction levels. They conclude that

> "Identification is the sine qua non of all estimation and especially of simultaneous equation estimation. ----
>
> Researchers who have employed simultaneous estimation techniques to study the deterrent effect of sanctions on crime have failed to recognize fully the importance of this issue. The restrictions that they (implicitly or explicitly) use to gain apparent identification have little theoretical or empirical basis.[3]"

Fisher and Nagin suggest that studies using aggregate non-experimental data

> "must have a time-series component in the data (i.e., pure time-series or a time-series, cross-section), and the estimation procedures must account for the possibility of serial correlation in the stochastic components of the specification.[4]"

The analysis pursued in this paper examines the issue of model complexity. Is the relationship between the homicide rate and variables such as the probability of imprisonment simultaneous, or is there one-way causality and, if so, which way? First, a theoretical framework is established which suggests three possible relationships between the offense rate and the probability of imprisonment: control (negative), definitional (negative), and cost minimizing response (positive). Second, the Granger-Newbold procedure is used to determine whether there is feedback, i.e., two way causality between the offense rate and the probability of imprisonment. This involves (1) fitting univariate Box-Jenkins models to each series, (2) cross correlating the innovations, (3) modelling the relation (or relations if two way causality) between the innovations, and (4) inferring the model between the offense rate and the probability of imprisonment based on the results from (1), (2) and (3). Lastly, multivariate transfer function models are estimated .

The analysis for homicide indicated that the causality was one way with the homicide rate depending inversely upon the probability of imprisonment. The estimation of the transfer function indicated a geometrically declining distributed lag. Extension of the analysis to multivarite transfer functions indicated no significant dependence of the homicide rate upon median time served in prison. Nor was there a significant dependence upon the unemployment rate for persons 16 years of age and over. However, the homicide rate was directly related to a geometric lag of the unemployment rate for 18 and 19 year old males. Finally, two measures of the death penalty were investigated: (1) the probability of execution given imprisonment, and (2) the probability of receiving the death sentence given imprisonment. The homicide rate was inversely related to either death penalty measure.

THEORY

The theory incorporates Stigler's [1970] recommendation that a model of rational agency behavior should be based on the objective criterion of minimizing the sum of damages plus control costs. This cost minimization is subject to the supply of offenses function and subject to the definition of the likelihood of imprisonment as the ratio of the number of felons committed to prison to the number of offenses. The conditions for cost minimization lead to a positive relationship between felons committed and the offense rate. The supply of offenses function and the definition of the likelihood of imprisonment both imply a negative relationship between the offense rate and the likelihood of imprisonment.

(i) Minimizing Damages to Victims and Public Control Costs

The optimal operation of the criminal justice system would require the minimization of the sum of damages to victims plus the public costs of operating the system. The damages to victims, D, is the sum over the various offenses of the product of the loss rate per offense, r_i, times the number of offenses, OF_i,

$$D = \sum_{i=1}^{m} r_i OF_i \; . \qquad (1)$$

The cost function specifies the technical relationship between inputs and outputs for the criminal justice system in terms of costs, C, as a function of a vector of outputs, $\vec{q}$, and a vector of input prices, $\vec{w}$,

$$C = C(\vec{q}, \vec{w}) \; . \qquad (2)$$

In principle, the cost function could be elaborated to include a number of measures of output such as a vector of arrests for the various crimes, $\vec{A}$, a vector of felony charges, $\vec{F}$, a vector of convictions, $\vec{V}$, a vector of imprisonments, $\vec{I}$, a vector of average time served, $\vec{TS}$, and a vector of other control measures such as probation $\vec{R}$:

$$C = C(\vec{A}, \vec{F}, \vec{V}, \vec{R}, \vec{I}, \vec{TS}, \vec{W}) \; . \qquad (3)$$

The supply of offense function for a given crime i could, in ideal form, be similarly elaborated to indicate the dependence of offenses, OF_i, upon the various control measures or outputs for that crime, as well as socioeconomic and demographic variables, x_j,

$$OF_i = g_i(A_i, F_i, V_i, R_i, I_i, TS_i, X_1, X_2, \ldots X_n). \qquad (4)$$

Substituting equation (4) into equation (1) and minimizing the sum of D and C, with respect to outputs such as imprisonments, one obtains

$$r_i \frac{\partial g_i}{\partial I_i} + \frac{\partial C}{\partial I_i} = 0 \; , \; i = 1, m, \qquad (5)$$

or with respect to outputs such as time served

$$r_i \frac{\partial g_i}{\partial TS_i} + \frac{\partial C}{\partial TS_i} = 0 \; , \; i = 1, m \; . \qquad (6)$$

Note that in taking the ratio of equations (5) and (6) one obtains the tangency condition that the marginal rate of substitution of the isocrime curve equals the marginal rate of transformation of the production possibility frontier:

$$\frac{\partial g_i}{\partial TS_i} \Big/ \frac{\partial g_i}{\partial I_i} = \frac{\partial C}{\partial TS_i} \Big/ \frac{\partial C}{\partial I_i} \tag{7}$$

which determines the optimal mix of output. The optimal level of the operation of the criminal justice system can be determined by evaluating equations such as (5) and (6) which equate the marginal costs of control to the marginal reduction in damages to victims.

Static Model

(i) A Simplified Illustrative Specification of the Cost Minimization Model

To illustrate the various behavioral and definitional relationships between the offense rate, OF_i , and imprisonments, I_i , that may be embedded in the data it is useful to simplify the general theory and use a concrete specification of functional forms in the model. The first simplification is to focus on the two controls of imprisonments, I , and average time served, TS , in the supply of offenses function. A log-linear functional form is assumed. This was the specification of the supply of offenses function used by Ehrlich (1973). The emphasis is on the control effect of imprisonment as a sanction. Thus equation (4) becomes

$$OF_i = kP_i^{\beta}\, TS_i^{\gamma}\, X_1^{\delta}\, X_2^{\varepsilon} \quad , \tag{8}$$

where the probability of imprisonment, P , is defined by

$$P_i = \frac{I_i}{OF_i} \, . \tag{9}$$

Equations (8) and (9) can be combined to eliminate P ,

$$OF_i = k^{\frac{\beta}{1+\beta}}\, I_i^{\frac{\beta}{1+\beta}}\, TS_i^{\frac{\gamma}{1+\beta}}\, X_1^{\frac{\delta}{1+\beta}}\, X_2^{\frac{\varepsilon}{1+\beta}} \quad , \tag{10}$$

where if crime is to be controllable with increases in expenditures on criminal justice and thereby increases in the outputs of imprisonments and time served, the parameter $\frac{\beta}{1+\beta}$ in equation (10) must be negative.

To simplify the cost function, we assume that costs are separable for crime i ,

$$C = C_i + \sum_{j \neq i} C_j \tag{11}$$

and can be expressed as a special case of the generalized Leontief cost function, the case which implies a linear transformation function between imprisonments and time served. The specification of outputs has been reduced to these two conformable to the supply of offenses function. (The assumption of a linear transformation function implies a constant ratio of imprisonments to time served, which is rather restrictive, but this simple form of the cost function is useful for illustration and could be complicated by generalization.) Thus, equation (3) becomes

$$C = (aTS_i + bI_i)(dw_p + ew_o) + \sum_{j \neq i} C_j , \qquad (12)$$

where w_p is the wage rate for police and w_o is the wage rate for other resources such as corrections plus the courts, etc.

Minimizing the damages from crime i, $r_i OF_i$, where OF_i is specified by equation (10), plus the costs of control C, where C is specified by equation (12), with respect to average time served, TS_i, one obtains,

$$\frac{\gamma}{1+\beta} r_i \frac{OF_i}{TS_i} = -a(dw_p + ew_o) \qquad (13)$$

and minimizing with respect to imprisonments for crime i, I_i, yields

$$\frac{\beta}{1+\beta} r_i \frac{OF_i}{I_i} = -b(dw_p + ew_o) \ . \qquad (14)$$

The ratio of equations (13) and (14) determines the optimal mix of outputs

$$\frac{\gamma}{\beta} \frac{I_i}{TS_i} = \frac{a}{b} \qquad (15)$$

which corresponds to equation (7).

Note that equation (14) indicates that imprisonments will vary positively with the offense rate. This is the system response to rising crime rates. Equation (8) specifies the inverse behavioral relationship between the offense rate and the probability of imprisonment and equation (9) specifies the inverse definitional relationship between these two variables.

There are five endogenous variables determined by this model: the costs of controlling crime i, C_i, holding the costs of controlling other crimes constant by assumption, as determined by equation (12), the offense rate for crime i, OF_i, as determined by equation (8), the probability of imprisonment, P_i, as defined by equation (9), the average time served, TS_i, as a function of the rate of imprisonments as determined by equation (15), and the rate of imprisonments, I_i, as determined by the offense rate in equation (14). The supply of offenses, equation (8), and the system response function, equation (14), could be estimated in principle, using the exogenous variables X_1, X_2, w_p, and w_o, and are identified, as the model is specified.

Of course the identification controversy raises the question of whether we are sufficiently knowledgeable about crime causation to properly choose the socioeconomic variables X_1 and X_2, and whether the simplifying assumptions used to derive the model from the theory are too restrictive to adequately model the complexity of the criminal justice system and human behavior. In any case, this model is useful in developing a path diagram of expected causal linkages between variables such as the offense rate and imprisonments. If some of the causal linkages occur with delay, it may be possible to determine the direction of causality without discovering the proper specification of socioeconomic variables X_j that would identify the formal model. If the model is recursive it may be much easier to identify. The causal path diagram is used to visually represent the economic model, and time series analysis is then utilized to see whether some of the causal relations may be recursive rather than simultaneous.

Path Diagram

(i) Causal Path Diagram of the Cost Minimization Model

The five equations (8,9,12,13, and 14) determining the five endogeneous variables of the offense rate, OF , average time served, TS , the imprisonment rate, I , the costs of control, C, and the probability of imprisonment, P , are represented by the path lines in Fig. 1. The arrow indicates the direction of causality, and the sign (+ or -) for the direction of impact. For example, the solid lines connecting average time served and the probability of imprisonment to the offense rate indicate the potential control effects of these variables on crime. However, these negative correlations between these two control variables and the crime rate may be obscured or offset by the positive relations running from the offense rate to these control variables indicated by the dotted lines. These dotted lines represent the responses of the system to changes in the crime rate and are occasioned in this model by cost minimizing behavior. The last leg of the positive connection between the offense rate and the probability of imprisonment is completed by the definition of the probability of imprisonment as shown in Fig. 1 by the dashed line. This definition also creates a negative link between the offense rate and the probability of imprisonment as illustrated in Fig. 1. The task is to devise a statistical procedure to distinguish these positive and negative correlations between the crime rate and the two control variables, i.e., the likelihood of imprisonment and the average time served. If there is a lag in behavioral response this may be possible using the Sims-Granger approach to causality.

Dynamic Model: Behavioral Lags

Suppose changes in the behavior of criminals lags behind changes in the control variables undertaken by criminal justice system authorities. These lags could be due to an elapse in time before the criminals realize the control variables have changed and/or due to lags in adjusting behavior to new information. In this case, the response of the crime rate to the probability of imprisonment may occur with a lag. For example, referring to equation (8),

$$OF(t) = kP^{\beta}\ TS^{\gamma}\ X_1^{\delta} \tag{16}$$

and taking logarithms one obtains

$$\ln OF(t) = \ln k + \beta \ln P(t) + \gamma \ln TS(t) + \delta \ln X_1(t) \tag{17}$$

As suggested above, the offense rate may depend upon the probability of imprisonment with a lag, i.e.,

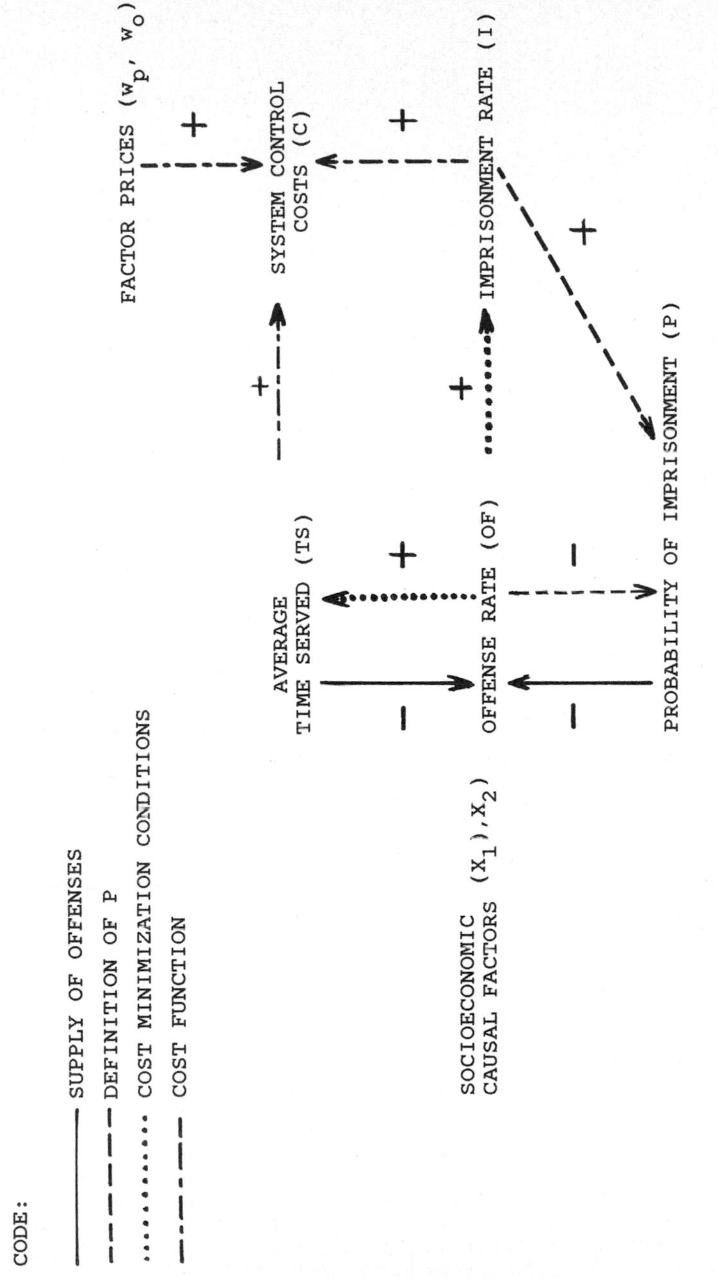

FIGURE 1: CAUSAL PATH DIAGRAM OF COST MINIMIZATION MODEL

$$\ln OF(t) = \ln k + \beta \ln P(t) + \beta\lambda_1 \ln P(t-1) + \beta\lambda_2 \ln P(t-2) + \gamma \ln S(t) + \delta \ln X_1(t) \qquad (18)$$

where we can define the lag operator Z^n

$$X_1(t-n) = Z^n X_1(t) \qquad (19)$$

and hence express equation (18) as

$$\ln OF(t) = \ln k + \beta[1 + \lambda_1 Z + \lambda_2 Z^2] \ln P(t) + \gamma \ln TS(t) + \delta \ln X_1(t) \qquad (20)$$

or

$$\ln OF(t) = \ln k + \beta\Lambda(Z) \ln P(t) + \gamma \ln TS(t) + \delta \ln X(t) \qquad (21)$$

where

$$\Lambda(Z) = [1 + \lambda_1 Z + \lambda_2 Z^2] \quad .$$

Of course, the offense rate may be a distributed lag, $\Phi(Z)$, of time served and a distributed lag, $H(Z)$, of the socioeconomic variables as well,

$$\ln OF(t) = \ln k + \beta\Lambda(Z) \ln P(t) + \gamma\Phi(Z) \ln TS + \delta H(Z) \ln X(t) \qquad (22)$$

The objective is to estimate this dependence of the offense rate upon the control variables and the socioeconomic variables. However, first we must consider confounding factors which could blur the proper identification of equation (22).

We recall that the system responds to varying crime rates as expressed in equations (13) and (14) above. Once again, this response may take place with a lag and consequently, for example, the probability of imprisonment may be a positive distributed lag of the offense rate,

$$\ln P(t) = gK(Z) \ln OF(t) \qquad (23)$$

where from the theory above, $g > 0$.

Lastly, from the definition of the probability of imprisonment in equation (9), we obtain a contemporaneous inverse dependence upon the offense rate.

$$\ln P(t) \equiv \ln I(t) - \ln OF(t) \quad . \qquad (24)$$

The three equations (22), (23), and (24) are simply an elaboration, in terms of lagged relationships, of the causal path diagram. Since the definitional or spurious inverse correlation between the offense rate and probability of imprisonment is contemporaneous, an empirical finding of an inverse lagged relationship between these two variables will reflect the behavioral control effect as postulated in equation (22). However, there is the possibility of two way causality as reflected by equation (23). The Granger-Newbold (1977) procedure is used to distinguish one-way from two-way causality.

<u>Time Series Methodology</u>

(i) Identifying Causality with Lagged Relationships: the Granger-Newbold Procedure.

Consider a simplified supply of offenses function (22) which focuses on the relationship between the offense rate and the probability of imprisonment, and where the other control and causal variables are represented by a residual $U_1(1)$, and where the contemporaneous definitional relationship (24) is now subsumed in this equation:

$$\ln OF(t) = \beta\Lambda(Z)\ln P(t) + U_1(t) \tag{25}$$

noting from the theory above that $\beta < 0$.

The response of the system to crime rates is likely to not be exact, but stochastic, and hence we add the residual or error time series $U_2(t)$ to equation (23), which may operate with a lag $M(Z)$, and where the errors $U_1(t)$ and $U_2(t)$ are presumed to be independent,

$$\ln P(t) = gK(Z)\ln OF(t) + M(Z)U_2(t) . \tag{26}$$

Solving for $\ln OF(t)$ and $\ln P(t)$ in terms of residuals,

$$[1- \beta g\, K(Z)\Lambda(Z)]\ln OF(t) = U_1(t) + \beta\Lambda(Z)M(Z)U_2(t) \tag{27}$$

$$[1 - \beta g\, K(Z)\Lambda(Z)]\ln P(t) = gK(Z)U_1(t) + M(Z)U_2(t) . \tag{28}$$

The cross-correlation function $\rho(v)$ of the filtered ($[1 - \beta gK(Z)\Lambda(Z)]$) offense rate and the filtered probability of imprisonment will, in general, be two sided, i.e., non-zero for both positive lags (v) and negative lags $(-v)$ indicating feedback or two-way causality.

Suppose, however, that the causality is one-sided because the criminal justice system does not respond to varying crime rates, i.e., $g = 0$, then from equation (26),

$$\ln P(t) = M(Z)\, U_2(t) \tag{29}$$

and fitting a univariate model to the probability of imprisonment we obtain the innovations $e_2(t)$

$$\frac{\ln P(t)}{M(Z)} = U_2(t) \equiv e_2(t) . \tag{30}$$

If $g = 0$, from equation (27) one obtains,

$$\ln OF(t) = \beta\Lambda(Z)\, M(Z)\, U_2(t) + U_1(t) \tag{31}$$

and fitting a univariate model to the offense rate, we obtain the innovations $e_1(t)$

$$\frac{\ln OF(t)}{\Lambda(Z)M(Z)} = \beta U_2(t) + \frac{U_1(t)}{\Lambda(Z)M(Z)} \equiv e_1(t) . \tag{32}$$

From equations (30) and (32), presuming the innovations $U_2(t)$ are white noise, as a result of the procedure of fitting a univariate model to the probability of imprisonment, and recalling the assumption that the error processes $U_1(t)$ and $U_2(t)$ are independent, the cross correlation function between the filtered offense rate and probability of imprisonment will be non zero and negative $(\beta < 0)$ only at lag (v) zero, i.e.,

$$\rho e_2,\ e_1(v) = \beta(\sigma_{e_2}/\sigma_{e_1})\sigma^v$$

In contradistinction, suppose that the relation is one-sided because there is no control effect and only spurious contemporaneous correlation, i.e., $\Lambda(Z) = \lambda$. Then from equation (25)

$$\ln OF(t) = \beta\lambda \ln(P)t + U_1(t)\ . \tag{34}$$

In addition, there is the positive response of the system to crime represented by equation (26). In this case, equations (26) and (34) can be collapsed to one equation where the probability of imprisonment is a positive distributed lag of the offense rate except at lag zero where it could be positive or negative.

Thus, the cross-correlation function of the innovations obtained from fitting univariate autoregressive integrated moving average processes to the offense rate and the probability of imprisonment, combined with the theoretical expectations of sign, should permit one to distinguish between two-way causality and the two different cases of one-way causality.

(ii) Modelling the One-Way Transfer Functions

In the case where the cross-correlation between the innovations is negative at lag zero and zero elsewhere, i.e., equation (33) holds, the relationship between the innovations can be modelled as

$$e_1(t) = \beta e_2(t) + U(t)\ . \tag{35}$$

If we were to substitute for the innovations from equations (30) and (32) we would, of course, obtain equation (25), the supply of offenses function. In general, following the Granger-Newbold procedure, we will have fitted two univariate integrated autoregressive moving average processes to the offense rate and the probability of imprisonment

$$A(Z)\Delta^d \ln OF(t) = B(Z)\ e_1(t) \tag{36}$$

where $\Delta = (1-Z)$ is the difference operator,

$$C(Z)\Delta^{d*} \ln P(t) = D(Z)\ e_2(t) \tag{37}$$

and estimated the parameters of $A(Z)$, $B(Z)$, $C(Z)$, $D(Z)$, d, and $d*$. Combining equations (35), (36), and (37) one obtains

$$\frac{A(Z)\Delta^d}{B(Z)} \ln OF(t) = \beta\ \frac{C(Z)\Delta^{d*}}{D(Z)} \ln P(t) + U(t) \tag{38}$$

or

$$\ln OF(t) = \beta\ \frac{B(Z)C(Z)\Delta^{d*}}{A(Z)D(Z)\Delta^d} \ln P(t) + \frac{B(Z)}{A(Z)\Delta^d}\ U(t) \tag{39}$$

where one can calculate

$$\Psi(Z) = (1 + \Psi_1 Z + \Psi_2 Z^2\ \ldots\ ) = \frac{B(Z)C(Z)\Delta^{d*}}{A(Z)D(Z)\Delta^d} \tag{40}$$

from the estimated parameters (see Box and Jenkins, Ch. 10, (1970). Alternatively, one can filter the probability of imprisonment to obtain the innovations $e_2(t)$,

$$\frac{C(Z)\Delta^{d*}}{D(Z)} \ln P(t) = e_2(t)$$

and use this same filter on the offense rate,

$$\frac{C(Z)}{D(Z)} \Delta^{d*} \ln OF(t) = \frac{C(Z)}{D(Z)} \frac{\Delta^{d*}}{\Delta^{d}} \frac{B(Z)}{A(Z)} e_1(t) \equiv e_1^*(t)$$

and from the cross-correlation between $e_2(t)$ and $e_1^*(t)$, observe the pattern of the transfer function. This observed transfer function can be approximated by a ratio of low order polynomials and estimated along with the residual according to the procedures of Box and Jenkins.

(iii) Estimating Multi-Input Transfer Functions

In the case where the Granger-Newbold procedure justifies one-way causality, multi-input transfer function models can be fitted using the procedure in Jenkins (1979). This involves (1) estimating univariate ARIMA models for the output and each input, (2) estimating bivariate transfer function models following the Box and Jenkins transfer function estimation procedure, and (3) combining the results and estimating the multivariate transfer function model. Thus, one can estimate equation (22) for the supply of offenses for homicide.

Homicide in California: Data

The annual number of felons committed to state prison is available for the major felonies, including the seven FBI index crimes for the years 1945 to 1979 in the various issues of *California Prisoners* and the *California Statistical Abstract*. Yearly values of median time served are available for the major felonies for the years 1945 to 1975 in *California Prisoners*. This data offers a long period of experience of the use of imprisonment as a sanction. Annual figures on reported offenses are available from the publication of *Crime in California*, which is entitled in recent years, *Crime and Delinquency in California*. In the case of homicide, the number of reported offenses tabulated by the California Bureau of Criminal Statistics can be checked against the number of homicides reported in the *Vital Statistics of the United States, Mortality Data*. Subsidiary information, such as the distribution of the number of offenses committed per offender is available for particular years in special studies such as *Willful Homicide in California*. There are four such studies for this particular crime. The number of executions is reported for the years 1945-1975. Felons received at state prison with a death sentence is reported in various issues of *California Prisoners* for the period 1950-1978.

The unemployment rate for persons 16 years of age and over is available from *Employment and Earnings* for the period since 1945. The unemployment rate for males 18 and 19 years of age is available from the same source since 1947.

Data on the number of willful homicides per capita was collected from the publications of the California Bureau of Criminal Statistics (BCS)[5] and compared to the number of homicides listed as a cause of death in the *Vital Statistics of the United States*. This comparison is illustrated in Fig. 2. Note the decline in the homicide rate

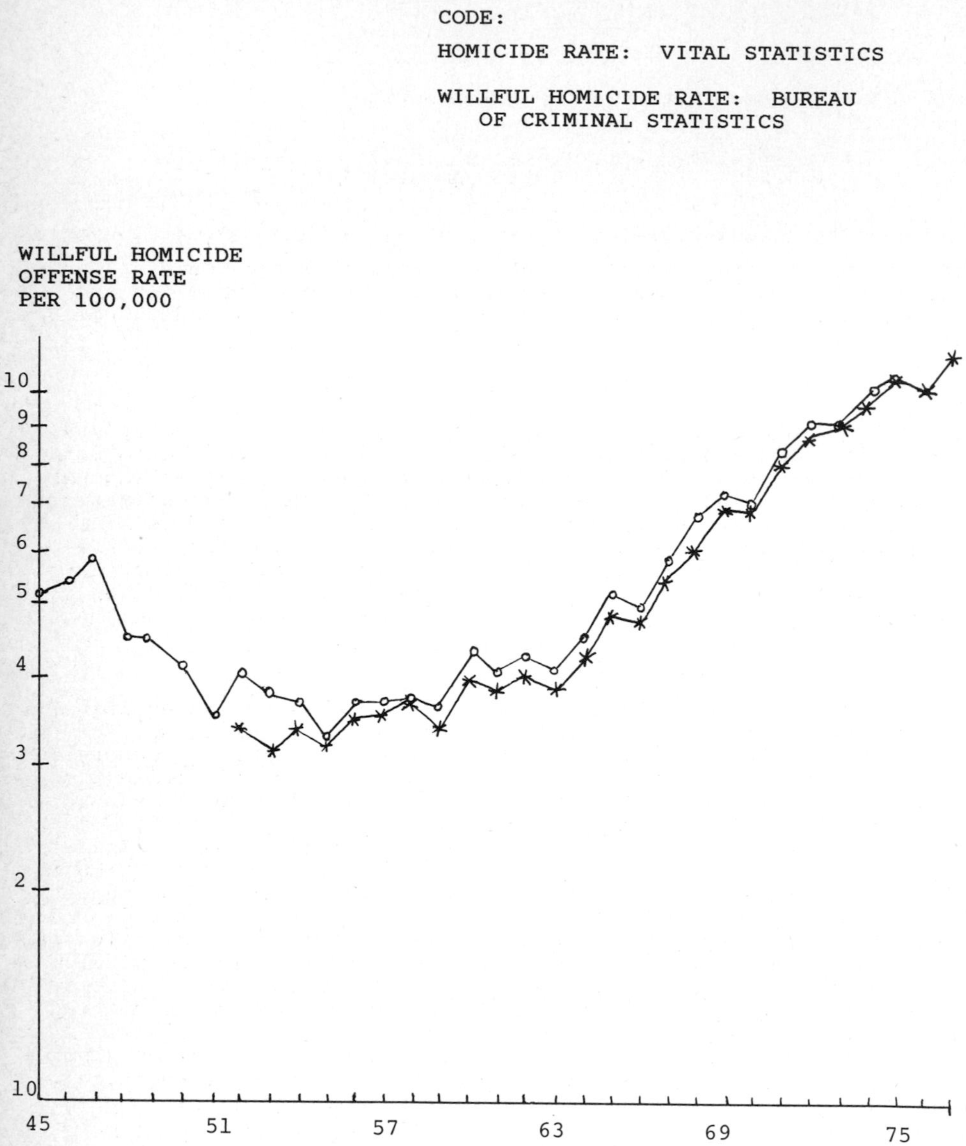

FIGURE 2: HOMICIDE RATE COMPARISON: VITAL STATISTICS AND REPORTED OFFENSES

for the ten year period 1945-1954. The ten year period 1955-64 was one of slow increase. The last ten years or so 1965-1977 was one of rapid increase. The probability of imprisonment was calculated from the number of imprisonments,[6] using, alternatively as a base, the number of reported offenses (BCS) and the number of homicides listed in the mortality statistics. The comparison is illustrated in Fig. 3. Note that the probability of imprisonment first increases and then decreases in this 30 year period. There is less variation in median time served. The median time served in years remained reasonably steady for the period 1945 to 1975, ranging from a low of 4.99 to a high of 6.65 with an average of 5.77.

<u>Determining Granger-Causality For Homicide</u>

(i) Estimating Univariate ARIMA models for $OF(t)$ and $P(t)$

Time series for the homicide offense rate and the probability of imprisonment were analyzed for the years 1950 to 1978, using homicide rates based on data from the <u>Vital Statistics of the United States</u> for the years 1950 to 1953 and data from the California Bureau of Criminal Statistics for subsequent years. A second order autoregressive model was estimated for the homicide offense rate:

$$\underset{(t=7.13)}{(1 - 1.2892Z} + \underset{(t=1.73)}{0.3243Z^2})[OF(t) - EOF(t)] = e_1(t) \qquad (41)$$

with a residual mean square of 0.3367. The t statistics for the estimated parameters are indicated. The mean of the offense rate series is denoted by its expected value, $EOF(t)$. The estimated parameters indicate that the homicide process is stationary, i.e., stable, although close to the boundary. The autocorrelation function of the residuals $e_1(t)$ indicate that they are approximately white noise. For 10 degrees of freedom, the Q statistic is 15.2, but for 22 degrees of freedom the Q statistic is 44.6. However, there is some structure remaining in $e_1(t)$.

A second order moving average model was estimated for the first difference of the probability of imprisonment:

$$(1 - Z)\,P(t) = (1 - \underset{(t=4.23)}{0.7525Z} + \underset{(t=1.85)}{0.3242Z^2})\, e_2(t) \qquad (42)$$

with a residual mean square of .001056. The estimated parameters indicate the process is invertible. The autocorrelation function of the residuals $e_2(t)$ indicates that they are white noise (Q statistic of 12.3 for 10 degrees of freedom).

(ii) Cross-Correlation of the Residuals $e_1(t)$ and $e_2(t)$

The cross-correlation function of the residuals is plotted in Fig. 4 with 95% confidence intervals around zero denoted by plusses (+) . The only significant correlation is at lag zero and is negative, indicating a lack of feedback and that the causality runs from the probability of imprisonment to the offense rate, as we argued on the basis of economic theory in the section <u>Time Series Methodology</u> (above).[7]

(iii) Modelling the Transfer Function between $P(t)$ and $OF(t)$

The estimated cross-correlation depicted in Fig. 4 can be modelled as (see equation 35),

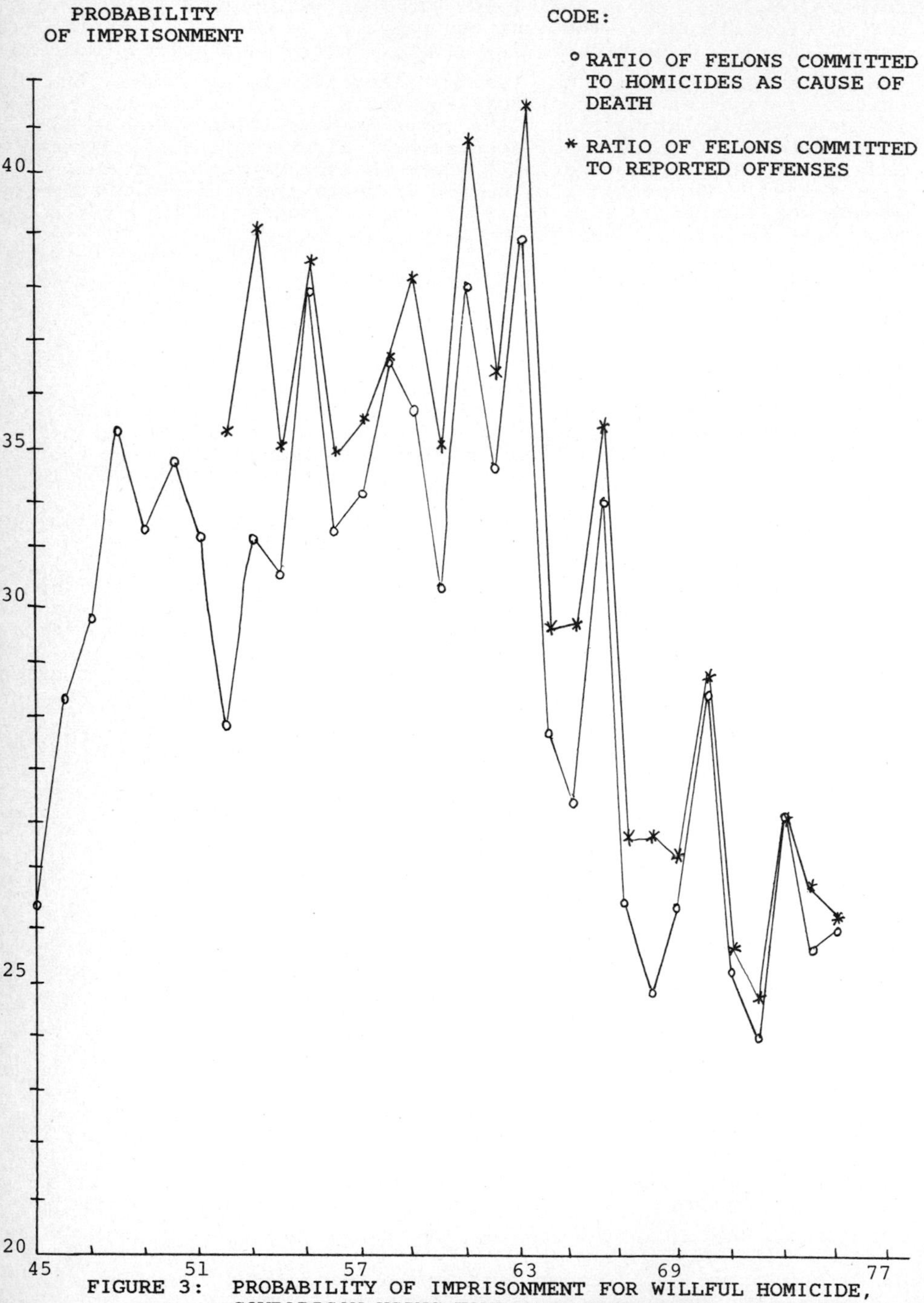

FIGURE 3: PROBABILITY OF IMPRISONMENT FOR WILLFUL HOMICIDE, COMPARISON USING HOMICIDES FROM VITAL STATISTICS AND REPORTED OFFENSES

```
                              CROSS-CORRELATION
           -1.0 -0.8 -0.6 -0.4 -0.2  0.0  0.2  0.4  0.6  0.8  1.0
             +----+----+----+----+----+----+----+----+----+----+
LAG
                                      I
-24  0.076        +                   IXX                 +
-23  0.019         +                  I                  +
-22  0.101          +                 IXXX              +
-21  0.063           +                IXX              +
-20  0.085            +               IXX             +
-19 -0.026             +              I              +
-18  0.122             +              IXXX           +
-17  0.033              +             IX            +
-16  0.258              +             IXXXXXX       +
-15 -0.134               +         XXXI            +
-14  0.264               +            IXXXXXXX     +
-13 -0.131                +        XXXI           +
-12  0.030                +           IX          +
-11 -0.076                +         XXI           +
-10 -0.068                 +        XXI          +
 -9 -0.082                 +        XXI          +
 -8 -0.027                 +          I          +
 -7 -0.280                  +  XXXXXXXI         +
 -6 -0.111                  +      XXXI         +
 -5 -0.001                  +         I         +
 -4 -0.213                  +    XXXXXI         +
 -3 -0.250                  +   XXXXXXI         +
 -2 -0.058                   +        I        +
 -1  0.212                   +        IXXXXX   +
  0 -0.569              XXXXX+XXXXXXXXI        +
  1  0.088                   +        IXX      +
  2  0.021                   +        IX       +
  3  0.073                  +         IXX       +
  4 -0.203                  +    XXXXXI         +
  5 -0.004                  +         I         +
  6  0.193                  +         IXXXXX    +
  7  0.011                  +         I         +
  8 -0.073                 +        XXI          +
  9  0.112                 +          IXXX       +
 10  0.077                 +          IXX        +
 11  0.166                +           IXXXX       +
 12 -0.080                +         XXI           +
 13  0.131                +           IXXX        +
 14  0.282               +            IXXXXXXX     +
 15  0.008               +            I            +
 16  0.005              +             I             +
 17  0.043              +             IX            +
 18  0.156             +              IXXXX          +
 19 -0.096             +            XXI              +
 20 -0.078            +             XXI               +
 21  0.076           +                IXX              +
 22 -0.001          +                 I                 +
 23 -0.060         +                  I                  +
 24 -0.014        +                   I                   +
--
```

CODE: 95% confidence intervals around zero.

FIGURE 4. CROSS-CORRELATION BETWEEN THE INNOVATIONS OF THE HOMICIDE OFFENSE RATE AND THE PROBABILITY OF IMPRISONMENT

$$e_1(t) = \beta\ e_2(t) + u(t) \qquad (43)$$

where the cross-correlation at lag zero, $\rho e_2,\ e_1(0)$, would be:

$$\rho e_2,\ e_1(0) = \frac{\beta\ \sigma^2 e_2}{\sigma e_1\ \sigma e_2} = \frac{\beta\ \sigma e_2}{\sigma e_1} \qquad (44)$$

and $\sigma^2 e$ is the residual mean square. The calculated value for β is -10.14.

The parameters of the lag structure of the transfer function from the probability of imprisonment to the homicide rate can be inferred by substituting for $e_1(t)$ and $e_2(t)$ from equations (41) and (42) into equation (43).[8] The lag structure can also be observed by cross-correlating the residuals obtained by filtering both the offense rate and the probability of imprisonment with the filter

$$(1-Z)/(1 - 0.7525Z + 0.3242Z^2)$$

as estimated in equation (42). The lag structure was approximately geometric and was fitted using a geometric transfer function,

$$\frac{(1-a)}{(1-aZ)}\ . \qquad (45)$$

Transfer Function Models of the Homicide Rate in Levels

(i) The Box and Jenkins Procedure for Estimating the Transfer Function

The homicide offense rate was estimated as a geometric distributed lag of the probability of imprisonment plus a residual using the procedure discussed in Box and Jenkins (1970, Chapter 11). The estimated model is

$$[OF(t) - EOF(t)] = \frac{\underset{(t=5.43)}{-11.728}}{\underset{(t=16.75)}{[1 - 0.8580Z]}}\ [P(t) - EP(t)] + \frac{U_1(t)}{\underset{(t=23.42)}{[1 - .9522Z]}} \qquad (46)$$

with a residual mean square of 0.1567. Comparing this residual mean square to that for the univariate model, equation (41), it is evident that the probability of imprisonment has more than halved the unexplained variance in the offense rate. The elasticity of the homicide rate with respect to probability of imprisonment was calculated at the means of the data for lag zero, i.e., the immediate impact, a value of -0.692. The elasticity for the sum of all lags, i.e., the total or long run impact, has a value of -4.565. This immediate impact elasticity compares to Ehrlich's two stage least square estimate for homicide of -0.892. However, the results reported here indicate that the total control effect is even greater when the impact at other lags is considered. The mean lag of the control effect estimated in equation (48) is 6 years, larger than one would expect from the lag structure inferred from the Granger-Newbold procedure (see footnote 8). There is still some structure in the residuals $U_1(t)$ suggesting the possibility of additional factors.

(ii) The Jenkins Procedure for Estimating Multi-Input Transfer Functions

(a) The median time served for homicide

The possible control effect of the length of sentence was investigated by estimating a multi-input transfer function model. The estimated model is

$$[OF(t) - EOF(t)] = \underset{(t=15.93)}{\overset{(t-5.74)}{\frac{-11.766}{[1-0.8566Z]}}} [P(t) - EP(t)]$$

$$\underset{(t-.55)}{-.0907} [TS(t)-ETS(t)] + \underset{(t=17.57)}{\frac{U_1(t)}{[1-0.9816Z]}} \qquad (47)$$

with a residual mean square of 0.1177. Median time served has the expected negative coefficient but it is not significant.[9] The residuals $U_1(t)$ are not completely orthogonal suggesting the presence of other factors.

(b) The Unemployment Rate of 18 and 19 Year Old Males

The possible effect of economic conditions on the homicide rate was investigated using unemployment rates. The unemployment rate for persons 16 years of age and over was not significant, controlling for the probability of imprisonment, but the unemployment rate for youths, μ_y, was significant. The latter has a great deal more variance over the business cycle. The estimated model is

$$[OF(t) - EOF(t)] = - \underset{(t=19.09)}{\overset{(t=9.27)}{\frac{14.9131}{[1 - .7507Z]}}} [P(t) - EP(t)]$$

$$+ \underset{(t=13.87)}{\overset{(t=6.86)}{\frac{.0983}{[1 - .7626Z]}}} [\mu_y(t) - E\mu_y(t)]$$

$$\underset{(t=0.58)}{- .0859} [TS(t) - ETS(t)] + \underset{(t=11.39)}{\overset{(t=5.33)}{\frac{[1 - 1.1409Z]}{[1 - 1.201Z]}}} U_1(t) \qquad (48)$$

with a residual mean square of .0622. Note that the addition of a causal variable significantly reduces the unexplained variance from that achievable with a control variable (see equation (48)), but the reduction is approximately half that of the control variable. Once again, median time served is not significant. The partial elasticity of the homicide rate with respect to the unemployment rate for youths is 0.226 at lag zero, the immediate impact, and is 0.911 for all lags, the long run impact. The mean lag is 3-1/3 years. Ehrlich (1975) reports a short run elasticity of about .068 for the unemployment rate for the civilian labor force. In our study, that unemployment rate is not significant but the rate for male youths is, and yields a higher elasticity.

(c) The Probability of Execution Given Imprisonment

The first approach was to estimate an intervention model, using a variable (DP(t)) that was 1 during years when there were executions (1945-1963 and 1966-1967) and 0 elsewhere (1964-1965 and 1968-1975). The estimation yielded the following model

$$[OF(t) - EOF(t)] = - \frac{\overset{(t=13.07)}{20.4167}}{\underset{(t=23.99)}{[1 - .7329Z]}} [P(t) - EP(t)]$$

$$+ \frac{\overset{(t=7.54)}{.1355}}{\underset{(t=10.65)}{[1 - .7242Z]}} [\mu_y(t) - E\mu_y]$$

$$+ \underset{(t=3.38)}{0.4986} \; DP(t) + (1 + \underset{(t=2.87)}{0.4985}Z) \, U_1(t) \; . \qquad (49)$$

The residual mean square was 0.1041. The results for youth unemployment and the probability of imprisonment are similar to those in equation (48). The intervention variable has a significant positive coefficient indicating either that executions are counterproductive to controlling homicide or that homicide rates were higher in the two decades prior to 1963, other causal and control variables equal. Note also that the residual mean square for this model is higher than for equation (50). To check this result for executions, the equation was estimated using the probability of executions given imprisonment for homicide, PX(t) . The estimated model is:

$$[OF(t) - EOF(t)] = - \frac{\overset{(t=6.83)}{12.9096}}{\underset{(t=20.32)}{[1 - 0.8488Z]}} [P(t) - EP(t)]$$

$$+ \frac{\overset{(t=3.19)}{.0559}}{\underset{(t=5.71)}{[1 - 0.8405Z]}} [\mu_y(t) - E\mu_y]$$

$$- \underset{(t=2.41)}{6.0108} \, [PX(t) - EPX(t)] + \frac{U_1(t)}{[1 - \underset{(t=3.01)}{0.6095}Z - \underset{(t=1.84)}{0.4284}Z^2]} \qquad (50)$$

with a residual mean square of 0.0778. The probability of execution given imprisonment has a significant immediate control effect. The elasticity at the means is -.032. This compares to the elasticity of execution given conviction ranging from -.039 to -.068 reported by Ehrlich (1975). It should be noted that the data sets and estimation techniques are different for the two studies, yet the results are similar.

(d) The Probability of Receiving the Death Sentence Given Imprisonment

The result for the impact of the death penalty was investigated using a different variable, death sentence. While executions ceased in California after 1967, prisoners continued to be sent to prison with a death sentence. Data was available beginning in 1950 and continuing through 1978, the period estimated. The model is:

$$[OF(t) - EOF(t)] = - \frac{\underset{(t=6.79)}{12.9883}}{\underset{(t=4.87)}{[1 - 0.6622Z]}} [P(t) - EP(t)]$$

$$+ \frac{\underset{(t=2.22)}{0.0899}}{\underset{(t=8.2)}{[1 - 0.824Z]}} [\mu_y(t) - E\mu_y(t)]$$

$$- \frac{\underset{(t=5.43)}{20.5425}}{\underset{(t=2.96)}{[1 - 0.5865Z]}} [PDS(t) - EPDS(t)] + \underset{(t=11.65)}{(1 - 0.941Z)} U_1(t) \qquad (51)$$

with a residual mean square of 0.2084. The death sentence has a significantly negative impact on the homicide rate with a geometrically declining lag. The mean lag is 1.4 years. The short run elasticity of the death sentence is -0.223 and the long run elasticity is -0.539. This result reinforces the significant control effect for the probability of execution. The mean lag of the probability of imprisonment in this model is 2 years, the short run elasticity is -0.72 and the long run elasticity is -2.13. We consider this to be the best model for the homicide rate in levels.

<u>Forecasts For the Transfer Function Model for Levels, Equation (51)</u>

The sensitivity of the homicide rate to various values of the unemployment rate for male youths and to various values of the probability of the death sentence is illustrated using equation (51) to make forecasts for the years 1979 through 1986. A univariate model was used to forecast the probability of imprisonment. Since equation (42) is a second order moving average model and has a memory of only two periods, a first order autoregressive model was estimated for differences in the probability of imprisonment. However, both models give nearly identical forecasts of the probability of imprisonment, 0.27 in 1979 and 0.28 thereafter. The observed value for 1979 is 0.27.

Forecasts for equation (51) are reported in Table 1. Values for the unemployment rate were based on forecasts from the UCLA Forecasting Project for unemployment rates for the U.S. for 1981-1983.[10] The ratio of the unemployment rate for males 18 and 19 to the unemployment rate for persons 16 and over in 1978 was used to extrapolate male youth unemployment rates from 1979 through 1983. The unemployment rate for 1984 through 1986 was set equal to the 1983 value. As is evident from the forecasts in Table 1, doubling the fraction of felons who receive a death sentence for homicide from about 1% to 2% would decrease homicide rates only 2 to 4 percent. However, increasing the probability of the death sentence dramatically to 7.8%, its mean value in the period 1950 to 1970, would decrease the homicide rate down to the levels observed in 1970 and 1971.

The effect of the unemployment rate for youths upon the homicide rate is illustrated in Table 2 where, for the first column of forecasts, the death sentence is .078 and the probability of imprisonment assumes the values as in Table 1, but the unemployment rate takes its actual values for 1978 and 1979 and then is decreased one percentage point every year thereafter. By 1986, the homicide rate has been decreased to 7.06 per 100,000.

TABLE 1

Forecasts of the Homicide Rate Per 100,000
Transfer Function Model in Levels - Equation (51)

YEAR	OBSERVED OF(t)	PDS(t)=0, all t P(t)[1], $\mu_y(t)$[2] FORECAST OF(t)	(σ)	PDS(t)=.009, all t P(t)[1], $\mu_y(t)$[2] FORECAST OF(t)	(σ)	PDS(t)=.018, all t P(t)[1], $\mu_y(t)$[2] FORECAST OF(t)	(σ)	PDS(t)=.078, all t P(t)[1], $\mu_y(t)$[2] FORECAST OF(t)	(σ)
1978	11.67								
1979	12.96	11.17	(.74)	10.98	(.74)	10.80	(.74)	9.56	(.74)
1980	14.32 (Prelim)	11.13	(1.10)	10.84	(1.10)	10.55	(1.10)	8.62	(1.10)
1981		11.41	(1.34)	11.05	(1.34)	10.70	(1.34)	8.36	(1.34)
1982		11.45	(1.56)	11.06	(1.56)	10.68	(1.56)	8.09	(1.56)
1983		11.43	(1.77)	11.02	(1.77)	10.61	(1.76)	7.89	(1.77)
1984		11.40	(1.96)	10.98	(1.96)	10.56	(1.96)	7.75	(1.96)
1985		11.37	(2.14)	10.95	(2.14)	10.52	(2.14)	7.67	(2.14)
1986		11.35	(2.30)	10.92	(2.30)	10.48	(2.30)	7.60	(2.30)

1,2 - Values of Probability of Imprisonment, P(t), and Unemployment Rate (18 & 19 aged males) $\mu_y(t)$, Used

YEAR	P(t) OBSERVED	P(t) VALUE	$\mu_y(t)$ OBSERVED	$\mu_y(t)$ VALUE
1978	.288		13.2	
1979	.266	.275	14.2	12.8
1980		.281	16.7	15.6
1981		.278		16.1
1982		.280		14.7
1983		.279		14.1
1984		.279		14.1
1985		.279		14.1
1986		.279		14.1

TABLE 2

Forecasts of the Homicide Rate Per 100,000
Transfer Function Model in Levels - Equation (51)

EFFECTIVE POLICIES

YEAR	FORECAST OF(t)	(σ)	P(t)	PDS(t)	$\mu_y(t)$	FORECAST OF(t)	(σ)	P(t)	PDS(t)	$\mu_y(t)$	FORECAST OF(t)	(σ)	P(t)	PDS(t)	$\mu_y(t)$
1979	9.70	(0.74)	.275	.078	14.2	11.04	(0.74)	.27	.009	12.8	11.15	(.74)	.27	.01	14.2
1980	8.83	(1.10)	.281	.078	16.7	10.90	(1.10)	.28	.009	15.6	10.87	(1.10)	.28	.02	16.7
1981	8.49	(1.34)	.278	.078	15.7	10.93	(1.34)	.29	.009	16.1	10.50	(1.34)	.29	.03	15.7
1982	8.20	(1.56)	.280	.078	14.7	10.72	(1.56)	.30	.009	14.7	9.87	(1.56)	.30	.04	14.7
1983	7.94	(1.77)	.279	.078	13.7	10.38	(1.77)	.31	.009	14.1	9.04	(1.77)	.31	.05	13.7
1984	7.67	(1.96)	.279	.078	12.7	10.02	(1.96)	.32	.009	14.1	8.09	(1.96)	.32	.06	12.7
1985	7.38	(2.14)	.279	.078	11.7	9.64	(2.14)	.33	.009	14.1	7.04	(2.14)	.33	.07	11.7
1986	7.06	(2.30)	.279	.078	10.7	9.25	(2.30)	.34	.009	14.1	5.91	(2.30)	.34	.08	10.7

PROBABLE SCENARIO

YEAR	FORECAST OF(t)	(σ)	P(t)	PDS(t)	$\mu_y(t)$
1979	11.04	(0.74)	.27	.009	12.8
1980	11.16	(1.10)	.26	.009	15.6
1981	11.63	(1.34)	.25	.009	16.1
1982	11.96	(1.56)	.24	.009	14.7
1983	12.26	(1.76)	.23	.009	14.1
1984	12.57	(1.96)	.22	.009	14.1
1985	12.91	(2.14)	.21	.009	14.1
1986	13.26	(2.30)	.20	.009	14.1

The effect of the probability of imprisonment is illustrated in Table 2 where it is increased a percentage point a year starting from 0.27 in 1979 and reaching 0.34 in 1986, comparable to its level in 1966. The unemployment rate takes the UCLA forecast values, as in Table 1, and the probability of receiving the death sentence is .009. The homicide rate falls from 11.0 in 1979 to 9.2 in 1986.

In 1979, the observed probability of imprisonment was .266, the unemployment rate for youths was 14.2% and the probability of receiving the death sentence given imprisonment for homicide was .026. In 1966, the corresponding values were .355, 10.2%, and .078 respectively. The effect on homicide rates of changing all of these variables back towards their mid-sixties levels is illustrated in Table 2, for the forecasts labeled "Effective Policies". In this case, homicide rates decrease to 5.91 per 100,000 by 1986.

A probable scenario is for the probability of imprisonment to continue to fall, the probability of receiving the death sentence to remain near current levels, and the unemployment rate to vary near the UCLA forecasts. In this case, the homicide rate rises to 13.2 by 1986, as illustrated in Table 2. Nonetheless, we expect this forecast of 13.2 for the homicide rate in 1986 will be a serious underestimate.

If equation (51) is used to forecast the homicide rate for 1979, conditional on the actual observed values of the probability of imprisonment equal .266, the probability of receiving the death sentence equal .026 and the male youth unemployment rate equal 14.2%, the forecast homicide rate is 10.87 with a standard deviation of 0.71. This is a significant underestimate of the observed value of 12.96.

The observed homicide rates appear to be rising more rapidly than the forecasts from the multivariate transfer function model of the rates in levels motivating the estimation of transfer function models for differences which could incorporate a constant, capturing trend, perhaps a more effective portmanteau variable for exogenous factors influencing homicides.

Transfer Function Models of the Homicide Rate in Differences

(i) The Probability of Imprisonment

Univariate Box-Jenkins models were estimated for the homicide rate and the probability of imprisonment, both in differences. The pattern of the cross-correlation function of the residuals was similar to that of Fig. 4, with a significant negative correlation at lag zero, and insignificant correlations elsewhere. Once again, the weights of the transfer function suggested a geometric approximation, and the following model was estimated

$$\Delta OF(t) = \underset{(t=4.24)}{1.0857} - \frac{\overset{(t=6.72)}{2.9963}}{\underset{(t=39.45)}{[1-.9407Z]}} \Delta P(t) + \frac{e(t)}{\underset{(t=4.57)}{[1+0.6668Z]}} \qquad (52)$$

where the constant term represents a trend in homicide rates and e(t) is the residual. The residual mean square is .0759. The mean lag is long, as was the case for the bivariate model in levels, equation (46).

(ii) The Probability of Receiving the Death Sentence Given Imprisonment

Inclusion in the model of the probability of receiving the death sentence given imprisonment, in differences, indicated a significant effect only at lag 0. The model is

$$\Delta OF(t) = \underset{(t=4.48)}{0.2943} - \frac{\overset{(t=4.54)}{7.9276}}{\underset{(t=4.52)}{[1-0.4418Z]}} \Delta P(t) - \underset{(t=1.97)}{4.8339} \Delta PDS(t) \qquad (53)$$

with a residual mean square of 0.1081. The short run elasticity of the death sentence is -0.0524, lower than in the case of the model in levels. The short run elasticity of the probability of imprisonment is -0.44 and the long run elasticity is -0.79. The residuals from this model are orthogonal.

(iii) The Unemployment Rate for 18 and 19 Year Old Males

The addition of the unemployment rate for youths, in differences, to equation (53) raises the residual mean square and the variable is not significant (t=0.15). It is noteworthy that this unemployment rate can be modelled as the sum of trend and white noise, i.e., the differenced variable is orthogonal. Thus, the influence of the variable is captured by the constant term in equation (53). If the constant term is dropped, the significance of the unemployment rate increases, but not by much. The model is,

$$\Delta OF(t) = \frac{\overset{(t=3.85)}{-7.8306}}{\underset{(t=2.28)}{[1-0.3285Z]}} \Delta P(t) - \underset{(t=2.01)}{6.0023} \Delta PDS(t) + \underset{(t=0.51)}{.0149} \Delta \mu y(t)$$

$$+ [1 + \underset{(t=1.24)}{0.2551}Z]\, e(t) \qquad (54)$$

with a residual mean square of .1851. We conclude that equation (53) is the best model of the homicide rate in differences. As in the case for levels, the variable median time served was not significant.

<u>Forecasts for the Transfer Function Model in Differences, Equation (53)</u>

The effect of exogenous factors causing homicide, captured by trend, as well as criminal justice system variables such as the probability of imprisonment and the probability of receiving the death sentence given imprisonment, are illutrated with forecasts for the period 1979 through 1986 in Table 3.

If the likelihood of imprisonment and the death sentence continue at 1978 levels, the homicide rate will rise from 11.67 in 1978 to 11.96 in 1979 and reach 14.02 in 1986, the trend effect of exogenous forces.

If the probability of the death sentence were increased by 1% per year beginning in 1980, i.e., from .009 in 1978 and 1979 to .019 in 1980 etc., reaching .079 in 1986, its value in 1964, then the homicide rate would continue to rise, reaching 13.69 in 1986.

If the probability of imprisonment, as well as the death sentence, were increased by 1 percentage point each year, beginning in 1980,

TABLE 3

Forecasts of the Homicide Rate: 1979-1986
Transfer Function Model in Differences - Equation (53)

	Observed			Effect of Trend				Effect of Raising the Death Sentence				Effect of Raising Both Sanctions			
YEAR	OF(t)	P(t)	PDS(t)	FORECAST OF(t)	(σ)	VALUE P(t)	VALUE PDS(t)	FORECAST OF(t)	(σ)	VALUE P(t)	VALUE PDS(t)	FORECAST OF(t)	(σ)	VALUE P(t)	VALUE PDS(t)
1978	11.67	.288	.009												
1979	12.96	.266	.026	11.88	(.49)	.288	.009	11.88	(.49)	.288	.009	11.88	(.49)	.288	.009
1980	14.36	(Prelim)		12.14	(.68)	.288	.009	12.10	(.68)	.288	.019	12.02	(.68)	.298	.019
1981				12.42	(.85)	.288	.009	12.33	(.85)	.288	.029	12.13	(.85)	.308	.029
1982				12.71	(.99)	.288	.009	12.57	(.99)	.288	.039	12.24	(.99)	.318	.039
1983				13.00	(1.12)	.288	.009	12.81	(1.12)	.288	.049	12.35	(1.12)	.328	.049
1984				13.29	(1.23)	.288	.009	13.05	(1.23)	.288	.059	12.45	(1.23)	.338	.059
1985				13.59	(1.33)	.288	.009	13.30	(1.33)	.288	.069	12.56	(1.33)	.348	.069
1986				13.88	(1.43)	.288	.009	13.54	(1.43)	.288	.079	12.66	(1.43)	.358	.079

SYMBOLS

OF(t) Homicide Rate per 100,000
P(t) Probability of Imprisonment for Homicide
PDS(t) Probability of Receiving the Death Sentence Given Imprisonment

i.e., from .288 in 1978 and 1979 to .298 in 1980 etc., reaching .358 in 1986, its approximate value in 1966, then homicide rates will continue to rise, but only reach 12.66 by 1986. Thus, there appear to be factors increasing homicides that will not be offset, even if the probabilities of sanctions are restored to their mid-sixties levels.

If the probability of imprisonment and the probability of receiving the death sentence take their observed values for 1979 and earlier years, the forecast of the homicide rate for 1979 is 11.98 per 100,000 with a standard error of 0.49. The observed homicide rate for 1979 is 12.96, at the boundary of a 95% confidence interval for the forecast. However, the preliminary value for the homicide rate in 1980 is 14.36, another large increase. It seems unlikely that the forecasting model, equation (53), will keep pace with the current surge in homicide rates. Nonetheless, equation (53) is capturing the upward trend in homicide rates and appears to be the best forecasting model currently available.

Conclusions

We find Hoenack and Weiler's econometric criticism that Ehrlich had not identified the supply of offenses function but had in fact reversed the causality and observed the system response to homicide, incorrrect. We find that causality runs from the probability of imprisonment to homicide.

Ironically, we find the criticism that not much is known about the exogenous variables determining homicide more telling. Using a forecasting model for homicide in differences, we observe that starting in 1980, even if the probabilities of imprisonment and receiving the death sentence given imprisonment were gradually restored to their mid-sixties values by 1986, homicide rates would continue to rise due to trends in exogenous factors. We conclude that if society is to control homicide, it must better understand the forces causing it.

Footnotes

[1] Hoenack and Weiler (1980), p. 337.

[2] Klein, Forst, and Filatov (1978), p. 347.

[3] Fisher and Nagin (1978), p. 396.

[4] Ibid., p. 397.

[5] The Bureau of Criminal Statistics defines criminal homicide to include all degrees of murder and all types of manslaughter, including vehicular.

[6] The imprisonment data used was total male felons newly received from court for homicide. The distribution is available for first degree murder, second degree murder, manslaughter and manslaughter by vehicle:

Commitments for Homicide

Year	Total	Murder 1st	Murder 2nd	Man-slaughter	Man-slaughter Vehicular
1950	152 (100%)	53 (34.9%)	51 (33.6%)	46 (30.3%)	2 (1.3%)
1960	218 (100%)	51 (23.4%)	67 (30.7%)	86 (39.5%)	14 (6.4%)
1970	417 (100%)	117 (28.1%)	138 (33.1%)	146 (35.0%)	16 (3.8%)
1975	579 (100%)	167 (28.8%)	200 (34.5%)	192 (33.2%)	20 (3.5%)

[7] If on the basis of an alternative theory the residuals were modelled, conversely to equation (43), as

$$e_2(t) = \beta' e_1(t) + u'(t)$$

where $\beta' < 0$, the parameters of the lag structure of the transfer function from the homicide rate to the probability of imprisonment can be inferred by substituting for $e_1(t)$ and $e_2(t)$ from equations (41) and (42). The parameters at lag 0,1, etc. are

$$\lambda_0 = 1$$

$$\lambda_1 = -1.04$$

$$\lambda_2 = 0.58$$

$$\lambda_3 = -.08$$

$$\lambda_4 = .02$$

The negative correlation at lag 0 (λ_o multiplied by β) could be explained by spurious correlation due to the definition of the probability of imprisonment as the ratio of imprisonment to the homicide rate, and the positive correlation at lag 1 could be explained by the positive (perhaps cost minimizing) response of imprisonments to the offense rate, but there is no explanation of the negative correlation at lag 2 within the theoretical framework of this paper. Also the bang-bang nature of the lag structure seems less plausible than the geometric (refer to footnote 8).

[8] Substituting for $e_1(t)$ and $e_2(t)$ from equations (41) and (42) into equation (43) one obtains

$$(1 - 1.2892Z + 0.3243Z^2)\ [OF(t) - EOF(t)] = \frac{-10.14\ (1-Z)\ P(t)}{(1 - .752Z + .3242Z^2)} + u(t)$$

or

$$[OF(t) - EOF(t)] = \frac{-10.14\ (1-Z)\quad P(t)}{(1-1.2892Z+1.3243Z^2)(1-.7525Z+.3242Z^2)}$$

$$+ \frac{U(t)}{(1-.752Z+.3242\ Z^2)}$$

and the transfer function $\Lambda\ (Z)$ is

$$(1 + \lambda_1 Z + \lambda_2 Z^2 + \ldots) = \Lambda(Z)$$

$$= \frac{(1-Z)}{(1-1.2892Z + 0.3243Z^2)(1-0.7525Z + 0.3242Z^2)}$$

and solving for the parameters λ_1 , λ_2 , etc., one obtains $\lambda_0 = 1$, $\lambda_1 = 1.0417$, $\lambda_2 = 0.5082$, $\lambda_3 = -0.0135$, $\lambda_4 = -0.2105$, $\lambda_5 = -0.2247$, $\lambda_6 = -0.1625$, $\lambda_7 = -0.1088$.

[9] The transfer function models incorporating time served, equations (47) and (48), the intervention variable for executions, equation (49), and the probability of execution given imprisonment, equation (50), were estimated using a time series from 1945 through 1975. The homicide rate was calculated from the data obtained from the Vital Statistics of the United States. The transfer function models incorporating the probability of receiving the death sentence given imprisonment (available in 1950 and subsequent years), as well as equations (41), (42), and (46) were estimated using a time series from 1950 through 1978. Homicide rates for 1950 through 1953 were calculated from data obtained from Vital Statistics of the United States, and subsequent years from data published by the California Bureau of Criminal Statistics. The two time series yield comparable models.

[10] The UCLA Forecasting Project's base forecasts for the unemployment rate as of March 1981 were:

Year	Unemployment Rate Forecast
1981	7.3
1982	6.7
1983	6.4

References

BLUMSTEIN, A., COHEN, J. and NAGIN, D., eds. (1978). *Deterrence and Incapacitation.* National Academy of Science, Washington, D.C.

BOX, G.E.P. and JENKINS, G.M. (1970). *Time Series Analysis, Forecasting, and Control.* Holden-Day, San Francisco.

EHRLICH, I. (1973). Participation in Illegitimate Activities: A Theoretical and Empirical Investigation, *Journal of Political Economy,* 81, No. 2, 521-565.

——— (1975). The Deterrent Effect of Capital Punishment: A Question of Life and Death, *American Economic Review*, 63, No. 3, 397-417.

FEDERAL BUREAU OF PRISONS (1970). *National Prisoner Statistics State Prisoners: Admissions Releases, 1970.* Washington, D.C.

——— (1960). *Prisoners Released from State and Federal Institutions, 1960.*

FISHER, F.M. and NAGIN, D. (1978). On the Feasibility of Identifying the Crime Function in a Simultaneous Model of Crime Rates and Sanction Levels. In *Deterrence and Incapacitation: Estimating the Effects of Criminal Sanctions on Crime Rates.* Eds: A. Blumstein, J. Cohen and D. Nagin. National Academy of Sciences, Washington.

GRANGER, C.W.J. (1969). Investigating Causal Relations by Econometric Models and Cross-Spectral Methods. *Econometrica*, 37, No. 3, 424-438.

——— and NEWBOLD, P. (1977). *Forecasting Economic Time Series.* Academic Press, New York.

HOENACK, S.A. and WEILER, W.C. (1980). A Structural Model of Murder Behavior, *American Economic Review*, 70, No. 3, 327-341.

JENKINS, G.M. (1979). Practical Experiences with Modelling and Forecasting Time Series. In *Forecasting.* Ed: O.D. Anderson, North-Holland, Amsterdam.

KLEIN, L.R., FORST, B. and FILATOV, V. (1978) The Deterrent Effect of Capital Punishment: An Assessment of the Estimates. In *Deterrence and Incapacitiation; Estimating the Effects of Criminal Sanctions on Crime Rates.* Eds: A. Blumstein, J. Cohen and D. Nagin, National Academy of Sciences, Washington.

SIMS, C.A. (1972). Money, Income and Causality. *American Economic Review*, 62, No. 4, 540-552.

STATE OF CALIFORNIA, Department of Corrections. *California Prisoners, 1945-49* (and subsequent issues).

———, Department of Justice, Bureau of Criminal Statistics. *Crime in California* (annual through 1964).

———. *Crime and Delinquency in California* (annual 1965-1977).

———. *Willful Homicide in California, 1963-65* (and subsequent issues).

STIGLER, G.J. (1970). The Optimal Enforcement of Laws. *Journal of Political Economy* 78, pp. 526-539 (1970).

UCLA FORECASTING PROJECT (1981). *The UCLA National Business Forecast, March 1981.*

APPLIED TIME SERIES ANALYSIS
O.D. Anderson & M.R. Perryman (eds.)

DISTRIBUTION OF PARAMETER ESTIMATES IN ARIMA(p,1,q) MODELS

S. E. Said and D. A. Dickey

Department of Statistics
North Carolina State University
Raleigh, North Carolina
U.S.A.

Let the time series Y_t satisfy

$$Y_o = 0$$

$$Y_t - \rho Y_{t-1} = e_t + \beta e_{t-1} \quad , \quad t \geq 1$$

where $|\beta| < 1$ and $\{e_t\}$ is a sequence of $NID(0,\sigma^2)$ random variables. The large sample distribution of the nonlinear regression estimates are established under the null hypothesis $H_o: \rho = 1$. The regression t-type statistic for testing the unit root is discussed. An illustrative example is given.

1. INTRODUCTION

In recent work several authors study distributions of estimates of coefficients in the autoregressive model (AR)

$$Y_o = 0$$

$$Y_t = \rho Y_{t-1} + Z_t \quad , \quad t \geq 1$$

$$Z_t = \alpha_1 Z_{t-1} + \dots + \alpha_p Z_{t-p} + e_t \quad , \quad \text{for all } t$$

where the e_t are $NID(0,\sigma^2)$ variates and the roots of $m^p - \alpha_1 m^{p-1} - \dots - \alpha_p = 0$ are all less than one in absolute value. Note that if $\rho = 1$ then Y_t can be written as

$$Y_t - Y_{t-1} = (\rho-1)Y_{t-1} + \alpha_1(Y_{t-1}-Y_{t-2}) + \dots + \alpha_p(Y_{t-p+1}-Y_{t-p}) + e_t$$

suggesting a regression of (Y_t-Y_{t-1}) on Y_{t-1}, $(Y_{t-1}-Y_{t-2}), \dots, (Y_{t-p+1}-Y_{t-p})$. Fuller and Dickey (1979) obtained the percentiles of the limit distribution of the coefficient and the regression "t-statistic" for Y_{t-1} under the assumption that $\rho = 1$. This may be used to test the null hypothesis that $\rho = 1$, that is, that the coefficient on Y_{t-1} is really zero. The distribution of this regression coefficient is not normal in the limit and the limit distribution depends on $\alpha_1, \alpha_2, \dots, \alpha_p$. Critical values for the limit distribution were calculated assuming that $\alpha_1 = \alpha_2 = \dots = \alpha_p = 0$ and are given in Fuller (1976) and Dickey (1976).

In the present paper we extend the distributional results to models with moving average terms. Since time is limited, we present the ARIMA(0,1,1) case. The general problem will be addressed in Said (1981).

2. MODEL AND ESTIMATION PROCEDURE

We now investigate the process Y_t defined by

$$Y_o = 0$$

$$Y_t - \rho Y_{t-1} = e_t + \beta e_{t-1} \quad , \quad t \geq 1 \tag{2.1}$$

where $|\beta| < 1$ and $\{e_t\}$ is a sequence of NID$(0,\sigma^2)$ random variables. Note that the process can be rewritten as

$$Y_t - Y_{t-1} = (\rho-1)Y_{t-1} + e_t + \beta e_{t-1} \tag{2.2}$$

Let $D_t = Y_t - Y_{t-1}$. The null hypothesis that $\rho=1$ is to be tested and, under this hypothesis, D_t is a simple invertible moving average process. The method of testing is based on the estimation procedure suggested in Fuller (1976) section 8.3, which is basically a nonlinear type of estimation.

The estimation begins by re-expressing model (2.1) as follows

$$e_t = Y_t - \rho Y_{t-1} - \beta e_{t-1} \quad , \quad t \geq 1 \tag{2.3}$$

or equivalently,

$$e_t = Y_t - \sum_{j=0}^{t-2}(-\beta)^j \ (\rho+\beta)\ Y_{t-1-j} + (-\beta)^t\ \delta \quad , \quad t \geq 1 \tag{2.4}$$

where $\delta = e_o$.

Equation (2.4) defines a function $e_t(\rho,\beta,\delta)$. For convenience, let $\theta' = (\rho,\beta,\delta)$ and $\hat{\theta}' = (\hat{\rho},\hat{\beta},\hat{\delta})$. Thus expanding $e_t(\theta)$ about $\hat{\theta}$ we obtain

$$e_t(\theta) = e_t(\hat{\theta}) - V_t(\hat{\theta})\cdot(\rho-\hat{\rho}) - W_t(\hat{\theta})\cdot(\beta-\hat{\beta}) - \Delta_t(\hat{\theta})\cdot(\delta-\hat{\delta}) + r_t \tag{2.5}$$

where r_t is the Taylor series remainder and V_t, W_t, Δ_t are the negative derivatives of $e_t(\theta)$ w.r.t. ρ, β and δ, respectively, evaluated at $\hat{\theta}$. Ignoring the remainder term and rearranging (2.5) as

$$e_t(\hat{\theta}) = V_t(\hat{\theta})\cdot(\rho-\hat{\rho}) + W_t(\hat{\theta})\cdot(\beta-\hat{\beta}) + \Delta_t(\hat{\theta})\cdot(\delta-\hat{\delta}) + e_t \tag{2.6}$$

suggests the regression of $e_t(\hat{\theta})$ on V_t, W_t, Δ_t as a way of estimating $\gamma' = (\rho-\hat{\rho},\beta-\hat{\beta},\delta-\hat{\delta})$. Assuming $\hat{\theta}$ is an initial estimator of θ we obtain an improved estimator $\hat{\theta} + \hat{\gamma}$.

The remainder term r_t involves second derivatives of $e_t(\theta)$ evaluated at a point $\ddot{\theta}$ between θ and $\hat{\theta}$. Under the null hypothesis, $H_o : \rho = 1$ we use $\hat{\rho} = 1$ as our initial estimator so that the first element of $\ddot{\theta}$ must be 1. This means that, in studying the null distribution of $\hat{\gamma}$, we may ignore remainder terms involving derivatives w.r.t. ρ (they get multiplied by 0). The second derivatives which will concern us are $T_t(\ddot{\theta})$, $U_t(\ddot{\theta})$ which are second derivatives of $e_t(\theta)$ in the order $\partial^2\beta$, $\partial\beta\partial\delta$, evaluated at $\ddot{\theta}$. The other second derivatives are either zero or get multiplied by zero. In table (1) a few values of the functions discussed thus far are given.

3. RECURSIVE FORMULAS

Several of our derivatives can be calculated recursively using difference equations obtained by differentiating both sides of equation (2.3). Table (1) serves to check these recursions and shows that appropriate starting values are obtained for each of these by assuming the function is zero for $t < 1$. The equations (for $t \geq 1$) are

$$V_t(\cdot) = Y_{t-1} - \hat{\beta}V_{t-1}(\cdot) \tag{3.1}$$

$$W_t(\cdot) = e_{t-1} - \hat{\beta}W_{t-1}(\cdot) \tag{3.2}$$

$$T_t(\cdot) = 2W_{t-1} - \ddot{\beta}T_{t-1}(\cdot) \tag{3.3}$$

Assuming that $|\hat{\beta}| < 1$ and $|\ddot{\beta}| < 1$, it is readily seen that the W_t and T_t series are behaving like stationary AR processes. For example, $T_t + \ddot{\beta}T_{t-1} = 2W_{t-1}$ so $T_t + (\ddot{\beta}+\beta)T_{t-1} + \beta\ddot{\beta}T_{t-2} = 2(W_{t-1}+\beta W_{t-2}) = 2e_{t-2}$ which means T_t behaves like an AR(2). The fact that the difference equations have initial value 0 means that the W_t and T_t processes deviate slightly from true AR's. In the appendix given in Dickey and Said (1981), we show that for each process, e.g. T_t, there is an AR process, say $\tilde{T}_t$, which is "exponentially close to T_t". By this we mean there exists M, $\lambda < 1$ such that $E\{\lambda^{-t}(T_t-\tilde{T}_t)\}^2 < M$ for all t. We then show that in

Table (1): First and Second Order Derivatives

t	$e_t(\hat\theta)$	$V_t(\hat\theta)$
1	$Y_1 - \hat\beta\hat\delta$	0
2	$Y_2 - (\hat\rho+\hat\beta)Y_1 + \hat\beta^2\hat\delta$	Y_1
3	$Y_3 - (\hat\rho+\hat\beta)Y_2 + \hat\beta(\hat\rho+\hat\beta)Y_1 - \hat\beta^3\hat\delta$	$Y_2 - \hat\beta Y_1$
4	$Y_4 - (\hat\rho+\hat\beta)Y_3 + \hat\beta(\hat\rho+\hat\beta)Y_2 - \hat\beta^2(\hat\rho+\hat\beta)Y_1 + \hat\beta^4\hat\delta$	$Y_3 - \hat\beta Y_2 + \hat\beta^2 Y_1$
5	$Y_5 - (\hat\rho+\hat\beta)Y_4 + \hat\beta(\hat\rho+\hat\beta)Y_3 - \hat\beta^2(\hat\rho+\hat\beta)Y_2 + \hat\beta^3(\hat\rho+\hat\beta)Y_1 - \hat\beta^5\hat\delta$	$Y_4 - \hat\beta Y_3 + \hat\beta^2 Y_2 - \hat\beta^3 Y_1$
	$W_t(\hat\theta)$	$\Delta_t(\hat\theta)$
1	$\hat\delta$	$\hat\beta$
2	$Y_1 - 2\hat\beta\hat\delta$	$-\hat\beta^2$
3	$Y_2 - (\hat\rho + 2\hat\beta)Y_1 + 3\hat\beta^2\hat\delta$	$\hat\beta^3$
4	$Y_3 - (\hat\rho + 2\hat\beta)Y_2 + (2\hat\beta\hat\rho+3\hat\beta^2)Y_1 - 4\hat\beta^3\hat\delta$	$-\hat\beta^4$
5	$Y_4 - (\hat\rho + 2\hat\beta)Y_3 + (2\hat\beta\hat\rho+3\hat\beta^2)Y_2 - (3\hat\beta^2\hat\rho+4\hat\beta^3)Y_1 + 5\hat\beta^4\hat\delta$	$\hat\beta^5$
	$T_t(\ddot\theta)$	$U_t(\ddot\theta)$
1	0	-1
2	$2\ddot\delta$	$2\ddot\beta$
3	$2Y_1 - 6\ddot\beta\ddot\delta$	$-3\ddot\beta^2$
4	$2Y_2 - (2\ddot\rho+6\ddot\beta)Y_1 + 12\ddot\beta^2\ddot\delta$	$4\ddot\beta^3$
5	$2Y_3 - (2\ddot\rho+6\ddot\beta)Y_2 + (6\ddot\beta\ddot\rho+12\ddot\beta^2)Y_1 - 20\ddot\beta^3\ddot\delta$	$-5\ddot\beta^4$

computing limits of sums of squares and cross products, the AR processes (like $\tilde T_t$) may be used in place of the actual derivatives defined in the previous recursive forumlas (like (3.2)). Thus for example, $n^{-1}\sum_{t=1}^{n} T_t T_{t+h}$ converges in probability to a continuous function of σ^2, β and $\ddot\beta$. If $\ddot\beta \xrightarrow{P} \beta$ then $n^{-1}\sum T_t T_{t+h} \xrightarrow{P} \gamma_T(h)$ where $\gamma_T(\cdot)$ is the autocovariance of the AR(2) process $T_t + 2\beta T_{t-1} + \beta^2 T_{t-2} = 2e_{t-2}$.

As for the V_t process, we first assume that $\hat\rho = 1$ and by (2.2) we have $D_t = e_t + \beta e_{t-1}$ where $D_t = Y_t - Y_{t-1}$. In a similar manner, we define $\dot V_t = V_t - V_{t-1}$ which implies by (3.1) that $\dot V_j + \hat\beta \dot V_{j-1} = e_{j-1} + \beta e_{j-2}$. Summing both sides as j goes from 1 to t we obtain $(1+\hat\beta)V_t = (1+\beta)\sum_{j=1}^{t-1} e_j + f_L(t)$ where $f_L(t)$ represents a linear combination of $\dot V_t$, $\dot V_o$, e_o, e_{-1}, e_{t-1}. Letting $Q_t = \sum_{j=1}^{t} e_j$, $d = (1+\hat\beta)^{-1}$ and $c = (1+\beta)d$ we get

$$V_t = cQ_{t-1} + d\, f_L(t) \tag{3.4}$$

When calculating limits of normalized sum of squares and cross products of series involving V_t, V_t may be replaced by cQ_{t-1}. This is justified by simply observing that $E\{f_L(t)\} = 0$ and $\mathrm{Var}\{f_L(t)\} = 0(1)$ which implies that $V_t = cQ_{t-1} + 0_p(1)$. It is shown in Dickey (1976) that

$$\left(n^{-1}\sum_{t=1}^{n} Q_{t-1}e_t,\ n^{-2}\sum_{t=1}^{n} Q_{t-1}^2 \right) \xrightarrow{\mathscr{L}} \left(\left(\sum_{i=1}^{\infty} \gamma_i Z_i\right)^2 - \frac{\sigma^2}{2},\ \sum_{i=1}^{\infty} \gamma_i^2 Z_i^2 \right) \equiv \left| \xi, \Gamma \right| \tag{3.5}$$

where Z_i is a NID(o,σ^2) sequence and $\gamma_i = (-1)^{i+1}\dfrac{2}{(2i-1)\pi}$.

4. LIMIT DISTRIBUTIONS

In this section we study the large sample distribution of the least square regression estimate of γ, which is given by $\hat{\gamma} = \{F'(\hat{\theta})F(\hat{\theta})\}^{-1}F'(\hat{\theta})E(\hat{\theta})$ where $F(\hat{\theta})$ is an $(n \times 3)$ matrix of derivatives, with rows given by $V_t(\hat{\theta})$, $W_t(\hat{\theta})$, $\Delta_t(\hat{\theta})$, $t = 1, 2, \ldots, n$ and $E(\hat{\theta}) = \left(e_1(\hat{\theta}), \ldots, e_n(\hat{\theta})\right)'$. Specifically, we wish to normalize and find the limit distribution of $\hat{\gamma} - \gamma$. That is, we will find the limit distribution of

$$D_n(\hat{\gamma}-\gamma) = \{D_n^{-1}F'(\hat{\theta})F(\hat{\theta})D_n^{-1}\}^{-1} \cdot \{D_n^{-1}F'(\hat{\theta})[E(\theta) + R(\theta)]\}$$

where $R(\theta) = (r_1, r_2, \ldots, r_n)'$ as in (2.5) and D_n is the diagonal matrix $D_n = \mathrm{diag}\{n, \sqrt{n}, 1\}$. We summarize the limiting results in the following two lemmas.

Lemma (1): Let $\bar{S}$ be a closed set containing $\theta' = (\rho, \beta, \delta)$ as an interior point. Let $\rho = \hat{\rho} = 1$, $\hat{\beta} - \beta = o_p(n^{-1/4})$, and $\hat{\delta} = O_p(1)$. Then $D_n^{-1}F'(\hat{\theta})F(\hat{\theta})D_n^{-1} \xrightarrow{\mathcal{L}} A$ where A is a diagonal matrix with element $A_{11} = \Gamma$, $A_{22} = (1-\beta^2)^{-1}\sigma^2$, $A_{33} = \beta^2(1-\beta^2)^{-1}$.

Proof: Observe

$$D_n^{-1}F'(\hat{\theta})F(\hat{\theta})D_n^{-1} = \begin{pmatrix} n^{-2}\Sigma V_t^2(\hat{\theta}) & n^{-3/2}\Sigma V_t(\hat{\theta})W_t(\hat{\theta}) & n^{-1}\Sigma V_t(\hat{\theta})\Delta_t(\hat{\theta}) \\ & n^{-1}\Sigma W_t^2(\hat{\theta}) & n^{-1/2}\Sigma W_t(\hat{\theta})\Delta_t(\hat{\theta}) \\ & & \Sigma \Delta_t^2(\hat{\theta}) \end{pmatrix}$$

Omitting the argument $\hat{\theta}$ for simplicity, we have $\sum_{t=1}^{n} \Delta_t^2 \longrightarrow \hat{\beta}^2(1-\hat{\beta}^2)^{-1}$ and since $\hat{\beta} \xrightarrow{P} \beta$ we obtain the result for A_{33}.

In the appendix given by Dickey and Said (1981) we show that $n^{-2}(\Sigma V_t^2 - c\Sigma Q_{t-1}^2) \xrightarrow{P} 0$ where $c = (1+\beta)/(1+\hat{\beta})$ so c converges to 1 when $\hat{\beta} \longrightarrow \beta$. By result (3.5) and Slutsky's theorem $A_{11} = \Gamma$.

W_t is exponentialy close to a stationary AR(1) process $\tilde{W}_t$ where $\tilde{W}_t(\theta) = \sum_{j=0}^{\infty} (-\beta)^j e_{t-1-j}$. The sum of squares of $W_t(\hat{\theta})$ (normalized by n^{-1}) converges to a continuous function of $\hat{\beta}$ and $\tilde{\beta}$. Noting that $\hat{\beta} \longrightarrow \beta$ we obtain the limit of the sum of squares for $W_t(\theta)$ as given by A_{22}.

The diagonality is shown as follows. (We show only $n^{-3/2}\sum_{t=1}^{n} V_t W_t \xrightarrow{P} 0$. The others follow similarly.) We first show $c\sum_{t=1}^{n} Q_{t-1}W_t = O_p(n)$ by simply observing that $E\left(\sum_{t=1}^{n} Q_{t-1}W_t\right) = \sum_{t=1}^{n} E\left(\sum_{i=1}^{t-1}\sum_{j=0}^{\infty} a_j e_i e_{t-j}\right) = O(n)$ where a_j is a sequence of weights which would be obtained by expressing (3.2) as an infinite MA. Also

$$\mathrm{Var}\left(\Sigma Q_{t-1}W_t\right) = \sum_{t=1}^{n}\sum_{s=1}^{n} \mathrm{Cov}\left(Q_{t-1}W_t, Q_{s-1}W_s\right)$$

$$= \Sigma\Sigma\, \{E(Q_{t-1}Q_{s-1}) \cdot E(W_tW_s) + E(Q_{t-1}W_s) \cdot E(W_tQ_{s-1})\}$$

But since $E(Q_{t-1}Q_{s-1}) \cdot E(W_tW_s) \le M \cdot \min(t-1, s-1) \cdot \lambda^{t-s}$, it follows that $\mathrm{Var}(\Sigma Q_{t-1}W_t) = O(n^2)$. Now, $\Sigma V_tW_t = \Sigma\{cQ_{t-1} + f_L(t)\}W_t$ where $f_L(t)$ is defined before (3.4). Since $\Sigma f_L(t)W_t = O_p(n)$, the proof is complete. □

Lemma (2): The vector $D_n^{-1}F'(\hat{\theta})\ E(\theta)$ converges in law to

$$\left(A_1,\ A_2,\ A_3\right)' \text{ where } \begin{pmatrix} A_2 \\ A_3 \end{pmatrix} \sim \text{MVN}\left[\begin{pmatrix} 0 \\ 0 \end{pmatrix}, \begin{pmatrix} \frac{\sigma^2}{1-\beta^2} & 0 \\ 0 & \frac{\beta^2\sigma^2}{1-\beta^2} \end{pmatrix}\right]$$

and $A_1 = \xi$ (see(3.5)).

Proof: Observe $D_n^{-1}F'(\hat{\theta})E(\theta) = \left(n^{-1}\Sigma\, \tilde{V}_t e_t,\ n^{-\frac{1}{2}}\Sigma\, W_t e_t,\ \Sigma\, \Delta_t e_t\right)'$. Let $\tilde{W}_t = \sum_{j=0}^{\infty}(-\beta)^j e_{t-1-j}$, $\tilde{V}_t = cQ_{t-1}$, $\tilde{\Delta}_t = -(-\beta)^t$. We have $D_n^{-1}\left(c\,\Sigma\, Q_{t-1}e_t,\ \Sigma\, \tilde{W}_t e_t,\ \Sigma\, \tilde{\Delta}_t e_t\right)' - D_n^{-1}F'(\hat{\theta})\ E(\theta) \xrightarrow{P} 0$. Now, by the theory of stationary time series $n^{-\frac{1}{2}}\Sigma\, \tilde{W}_t\ e_t,\ \Sigma\, \tilde{\Delta}\, e_t$ converge to the stated distribution (see Fuller, pg. 348 for example). Note that $-\sum_{t=1}^{n}(-\hat{\beta})^t e_t$ converges for $\hat{\beta}\ \varepsilon\ \bar{S}$, and e_t normal, to $N\left[0, \hat{\beta}^2(1-\hat{\beta}^2)^{-1}\sigma^2\right]$. The covariance between $n^{-\frac{1}{2}}\sum_{t=1}^{n}\tilde{W}_t e_t$ and $\sum_{t=1}^{n}(-\hat{\beta})^t e_t$ is $n^{-\frac{1}{2}}\sum_{t=1}^{n}\sum_{s=1}^{n}(-\hat{\beta})^s\ E(\tilde{W}_t e_t e_s) = 0$ since $\tilde{W}_t$ is a function of only $e_{t-1}, e_{t-2}, \ldots$, etc. It is shown in the appendix given in Dickey and Said (1981) that $n^{-1}\sum_{t=1}^{n}V_t e_t - cn^{-1}\sum_{t=1}^{n}Q_{t-1}e_t \xrightarrow{P} 0$. Recall from (3.5) that $n^{-1}\sum_{t=1}^{n}Q_{t-1}e_t \xrightarrow{\mathcal{L}} \xi$. As $\hat{\beta} \longrightarrow \beta$, we have $c \longrightarrow 1$. □

Combining lemmas (1) and (2) we see that, if the remainder r_t in (2.5) can be ignored then $D_n(\hat{\gamma}-\gamma) \xrightarrow{\mathcal{L}} (B_1,B_2,B_3)$ where $B_1 = \xi/\Gamma$ (this is the limit distribution of $n(\hat{\rho}-1)$ in Dickey and Fuller (1979)) and where

$$\begin{bmatrix} B_2 \\ B_3 \end{bmatrix} \sim \text{MVN}\left[\begin{pmatrix} 0 \\ 0 \end{pmatrix}, \begin{pmatrix} (1-\beta^2) & 0 \\ 0 & \frac{1-\beta^2}{\beta^2}\sigma^2 \end{pmatrix}\right] .$$

Thus we get the same limiting behavior for the coefficient $\hat{\beta}$ as in the stationary case and the same limit distribution for the unit root estimator as in the AR(1) case of Dickey and Fuller (1979).

5. THE REMAINDER TERM

To justify the limiting distributions obtained in the previous section, we need to show that the remainder effect can be ignored, that is, $D_n^{-1}F'(\hat{\theta})\ \ R(\ddot{\theta}) = o_p(1)$. Now observe the tth row of $R(\ddot{\theta})$ is $\frac{1}{2}T_t(\ddot{\theta})\ (\beta-\hat{\beta})^2 + U_t(\ddot{\theta})\ (\beta-\hat{\beta})\ (\delta-\hat{\delta})$. This means we must show that all of the following expressions converge to zero:

1) $n^{-1}\ (\beta-\hat{\beta})^2\ \Sigma\ V_t(\hat{\theta})\ T_t(\ddot{\theta})$
2) $n^{-\frac{1}{2}}\ (\beta-\hat{\beta})^2\ \Sigma\ W_t(\hat{\theta})\ T_t(\ddot{\theta})$
3) $(\beta-\hat{\beta})\ \Sigma\ \Delta_t(\hat{\theta})\ T_t(\ddot{\theta})$
4) $n^{-1}(\beta-\hat{\beta})\ (\delta-\hat{\delta})\ \Sigma\ V_t(\hat{\theta})\ U_t(\ddot{\theta})$
5) $n^{-\frac{1}{2}}(\beta-\hat{\beta})\ (\delta-\hat{\delta})\ \Sigma\ W_t(\hat{\theta})\ U_t(\ddot{\theta})$
6) $(\beta-\hat{\beta})\ (\delta-\hat{\delta})\ \Sigma\ \Delta_t(\hat{\theta})\ U_t(\ddot{\theta})$

Proof: In the proof of lemma (1) it is shown that $\Sigma\ V_t(\hat{\theta})\ W_t(\hat{\theta}) = 0_p(n)$. Since $T_t(\ddot{\theta})$, like $W_t(\hat{\theta})$, behaves as a stationary AR, the same logic shows $\Sigma\ V_t(\hat{\theta})\ T_t(\ddot{\theta}) = 0_p(n)$. Now $n^{-1}(\beta-\hat{\beta})^2$ is $o_p\left(n^{-3/2}\right)$. This proves 1.

Expression 2 is easily shown since it involves two series which behave like stationary AR's so the sum of their cross product is $O_p(n)$ and since $n^{-\frac{1}{2}}(\beta-\hat{\beta})^2 = o_p(n^{-1})$, the result is obtained.

Now $\ddot{U}_t(\theta)$ is bounded by an exponentially decreasing function in t as is $\Delta_t(\hat{\theta})$. Thus 3, 5, 6 are obviously converging to zero.

Expression 4 involves $\Sigma V_t(\hat{\theta})\ddot{U}_t(\theta)$. Now there exists M, $\lambda < 1$ such that $\ddot{U}_t(\theta) < M\lambda^t$. Also $E\{V_t(\hat{\theta})\} = 0$ and

$$\mathrm{Var}\{\sum_{t=1}^{n}\ddot{U}_t(\theta)V_t(\hat{\theta})\} = \sum_{t=1}^{n}\sum_{s=1}^{n}\ddot{U}_t(\theta)\ddot{U}_s(\theta)\,\mathrm{Cov}(V_t(\hat{\theta}), V_s(\hat{\theta}))$$

$$\le \sum_{t=1}^{n}\sum_{s=1}^{n} M^2\lambda^{t+s}\{c^2\min(t-1,s-1)\sigma^2 + E(Q_{s-1}f_L(t)) + E(Q_{t-1}f_L(s)) + E(f_L(t)f_L(s))\} = O(1)$$

Since $n^{-1}(\delta-\hat{\delta})(\beta-\hat{\beta}) = o_p(n^{-5/4})$, we are done.

6. HYPOTHESIS TESTING.

We first show that the improved estimator $\hat{\theta}+\hat{\gamma}$ leads to errors $e_t(\hat{\theta}+\hat{\gamma})$ for which $n^{-1}\sum_{t=1}^{n}e_t^2(\hat{\theta}+\hat{\gamma}) \xrightarrow{P} \sigma^2$ so that $n^{-1}\sum_{t=1}^{n}e_t^2(\hat{\theta}+\hat{\gamma})$ may be used in practice to estimate σ^2. Recall from (2.1), $e_t = Y_t - \rho Y_{t-1} - \beta e_{t-1}$. Let $D_t = Y_t - Y_{t-1}$. Under the null hypothesis, $H_o: \rho = 1$, we have $e_t = D_t - \beta e_{t-1}$ which implies that the D_t series behaves as a simple MA(1). Note the D_t process is readily obtained from the Y_t process. By results of Fuller (1976), pg. 348 we have $n^{-1}\sum_{t=1}^{n}\hat{e}_t^2 \xrightarrow[n]{P} \sigma^2$. To obtain an initial estimate of β, simply set $(1+\hat{\beta}^2)^{-1}\hat{\beta} = \hat{\rho}_D(1)$ where $\hat{\rho}_D(1) = \sum_{t=1}^{n}D_tD_{t-1}/\sum_{t=1}^{n}D_t^2$. We choose the solution $\hat{\beta}$ such that $|\hat{\beta}| < 1$.

Now to test for unit root, that is, $H_o: \rho = 1$ we use the regression "t-type statistic" defined by

$$\tau = \frac{\hat{\rho}-1}{\sqrt{c_1 S_e^2}} \tag{6.1}$$

where $S_e^2 = n^{-1}\sum_{t=1}^{n}e_t^2(\hat{\theta}+\hat{\gamma})$ and c_1 is the element of $\{F'(\hat{\theta})F(\hat{\theta})\}^{-1}$ associated with $\hat{\rho}$.

Theorem: Under the assumptions of model (2.1) and defining τ as in (6.1) we have $\tau \xrightarrow{\mathcal{L}} \xi/\sqrt{\Gamma}$ (6.2)

where ξ and Γ are defined as in (3.5).

Proof: We note that the normalized denominator for $n(\hat{\rho}-1)$ is $(n^2\sigma^2)^{-1}\sum_{t=1}^{n}V_t^2(\hat{\theta})$ and from (6.1)

$$\tau - \frac{n(\hat{\rho}-1)\sqrt{\Sigma V_t^2(\hat{\theta})/n^2\sigma^2}}{\sqrt{\Sigma\,\hat{e}_t^2/n\,\sigma^2}} \xrightarrow{P} 0$$

Now $(n^2\sigma^2)^{-1}\Sigma V_t^2(\hat{\theta}) \xrightarrow{\mathcal{L}} \Gamma$ and $(n\sigma^2)^{-1}\Sigma\hat{e}_t^2 \xrightarrow{P} 1$. Therefore by Slutsky's theorem $\tau \longrightarrow \dfrac{\frac{\xi}{\Gamma}\cdot\sqrt{\Gamma}}{1} \equiv \xi/\sqrt{\Gamma}$ □

The above limiting distribution for τ is the same as the one obtained by Dickey (1976). Empirical percentiles for the τ-statistic are given in Dickey (1976).

7. EXAMPLE. (Box and Jenkins Series A)

Box and Jenkins (1970, pg. 525) list 197 consecutive hourly readings of concentration in a chemical process. We used the observations and our suggested fitting technique to obtain the following fitted equation

$$\widehat{Y_t - Y_{t-1}} = \underset{(.0359)}{-.0992}\, Y_{t-1} + e_t - \underset{(.0706)}{.5468}\, e_{t-1}$$

$$n(\hat{\rho}-1) = -19.44$$

$$\tau = -2.76$$

Both $n(\hat{\rho}-1)$ and τ are significant when compared to their corresponding 1% critical values which are given in Dickey (1976). This result implies that the Box and Jenkins Series A behaves as a stationary ARMA process and hence the series should not be differenced.

REFERENCES:

BOX, G.E.P. and JENKINS, G.M. (1970). *Time Series Analysis Forecasting and Control.* Holden-Day, San Francisco.

DICKEY, D.A. (1976). "Estimation and Hypothesis Testing in Non-Stationary Time Series", Ph.D. thesis, Iowa State University.

DICKEY, D.A. and FULLER, W.A. (1979). "Distribution of the Estimators for Autoregressive Time Series with a Unit Root", *Journal of the American Statistical Association*, v. 74, 427-431.

DICKEY, D.A. and SAID, S.E. (1981). "Testing ARIMA (p,1,q) versus ARIMA (p+1,q)", to appear in the 1981 ASA proceedings, Business & Economics Statistic Section.

FULLER, W.A. (1976). *Introduction to Statistical Time Series.* Wiley, New York.

SAID, S.E. "Testing for Unit Roots in ARMA Models, Ph.D. thesis, North Carolina State University. (To appear 1981.)

APPLIED TIME SERIES ANALYSIS
O.D. Anderson & M.R. Perryman (eds.)

A DYNAMIC RATIONAL EXPECTATIONS MODEL OF MEXICO

Javier Salas
Banco De Mexico, S. A.

and

George Tauchen
Duke University

This paper uses post-WWII annual data to estimate the parameters of a dynamic rational expectations model of Mexico. The estimation technique is a version of Hatanaka's method for estimating dynamic simultaneous equations models with autocorrelated errors. Sims-Granger causality tests are used to check the assumptions about exogeneity. Dynamic simulations are performed to determine the system's response to movements in macro policy variables and to random shocks in aggregate supply and demand.

INTRODUCTION

In this paper we present estimates of the parameters of a small-scale classical rational expectations model of Mexico. Our major findings are as follows. First, the model is consistent with the data. In fact, we find it remarkable that application of the model to a less industrialized nation gives parameter estimates that are statistically significant and have the correct signs. Second, the point estimates of the price elasticities of the aggregate supply and demand schedules are well below unity in absolute value. These low elasticities indicate that the price effect of a shift in either aggregate demand or supply is quite large while the output effect is relatively small. Furthermore, the small elasticity of aggregate demand indicates that Lucas's (1973) assumption of a unitary elastic aggregate demand schedule is untenable. Although his assumption is convenient -- it reduces the data requirements by allowing the researcher to identify nominal income with nominal aggregate demand -- it is not true empirically, and consequently it is important in empirical work to take account of all the variables shifting aggregate demand. Third, and finally, our results indicate that the position of the Mexican aggregate supply schedule is quite sensitive to the level of agricultural employment in the southwestern United States. Specifically, the estimates indicate that a ten percent increase in the agricultural employment variable is associated with a four percent leftward shift of the Mexican aggregate supply schedule.

The two-equation model consists of an aggregate demand schedule and a Lucas-type aggregate supply schedule. Policy variables together with export demand determine the position of the aggregate demand schedule; the U.S. agricultural employment variable determines the position of the aggregate supply schedule. The two schedules jointly determine the current price level and the level of aggregate output conditional on the history of the endogenous variables and the values of the exogenous variables. Time variation in the exogenous variables and the structural disturbances drive the system.

Section II of this paper describes the model in more detail. Section III reports the point estimates and associated test statistics. Section IV briefly reports the results of a battery of tests designed to check our assumptions about exogeneity. Readers uninterested in this kind of testing can skip directly to the dynamic simulations reported in Section V.

MODEL

The estimated model of this paper is a demand and supply system determining the current level of output and the price level:

$$(1) \quad y_t = \alpha_{10} + \alpha_{11}t + \lambda_1 y_{t-1} + \beta_1 p_t + \gamma_{11} m_t + \gamma_{12} g_t + \gamma_{13} x_t + u_{1t}$$

$$(2) \quad y_t = \alpha_{20} + \alpha_{21}t + \lambda_2 y_{t-1} + \beta_2 (p_t - \hat{p}_{t-1,t}) + \gamma_{24} n_t + u_{2t}$$

where

- y_t : log of real GNP
- p_t : log of the price level
- $\hat{p}_{t-1,t}$: the expected value of p_t conditional on information available at time t-1
- m_t : log of the nominal money stock (currency + demand deposits)
- g_t : log of real government expenditures
- x_t : log of real exports
- n_t : log of U.S. agricultural employment in fifteen southwestern states
- t : time, in years (1950-77)

Equation (1) is a logarithmic approximation to a standard aggregate demand schedule in the price-output plane. Theory suggests a downward sloping demand curve ($\beta_1 < 0$); for, an increase in the price level deflates the real value of nominal assets thus causing a decline in aggregate demand through an interest rate effect and through any real-wealth effect on consumption. The policy variables and exports shift aggregate demand outwards ($\gamma_{11}, \gamma_{12}, \gamma_{13} > 0$). The lagged output term detects any dependence of aggregate demand on lagged output through, say, an accelerator effect. In earlier versions of this work equation (1) included lagged real balances, $m_{t-1} - p_{t-1}$, but the coefficient was always small and statistically insignificant. Finally, the constant, trend and random disturbance u_{1t} reflect all other influences on aggregate demand.

Equation (2) is a Lucas-type (1973) aggregate supply schedule. It links output only to the surprise term, $p_t - \hat{p}_{t-1,t}$, the unforecastable component of prices. Thus, an unforeseen increase in the price level induces an upward movement along the aggregate supply schedule ($\beta_2 > 0$), while a foreseen increase in the price level generates no supply response. The lagged output term in (2) reflects the presence of frictional (adjustment cost) terms in the objective functions of the suppliers of goods and services (Sargent, 1978). The trend term accounts jointly for secular growth in productivity, population, and the capital stock. In earlier work, equation (2) included a separate population term which was always statistically insignificant.

An important explanatory variable in (2) is U.S. agricultural employment in fifteen southwestern states. Expansion of U.S. job opportunities for legal and illegal Mexican labor reduces Mexican labor supply which in turn reduces Mexican aggregate supply ($\gamma_{24} < 0$). Statistical work guided the choice of this particular U.S. variable. Equations (1)-(2) were first estimated with no U.S. variables in the system. Estimates of the supply disturbances u_{2t} were then related to three U.S. employment variables.[1] The chosen variable gave by far the best fit, and so it was used in the reported estimation.

The strictly exogenous variables in (1) and (2) are money, m_t, government expenditures, g_t, exports, x_t, and the U.S. agricultural employment variable, n_t. Here strictly exogenous means that the variables are uncorrelated with the disturbance terms, u_{1t} and u_{2t}, at all leads and lags. Although the disturbances are unrelated to the exogenous variables, they can display serial correlation which is corrected for in the estimation.

It is now well known that the system (1)-(2) allows no role for systematic demand management for the purpose of output stabilization. Taking the expected value of (2) conditional at t-1 gives

$$\hat{y}_{t-1,t} = \alpha_{20} + \alpha_{21}t + \lambda_2 y_{t-1} + \gamma_{24}\hat{n}_{t-1,t} + \hat{u}_{2,t-1,t}$$

which is independent of the stochastic processes generating the demand shift variables in the sense discussed by Sargent (1976b). Of course the system does allow a role for demand management policies for the purpose of price-level stabilization.

RESULTS

Estimation requires an observable proxy for the unobserved price-level forecast, $\hat{p}_{t-1,t}$, appearing in the supply equation (2). Here we use Sargent's (1976a) technique. The price level is regressed on a vector of variables dated t-1 and earlier

$$\text{(3)} \quad \hat{p}_{t-1,t} = \underset{(3.268)}{-8.512} - \underset{(.012)}{.035t} + \underset{(.188)}{.490p_{t-1}} + \underset{(.092)}{.775m_{t-1}} - \underset{(.210)}{.125m_{t-2}} + \underset{(.276)}{.761n_{t-1}} - \underset{(.248)}{.397n_{t-2}}$$

with standard errors in parentheses. Other variables insignificant at 10 percent have been deleted from (3). The significance of p_{t-1} in (3) is evidence of serial correlation in the disturbances of the structure (1)-(2); a serially correlated error structure is the only way the lagged price level can enter (3). The sample residual from (3) is used as the endogenous price surprise variable in (2). This method does produce an error-ridden proxy for the true surprise variable. Nonetheless, the variable's error is uncorrelated with the exogenous variables, and, though the point is not particularly relevant given the small sample in this work, the variable's error vanishes in large samples.

Table 1 displays two stage least squares estimates of the aggregate demand and supply equations. The table reports results both with and without the limited information version of Hatanaka's Method (a) (1976, p. 194) to correct for serial correlation in structural equations. Each error is assumed to follow a first order univariate autoregression, $u_{it} = \rho_i u_{i,t-1} + \eta_{it}$ $(i = 1,2)$. The method produces both intermediate and final estimates of ρ_i. Table 1 reports the final estimates along with their standard errors. There is very little evidence for serial correlation in the demand equation; on the other hand, there is rather strong evidence for serial correlation in the supply equation.

Inspection of the corrected aggregate demand equation in Table 1 reveals several points worth noting. First, the coefficients of government expenditures and exports are roughly equal to the variables' shares in real GNP, indicating an autonomous expenditures demand multiplier of about unity. Second, the small and insignificant coefficient on money suggests a weak monetary demand effect. In interpreting this result, remember that the coefficient of money detects only portfolio effects since fiscal policy is held constant. Third, the coefficient on lagged income gives a mean lag of $.420/(1 - .420) = .724$ years for the response of aggregate demand to a random shock. And fourth, the point estimate $\hat{\beta}_1 = -.182$ indicates a highly price inelastic aggregate demand schedule. This

TABLE 1 a/

<u>Uncorrected for serial correlation</u> (Period: 1949-77)

Aggregate demand (1):

$$y_t = \underset{(1.79)}{3.576} + \underset{(0.011)}{0.030}\, t + \underset{(0.159)}{0.565}\, y_{t-1} - \underset{(0.066)}{0.179}\, p_t + \underset{(0.076)}{0.023}\, m_t + \underset{(0.039)}{0.036}\, g_t + \underset{(0.043)}{0.069}\, x_t + u_{1t}$$

Aggregate supply (2):

$$y_t = \underset{(2.074)}{6.763} + \underset{(0.009)}{0.019}\, t + \underset{(0.156)}{0.574}\, y_{t-1} + \underset{(0.186)}{0.200}\, (p_t - \hat{p}_{t-1,t}) - \underset{(0.086)}{0.255}\, n_t + u_{2t}$$

<u>Corrected for serial correlation</u> (Period: 1950-77)

Aggregate demand (1):

$$y_t = \underset{(2.019)}{4.757} + \underset{(0.014)}{0.039}\, t + \underset{(0.239)}{0.420}\, y_{t-1} - \underset{(0.063)}{0.182}\, p_t + \underset{(0.079)}{0.043}\, m_t + \underset{(0.036)}{0.063}\, g_t + \underset{(0.039)}{0.065}\, x_t + u_{1t}$$

$$u_{1t} = \underset{(0.333)}{0.403}\, u_{1t-1} + v_{1t}$$

Aggregate supply (2):

$$y_t = \underset{(2.575)}{11.981} + \underset{(0.015)}{0.037}\, t + \underset{(0.261)}{0.208}\, y_{t-1} + \underset{(0.222)}{0.406}\, (p_t - \hat{p}_{t-1,t}) - \underset{(0.127)}{0.405}\, n_t + u_{2t}$$

$$u_{2t} = \underset{(0.319)}{0.685}\, u_{2t-1} + v_{2t}$$

a/ standard errors in parentheses

last result is in contrast to Lucas's (1973) assumption of unity for the price elasticity of aggregate demand. The Lucas (1977)-Arak (1977) exchange discusses some of the issues involved with this elasticity.

One interesting hypothesis regarding aggregate demand is whether or not the schedule is homogenous of degree zero in the two nominal variables, money and the price level. This will be the case if the liquidity preference (portfolio balance) schedule is homogeneous of degree zero and if only real wealth enters the consumption function. A simple t test for the null hypothesis $\beta_1 + \gamma_{11} = 0$ gives $t(21) = -1.04$, insignificant at conventional levels.

Consider now the corrected aggregate supply equation in Table 1. The point estimate $\hat{\beta}_2 = .406$ is the elasticity of aggregate supply with respect to unforeseen movements in the price level; the $t(23) = 1.83$ is significant at the eight percent level. Furthermore, the small and insignificant coefficient on lagged output suggests that the output effects of a nominal shock persist for no longer than one year. Finally, the coefficient of n_t indicates that a 10 percent increase in southwestern agricultural employment leads to a 4.05 percent leftward shift of Mexico's aggregate supply schedule.

The only apparent paradox in the system is why equation (3) shows such strong feedback from money into prices and yet the fit of equation (1) suggests only a weak portfolio effect of money. Three explanations come to mind. First, lagged money could be a good predictor of future levels of government expenditures, and the autoregressions summarized in the following sections suggest that this is the case. Alternatively, money's role in the exchange process might not be correctly modeled in the two-equation demand and supply system. More complete models of the exchange process might in fact have monetary disturbances entering the supply equation. Lastly, accepting the null hypothesis $\gamma_{11} = 0$ in equation (1) could very well be a simple Type II error.

EXOGENEITY

The fit of any simultaneous equations system is only as valid as the maintained assumptions about exogeneity, and thus the assumptions should be routinely checked whenever possible. The Granger (1969) and Sims (1972, 1977) methodology provides a convenient framework for testing exogeneity assumptions in a time series context. If the putative exogenous variables are indeed strictly exogenous, then as a block these variables should Granger cause the endogenous variables and in turn should not be Granger caused by the endogenous variables.

Table 2 summarizes the outcome of a battery of Granger feedback tests. Each

TABLE 2
Exogeneity Tests

Dependent Variable	Null Hypothesis Coef.'s of indicated variables = 0	Statistic (Prob-val in parenthesis)
y_t	m_{t-1}, g_{t-1}, x_{t-1}, n_{t-1}	F(4,20) = 2.57(0.69)
p_t	m_{t-1}, g_{t-1}, x_{t-1}, n_{t-1}	F(4,20) = 7.04(.001)
m_t	y_{t-1}, p_{t-1}	F(2,20) = .96(.401)
g_t	y_{t-1}, p_{t-1}	F(2,20) = .54(.591)
x_t	y_{t-1}, p_{t-1}	F(2,20) = 1.16(.331)
n_t	y_{t-1}, p_{t-1}	F(2,20) = 2.01(.161)
n_t	m_{t-1}, g_{t-1}, x_{t-1}, y_{t-1}, p_{t-1}	F(5,20) = .93(.482)

variable in the system is regressed on a constant, trend, two lags of itself and, in the interest of parsimony, one lag each of the other five variables in the system. The first two tests indicate feedback at the 7 percent level or lower from the block of exogenous variables into the endogenous variables. The next four tests reveal no feedback at the 15 percent level from the block of endogenous variables into the exogenous variables. The last test indicates that as a group the Mexican variables do not contribute to the explanation of the n_t ; i.e., the southwestern agricultural employment variable is causally prior with respect to all variables in the system. These tests do not prove the exogeneity assumptions -- no test can -- but clearly they suggest that the maintained exogeneity assumptions are not wildly in conflict with the data.

SIMULATIONS

In order to illustrate the usefulness of the model for forecasting and policymaking, we next calculate the response of the estimated structure (1)-(2) to three forms of shocks. The first two shocks are random pulses to the disturbance terms in the aggregate demand and supply schedules. The third shock is a random pulse to the southwestern agricultural employment variable. The third simulation is carried out under two scenarios. One version has the policy variables turned off and the other has the policy variables responding to the shock according to estimated feedback equations. The first scenario can be thought of as a partial effect, the second as a total effect. In all simulations, we assume that in the initial period the price level is unknown (i.e., we solve for y_t and $p_t - \hat{p}_{t-1,t}$ simultaneously in (1) and (2) above) and that in subsequent periods the price level is known (i.e., we set $p_t - \hat{p}_{t-1,t}$ to zero). Further, the starting conditions are all variables equal zero and all results are relative to trend. And finally, we use the corrected equations in Table 1.

Table 3 reports the results of a one unit movement in the demand disturbance u_{1t}.

TABLE 3
Response to a Demand Disturbance

Year	Prices (p)	Output (y)	Disturbance (u_1)
0	1.701	.690	1.000
1	.805	.143	0
2	.167	.029	0
3	.034	.006	0
4	.007	.001	0
5	.001	.000	0
6	.000	.000	0

This can be thought of as an innovation[2] in, say, autonomous consumption or fiscal policy sufficiently large to produce a one percent rightward shift in aggregate demand. The results need little explanation. The initial output effect is large but subsequent movements are heavily damped. The price effect is well above unity (both schedules are inelastic), and it is less damped than the output effect. The difference in damping rates is attributable to the difference in coefficients on lagged output. With $\hat{\lambda}_1 = .420$ versus $\hat{\lambda}_2 = .208$ the demand schedule relaxes to the stationary position more slowly than the supply schedule.

Table 4 displays the results of a one unit pulse in the supply disturbance u_{2t}. This can be thought of as, say, a productivity gain relative to trend or as a fall in imported raw materials prices sufficiently large to produce a one percent rightward shift in aggregate supply. Here the effects are more spread out over time, which is to be expected because the shock continues to feed into the supply equation through the shock's own autoregression. Note the size of the price effect, it bottoms out in period 1 at 3.394 percent below trend. The output effect is smaller, peaking in period 1 at 0.749 percent above trend. Again the

TABLE 4
Response to a Supply Disturbance

Year	Prices (p)	Output (y)	Disturbance (u_2)
0	-1.701	0.310	1.000
1	-3.394	0.749	0.685
2	-1.698	0.625	0.469
3	-1.033	0.451	0.321
4	-0.681	0.314	0.220
5	-0.461	0.216	0.151
6	-0.314	0.148	0.103
7	-0.215	0.102	0.071
8	-0.147	0.070	0.048

price effect is more slowly damped than the output effect.

Panel A of Table 5 displays the system's response to a one percent movement in the U.S. variable, n , assuming no response by policymakers. Since n is serially correlated, the initial pulse sets off subsequent movements in n . We account for this by treating the initial pulse as a one unit movement in the residual of

TABLE 5
Response to a Shock in U.S. Agricultural Employment Variable

Panel A: No Policy Response

Year	Prices(p)	Output(y)	U.S. Vbl.(n)	Govt.Exp.(g)	Money(m)
0	0.688	-0.126	1.000	0.000	0.000
1	2.813	-0.566	1.334	0.000	0.000
2	2.045	-0.611	1.220	0.000	0.000
3	1.239	-0.483	0.880	0.000	0.000
4	0.525	-0.299	0.491	0.000	0.000
5	0.011	-0.128	0.162	0.000	0.000
6	-0.279	-0.003	-0.059	0.000	0.000
7	-0.378	0.068	-0.169	0.000	0.000
8	-0.348	0.092	-0.193	0.000	0.000

Panel B: With Policy Response

Year	Prices(p)	Output(y)	U.S. Vbl.(n)	Govt.Exp.(g)	Money(m)
0	0.796	-0.082	1.000	0.898	0.154
1	3.659	-0.557	1.334	1.930	0.543
2	2.983	-0.609	1.220	1.911	1.126
3	2.172	-0.482	0.880	1.583	1.618
4	1.454	-0.299	0.491	1.418	1.848
5	0.851	-0.128	0.162	1.171	1.838
6	0.395	-0.003	-0.059	0.822	1.646
7	0.127	0.068	-0.169	0.549	1.331
8	0.011	0.092	-0.193	0.369	0.975

the following autoregression

$$(4) \qquad n_t = \underset{(0.728)}{1.821} - \underset{(0.003)}{0.006}\, t + \underset{(0.173)}{1.334}\, n_{t-1} - \underset{(0.181)}{0.560}\, n_{t-2} ,$$

and then feeding the output of (4) into the supply equation. The results indicate that prices peak in period 1 at 2.813 percent above trend; output troughs in period 2 at 0.611 percent below trend. It appears that the Mexican economy experiences large and sustained effects from movenets in border-state agricultural employment.

In panel B we report the results for the case in which the policy variables are allowed to vary in response to the shock. To find the policy response, we estimated the following two equations

$$g_t = \underset{(3.348)}{-4.385} + \underset{(0.032)}{0.064}\, t - \underset{(0.194)}{0.105}\, g_{t-1} - \underset{(0.180)}{0.386}\, g_{t-2} + \underset{(0.248)}{0.778}\, m_{t-1}$$

(5)

$$- \underset{(0.187)}{0.174}\, x_{t-1} + \underset{(0.703)}{0.898}\, n_t + \underset{(0.782)}{0.707}\, n_{t-1}$$

$$m_t = \underset{(0.221)}{-0.586} + \underset{(0.597)}{0.015}\, t + \underset{(4.407)}{1.246}\, m_{t-1} - \underset{(1.697)}{0.475}\, m_{t-2} + \underset{(0.143)}{0.022}\, g_t$$

$$+ \underset{(1.343)}{0.197}\, g_{t-1} - \underset{(0.087)}{0.013}\, x_{t-1} + \underset{(0.235)}{0.134}\, n_t - \underset{(0.080)}{0.047}\, n_{t-1}$$

An F test on n_t, n_{t-1} in (5) is $F(2,21) = 5.88$, significant at the one percent level; in (6) the statistic is $F(2,20) = 0.04$, significant at only the 96 percent level. It appears that the Mexican government uses fiscal policy to offset movements in the U.S. variable. The simulations in Panel B are obtained as follows. As in Panel A the agricultural employment variable, the output of (4), is fed into the aggregate supply equation. Since in this case we allow the policy variables to respond to the shock, the output of (4) is also fed into the policy equations (5) and (6). The policy outputs from (5) and (6) are then fed into the aggregate demand equation. These calculations attribute all of the contemporaneous correlation between n and the policy variables to the policy reaction function. Furthermore, the calculations implicitly assume that the policymaker "sees" the shock before the public does. To the extent that the public is aware of the shock (and the policy equations), the simulations overstate the ability of demand policy to offset the shock in initial period, and thus the results should be interpreted with circumspection.

Mindful of these caveats, a comparison of Panel B with Panel A of Table 5 suggests that demand policy eliminates roughly thirty-five percent of the output drop in the initial period. This is accomplished, of course, at the expense of a higher price level in the future. Apparently, the policy objective function places relatively more weight on output fluctuations than on price level fluctuations.

In a minor way Lucas's point about noninvariance of parameters with respect to policy shifts applies to the calculations presented in Panel A of Table 5. For, the slope parameter β_2 in the aggregate supply schedule (2) depends upon the variance of the forecast error $p_t - \hat{p}_{t-1,t}$.[3] If the variation in the policy variables is eliminated, then the variance of $p_t - \hat{p}_{t-1,t}$ can change, though in which direction is not immediately obvious. There are plausible arguments for the variance either to decrease or to increase. If in fact the variance decreases, then the true β_2 in (2) would be larger than the estimate reported in Table 1. In this case the initial decline in output would be smaller than calculated in Panel A. Nonetheless, the calculations correctly derive the output and price effects from period 1 onwards.[4]

SUMMARY AND CONCLUSION

We have set forth and estimated a small-scale classical rational expectations model of the Mexican economy. Each parameter has the theoretically correct sign and in most cases is statistically significant at conventional levels. The model is suitable for tracing through the system the effects of a wide variety of exogenous shifts to aggregate demand and supply. Our demand-shift simulations indicate that the effect of monetary and/or fiscal policy on real national income (and thus employment) are concentrated within one year. In contradistinction, the effect of policy on inflation persists for two or three years.

FOOTNOTES

1. The other two series were total U.S. agricultural employment and total U.S. employment.

2. The innovation in a variable is that part of the series that cannot be linearly forecasted from the past values of all series in the system.

3. See Lucas's (1973, p. 328) derivation of the slope coefficient. His term $\theta\gamma$ is our β_2, and his σ^2 is Var $(p_t - \hat{p}_{t-1,t})$.

4. Of course there is still the surely negligible error in the contribution of $\lambda_i y_{t-1} (i = 1,2)$ to the determination of the positions of the aggregate demand and supply schedules.

APPENDIX

MONEY (m): We use the definition of M1 which includes coins and banknotes held by the public and checking accounts on December 31st of each year. The figures are not seasonally adjusted. The sources are: Statistics on the Mexican Economy published by Nacional Financiera S. A. and Indicadores Economicos published by Banco de México S. A.
PRICES (p): We use the implicit GNP deflator. The source is: Informe Anual published by Banco de México S. A.
REAL GNP (y): We use the values of nominal GNP deflated by the implicit GNP deflator. The source is: Informe Anual published by Banco de México S.A.
REAL GOVERNMENT EXPENDITURES (g): We use the data on nominal government expenditures deflated by the implicit GNP deflator. The series includes expenditures other than for final goods and services. Nonetheless, it surely is highly correlated with the theoretically correct figure. The figures on nominal government expenditure include: Wages and salaries, interests and payments on the public debt, transfers, physical investment, financial investment and other capital expenditures. The sources are: Anuario Estadístico de los Estados Unidos Mexicanos published by Secretaría de Industria y Comercio (Dirección General de Estadística) and Indicadores Económicos published by Banco de México S. A.
REAL EXPORTS (x): We use data on nominal exports deflated by the implicit GNP deflator. The figures on nominal exports include: agricultural goods, cattle, fish, products from extractive industries and products from manufacturing industries. The source is: International Financial Statistics published by the International Monetary Fund.
U.S. AGRICULTURAL EMPLOYMENT (n): The series is total farm employment (hired workers plus family workers) in fifteen southwestern states. The figures are for: Montana, Idaho, Wyoming, Colorado, New Mexico, Arizona, Utah, Nevada, Washington, Oregon, California, Arkansas, Louisiana, Oklahoma and Texas. The source is: Agricultural Statistics published by the United States Department of Agriculture.

REFERENCES

[1] Arak, M. (1977). "Some International Evidence on Output-Inflation Tradeoffs: Comment," American Economic Review 67, pp. 728-30.

[2] Banco de México S. A. Indicadores Económicos, Issues from July 1976 to May 1978.

[3] ——. Informe Anual, selected issues from 1947-1978.

[4] Granger, C.W.J. (1969). "Investigating Causal Relations by Econometric Models and Cross Spectral Methods," Econometrica 37, pp. 424-38.

[5] Hatanaka, M. (1976). "Several Efficient Two-Step Estimators for the Dynamic Simultaneous Equations Model with Autoregressive Disturbances." Journal of Econometrics 4, pp. 189-204.

[6] International Monetary Fund, International Financial Statistics, selected issues from 1950-1978.

[7] Lucas, R.E. (1973). "Some International Evidence on Output-Inflation

Tradeoffs." American Economic Review 63, pp. 326-34.

[8] ——— (1977). "Some International Evidence on Output-Inflation Tradeoffs: Reply." American Economic Review 67, pp.

[9] Nacional Financiera S. A. (1966). Statistics on the Mexican Economy.

[10] Sargent, T.J. (1976a). "A Classical Macroeconometric Model for the United States." Journal of Political Economy 84, pp. 207-37.

[11] ——— (1976b). "The Observational Equivalence of Natural and Unnatural Rate Theories of Macroeconomics." Journal of Political Economy 84, pp. 631-40.

[12] ——— (1978). "Estimation of Dynamic Labor Demand Schedules Under Rational Expectations." Journal of Political Economy 86, pp. 1009-44.

[13] Secretaría de Industria y Comercio (Dirección General de Estadística), Anuario Estadístico de los Estados Unidos Mexicanos, selected issues from 1957-1965.

[14] Sims, C.A. (1972). "Money, Income and Causality." American Economic Review 62, pp. 540-52.

[15] ——— (1977). "Exogeneity and Causal Ordering in Macroeconomic Models," in New Methods in Business Cycle Research: Proceedings from a Conference, pp. 23-44.

[16] United States Department of Agriculture, Agricultural Statistics, selected issues from 1948-1978.

APPLIED TIME SERIES ANALYSIS
O.D. Anderson & M.R. Perryman (eds.)

A TRANSFER FUNCTION MODEL OF A STOCK AND ITS OPTION PRICES: A MICROCOMPUTER ANALYSIS

Joseph B. Seif

Department of Quantitative Analysis
College of Business Administration
St. Johns University
Jamaica, N.Y.
U.S.A.

Box and Jenkins (1970) have specified a procedure for the development of a "transfer function model", a model which expresses the interrelationships between two time series. In this paper we present an application of the identification method as programmed by the author on a microcomputer. The series studied are prices of a stock and its options. The essentially random walk behavior of stock prices simplifies the estimation procedure which, in fact, turns out to be equivalent to an OLS regression of the differenced series.

METHOD AND RESULTS

Suppose we have a series Y(t), which we consider as an output series controlled by an input series X(t). We would like a model relating the values of Y(.) to those of X(.). Ideally, Y(t) would be related only to lagged values of X(.), i.e. X(t-k), k>0, so that knowledge of X(t-1),X(t-2), ... could be used to forecast the value of Y(t). That this expectation is too ambitious in the case of stock prices is widely believed and is also a result of this study. Using the Box-Jenkins identification procedure we develop a transfer function model between X(t) and Y(t), but unfortunately the model does not contain any lags.

We are seeking a model of the form

$$Y(t) = v(B)X(t) + N(t)$$

where B is the backward shift operator, BX(t) = X(t-1), and N(t) is noise. Thus v(B)X(t) represents a (possibly infinite) linear combination of X(t) and its previous values. The prodedure for identifying such models is (Anderson (1975), Box and Jenkins (1970) and Pack (1977)) as follows. If the series X(t) and Y(t) are not stationary, they must first be differenced d times until they result in stationary series x(t) and y(t). (Usually d=1 as is the case in our example.) Then identify the ARIMA model of the input series, X(t) to obtain

$$\theta^{-1}(B)\ \phi(B)x(t) = a(t)$$

where a(t) is a white noise series. We call this "prewhitening" and refer to a(t) as the prewhitened input series. We then apply the same prewhitening operator to the output series

$$b(t) = \theta^{-1}(B)\,\phi(B)\,y(t),$$

b(t) is the prewhitened output series.

In our example, the differenced series x(t) is already a white noise series. This is a well known property of stock prices (see [3]), and we verify it by examination of its autocorrelation and partial autocorrelation functions. We then seek a transfer function model of the form

$$y(t) = v_0\,x(t) + v_0\,x(t-1) + \ldots + n(t)$$

where n(t) is white noise.

Box and Jenkins show that a rough estimate of the weights v may be obtained by

$$v_{ab} = \frac{r_{ab}(k)\,s_b}{s_a}$$

where $r_{ab}(k)$ is the k-th cross-correlation coefficient between the prewhitened series a(t) and b(t),

$$r_{ab}(k) = \frac{c_{ab}(k)}{s_a\,s_b}$$

and

$$c_{ab}(k) = \frac{1}{n}\sum_{t=1}^{n-k} a(t)b(t+k) \qquad \text{for } k = 0,1,2,\ldots.$$

$$c_{ab}(k) = c_{ab}(-k) \qquad \text{for } k = 0,-1,-2,\ldots$$

The cross-correlations are insignificantly different from zero at all lags except except k=0 thus leading to the (not surprising) interpertation that Y(t) may be predicted from the current values of X(t) but not from its past values. In general this estimate of v is inefficient and Box and Jenkins suggest that its main value is in identifying the model and a non-linear least squares procedure is to be used for estimation of the coefficients. In our example, however, it is easily seen that the v given above is equal to the OLS estimate of simple linear regression of y(t) on x(t).

To complete the model we must specify the structure of the noise term, n(t). This is accomplished by calculating the autocorrelation function of the residuals, y(t)−v x(t). It is seen that n(t) can be considered to be a white noise series, again justifying the OLS estimate of v .

Finally, we perform this OLS estimate and obtain the equation

$$y(t) = -0.041 + \underset{(23.265)}{0.469}\, x(t)$$

with v = 0.469 and a t-statistic of 23.265. The value of R^2 is 0.801 and the Durbin-Watson statistic is 2.19.

This is the equation for the differenced series. In terms of the original prices the model becomes

$$Y(t) = Y(t-1) - 0.041 + 0.469(X(t)-X(t-1)) + n(t)$$

where n(t) is white noisewith standard deviation 0.174 and the constant term should probabely be ignored. Actually, the value we get for v is 0.481 the difference being due to setting the mean of the prewhitened series to zero in calculating the crosscorrelations.

THE COMPUTATIONS

The data was obtained from a time sharing service and downloaded into a personal microcomputer. We have the closing daily prices of IBM common stock and the closing prices of IBM call option at the striking price of $60 and expiration date of October 21, 1981. Denoting the stock prices by X(t) we first calculate the autocorrelation function X(t) i.e.

$$c(k) = \frac{1}{n} \sum_{t=0}^{n-k} [X(t)-X][X(t+k)-X]$$

K :	1	2	3	4	5	6	7	8	9	10	11	12	13	14	15
a.c.:	.94	.89	.84	.80	.77	.74.	.71	.69	.66	.63	.61	.59	.56	.53.	.50

s.e. of ac=.382 white noise chi-square = 1163.7

Table 1. Autocorrelation function of 137 daily closing prices of IBM call options. Series X(t).

The results are shown in Table 1 and since the autocorrelation fail to damp at long lags, first differencing is indicated.

Forming the first differences

$$x(t) = X(t) - X(t-1)$$

we again compute the autocorrelations and also the partial autocorrelations. The results are shown in Tables 2 and 3. Under the hypothesis of no autocorrelation, the 5% significance critical values are approximated by 2/n = 0.17 and we accept the hypothesis that the differenced series is white noise, i.e. x(t) = a(t).

K	: 1	2	3	4	5	6	7	8	9	10
a.c.	:-.03	-.02	-.15	-.19	.01	-.07	.12	.02	-.03	-.09

s.e. of ac = .153, white noise chi-square = 12.25

Table 2. Autocorrelations of x(t), first differences of X(t).

K	: 1	2	3	4	5	6	7	8	9	10
pac.	:-.03	-.02	-.16	-.20	.02	-.12	.05	-.01	-.06	-.11

s.e. of pac = .153

Table 3.Partial autocorrelations of x(t), first differences of X(t).

Thus no further prewhitening is necessary. We perform the same differencing on the Y(t) series of option prices and obtain the series y(t). We now calculate the cross correlation function between x and y (Table 4), and these estimates combined with those of the standard deviations of the x and y series provide the estimates of the transfer model weights v .

K	: -7	-6	-5	-4	-3	-2	- 1	0	1	2	3	4	5	6
cc.	:.08	.03	.04	-.16	-.17	-.05	.08	.92	.05	-.08	-.16	-.16	.00	.05

Table 4.Cross correlations between x(t) and y(t).

We now calculate the autocorrelation function of the residuals (not shown) and this indicates the identification of n(t) as a white noise process.

CONCLUSIONS and LIMITATIONS

We have applied the Box-Jenkins transfer model porocedure to the identification and estimation of a linear relationship between a stock and its options.

The major motivation for this work was to develop microcomputer packages for the Box-Jenkins univariate and transfer function models. This has been accomplished as far as the identification of these models. We have not yet developed the routines for non-linear least-sqares estimation of the general case, nor have we addressed the case of seasonal models. The example used in this paper was chosen to accommodate these limitations.

Also, we have not made use of available finance theory. A more accurate theory of option prices would include considerations of the time-to-expiration value of the option and also account for differences between in-the-money and out-of-the-money calls (see Black and Scholes (1973)).

REFERENCES

ANDERSON, O.D. (1975). *Time Series Analysis and Forecasting: The Box-Jenkins Approach.* Butterworths, London.

BLACK, F. and SCHOLES, M. (1973). Options and Liabilities, *Journal of Political Economy 81.*

BOX G.E.P. and JENKINS, G.M. (1970). *Time Series Analysis, Forecasting and Control.* Holden-Day, San Francisco.

COOTNER, P. (1964). *The Random Character of Stock Prices.* M.I.T. Press, Cambridge, Massachusetts

PACK, D.J. (1977). Reavealing Time Series Interrelationships. *Decision Sciences 8,* 377-402.

APPLIED TIME SERIES ANALYSIS
O.D. Anderson & M.R. Perryman (eds.)

LEAST SQUARE SPECTRAL ESTIMATION OF TRUNCATED SINUSOIDS

Piyare L. Sharma
C. S. Chen

Electrical Engineering Department
The University of Akron
Akron, Ohio
U.S.A.

It is shown analytically that for a single noise-free sinusoid, LS spectral estimation by second order prediction filter produces no frequency bias. The results have been shown to be independent of the data length, initial phase and frequency of the sinusoid. It has also been shown that identical results are obtained by taking only a unidirectional prediction error filter (forward or backward) for the single sinusoid case. Finally a computer simulation has been presented which agrees with the theoretical derivation of LS bidirectional and unidirectional prediction error filters.

INTRODUCTION

Frequency error due to data truncation, noise and the initial phase in a single real sinusoidal signal has been of considerable interest. Chen and Stegen [3] presented a computer simulation of Burg's maximum entropy method for a noisy sinusoid with varying initial phase and length of the sinusoid. Ulrych [2], Ulrych and Clayton [7] compared the results of maximum entropy method and least square spectral estimation and found that least square error spectral estimation method is much less sensitive to the varying initial phase and noise. Recently Swingler [4,8] derived simple expressions for the frequency error due to data truncations, sampling rate and initial phase effects for maximum entropy and discrete fourier transform techniques, for a single real sinusoid. He concluded the DFT exhibits zero error independent of initial phase if the data length contains an odd number of quarter cycles and the error drops inversely as the square of the data length. Whereas, Burg's MEM method gives zero error for all initial phases if the data contains (approximately) even numbers of quarter cycles and the error drops off more slowly as the inverse of the number of data points.

In this paper, it is shown analytically that for a single noise-free sinusoid, least square error spectral estimation by second order prediction filter produces no frequency bias. The results have been shown to be independent of the data length, initial phase and frequency of the sinusoid. It has also been shown that identical results are obtained by taking only a unidirectional prediction error filter (forward or backward) for the single sinusoid case.

THEORY

Consider (N+1) point data segment of a real sinusoid

$$X(K)=\cos(\theta K+\phi) \; ; \; K=0,\ 1,\ 2,\ 3,\ ---,\ N$$

where $\theta=\omega T$, ω is the angular frequency in radian, T is the sampling period in second and ϕ is the initial phase. It is noted that the "correct" prediction filter coefficient is (1, -2cosθ, 1) [8].

Let the second order prediction filter be $(1, -g_1, -g_2)$, mean square forward error e_f^2 and mean square backward error e_b^2 are,

$$e_f^2 = \frac{1}{(N-2)} \sum_{t=o}^{N-2} (X_{t+2} - g_1 X_{t+1} - g_2 x_t)^2$$

and

$$e_b^2 = \frac{1}{(N-2)} \sum_{t=o}^{N-2} (X_t - g_1 X_{t+1} - g_2 X_{t+2})^2$$

respectively.

The minimization of $(e_f^{\,2} + e_b^{\,2})$ with respect to g_1 and g_2 yields,

$$g_1 (2 \sum_{t=o}^{N-2} X^2{}_{t+1}) + g_2 \sum_{t=o}^{N-2} (X_t X_{t+1} + X_{t+1} X_{t+2}) = \sum_{t=o}^{N-2} (X_t \cdot X_{t+1} + X_{t+1} X_{t+2}) \tag{1}$$

and

$$g_1 \sum_{t=o}^{N-2} (X_t \cdot X_{t+1} + X_{t+1} X_{t+2}) + g_2 \sum_{t=o}^{N-2} (X_t^{\,2} + X^2{}_{t+2}) = 2 \sum_{t=o}^{N-2} X_t \cdot X_{t+2} \tag{2}$$

After further simplification, eqs. (1) and (2) become

$$Ag_1 + Bg_2 = B \tag{3}$$

$$Bg_1 + Cg_2 = D \tag{4}$$

where

$$A=N-1+\cos(N\theta+2\phi)\frac{\sin(N-1)\theta}{\sin\theta} \tag{5}$$

$$B=(\cos\theta)\{(N-1)+\cos(N\theta+2\phi)\frac{\sin(N-1)\theta}{\sin\theta}\} \quad (6)$$

$$C=N-1+\cos(N\theta+2\phi)\cdot\cos 2\theta\,\frac{\sin(N-1)\theta}{\sin\theta} \quad (7)$$

$$D=(N-1)\cos 2\theta+\cos(N\theta+2\phi)\frac{\sin(N-1)\theta}{\sin\theta} \quad (8)$$

The solution to eqs. (3) & (4) is

$$g_1=2\cos\theta,\ g_2=-1$$

which is identical to the "correct" filter coefficient. The minimization of e_f^2 or e_b^2 lead to the identical solutions.

COMPUTER SIMULATION

(A) The bidirectional and unidirectional spectral estimation of truncated sinusoidal signals are presented in Figs. 1-4.

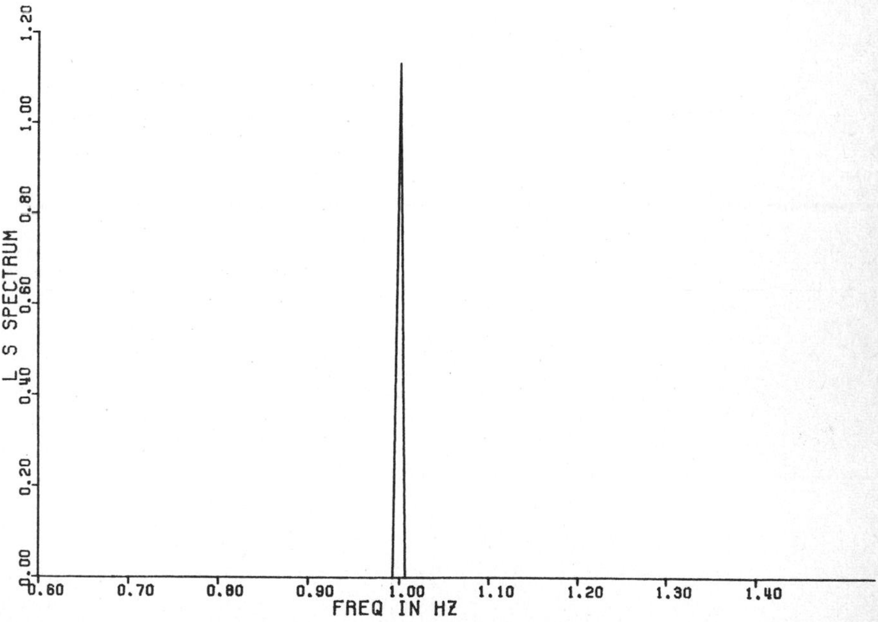

Fig. 1 Bidirectional LS Estimation

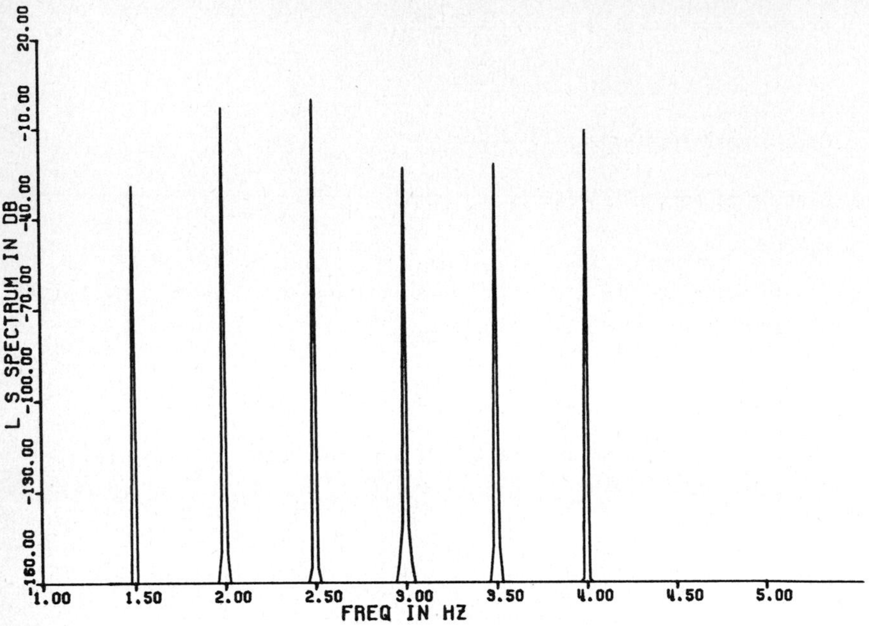

Fig. 2 Bidirectional LS estimation is independent of frequency

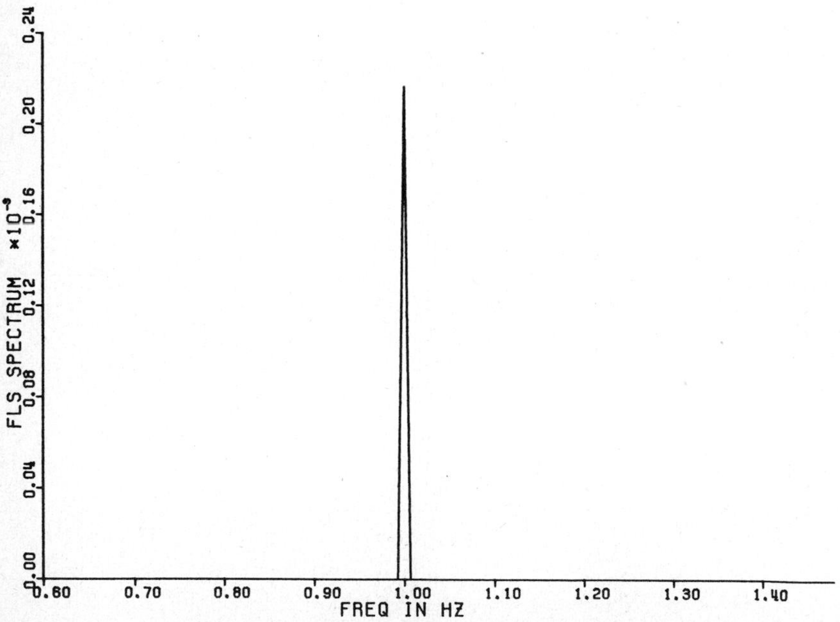

Fig. 3 Forward LS Estimation

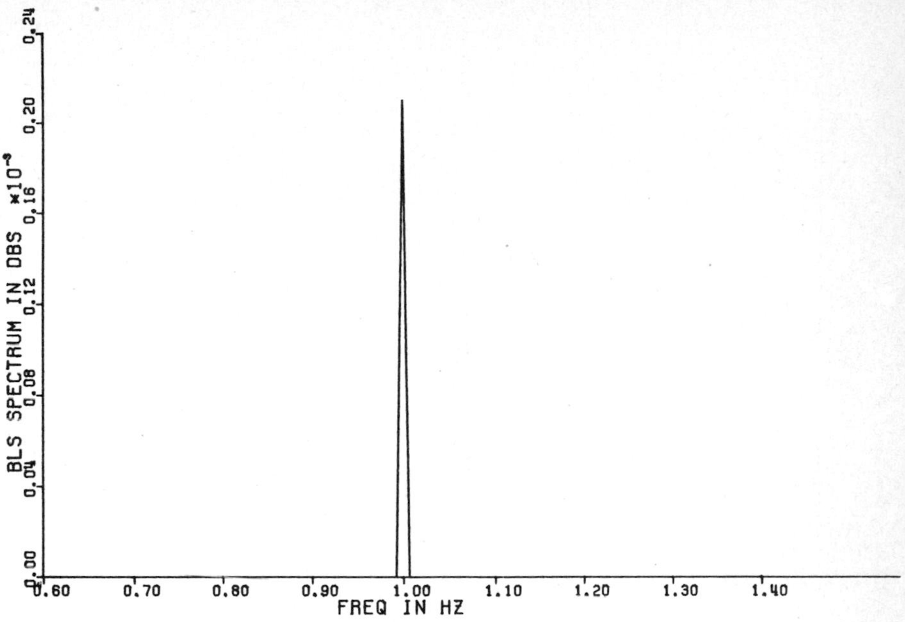

Fig. 4 Backward LS Estimation

The data points used in the simulation are 30 for a 1 Hz sinusoid and the sampling period is 0.05 sec. No frequency bias is observed. For comparison, the maximum entropy estimation error Δ_{MEM} and discrete Fourier transform estimation error Δ_{DFT} are given below [8],

$$\Delta\theta_{MEM} = \left(\frac{-1}{N-1}\right)\cos\left[2\phi+\theta(N-1)\right]\sin\,(N-1)\theta$$

and

$$\Delta\theta_{DFT} = \frac{6\cos\left[2\phi+\theta(n-1)\right]\cos N\theta}{N^2\sin\theta}$$

The result shown in Fig. 2 indicates that the LS estimation is independent of frequency.

(B) The least square method is independent of data length and the initial phase as shown in Figs. 5 and 6. The data point used in the simulated result of Fig. 6 varies from 15 to 40.

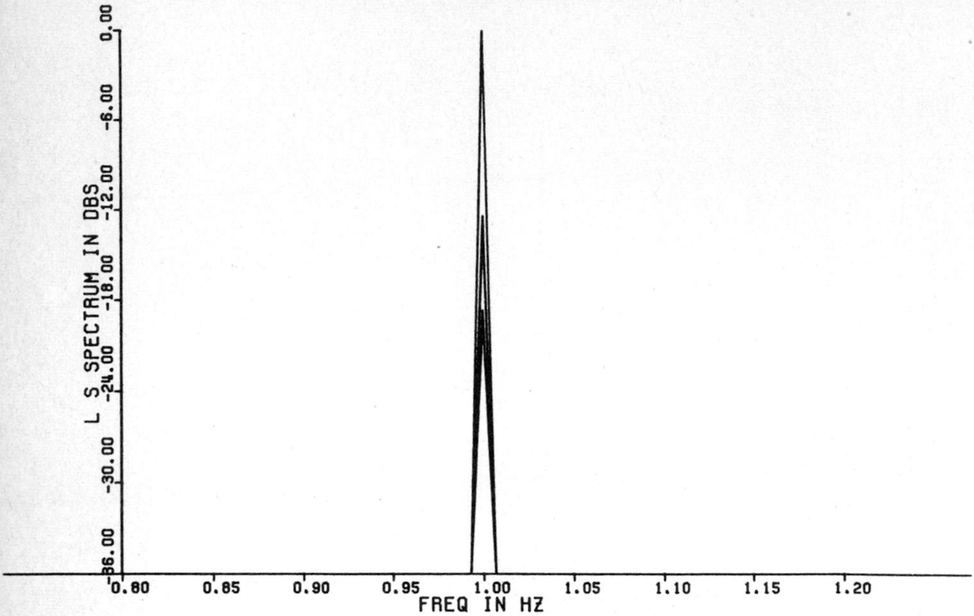

Fig. 5 Bidirectional LS estimation is independent of phase

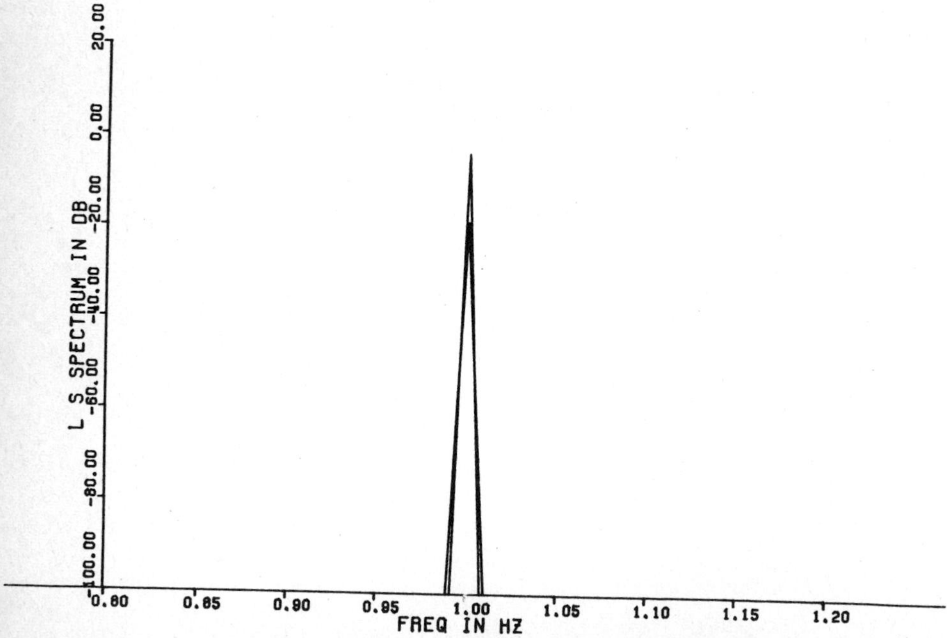

Fig. 6 Bidirectional LS estimation is independent of data length

(C) If the prediction filters of order other than two are used for sinusoids with additive noise, the spectrum becomes smoother and at the same time there is bias in frequency estimation. Fig. 7 shows results of fourth to seventh order filter estimation and this effect is inversely proportional to the filter order.

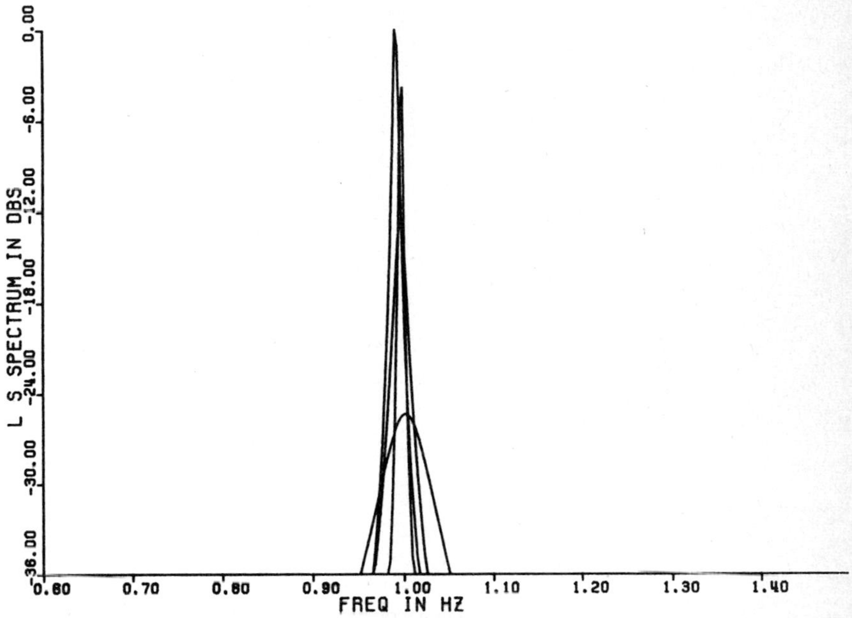

Fig. 7 Spectral estimation for 4th to 7th order filter

(D) If the data contain white noise, the spectrum becomes smooth and broad. Fig. 8 shows the noise effect for signal-to-noise ratio varying from 6 to 400. Interestingly, the noise effect on spectral broadening is much more severe on unidirectional estimation (Fig. 9) than on the bidirectional estimation (Fig. 8).

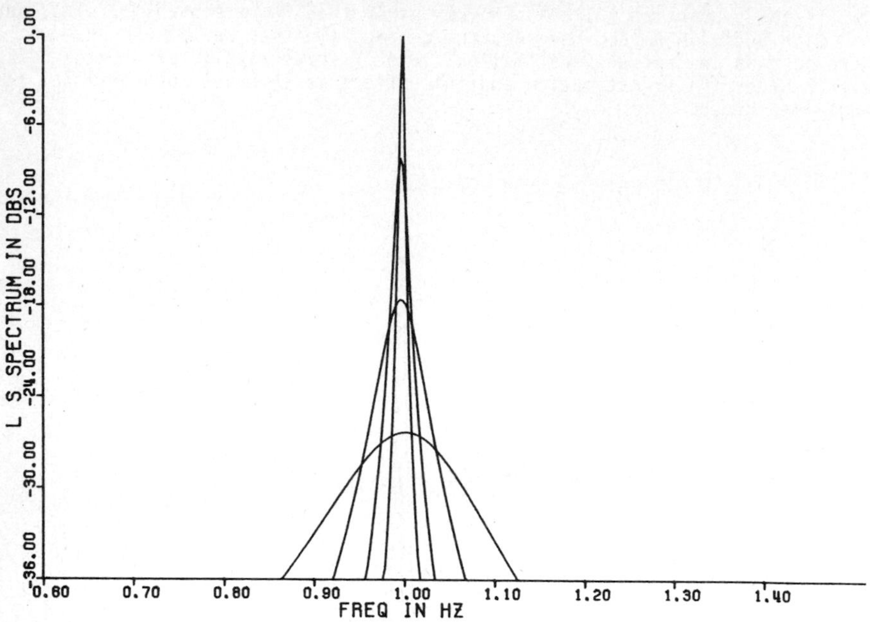

Fig. 8 Effect of white noise in bidirectional estimation

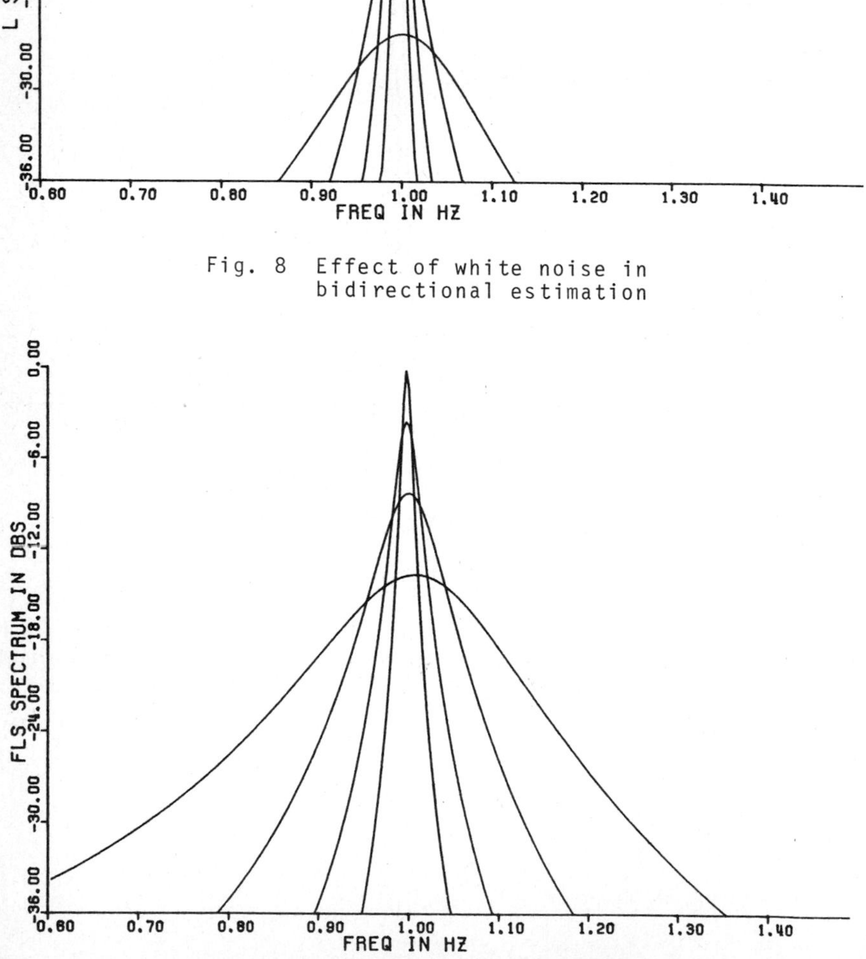

Fig. 9 Effect of white noise in forward estimation

SUMMARY

It is derived analytically that for a single truncated, noise-free, real sinusoidal signal, least square second order filter spectral estimation produces no frequency shifting. This result is independent of frequency, phase, data length, bidirectional or unidirectional estimation. For filter order other than two, simulation results indicate larger frequency bias for lower order filter prediction.

If white noise is added to the data, the result is broadening and reduction in magnitude of the spectrum, and the effect is more severe on unidirectional estimation than on bidirectional estimation.

REFERENCES

[1] Burg, J. P., Maximum entropy spectral analysis, Ph.D. Thesis, Stanford Univ., Stanford, California (1975).

[2] Ulrych, T. J., Maximum entropy power spectrum on truncated sinusoids, J. Geophysical Res., Vol. 77, No. 8 (March 1972) 1396-1400.

[3] Chen, W. Y. and Stegen, G. R., Experiments with maximum entropy power spectrum of truncated sinusoids, J. Geophysical Res., Vol. 79, No. 20 (July 1974) 3019-3022.

[4] Swingler, D. N., Frequency error in MEM Processing, IEEE Trans. Acoust. Speech and Signal Processing, Vol. ASSP-28, No. 2 (April 1980) 257-259.

[5] Fougere, P. F., A solution to the problem of spontaneous line splitting in maximum entropy power spectrum analysis, J. Geophysical Res., Vol. 82, No. 7 (March 1977) 1051-1054.

[6] Kay, Steven M., Maximum entropy spectral estimation using analytical signal, IEEE Trans. on Acoust. Speech and Signal Processing, Vol. ASSP-26, No. 5 (October 1978) 467-469.

[7] Ulrych, T. J. and Clayton, R. W., Time series modelling and maximum entropy, Phys. Earth Planet Interiors, Vol. 12 (1976) 188-200.

[8] Swingler, D.N., A comparison between Burg's maximum entropy method and a non-recursive technique for the spectral analysis of deterministic signal, J. of Geophysical Res., Vol. 84, No. B2 (February 1979) 679-685.

[9] Jackson, P. L., Truncation and phase relationship of sinusoids, J. of Geophysical Res., Vol. 72, No. 4 (1967) 1400-1403.

[10] Swingler, D. N., Burg's maximum entropy algorithm versus the discrete Fourier transform as a frequency estimator for truncated real sinusoids, J. of Geophysical Research, Vol. 85, No. B3 (March 1980) 1435-1438.

APPLIED TIME SERIES ANALYSIS
O.D. Anderson & M.R. Perryman (eds.)

CONSISTENT ESTIMATION OF THE TIME VARYING PARAMETER MODEL: A STATE SPACE REPRESENTATION

Tsunemasa Shiba
and
Hiroki Tsurumi

Department of Economics, Rutgers University
New Brunswick, N.J.
U.S.A.

In 1976, Cooley and Prescott proposed a particular estimation method for a time varying parameter model. In an earlier paper (Tsurumi and Shiba, 1981) we have proven that Cooley and Prescott's estimator is inconsistent and lacks robustness. In this paper we shall reformulate Cooley and Prescott's model in state space representation to prove strong consistency of the maximum likelihood estimator of the parameters in the model. The likelihood function is written with the Kalman filter innovation process. This enables us to use various asymptotic properties of the Kalman filter.

INTRODUCTION

Among various time varying parameter models the Cooley-Prescott (CP hereafter) model (CP (1976)) has been used in many applied works (for example Roll (1972), Laumas and Mehra (1976), Rausser and Laumas (1976), Cooley and DeCanio (1977), and Mullineaux (1980), among others). In the studies cited above the model is defined to be stable if the regression coefficients, β_t's, do not vary over time. Since the CP model reduces to the classical linear regression model if the measure of parameter variation, γ, is zero, the test of the model stability is made by conducting a hypothesis test: null hypothesis $\gamma=0$ against $\gamma\neq 0$. The test relies on the asymptotic distribution of the maximum likelihood estimator of γ derived in CP (1976).

In our earlier paper (Tsurumi and Shiba (1981)) we pointed out that CP's formulation of their model leads to an unusual likelihood function, and their estimator of parameters is not consistent and asymptotically normal. As Pagan (1980) points out one may also note that the CP model has difficulty in identifying fixed part of regression coefficient. An interesting question, then, will be whether or not a (strong) consistency of γ can be established if one formulates a likelihood function differently from the way CP did.

In this paper we shall establish a strong consistency for the maximum likelihood estimators (MLE) of γ and σ^2 based on a state space representation. We use some of the well-known asymptotic properties of the Kalman filter (Jazwinski (1970) and McGarty (1974) among others), and employ the line of argument first suggested by Jennrich (1969). In a nonlinear regression model Jennrich (1969) presented a general method of proving a consistency for the least squares estimator. Then Amemiya (1973) used Jennrich's method to establish a consistency for the MLE based on a truncated normal distribution. Frydman (1980) summarized and gave a clear proof of Jennrich's argument, and then he used it to show a strong consistency for the MLE of a general nonlinear model. We may note that, with this method, we do not need to compute derivatives of the likelihood function to show a consistency of the MLE.

The plan of this paper is as follows. In section II we present the model and assumptions. We prove a strong consistency of the MLE of parameters in the model in section III.

THE MODEL AND ASSUMPTIONS

CP's model is reproduced below.

$$(1)\qquad \begin{aligned} y_s &= x_s'\beta_s \\ \beta_s &= \beta_s^p + u_s \\ \beta_s^p &= \beta_{s-1}^p + v_s \end{aligned}$$

where u_s and v_s are independently distributed normal random variables with $N(0,(1-\gamma)\sigma^2\Sigma_u)$ and $N(0,\gamma\sigma^2\Sigma_v)$, respectively. y_s is a scalar, x_s is a kx1 vector, and β_s^p and β_s are kx1 time varying parameters. We can rewrite the above system as

$$(2)\qquad y_s = x_s'\beta_s^p + x_s'u_s$$

$$(3)\qquad \beta_s^p = \beta_{s-1}^p + v_s$$

where $x_s'u_s \sim N(0,(1-\gamma)\sigma^2 x_s'\Sigma_u x_s)$. In the CP model Σ_u and Σ_v are assumed to be known. And also one realization of β_s^p, β_{T+1}^p, is regarded as a fixed, nonrandom parameter.

In this paper we shall reformulate equations (2) and (3) slightly to conform to the conventional state space representation.

$$(4)\qquad y_s = x_s'\beta_s + x_s'u_s$$

$$(5)\qquad \beta_{s+1} = \beta_s + v_s$$

where $x_s'u_s \sim N(0,(1-\gamma)\sigma^2 x_s'\Sigma_u x_s)$ and $v_s \sim N(0,\gamma\sigma^2\Sigma_v)$ are independently distributed.

The main difference between our model and the CP model is in the next assumption, where we do not assume any β_s to be fixed.

<u>Assumption 1</u>: $E(\beta_o)=b_o$ and $E(\beta_o-b_o)(\beta_o-b_o)' = \bar{\Sigma}$ are known <u>a priori</u> for the stochastic process $\{\beta_s\}$.

Given the above assumption and assuming that Σ_u and Σ_v are known (as CP do) we can write down the Kalman filter equations as

$$(6)\qquad b_{s+1} = b_s + P_{s+1}(y_{s+1} - x_{s+1}'b_s)$$

$$(7) \qquad P_{s+1} = (\Pi_s+\gamma\sigma^2\Sigma_v)x_{s+1}(x'_{s+1}(\Pi_s+\gamma\sigma^2\Sigma_v + (1-\gamma)\sigma^2\Sigma_u)x_{s+1})^{-1}$$

$$(8) \qquad \Pi_{s+1} = (I - P_{s+1}x'_{s+1})(\Pi_s +\gamma\sigma^2\Sigma_v)$$

where P_s= filter gain vector of b_s, Π_s= $E(\beta_s-b_s)(\beta_s-b_s)'$ = error covariance matrix of b_s.

We shall use some well-known asymptotic properties of the Kalman filter. They are given in the following two remarks.

Remark 1: If a linear system is uniformly completely observable (UCO) and uniformly completely controllable (UCC) then its filter is uniformly asymptotically stable (UAS) and its error covariance matrix is uniformly bounded.

Remark 2: If an UAS filter has bounded inputs then the filter is uniformly bounded.

Definitions of UCO and UCC are found in Jazwinski (1970, pp.232-233) for a general linear system. Remark 1 is found in Jazwinski (1970, p.234 and p.240), and Remark 2 in Jazwinski (1970, p.240). The following Assumptions 2.1 and 2.2 give UCO and UCC conditions, respectively, for our model.

Assumption 2: 2.1 x'_s s are uniformly bounded.
2.2 Σ_v is a positive definite matrix.

That x'_ss are bounded is frequently assumed in the time varying parameter literature, e.g., CP (1976) and Pagan (1980) among others. A direct consequence of the above assumption is given below.

Remark 3: b_s and Π_s in (6) and (8) are uniformly bounded.

Finally we shall need the following two assumptions.

Assumption 3: The parameter space Ξ is compact, does not contain $\sigma^2 \leq 0$ region, and contains an open neighborhood of $\theta_o = (\gamma_o, \sigma_o^2)$, where θ_o = true parameter vector.

Assumption 4: The innovation process, e_s, has the smallest variance, h_s, at the true parameter vector, θ_o, i.e., $h_s(\theta_o) \leqq h_s(\theta)$ for $\theta \in \Xi$ where $h_s(\theta)$ = variance of the innovation $e_s(\theta)$ (see (9) below for the definition of $e_s(\theta)$).

STRONG CONSISTENCY OF MLE OF γ AND σ^2

We shall derive a likelihood function for the system (4) and (5) using the Kalman filter equations (6) to (8) and the innovation process, $e_s(\theta)$,

(9) $$e_s(\theta) = y_s - x_s' b_{s-1}(\theta)$$

where $b_s(\theta)$ = Kalman filter generated by parameter vector θ. Noting that e_s's are independently normally distributed with zero mean (Schweppe (1965) proves this) and variance $h_s(\theta)$

(10) $$\begin{aligned} h_s(\theta) &= E(y_s - x_s' b_{s-1}(\theta); \mathcal{T}^{s-1})^2 \\ &= x_s'(\Pi_{s-1} + \gamma\sigma^2\Sigma_v + (1-\gamma)\sigma^2\Sigma_u)x_s \end{aligned}$$

where $\mathcal{T}^{s-1}$= information up to (s-1), we obtain the following joint probability density function of $e_1(\theta), \ldots, e_T(\theta)$.

(11) $$\prod_{s=1}^{T} \frac{1}{\sqrt{2\pi h_s(\theta)}} \exp(-\frac{1}{2} \Sigma \frac{1}{h_s(\theta)} (y_s - x_s' b_{s-1}(\theta))^2) .$$

The likelihood function of the system (4) and (5) is given by

(12) $$L_T^*(\theta; y) = \S(y_1) \prod_{s=2}^{T} \S(y_s; \mathcal{T}^{s-1})$$

where $\S(y_1)$ = probability density function of y_1, and $\S(y_s; \mathcal{T}^{s-1})$ = conditional probability density function of y_s given information up to (s-1). This representation of the likelihood function has been useful in evaluating MLE for similar models to ours (Schweppe (1965) and Pagan (1980) among others). It is easy to see that the Jacobian of the transformation from $y_1, \ldots, y_T$ to $e_1(\theta), \ldots, e_T(\theta)$ is 1, hence the log likelihood function, L_T, in view of (11) and (12) becomes

(13) $$L_T(\theta) = -\frac{T}{2} \log(2\pi) - \frac{1}{2} \Sigma \log h_s(\theta) - \frac{1}{2} \Sigma\, e_s(\theta)^2/h_s(\theta).$$

Noting that the model expressed in terms of θ_o is given by $y_s = x_s' b_{s-1}(\theta_o) + e_s(\theta_o)$, the third term in (13) becomes

(14) $$\Sigma\, e_s(\theta)^2/h_s(\theta) = \Sigma\, (y_s - x_s' b_{s-1}(\theta))^2/h_s(\theta)$$

$$= \Sigma(x_s'(b_{s-1}(\theta_o) - b_{s-1}(\theta)) + e_s(\theta_o))^2/h_s(\theta)$$
$$= \Sigma q_s(\theta,\theta_o)^2 h_s(\theta) + \Sigma 2q_s(\theta,\theta_o)e_s(\theta_o) + \Sigma e_s(\theta_o)^2/h_s(\theta)$$

where $q_s(\theta,\theta_o) = x_s'(b_{s-1}(\theta_o) - b_{s-1}(\theta))/h_s(\theta)$.

We are now in a position to use the line of argument developed by Jennrich (1969) and later used by Amemiya (1973) and Frydman (1980) among others.

Lemma 1: If

(1) $Q_T(\theta) \to Q(\theta)$ almost everywhere (a.e.) uniformly in θ,

(2) $Q(\theta) \leqq Q(\theta_o)$ for all $\theta \in \Xi$

(3) $Q(\theta) = Q(\theta_o)$, then $\theta = \theta_o$

hold, then MLE of θ is strongly consistent, where $Q_T = \frac{1}{T} L_T(\theta)$ and $Q(\theta)$ = limiting value of $Q_T(\theta)$.

Proof: Frydman (1980, pp.854-855) gives a clear proof of this lemma that is suggested in Jennrich's (1969) paper.

We show that the above (1), (2), and (3) are satisfied in our model in Lemmas 2, 3, and 4, respectively.

Lemma 2: $Q_T(\theta) \to Q(\theta)$ a.e. uniformly in θ.

Proof: Combining (13) and (14), $Q_T(\theta)$ becomes

$$(15) \qquad Q_T(\theta) = -\frac{1}{2}\log(2\pi) - \frac{1}{2T}\Sigma \log h_s(\theta) - \frac{1}{2}\Big(\frac{1}{T}\Sigma q_s(\theta,\theta_o)^2 h_s(\theta) + \frac{2}{T}\Sigma q_s(\theta,\theta_o)e_s(\theta_o) + \frac{1}{T}\Sigma e_s(\theta_o)\Big)^2/h_s(\theta).$$

We shall show that each term in (15) converges to a constant under assumptions 1 to 4.

Let us take up the second term in (15). Jazwinski (1970, Lemma 7.1 and 7.2) shows $h_s(\theta)$ is bounded, hence we may define a limiting value $\tilde{h}(\theta)$ as

$$(16) \qquad \lim \frac{1}{T}\Sigma \log h_s(\theta) = \tilde{h}(\theta).$$

The third term also converges to a constant, say $\delta_b(\theta-\theta_o,\theta)$, since $b_s(\theta_o)-b_s(\theta)$ is bounded with small order of 1/T by Remark 3, and $h_s(\theta)$ and x_s's are also bounded.

(17) $\lim \frac{1}{T} \Sigma q_s(\theta,\theta_o)^2 h_s(\theta) = \delta_b(\theta-\theta_o,\theta)$

where $\delta_b(\theta-\theta_o,\theta)$ is a function of $\theta - \theta_o$ and θ because derivatives of $b_s(\cdot)$ are continuous, and $\delta_b(0,\theta)=0$. The fourth term is a weighted average of $e_s(\theta_o)$ where weights are $q_s(\theta,\theta_o) < \infty$ (by UCO and UCC), hence using the strong law of large numbers

(18) $\lim \frac{1}{T} \Sigma q_s(\theta,\theta_o) e_s(\theta_o) = 0$ a.e. uniformly.

The fifth term becomes

(19) $\frac{1}{T} \Sigma e_s(\theta_o)^2/h_s(\theta) = \frac{1}{T} \Sigma \eta_s(\theta,\theta_o)(e_s(\theta_o)^2/h_s(\theta_o))$

where $\eta_s(\theta,\theta_o) = h_s(\theta_o)/h_s(\theta)$. Note that (19) converges to $\tilde{\eta}(\theta,\theta_o)$ where $\tilde{\eta}(\theta,\theta_o) = \lim \frac{1}{T} \Sigma \eta_s(\theta,\theta_o)$ and $\eta(\theta_o,\theta_o) = 1$, because $e_s^2(\theta_o)/h_s(\theta_o) \sim \chi^2(1)$.

We have thus shown that $Q_T(\theta)$ converges to $Q(\theta)$ a.e. uniformly in $\theta \in \Xi$ where

(20) $Q(\theta) = -\frac{1}{2}\log(2\pi) - \frac{1}{2}(\tilde{h}(\theta) + \delta_b(\theta-\theta_o,\theta) + \tilde{\eta}(\theta,\theta_o))$.

<u>Lemma 3</u>: $Q(\theta) \leqq Q(\theta_o)$ for $\theta \in \Xi$.

<u>Proof</u>: From (20) we obtain

(21) $Q(\theta_o) = -\frac{1}{2}\log(2\pi) - \frac{1}{2}\{h(\theta_o) + \delta_b(0,\theta_o) + \tilde{\eta}(\theta_o,\theta_o)\}$.

In view of Assumption 4, we have $\tilde{\eta}(\theta_o,\theta_o) = 1 \leqq \tilde{\eta}(\theta,\theta_o)$. Hence each term in $\{\cdot\}$ above is smaller than that of (20), which proves the lemma.

<u>Lemma 4</u>: If $Q(\theta) = Q(\theta_o)$, then $\theta = \theta_o$.

<u>Proof</u>: Note that $Q(\theta_o) - Q(\theta) = 0$ implies

(22) $(\tilde{h}(\theta) - \tilde{h}(\theta_o)) + (\tilde{\eta}(\theta,\theta_o) - 1) + \delta_b(\theta-\theta_o,\theta) = 0$.

But the three terms in (22) are all greater than or equal to zero, hence the equality holds only when they are all zero's. The above discussion has indicated that this is true when $\theta = \theta_o$.

We have thus shown Lemma 1 to hold, hence proved the following.

Theorem: Under the Assumptions 1 to 4, the maximum likelihood estimator of θ, $\hat{\theta}_T$, is strongly consistent, where $\hat{\theta}_T$ is defined as $L_T(\hat{\theta}_T) \geqq L(\theta)$ for $\theta \in \Xi$.

CONCLUSION

We showed a strong consistency for the MLE of parameters of a time varying parameter model which is frequently used in econometrics. The state space representation of the likelihood function was useful since various asymptotic properties of the Kalman filter became applicable. In proving the consistency of the MLE we closely followed the line of argument developed by Jennrich (1969).

After the proof of a (strong) consistency of the MLE one may be interested in establishing asymptotic normality of the estimator. This requires calculation of derivatives of the likelihood function, and becomes rather involved. Thus, the proof of asymptotic normality of the estimator, which uses martingale central limit theorem by McLeish (1974), is presented in another paper by the authors (Shiba and Tsurumi (1980)).

REFERENCES

(1) Amemiya, T., Regression analysis when the dependent variable is truncated normal, Econometrica 41 (1973) 997-1016.

(2) Cooley, T. and DeCanio, J., Rational expectations in American agriculture, 1867-1914, Review of Economics and Statistics 59 (1977) 9-17.

(3) Cooley, T. and Prescott, E., Estimation in the presence of stochastic parameter variation, Econometrica 44 (1976) 167-184.

(4) Frydman, R., A proof of the consistency of maximum likelihood estimators of nonlinear regression models with autocorrelated errors, Econometrica 48 (1980) 853-860.

(5) Jazwinski, A.H., Stochastic processes and filtering theory (Academic Press, New York, 1970).

(6) Jennrich, R.I., Asymptotic properties of nonlinear least squares estimators, Annals of Mathematical Statistics 40 (1969) 633-643.

(7) Laumas, G.S. and Mehra, Y.P., The stability of the demand for money function: the evidence from quarterly data, Review of Economics and Statistics 58 (1976) 463-468.

(8) McGarty, T.P., Stochastic systems and state estimation (John Wiley and Sons, New York, 1974)

(9) McLeish, D.L., Dependent central limit theorems and invariance principles, Annals of Probability 2 (1974) 620-628.

(10) Mullineaux, D.J., Inflation expectation and money growth in the United States, American Economic Review 70 (1980) 149-161.

(11) Pagan, A., Some identification and estimation results for regression models with stochastically varying coefficients, Journal of Econometrics 13 (1980) 341-363.

(12) Rausser, G.C. and Laumas, P.S., The stability of the demand for money in Canada, Journal of Monetary Economics 2 (1978) 367- 380.

(13) Roll, R., Interest rates on monetary assets and commodity price index changes Journal of Finance 27 (1972) 241-277.

(14) Schweppe, F.C., Evaluation of likelihood functions for Gaussian signals, I.E.E.E. Transactions on Information Theory, 11 (1965) 61-70.

(15) Shiba, T. and Tsurumi, H., A state space representation of the time varying parameter model, Discussion Paper No.80-24 Bureau of Economic Research Rutgers University (January 1980)

(16) Tsurumi, H. and Shiba, T., On Cooley and Prescott's time varying parameter model, Economic Studies Quarterly 32 (1981) 176-180

APPLIED TIME SERIES ANALYSIS
O.D. Anderson & M.R. Perryman (eds.)

THE EFFECT OF MONETARY VARIABLES ON RELATIVE PRODUCER PRICES*

Houston H. Stokes

Professor of Economics
University of Illinois at Chicago Circle
Chicago, Illinois
U.S.A.

INTRODUCTION

Economic theory predicts that an anticipated change in the money supply in a country is reflected in equal and offsetting changes, in all prices.[1] If the change in the money supply is not fully anticipated, not all prices adjust at the same speed, resulting in relative price movements. Various explanation for these relative price movements have included differences in the supply and demand elasticities of the products and, in more recent work in the period 1957-1971 (Bordo [1980]), the effect of contract length on the speed of adjustment. Bordo's study assumed the line of causation went from monetary influence to price movements. His study estimated constrained Almon weights for separate models relating price movements by industry group to differences in the natural log of the money supply. The speed of adjustment was inferred by inspection of the weight patterns.

One problem with Bordo's study (acknowledged in his footnotes 37 and 38) is that income effects are ignored and feedback is not explicitly modeled. The present study uses a vector autoregressive moving average model[2] to explicitly estimate the dynamic relationship between velocity and (1) the ratio of farm to industrial prices, (2) the ratio of crude to finished prices, and (3) the ratio of durable to nondurable prices. The dynamic relationship between lagged velocity and the price ratio variables will indicate which of the two price series adjusts first to a monetary change, while the dynamic relationship between the lagged ratio of price and the velocity series will measure the feedback of monetary growth (via inflation) on income. After a brief look at the literature, alternative statistical procedures will be discussed. The vector autoregressive moving average model used is outlined and the results will be presented.

SURVEY OF THE LITERATURE

Bordo [1980], building on the pioneering work of Cairnes [1873], Graham [1930], Telser (1975), and others, argued that there are a number of reasons why some price series might adjust to monetary and other changes before other price series. The traditional model (Cairnes [1873]) emphasized the role of the elasticity of demand and supply in the respective markets, while an alternative view emphasized the role of long-term contracts in slowing adjustment of some price series (Wachter and Williamson [1978]). In the Cairnes model, there were two mechanisms by which an increase in the money supply (gold in his model) could increase prices. The direct effect resulted from a shift in the demand curve of initial recipients of the monetary increase, while the indirect effect arose from the wages in the initially affected industries rising and causing wages in other industries to increase. The direct effect will be larger the less the elasticity of supply of the good in question, while the indirect effect will be larger the less elastic the supply of labor. Extensions to the Cairnes model include the effect of inventories in slowing adjustment in prices, the effect of governmental intervention in agricultural marketing boards and farm supports dampening price adjustment, and the effect of commodity exchanges in increasing the speed of adjustment.

Using data from the German hyperinflation (1921-1923) Graham [1930] found that the prices of raw materials tended to adjust faster than the prices of finished goods. Following the Tesler [1975] model, one would expect that durables, because they are relatively more standardized and hence relatively more profitable to hold in inventory, would adjust at a slower rate than nondurables.

The contract length approach argues that "the length of fixed-price contract negotiated by an industry for given transactions costs and degree of risk aversion will be inversely related to the variance of the industries relative prices" (Bordo [1980], pp 1092-1093).[3] Its effects will be superimposed on the effects generated by the fundamental approach and, for the sectors under study, in the same direction. According to Bordo, the traditional approach predicts that a change the money supply will cause: a faster response for crude prices than manufactured (finished) product prices, a faster response for farm prices than industrial prices and a faster response for nondurable prices than durable prices. In the next section, various alternative statistical techniques to test these effects will be discussed.

APPROPRIATE STATISTICAL PROCEDURE

The Bordo [1980] approach involved estimating

$$(1-B)\ \ln P_{it} = A_i + \sum_{j=0}^{n} C_{jt}\ (1-B)\ \ln M_{t-j} + e_i \tag{1}$$

using the Almon (polynomial distributed lag -PDL) procedure and inspecting the weight patterns (C_{jt}) for the different sectors.[4] Although such a procedure involves little computational burden, there are a several problems with the PDL model. The greatest disadvantage of the PDL procedure is that it is not a test for causality in the sense of Granger [1969] because the effects of possible autocorrelation in dependent variable are not explicitly modeled. If both P and M are highly autocorrelated and are generated from the same process, equation (1) will indicate that they are related, while both the Box-Jenkins two-filter causality test and transfer function model will correctly reject causality.[5] If theory sugests that the direction causality runs from the monetary variable to the price variable, a transfer function model of the form

$$(1-B)\ \ln P=(W(B)/\delta(B)\ )\ (1-B)\ln M + (\theta(B)/\phi(B)\ )\ e_t \tag{2}$$

where W(B) and $\delta(B)$ are finite polynomials in the lag operator B for the input series and $\theta(B)$ and $\phi(B)$ are finite polynomials in the lag operator B for the noise model would be the preferred method of analysis. In a transfer function the effects of omitted variables are captured in the noise model ($\theta(B)/\phi(B)$).

Unlike the PDL technique, where there is no structured procedure to determine the form of the lag, Box and Jenkins [1970] have provided detailed estimation procedures to identify, estimate, and further verify the appropriate specification of the transfer function model. The transfer function model will not be the appropriate estimation technique if there is feedback from the left-hand side variable to the right-hand side variable. In the next section, it is argued that if Bordo's model is modified to take into account real factors, the resulting feedback will require that a vector autoregressive-moving average model be used in place of the transfer function model.[6]

If it is assumed that the economy is in a long-run equilibrium, where the increase in the money supply is equal to the growth rate, there is no reason for prices to move, unless there is a structural change in the economy that would be reflected in a change in velocity. Assume two cases: in the first, the rate of growth of the money supply increases and the rate of growth of real output remains unchanged. In the second, the rate of growth of the money supply increases and the rate of growth of production also increases. Bordo's model does not distinguish between these two cases, since real variables are not accounted for in his model although in the first case prices increase more than in the second. In this study, two important changes have been made in Bordo's model: (1) the monetary variable has been replaced by velocity, and (2) the price variable has been replaced by a price-ratio variable.[7] The use of velocity will allow a distinction to be made between the first case, where velocity would change, and the second case, where velocity would not change. Unless there was a shift in the equilibrium velocity, the revised model is able to distinguish between monetary changes that would be reflected in price changes and monetary changes that would not be reflected in price changes since there were offsetting changes in production.[8]

Unlike Bordo's model, in this study feedback from prices to velocity can be explicitly modeled. Assuming an initial equilibrium, the model can test whether an increase in the money supply might first be reflected in price movements, later in income effects.[9] The second major change in the Bordo model replaces the price of one sector with a ratio of prices. Since the question under study is the effect of monetary disturbances on relative prices, not the effect of monetary disturbances on price levels, the proposed change obviates the necessity of comparing two lag-weight streams (the C_{jt}'s from equation [1]).

We will now discuss how the above tests can be made with the vector autoregressive moving average (VARMA) procedure. Quenouille [1957] indicated that a dynamic simultaneous equation system could be written in the form

$$H(B)\ X_t = F(B)\ e_t \tag{3}$$

Where X_t' is a suitable differenced (to achieve stationarity) row vector of t th observation of the two-series, velocity and relative prices.[10] H(B) and F(B) are each 2-by-2 matrices whose elements are finite polynomials in the lag operator B. Equation [3] assumes $|H(B)|$ and $|F(B)|$ are on or outside the unit circle (the invertibility condition), and the expected value of the error vector e_t is a null vector ($Ee_t = 0$).[11] H(B) and F(B) can be written

$$H(B) = h_0 + h_1B + \ldots + h_pB^p \tag{4}$$

$$F(B) = f_0 + f_1B + \ldots + f_qB^q \tag{5}$$

where h_i is a 2-by-2 matrix of the ith order elements of the matrix H(B) and f_j is a 2-by-2 matrix of the jth order element of matrix F(B). In H(B) the maximum order polynomial is p, while in F(B) the maximum order polynomial is q. Without loss of generality, equation [3] can be expressed in a number of forms. In subsequent estimation, what Granger-Newbold [(1977), 223] have called Model A has been used. Model A involves assuming that the zero order elements in H(B) and F(B) are an identity matrix, i.e., $h_0 = I$ and $f_0 = I$. In this formulation, instantaneous causality, if present, will be seen as significant off-diagonal cross correlations between the error vectors, since Model A assumes the error covariance matrix $\sum_e$ is not a diagonal matrix.

Granger-Newbold [1977], pp 223-224) use matrix P to diagonalize the covariance matrix $\sum e$ to $\sum u$.

$$P\textstyle\sum_e P' = \sum_u \tag{6}$$

Matrix P is used to transform h_o (see equation [4]) to a lower triangular matrix h^*_0,

$$h^*_0 = Ph_0. \tag{7}$$

Off-diagonal elements in h^*_0 represent the instantaneous relationship between the variables in the vector Z_t. The advantage of estimating the model in the vector autoregressive form (equation [3]) is that it is a very parsimonious representation. The fewer the parameters that have to be estimated, the easier the computational burden. The difficulty with estimating the model in the VARMA form is that it is sometimes harder to interpret the coefficients. Assuming that $|H(B)|$ lies outside the unit circle, equation [3] can be written in vector moving average form (VMA) as

$$Z_t = (H(B))^{-1}F(B)\ e_t \tag{8}$$

which can be simplified to[12]

$$Z_t = M(B)\ e_t \tag{9}$$

Expanding equation [3] gives

$$\begin{bmatrix} H_{11}(B) & H_{12}(B) \\ H_{21}(B) & H_{22}(B) \end{bmatrix} (1-B) \begin{bmatrix} \text{Ln } V \\ \text{Ln}(P_i/P_j) \end{bmatrix} = \begin{bmatrix} F_{11}(B) & F_{12}(B) \\ F_{21}(B) & F_{22}(B) \end{bmatrix} \begin{bmatrix} e_{1t} \\ e_{2t} \end{bmatrix}, \tag{10}$$

and expanding equation [9] results in

$$(1-B) \begin{bmatrix} \text{Ln } V \\ \text{Ln}(P_i/P_j) \end{bmatrix} = \begin{bmatrix} M_{11}(B) & M_{12}(B) \\ M_{21}(B) & M_{22}(B) \end{bmatrix} \begin{bmatrix} e_{1t} \\ e_{2t} \end{bmatrix}. \tag{11}$$

If elements $H_{12}(B)$ and $F_{12}(B)$ are equal to zero, equation [10] can be reduced to a transfer function model of the form of equation [2], and an ARIMA model on LnV since there is no feedback in the system. If only the diagonal elements in H(B) and F(B) are significant, there is no relationship between the two series. Each series is best predicted by lagged values of itself, a VARIMA model, and no further information can be obtained by adding the other series. Bordo's PDL model can be written in terms of equation [10] (assuming V is replaced by M and P_i/P_j is replaced by P_i) if (1) all elements in F(B) are assumed to equal zero except $F_{22}(B)$, which is assumed to equal one, and (2) all elements in H(B) are assumed equal to zero, except $H_{22}(B)$, which is assumed to equal 1, and (3) $H_{21}(B)$, is assumed to equal the PDL weights.[13]

Equation [11] provides another look at the process that generates the two series LnV and Ln (P_i/P_j). If only the diagonal elements in M(B) were significant, the interpretation is that the series are independent. In this case each series is best written as a moving average model on past error processes of that series alone. If off-diagonal elements as well as on-diagonal elements are significantly different from zero,[14] the series in question is best represented as a moving average process on errors of the other series as well as past errors of the series in question.

THE RESULTS

Equation [10] has been estimated assuming P_i/P_j is farm prices/industrial prices and the results are reported in Table 1. The vector moving average form of the model (equation [11] is also given. Because the off-diagonal elements of the residual correlation matrix are small, there is little evidence of instantaneous causality, and as a consequence the off-diagonal elements of the P matrix are small (see equation [6] and [7]). Off-diagonal elements in H(B) and F(B) are significant, indicating a dynamic relationship between the two series. If only $H_{21}(B)$ was significant, it could be inferred that a positive causal relationship went from velocity to price with a 10-month lag.[15] However, terms in $F_{12}(B)$ and $F_{21}(B)$ are significant. To interpret the results, the vector moving average matrix M(B) must be inspected. M(B) was calculated as $(H(B))^{-1}F(B)$, assuming the

TABLE 1

Coefficients of Vector Autoregressive - Moving Average Model of the Form $H(B)(1-B)X_t=F(B)e_t$ for X_1 = Velocity, X_2 = Farm/Industrial

Elements of Vector Autoregressive Matrix H(B)

$H_{11} = 1 -.3099B -.1675B^3 -.0851B^9 +.0887B^{10}$
(6.66) (3.60) (1.70) (-1.80)

$H_{12} = 0$

$H_{21} = 0 -.1650B^{10}$
(2.70)

$H_{22} = 1 -.1590B^2 +.1795B^4 -.1056B^6$
(3.26) (-3.71) (2.18)

Note: Series 1 is velocity

Elements of Vector Moving Average Matrix F(B)

$F_{11} = 1 -.0751B^5 +.0936B^{12} -.2391B^{14} +.0781B^{18}$
(1.51) (-1.84) (4.78) (-1.55)

$F_{12} = 0 +.0858B^8 +.0580B^9 -.0692B^{16} -.1415B^{17}$
(-2.46) (-1.66) (1.98) (4.01)

$F_{21} = 0 -.1651B^8$
(2.33)

$F_{22} = 1 +.14435B^{11} -.1433B^{13} -.1097B^{16}$
(-2.93) (2.92) (2.22)

Residual Covariance Matrix

.0001423	
.0000025	.0002914

Residual Correlation Matrix

1.00000	
.01175	1.00000

Transformation to diagonalize Covariance Matrix = P

1.00000	
-.01681	1.00000

Diagonal elements of transformed Covariance Matrix

.00014233 .00029139

TABLE 1 (Cont'd)

Vector Moving Average Form of Model Matrix M(B)				
lag	M_{11}	M_{12}	M_{21}	M_{22}
1	1.31	.00	.00	1.00
2	1.41	.00	.00	1.16
3	1.60	.00	.00	1.16
4	1.72	.00	.00	1.00
5	1.69	.00	.00	1.00
6	1.72	.00	.00	1.06
7	1.74	.00	.00	1.06
8	1.75	.09	-.17	1.11
9	1.84	.17	-.17	1.11
10	1.81	.20	-.03	1.09
11	1.78	.22	.02	1.24
12	1.89	.24	.09	1.23
13	1.91	.25	.13	1.11
14	1.66	.26	.12	1.12
15	1.61	.26	.11	1.07
16	1.59	.20	.12	.96
17	1.55	.05	.12	.99

For ease of notation, $H_{ij}(B)$, $F_{ij}(B)$ and $M_{ij}(B)$ are written without (B).

maximum lag was 17 periods. The elements $M_{11}(B)$ and $M_{22}(B)$ control for the effect of past errors on the series. The positive elements in $M_{12}(B)$, starting at period 8 and building to a maximum in period 14, indicate that shocks lagged 14 periods in the process generating the relative price series are negatively related to the velocity series. The negative elements in $M_{21}(B)$ at period 8 and 9 of -.17 indicate that lagged shocks in the process generating the velocity series are positively related to the relative price series.

The results are consistent with the following mechanism. If there is an incraese in the level of economic activity, without an increase in the money supply, the ratio of farm prices to industrial prices will rise with an 8 to 9 period lag. This is consistent with the theory discussed earlier. This increase, in turn, will cause a negative change in velocity (via industrial production falling) starting in 9 periods and peaking in period 14. This feedback would not be observed if the money supply had been increased as industrial production went up. It appears that we have witnessed the self-limitation of a boom.

Table 2, which relates velocity to the ratio of crude to finished prices, and Table 3, which relates velocity to the ratio of durable to nondurable prices, illustrate a similar picture.[16] In Table 2, there appears to be less feedback than was observed in Table 1. The negative sign at the 17th period is hard to interpret and has been ignored.[17] In Table 2 there is more effect from shocks in the velocity series on the relative price series than was seen in Table 1. If there were un-expected increases in production relative to the money supply (as would be the case in the start of a business boom or a credit crunch), there would be an increase in the ratio of crude to finished goods prices after 8 periods.

Table 3, in contrast to Table 2 but similar to the pattern observed in Table 1, shows relatively more feedback from relative prices to velocity than vice versa. The signs of the effects in both directions are similar to those in the previous two tables.

TABLE 2

Coefficients of Vector Autoregressive-Moving Average Model of the Form
$H(B)(1-B)X_t=F(B)e_t$ X_1 = Velocity, X_2 = Crude/Finished

Elements of Vector Autoregressive Matrix H(B)

Constant Series #1 -.00097749
(-2.14)

$H_{11} = 1-.2783B - .1858B^3 + .1258B^5 - .1027B^9 + .1289B^{10}$
(5.99) (4.06) (2.79) (2.10) (-2.65)

$H_{12} = 0-.0606B$
(2.07)

$H_{21} = 0$

$H_{22} = 1 + .1706B^4 - .1407B^{11}$
(-3.56) (2.97)

Note: Series 1 is velocity

Elements of Vector Moving Average Matrix F(B)

$F_{11} = 1-.2169B^{14}$
(4.26)

$F_{12} = 0-.0699B^{15} - .0868B^{16} - .0921B^{17}$
(2.30) (2.89) (3.03)

$F_{21} = 0-.2266B^8$
(2.70)

$F_{22} = 1 - .12795B^{13} - .1518B^{16}$
(2.61) (3.10)

Residual Covariance Matrix		Residual Correlation Matrix	
.0001438		1.00000	
.0000131	.0004111	.05401	1.00000

Transformation to diagonalize Covariance Matrix = P

1.00000	
-.09134	1.00000

Diagonal elements of transformed Covariance Matrix

.000143758 .000409926

Table 2 (Cont'd)

Vector Moving Average Form of Model Matrix M(B)				
lag	M_{11}	M_{12}	M_{21}	M_{22}
1	1.28	.06	.00	1.00
	.36	.08	.00	1.00
3	1.56	.08	.00	1.00
4	1.67	.09	.00	.83
5	1.59	.09	.00	.83
6	1.57	.08	.00	.83
7	1.58	.08	.00	.83
8	1.54	.08	-.23	.86
9	1.60	.08	-.23	.86
10	1.53	.08	-.23	.86
11	1.47	.08	-.23	1.00
12	1.48	.08	-.19	.99
13	1.46	.09	-.19	.87
14	1.20	.08	-.19	.87
15	1.15	.01	-.19	.82
16	1.14	-.10	-.19	.67
17	1.08	-.24	-.19	.69

SUMMARY

Bordo's study (1980) found a positive relationship between monetary increases and prices of different sectors of the economy, with certain price series adjusting faster than others. This study differed from the prior one in a number of important details. The first difference of the natural log of velocity was used in place of the natural log of the money supply to partially control for real effects. In place of one price series at a time, first differences of the natural log of a ratio of price series was used. In order to allow for the possiblity of feedback, the VARMA and VMA procedures were used instead of the more restrictive PDL technique.

Velocity was found to be significantly related to relative prices. An unexpected increase in industrial production, relative to the money supply (i.e., an increase in velocity) was found to cause farm prices to rise relative to industrial prices, crude product prices to rise relative to finished product prices, and nondurable good prices to rise relative to durable good prices. Conversely, if there are unexpected increases in the money supply without an increase in industrial production, industrial prices will rise relative to farm prices, finished product prices will rise relative to crude product prices, and durable good prices will rise relative to nondurable good prices.

These effects are in the reverse direction to those predicted by Bordo. This can be explained if we note that Bordo did not separate out changes in the money supply that were associated with a greater increase in industrial production from changes in the money supply that were associated with a lesser increase in indusinddustrial production.

Negative feedback was found from relative prices to velocity. Making the usual assumption that the money supply is exogenous, this is consistent with inflactionary forces slowing the growth of industrial production. Since other variables are not in the model, a definitive determination cannot be made at this time.

TABLE 3

Coefficients of Vector Autoregressive-Moving Average Model of the Form
$H(B)(1-B)X_t=F(B)e_t$ X_1 = Velocity, X_2 = Durable/Nondurable

Elements of Vector Autoregressive Matrix H(B)

$H_{11} = 1-.3223B -.1666B^3 +.0852B^9 +.0756B^{10}$
(6.78) (3.52) (1.69) (-1.49)

$H_{12} = 0$

$H_{21} = 0+.0627B^{10}$
(-2.46)

$H_{22} = 1-.1343B -.2027B^2 +.1344B^4 -.1086B^5 -.2127B^6 +.1064B^7 -.1152B^{12}$
(2.76) (4.23) (-2.78) (2.25) (4.31) (-2.17) (2.46)

Elements of Vector Moving Average Matrix F(B)

$F_{11} = 1-.0924B^5 -.0975B^8 -.1896B^{14} +.0814B^{18}$
(1.81) (1.91) (3.69) (-.1.59)

$F_{12} = 0-.1019B -.1639B^9 +.2407B^{16} +.1757B^{17}$
(1.18) (1.84) (-2.77) (-1.96)

$F_{21} = 0$

$F_{22} = 1$

Residual Covariance Matrix		Residual Correlation Matrix	
.000148011		1.00000	
.000001597	.00005003	.01855	1.0000

Transformation to diagonalize Covariance Matrix = P

1.0000000	
-.0107864	1.00000000

Diagonal elements of transformed Covariance Matrix

.000148011 .00005000966

TABLE 3 (Cont'd)

Vector Moving Average Form of Model Matrix M(B)				
lag	M_{11}	M_{12}	M_{21}	M_{22}
1	1.32	-.10	.00	1.13
2	1.43	-.13	.00	1.36
3	1.63	-.15	.00	1.41
4	1.74	-.17	.00	1.33
5	1.71	-.18	.00	1.42
6	1.73	-.18	.00	1.61
7	1.76	-.19	.00	1.60
8	1.66	-.19	.00	1.68
9	1.72	-.36	.00	1.66
10	1.69	-.42	-.06	1.63
11	1.65	-.44	-.09	1.68
12	1.66	-.47	-.11	1.82
13	1.65	-.49	-.14	1.85
14	1.44	-.50	-.14	1.93
15	1.38	-.51	-.15	1.93
16	1.36	-.27	-.17	1.93
17	1.31	-.02	-.17	1.96

APPENDIX

Data Sources

All data have been taken from the August 1980 version of the NBER/City Bank Data Bank for the period 1947-1979.

Velocity is defined as industrial production: total index (1967 = 100 SA) divided by Money Stock (M-2).

The price ratio series all use various components of the producer price index PPI (formerly the wholesale price index) 1967 = 100.

Farm/industrial is defined as PPI farm products, processed foods, and feeds divided by PPI industrial commodities.

Crude/finished is defined as PPI crude materials for further processing divided by PPI finished consumer goods (except food).

Durable/nondurable is defined as PPI durable manufactured goods divided by nondurable manufactured goods.

FOOTNOTES

*Computer time for this study was provided by the University of Illinois at Chicago Circle. Diana Stokes provided editorial assistance. I am responsible for any errors or omissions.

[1]A classic example is the 1958 decision of the French government to exchange 100 old francs for one new franc. When the decision was announced, all prices fell to one hundredth of their former value and were quoted in new francs. There

FOOTNOTES (cont'd)

was no change in relative prices and apart from a few ignorant people who were concerned over the reduction in their bank balance now quoted in the new franc, no real effects on the economy were observed.

[2]To be defined later.

[3]For a more complete discussion and mathematical derivation, see Bordo [1980], pp. 1093-1096.

[4]In Bordo's model, B is the differencing operator defined such that $B^k X_t = X_{t-k}$, P_{it} is the price level for the ith sector in period t, A_i is the intercept for the ith sector, M_t is the level M_2 in period t, and e_i is the error vector for the ith sector.

[5]The literature on causality testing, using the Granger [1969] definition, is too large to summarize here. According to Granger [1969], a series X_t causes a series Y_t if and only if a model predicting Y_t as function of past values of Y is inferior, (in the sense of minimizing the sum of squared errors) to a model estimating Y_t as a function of lagged vaues of X_t. While the two-filter causality testing method is downwardly biased, if there is no feedback, the transfer function approach is the appropriate procedure. For a more complete discussion, see Pierce and Haugh [1977] and Stokes and Neuburger [1979]. Since we cannot reject feedback, we cannot use the transfer function procedure in this study.

[6]It should be noted that the transfer function is a special case of the vector auto-regressive moving average model. (See Tiao-Box-Grupe-Hudak-Bell-Chang [1980].)

[7]Monthly data (1947-1979) for this study has been taken from the NBER/City Bank Data Bank. For further details, see Appendix. All calculations have been made using the B34S Program (Stokes [1980]), which contains a modified version of the Wisconsin Multiple Time Series program (WMTS-1) (See Tiao-Box-Grupe-Hudak-Bell-Chang [1980]) as an option. All data series have not been seasonally adjusted, with the exception of the industrial production series, which is not readily available in nonseasonally adjusted form.

[8]In the analysis that follows, it is assumed that the effect of a change in velocity on relative prices remains stable. While in theory this assumption is well founded, there are some possible institutional problems that are worth mentioning. In late 1980 there were major changes in the composition of the money supply as demand deposits became NOW accounts and were counted as part of broader definitions in the money supply. If this period were studied it would be important to be able to distinguish carefully between shifts in measured velocity that really reflect structural changes in the economy and changes in velocity that reflect shifting composition of money holdings. The proper procedure would be to define velocity using a very broad definition of money. However, if this is done an implicit assumption is made that the price effects of one dollar of M-1 is the same as price effects of one dollar of one of the components of M-5 (not found in M-1) Another research task would be to determine whether prices or relative prices are differentially related to velocity measured using alternative money definitions.

[9]If feedback were found, it would imply that either money was not neutral in its effect on real magnitudes (something that rational expectations school assumes could happen only if the money supply increase was unexpected) or the shock to the economy came from the fiscal side in the form of a change in production. A fiscal shock would shift the demand curve for the product, be reflected in the price of the product (or relative price of the product) and possibly later be reflected in

FOOTNOTES (cont'd)

will not distinguish between price shocks that originated from prior monetary disturbances or price shocks that originated from fiscal or other unspecified shocks to the economy.

[10]In the empirical section, three models are estimated, each with a different relative price series. The discussion of the VARMA procedure is written as if only one model was estimated.

[11]For further discussion of the importance of these assumptions, see Tiao-Box-Grupe-Hudak-Bell-Chang [1980] and Granger-Newbold [1977]. If $|H(B)|$ lies outside the unit circle, equation [3] can be written in vector moving average form, while if $|F(B)|$ lies outside the unit circle, equation [3] can be written in vector autoregressive form.

[12]For a discussion of the interpretation of the VMA form of the model, see Sims [1980]. Another way to get equation [9] is to estimate the model as a vector autoregressive process of the form

$$A(B)\ Z_t = e_t$$

using OLS and, provided the $|A(B)|$ lies outside the unit circle, transform the model to a VMA model by the transformation $M(B) = (A(B))^{-1}$. The latter procedure was used by Sims [1980], who estimated A(B) without constraining insignificant parameters to zero or testing for significance of any parameters. In contrast, in the estimation procedure used in this study insignificant terms in H(B) and F(B) have been constrained to zero and significance tests have been used for the remaining parameters. Cross correlations of the error terms have been inspected to test for the adequacy of the form of the model. By estimating fewer parameters, multicollinearity among parameters is reduced. Multicollinearity of the estimated parameters has been checked via inspection of the variance-covariance matrix of the parameters.

[13]The PDL model will not allow testing for Granger causality because of the assumption that all parameters of an order greater than 1 in $H_{22}(B)$ are equal to zero.

[14]If equation (11) is estimated by first estimating equation (10) and then forming M(B) as $(H(B))^{-1}F(B)$ (see equation (8)) the calculation of standard errors on the terms in each element in M(B) is not possible.

[15]The terms in H(B) are reported with a negative sign because all the variables are on the left-hand side. To interpret the results, one must shift the sign.

[16]In Tables 1 and 2, the numerator of the relative price series is postulated to adjust first. For this reason, the off diagonal elements in M(B) in Tables 1 and 2 are the same sign, while the off-diagonal elements in Table 3 are reversed in sign.

[17]I am troubled by the significant constant for series 1 in Table 2 in view of Granger [1981], who noted that the spectrum of both series in a model should be consistent. If two constants are entered in the equation, neither are significant (the t statistic for the constant series 1 falls to -1.69). When constants were tried in all three tables, the pattern of the lag structure remained the same.

BIBLIOGRAPHY

1. Bordo, Michael D., "John E. Cairnes on the Effects of the Austrian Gold Discoveries, 1851-73: An Early Application of the Methodology of Positive Economics", History of Political Economics No. 7 (Fall 1975), pp. 337-359.

2. Bordo, Michael D., "The Effects of Monetary Change on Relative Commodity Prices and the Role of Long-Term Contracts", The Journal of Political Economy Vol 88 Number 6 (December 1980) pp. 1088-1109.

3. Box, G. E. P. and G. M. Jenkins, Time Series Analysis, Forecasting and Control, San Francisco: Holden-Day, 1970.

4. Cairnes, John, Essays on Political Economy: Theoretical and Applied, 1873 Reprint., New York: Kelly, 1965.

5. Graham, Frank D., Exchange, Prices, and Production in Hyperinflation: Germany 1920-1923. Princeton, N.J.: Princeton University Press, 1930.

6. Granger, C. W. J., "Investigating Causal Relationships by Econometric and Cross Spectral Methods," Econometrica 37 (May 1969).

7. Granger, C. W. J., and P. Newbold, Forecasting Economic Time Series (New York: Academic Press, 1977).

8. Granger, C. W. J. "Some Properties of Time Series Data and Their Use in Econometric Model Specification" Journal of Econometrics Vol. 16 (1981) pp. 121-130.

9. Pierce, David and Larry Haugh, "Casuality in Temporal Systems: Characterizations and a Survey," Journal of Econometrics Vol. 5 (1977), pp. 265-293.

10. Quenouille, M. H., The Analysis of Multiple Time Series, 1957, Griffin: London.

11. Sims, Christopher, "Macroeconomics and Reality", Econometrica Vo. 48, No. 1 Janury 1980 pp. 1-48.

12. Stokes, Houston H., and Neuburger, Hugh "The Effect of Monetary Changes on Interest Rates: A Box-Jenkins Approach," The Review of Economics and Statistics Vol. LXI, No. 4 (November 1979) pp. 534-548.

13. Stokes, Houston H., "The B34S Data Analysis Program: A Short Writeup," Report FY 77-1 (Revised 14 July 1980) College of Business Administration Working Paper Series. University of Illinois at Chicago Circle.

14. Telser, Lester G., "When are Prices More Stable Than Purchase Rates?" In The Competitive Economy: Selected Readings, edited by Yale Brozen. Morristown, N.J.: General Learning Press, 1975.

15. Tiao, George, G. E. P. Box, M. R. Grupe, G. B. Hudak, W. R. Bell and I Chang, "The Wisconsin Multiple Time Series (WMTS-1) Program: A Preliminary Guide," University of Wisconsin, Department of Statistics, 1980, pp. 1-75.

16. Wachter, Michael and O. Williamson, "Obligational Markets and the Mechanics of Inflation." Bell Journal of Economics Vol. 9 (Autumn 1978) pp. 549-571.

APPLIED TIME SERIES ANALYSIS
O.D. Anderson & M.R. Perryman (eds.)

MAXIMUM LIKELIHOOD ESTIMATION OF STOCHASTIC LINEAR DIFFERENCE EQUATIONS WITH AUTOREGRESSIVE MOVING AVERAGE ERRORS AND WITH MISSING OBSERVATIONS AND OBSERVATIONAL ERRORS

Suan-Boon Tan

Department of Mathematics and Statistics
University of Nebraska
Lincoln, Nebraska
U.S.A.

This paper extends the work of Jones (1980) to stochastic difference equations with autoregressive moving average errors which are shown to have a minimal Markovian representation that is suitable for the application of Kalman filtering techniques. This state space approach involves matrices and vectors with dimension equal to Max(p+r,q+1) rather than those with dimension equal to the number of observations.

1. INTRODUCTION

Recently the use of state space methodology has been quite extensive. Mehra (1974) and Caines and Rissanen (1974) used Kalman recursion estimation to calculate the exact likelihood of an autoregressive moving average (ARMA) process. Harvey and Phillips (1979) extended the idea to regression model with ARMA disturbances, and Jones (1980) used Akaike's representation and Kalman recursive technique to obtain the exact likelihood function of a stationary ARMA process with both the observational error and missing observations. In this paper we extend the work of Jones (1980) to stochastic linear difference equations with ARMA errors and with the observational errors and missing observations. Maximum likelihood estimation of stochastic linear difference equations with ARMA errors has previously been considered by Riensel (1976), whose method of estimation is closely related to that of Anderson (1975).

2. MODEL AND MINIMAL MARKOVIAN REPRESENTATION

Consider the model

(1)
$$y_t - \sum_{i=1}^{r} \alpha_i y_{t-i} = \sum_{i=1}^{k} \beta_i x_{ti} + u_t$$

$$u_t - \sum_{i=1}^{p} \phi_i u_{t-i} = \epsilon_t + \sum_{i=1}^{q} \gamma_i \epsilon_{t-i}$$

$$= \sum_{i=0}^{q} \gamma_i \epsilon_{t-i} \qquad (t=\ldots,-1,0,1,\ldots).$$

where $\gamma_0 = 1$.

We shall assume throughout that the following assumptions hold:

(i) The ϵ_t's are i.i.d. $n(0,\sigma^2)$

(ii) all roots of $A(z) = 0$, $\phi(z) = 0$ and $\Gamma(z) = 0$ are greater than 1 in absolute value and there are no roots common to the three equations where

$$A(z) = 1 - \sum_{i=1}^{r} \alpha_i z^i$$

$$\phi(z) = 1 - \sum_{i=1}^{p} \phi_i z^i$$

$$\text{and} \quad \Gamma(z) = 1 + \sum_{i=1}^{q} \gamma_i z^i$$

(iii) $x_{ti}(i=1,\ldots,k)$ are known for all t.

Model (1) includes many models that have been considered in the literature. When $\beta_i = 0$, $i=1,\ldots,k$, and $\phi_i = 0$, $i=1,\ldots,p$, the process is called an ARMA process. When only $\phi_i = 0$, $i=1,\ldots,p$, the process is called an ARMAX (autoregressive moving average with exogenous variables) process. When $\alpha_i = 0$, $i=1,\ldots,r$, we have the process considered by Harvey and Phillips (1979)[1].

The work by Jones depends heavily on Akaike minimal Markovian representation for an ARMA process (see Akaike (1974)). In this section we show that such a minimal representation can also be obtained for model (1). Let $R(y_1,y_2,\ldots)$ denote the closure in the sense of mean square of the linear space of finite linear combination of $y_1,y_2,\ldots$, and let $y(t+j|t)$ denote the projection of y_{t+j} onto $R(y_t,y_{t-1},\ldots)$. Obviously $y(t|t) = y(t)$. The one-step prediction is

$$y(t+1|t) = \sum_{i=1}^{r} \alpha_i y_{t+1-i} + \sum_{i=1}^{k} \beta_i x_{t+i,i} + u(t+1|t) \tag{3}$$

where $u(t+j|t)$ is similarly defined as $y(t+j|t)$. Equation (3) follows because $R(y_t,y_{t-1},\ldots) = R(u_t,u_{t-1},\ldots)$ which is a consequence of the fact that $A(z) = 0$ has roots outside of unit disc. The expression for $u(t+1|t)$ is (see Jones (1980), (2.6))

$$u(t+1|t) = \sum_{i=1}^{p} \phi_i u_{t+1-i} + \sum_{i=1}^{q} \gamma_i \epsilon_{t+1-i} \tag{4}$$

The j step prediction is given by

(5) $$y(t+j|t) = \sum_{i=j}^{r} \alpha_i y_{t+j-i} + \sum_{i=1}^{j-1} \alpha_i y(t+j-i|t) + \sum_{i=1}^{k} \beta_i x_{t+j,i} + u(t+j|t) \text{ with}$$

(6) $$u(t+j|t) = \sum_{i=j}^{p} \phi_i u_{t+j-i} + \sum_{i=1}^{j-1} \phi_i u(t+j-i|t) + \sum_{i=j}^{q} \gamma_i \epsilon_{t+j-1}$$

where the various summations are eliminated if the indices are not in the proper range. Similarly,

(7) $$y(t+j|t+1) = \sum_{i=j}^{r} \alpha_i y_{t+j-i} + \sum_{i=1}^{j-1} \alpha_i y(t+j-i|t+1) + \sum_{i=1}^{k} \beta_i x_{t+j,i} + u(t+j|t+1) \text{ with}$$

(8) $$u(t+j|t+1) = \sum_{i=j}^{p} \phi_i u_{t+j-i} + \sum_{i=1}^{j-1} \phi_i u(t+j-i|t+1) + \sum_{i=j-1} \gamma_i \epsilon_{t+j-1}$$

From (5)-(8), we have

(9) $$y(t+j|t+1) - y(t+j|t) = \sum_{i-1}^{j-1} \alpha_i [y(t+j-i|t+1) - y(t+j-i|t)] + \sum_{i=1}^{j-1} \phi_i [u(t+j-i|t+1) - u(t+j-i|t)] + \gamma_{j-1} \epsilon_{t+1}$$

This equation gives a recursion involving only the random input at time t + 1 because of assumption (ii) which implies $R(y_t, y_{t-1}, \ldots) = R(u_t, u_{t-1}, \ldots) = R(\epsilon_t, \epsilon_{t-1}, \ldots)$. Let

(10) $$y(t+j|t+1) - y(t+j|t) = g_j \epsilon_{t+1} \text{ and}$$
$$u(t+j|t+1) - u(t+j|t) = h_j \epsilon_{t+1}.$$

Then from (9) and (10) we get

(11) $$y(t+j|t+1) = y(t+j|t) + \left[\sum_{i=1}^{j-1} (\alpha_i g_{j-i} + \phi_i h_{j-i}) + \gamma_{j-1}\right]\epsilon_{t+1}$$

or

$$y(t+j|t+1) = y(t+j|t) + g_j \epsilon_{t+1}$$

where the g's and h's are generated by the recursion (see Jones (1980), (2.13))

$$h_1 = 1$$

$$h_j = \gamma_{j-1} + \sum_{i=1}^{j-1} \phi_i h_{j-i}$$

$$g_1 = 1$$

$$g_j = \gamma_{j-1} + \sum_{i=1}^{j-1} (\alpha_i g_{j-i} + \phi_i h_{j-i})$$

$$= \sum_{i=1}^{j-1} \alpha_i g_{j-i} + h_j$$

where

$$\alpha_j = 0 \text{ for } j > q$$

$$\phi_i = 0 \text{ for } i > p$$

$$\gamma_i = 0 \text{ for } i > q$$

By (1) and (10) and some algebra, for $c \geq q + 1$,

(12) $$y(t+c|t+1) = y(t+c|t) + g_c \epsilon_{t+1}$$

$$= \sum_{i=1}^{r} \alpha_i y(t+c-i|t) + \sum_{i=1}^{k} \beta_i x_{t+c,i} + u(t+c|t) + g_c \epsilon_{t+1}$$

$$= \sum_{i=1}^{r} \alpha_i y(t+c-i|t) + \sum_{i=1}^{k} \beta_i x_{t+c,i} + \sum_{i=1}^{p} \phi_i u(t+c-i|t)$$

$$+ \sum_{i=1}^{q} \gamma_i \epsilon(t+c-i|t) + g_c \epsilon_{t+1}$$

$$= \sum_{i=1}^{p+r} \theta_i y(t+c-i|t) + \sum_{i=1}^{k} \beta_i \phi(L) x_{t+c,i} + g_c \epsilon_{t+1}$$

where

$$
\begin{aligned}
\theta_i &= \alpha_1 + \phi_1, && i = 1 \\
&= \alpha_i + \phi_i - \sum_{j=1}^{i-1} \alpha_j \phi_{i-j}, && i = 2,\dots, \min(p,r). \\
&= \alpha_i - \sum_{j=1}^{p} \phi_j \alpha_{i-j}, && i = p+1,\dots,r, \text{ if } p < r \\
&= \phi_i - \sum_{j=1}^{r} \alpha_j \phi_{i-j}, && i = r+1,\dots,p \text{ if } p > r, \\
&= -(\phi_{i-r}\alpha_r + \dots + \phi_p \alpha_{i-p}), && i = \max(p,r)+1,\dots,p+r \\
&= 0 && \text{otherwise,}
\end{aligned}
$$

and $\phi(L)x_{t+c,j} = x_{t+c,j} - \sum_{i=1}^{p} \phi_i x_{t+c-i,j}$. We have used the fact that $\varepsilon(t+c-i|t) = 0$ for $i=0,\dots,q$ since $c \geq q + 1$. Therefore a minimal Markovian representation is given by

$$
(13) \quad
\begin{bmatrix} y(t+1|t+1) \\ y(t+2|t+1) \\ \vdots \\ y(t+m|t+1) \end{bmatrix}
=
\begin{bmatrix} 0 & 1 & 0 & 0 \dots 0 \\ 0 & 0 & 1 & \dots 0 \\ & & \vdots & \\ \theta_m & \dots & \theta_2 & \theta_1 \end{bmatrix}
\begin{bmatrix} y(t|t) \\ y(t+1|t) \\ \vdots \\ y(t+m-1|t) \end{bmatrix}
+
\begin{bmatrix} g_1 \\ g_2 \\ \vdots \\ g_m \end{bmatrix} \epsilon_{t+1}
$$

$$
+ \begin{bmatrix} 0 \\ \dots\dots\dots\dots \\ \phi(L)x_{t+m,1},\dots,\phi(L)x_{t+m,k} \end{bmatrix}
\begin{bmatrix} \beta_1 \\ \vdots \\ \beta_k \end{bmatrix}
$$

where $m = \max(p+r, q+1)$. In matrix notation

$$
\underset{\sim}{Z}(t+1) = \underset{\sim}{F}\, \underset{\sim}{Z}(t) + \underset{\sim}{G}\, \epsilon(t+1) + \underset{\sim}{\Gamma}(t) \underset{\sim}{\beta}
$$

where $\underset{\sim}{Z}(t) = [y(t|t), y(t+1|t),\dots,y(t+m-1|t)]'$

$$\underset{\sim}{F} = \begin{bmatrix} 0 & 1 & 0 & \dots & 0 \\ 0 & 1 & 1 & \dots & 0 \\ & & \vdots & & \\ \theta_m & \cdot & \cdot\cdot & & \theta_1 \end{bmatrix}$$

$$\underset{\sim}{G} = [g_1, \ . \ . \ . \ , \ g_m]'$$

$$\underset{\sim}{\Gamma}(t) = \begin{bmatrix} & 0 & \\ \dots\dots\dots & \dots\dots\dots & \dots\dots\dots \\ \phi(L)x_{t+m,1} & \dots & \phi(L)x_{t+m,k} \end{bmatrix}$$

and
$$\underset{\sim}{\beta} = [\beta_1 \ , \ . \ . \ . \ . \ , \ \beta_k]'$$

The second equation in the state space representation of the process is the observational equation

(14)
$$w(t) = [1,0,\dots 0] \begin{bmatrix} y(t|t) \\ \vdots \\ y(t+m-1|t) \end{bmatrix} + v(t)$$

i.e. $w(t) = \underset{\sim}{H}'\underset{\sim}{Z}(t) + v(t)$

where $v(t)$ is a random variable denoting observational error which is uncorrelated at different times, uncorrelated with the ϵ's and

(15)
$$E[v(t)] = 0$$
$$E[v(t)^2] = R$$

3. EVALUATION OF THE LIKELIHOOD FUNCTION

Assuming that the errors are normally distributed, the likelihood function can be calculated using Kalman recursive estimation technique for given values of α_i, ϕ_i, γ_i, β_i and R. The procedure starts by specifying the initial state vector $\underset{\sim}{Z}(0|0)$ and covariance matrix of $\underset{\sim}{Z}(0|0)$, $\underset{\sim}{P}(0|0)$. We will assume that $\underset{\sim}{Z}(0|0)$ is known. The equation for generating $\underset{\sim}{P}(0|0)$ will be derived in the next section. This section is almost the same as Section 3 of Jones but it is given here for the sake of completeness.

The recursion first calculates a one-step prediction

(16)
$$\underset{\sim}{Z}(t+1|t) = \underset{\sim}{F}\ \underset{\sim}{Z}(t|t) + \underset{\sim}{\Gamma}(t)\underset{\sim}{\beta}$$

where $\underset{\sim}{Z}(t+j|t) = E[\underset{\sim}{Z}(t+j|\omega(s), s \le t]$, the projection of $\underset{\sim}{Z}(t+j)$ on the observations up to time t. The covariance matrix of this prediction is

(17) $$\underset{\sim}{P}(t+1|t) = \underset{\sim}{F}\,\underset{\sim}{P}(t|t)\underset{\sim}{F}' + \sigma^2\underset{\sim}{G}\,\underset{\sim}{G}'.$$

The predicted value of the next observation is

(18) $$\omega(t+1|t) = \underset{\sim}{H}'\,\underset{\sim}{Z}(t+1|t)$$

Using the next observation, the state vector estimate is updated

(19) $$\underset{\sim}{Z}(t+1|t+1) = \underset{\sim}{Z}(t+1|t) + \underset{\sim}{\Delta}(t+1)[\omega(t+1)-\omega(t+1|t)]$$

where

(20) $$\underset{\sim}{\Delta}(t+1) = \underset{\sim}{P}(t+1|t)\underset{\sim}{H}[\underset{\sim}{H}'\underset{\sim}{P}(t+1|t)\underset{\sim}{H} + R]^{-1}.$$

Note that $\underset{\sim}{P}(t+1|t)\underset{\sim}{H}$ is the first column of $\underset{\sim}{P}(t+1|t)$ and $\underset{\sim}{H}'\underset{\sim}{P}(t+1|t)\underset{\sim}{H} = P(t+1|t)$, the upper left-hand element of the $\underset{\sim}{P}(t+1|t)$ matrix. Finally, the covariance matrix of the updated state is

(21) $$\underset{\sim}{P}(t+1|t+1) = \underset{\sim}{P}(t+1|t) - \underset{\sim}{\Delta}(t+1)\underset{\sim}{H}'\underset{\sim}{P}(t+1|t).$$

Let

(22) $$\tilde{\omega}(t+1) = \omega(t+1) - \omega(t+1|t)$$

Then $\tilde{\omega}(t+1)$ is orthogonal to the previous observations and it has variance

(23) $$\begin{aligned}\tilde{V}(t+1) &= \underset{\sim}{H}'\underset{\sim}{P}(t+1|t)\underset{\sim}{H} + R \\ &= P_{11}(t+1|t) + R\end{aligned}$$

the one-step prediction error variance of the process at time t+1 plus the observational error variance. The variance σ^2 can be removed from the nonlinear estimation problem by dividing R by σ^2. Therefore, the log likelihood function for $\omega(i)$, i=1,...,n, to be maximized with respect to α_i, ϕ_i, γ_i, β_i and R/σ^2, after dropping constants, is given by (see Jones (1980) 3.12-3.15, and Mehra (1974)).

(24) $$\ell = -\sum_{i=1}^{n} \ell n\, V(i) - n\ell n \sum_{i=1}^{n} \tilde{\omega}^2(i)/\tilde{V}(i)$$

and the maximum likelihood estimator for σ^2 is

(25) $$\hat{\sigma}^2 = \frac{1}{n} \sum_{i=1}^{n} \tilde{\omega}^2(i)/\tilde{V}(i).$$

As mentioned in Jones, σ^2 can be set equal to unity first while maximizing the likelihood function with respect to other parameters.

4. THE INITIAL STATE COVARIANCE MATRIX.

The quantities involved are of the form $\mathrm{cov}[y(t|t),y(t+j|t)]$ and $\mathrm{cov}[y(t+i|t),y(t+j|t)]$ where $i, j > 0$. These can be calculated by a method due to Akaike as shown in Jones. Let $\lambda_j = \mathrm{cov}[y_{t+j},\epsilon_t]$. Then (see Jones 4.2-4.4)

(26) $$\lambda_0 = \mathrm{cov}[y_t,\epsilon_t] = \mathrm{cov}[u_t,\epsilon_t] = \sigma^2$$

(27) $$\lambda_j = \mathrm{cov}[y_{t+j},\epsilon_t] = E[(\sum_{i=1}^{r} \alpha_i y_{t+j-i} + \sum_{i=1}^{k} \beta_i x_{t+j,i} + u_{t+j})\epsilon_t]$$

$$= \sum_{i=1}^{r} \alpha_i \lambda_{j-i} + \sigma^2 h_{j+1},\ j > 0,$$

where

(28) $$\lambda_j = 0 \quad \text{if } j < 0.$$

From (11) and (26)-(28), if we let $\bar{\lambda}_j = \lambda_j/\sigma^2$, the g's are the same as the $\bar{\lambda}$'s in equation (26)-(28) shifted by one. i.e.

(29) $$g_j = \bar{\lambda}_{j-1}$$

Thus y_t can be written as

(30) $$y_t = \sum_{j=0}^{\infty} \bar{\lambda}_j \epsilon_{t-j} + \sum_{i=1}^{k} \beta_j A(L)^{-1} x_{t,i}$$

$$= \sum_{j=0}^{\infty} g_{j+1} \epsilon_{t-j} + \sum_{i=1}^{k} \beta_j A(L)^{-1} x_{t,i}$$

where L is the lag operator such that $Lx_{t,i} = x_{t-1,i}$. For $j \geq 0$,

(31) $$y(t+j|t) - E[y(t+j|t)] = \sum_{\ell=j}^{\infty} g_{\ell+1}\epsilon_{t+j-\ell}$$

and for $j > 0$,

(32) $$y(t+j) - E[y(t+j)] = \sum_{\ell=0}^{\infty} g_{\ell+1}\epsilon_{t+j-\ell}$$

$$= y(t+j|t) - E[y(t+j|t)] + \sum_{\ell=0}^{j-1} g_{\ell+1}\epsilon_{t+j-\ell}$$

Since the ϵ's are uncorrelated with the past y's, it follows from (32) that

(33) $$\text{cov}[y(t),y(t+j)] = \text{cov}[y(t|t),y(t+j|t)]$$

and for $j \geq i > 0$,

(34) $$\text{cov}[y(t+i),y(t+j)] = \sigma^2 \sum_{\ell=0}^{i-1} g_{\ell+1+j-i}g_{\ell+1}$$

$$+ \text{cov}[y(t+i|t),y(t+j|t)].$$

From (1) the covariances satisfy, for $j \geq 0$

(35) $$c_j = \text{cov}[y_{t+j},y_t]$$

$$= \text{cov}[\sum_{i=1}^{r} \alpha_i y_{t+j-i} + u_{t+j} + \sum_{i=1}^{k} \beta_i x_{t+j,i}, y_t]$$

$$= \sum_{i=1}^{r} \alpha_i c_{j-i} + c_{u,y}(j)$$

where $c_{u,y}(y) = \text{cov}[u_{t+j},y_t]$ in turn satisfies

(36) $$C_{u,y}(j) = \text{cov}[u_{t+j}, y_t]$$

$$= \text{cov}[\sum_{i=1}^{p} \phi_i u_{t+j-i} + \sum_{i=0}^{q} \gamma_i \epsilon_{t+j-i}, y_t]$$

$$= \sum_{i=1}^{p} \phi_i c_{u,y}(j-i) + \sum_{i=j}^{q} \gamma_i \lambda_{i=j}$$

where $\gamma_i = 0$ for $i > q$ and $\lambda_i = 0$ for $i < 0$. Note that $C_j = C_{-j}$ but $C_{u,y}(j) \neq C_{u,y}(-j)$. In fact for $j \geq 0$,

(37) $$C_{u,y}(j) = \text{cov}[u_{t+j}, y_t]$$

$$= E[\sum_{k=0}^{\infty} h_{k+1} \epsilon_{t+j-k})(\sum_{\ell=0}^{\infty} g_{\ell+1} \epsilon_{t-\ell})]$$

$$= \sigma^2 \sum_{k=0}^{\infty} h_{k+1+j} g_{k+1}$$

$$= \sigma^2 \sum_{k=1}^{\infty} h_{k+j} g_k$$

and similarly,

(38) $$C_{u,y}(-j) = \text{cov}[u_{t-j}, y_t]$$

$$= \sigma^2 \sum_{k=1}^{\infty} h_k g_{j+k}$$

Since equation (36) is a pth-order difference equation, $c_{u,y}(j)$, $j \geq 0$ can be obtained from (36) for a given set of γ_i, λ_i and ϕ_i and once $c_{u,y}(j)$, $j=-1,\ldots,-p$ are known. As $|h_k|$'s and $|g_k|$'s are decreasing to zero at an exponential rate, $c_{u,y}(-j)$ can be approximated as accurately as desired from (38) by using a finite number of terms. Therefore for a given set of γ_i, λ_i and ϕ_i, c_j, $j=0,\ldots,r$ can be computed using (35) and (36) as in Jones 4.10-4.11. The remaining c_j's necessary to calculate the state covariance matrix can be calculated from (35).

5. MISSING OBSERVATIONS

Proceeding as in Jones, if an observation, $\omega(t+1)$, is missing, the recursion proceeds through equations (16)-(18) noting that σ^2 can be set equal to one and estimated later. Equations (19) and (22) are replaced by

$$\underset{\sim}{Z}(t+1)|t+1) = \underset{\sim}{Z}(t+1|t)$$

and

$$\underset{\sim}{P}(t+1|t+1) = \underset{\sim}{P}(t+1|t)$$

The corresponding term in equations (24) and (25) is skipped. This procedure works for any number of observations missing in a row. The above replacement is made to account for the fact that no new information is obtained from the missing observation.

6. REPARAMETRIZING FOR STABILITY

To ensure that the stability condition made in assumption (ii) is met in the process of maximizing the likelihood function, we have to know exactly where the α_i, ϕ_i and γ_i are allowed to assume their values. Let us consider just the condition imposed by $A(z) \neq 0$ for $|z| \leq 1$. The conditions imposed by $\phi(z) \neq 0$ for $|z| \leq 1$ and $\Gamma(z) \neq 0$ for $|z| \leq 1$ are similar. Let

$$A = \{(\alpha_1,\ldots,\alpha_p) \in R^p | A(z) \neq 0 \text{ for } |z| \leq 1\} \text{ and}$$

let π_j denote the j-th partial autocorrelation. It follows from Theorem (2) of Barndorff-Nielsen and Schou (1973) that the transformation

$$\phi : (-1,1)^p \to A$$

such that $\phi(\pi_1,\ldots,\pi_p) = (\alpha_1,\ldots,\alpha_p)$ is one-to-one, and both ϕ and ϕ^{-1} are continuously differentiable. Such a transformation is explicitly given in Durbin (1960) in a recursive form. Obviously $\mathbb{R}^p$ is diffeomorphic to $(-1,1)^p$. Therefore it would be easier to maximize the log likelihood function with respect to elements in $\mathbb{R}^p$ than to do it directly with respect to elements in A. This is exactly what is done in Section 6 of Jones (1980). We shall therefore not repeat ourselves here. See Jones for details. After reparameterization, the maximization can be performed using a non-linear optimization program.

Remark: As for model identification, one can use Akaike information criterion, AIC, as done in Jones.

REFERENCES

Akaike, H. (1974), Markovian representation of stochastic processes and its application to the analysis of autoregressive moving average process. Annals of the Institute of Statistical Mathematics, 26, 363-387.

Anderson, T. W. (1975), Maximum likelihood estimation of parameters of autoregressive processes with moving average residuals and other covariance matrices with linear structure. The Annals of Statistics 3, 1283-1304

Barndorff-Nielsen, O. and Schow, G. (1973), On the parametrization of autoregressive models by partial autocorrelation. Journal of Multivariate Analysis, 3, 408-419.

Caines, P. E. and Rissaner, J. (1974), Maximum likelihood estimation of parameters in multivariate Gaussian stochastic processes. IEEE Transactions on Information Theory. IT-20, 102-104.

Durbin, J. (1960), The fitting of time series models. Review of the International Institute of Statistics 28, 233-244.

Harvey, C. C. and Phillips, G. D. (1979), Maximum likelihood estimation of regression on models with autoregressive-moving average disturbances. Biometrika, 66, 1, 49-58.

Jones, R. H. (1980), Maximum likelihood fitting of ARMA models to time series with missing observations. Technometrics, 22, 389-395.

Mehra, R. (1974), Identification in control and econometrics; similarities and differences. Annals of Economics and Social Measurements, 3, 21-47.

Riensel, G. (1976), Maximum likelihood estimation of stochastic linear difference equations with autoregressive moving average errors. Technical Report No. 112, Carnegie-Mellon University.

APPLIED TIME SERIES ANALYSIS
O.D. Anderson & M.R. Perryman (eds.)

A MULTIPLE TIME SERIES ANALYSIS OF COTTON-POLYESTER MARKET COMPETITION

Lucille M. Terry

Bowling Green State University
Bowling Green, Ohio

W. Robert Terry

University of Toledo
Toledo, Ohio

Cotton and polyester compete for a share of the fiber market. This paper utilized state space analysis to develop a model of the interaction between price and consumption for cotton and polyester. The resulting model indicated that the prices of both fibers follow a random walk. This model also suggests that price competition exists between cotton and polyester.

1. INTRODUCTION

This paper utilizes the state space approach to stochastic multivariate model building developed by Akiake (1976) to develop a model of the interfiber competition between cotton and polyester. In particular, price and consumption time series for cotton and polyester are regarded as a multivariate time series system. The objective of the state space analysis is to develop a simultaneous system of stochastic difference equations which best describe the multivariate stochastic process which generated the observed price and consumption time series for cotton and polyester.

A number of researchers have utilized a multiple time series approach to model commodity systems. Quenouille (1957) utilized a first order vector autoregressive process to model the relationships between: hog supply, hog price, corn supply, corn price and farm wages. Box and Tiao (1977) developed a canonical transformation for a stationary vector autoregressive process and applied it to Quenouille's data. Their analysis yielded a model of the transformed process which could be more readily interpreted than the Quenouille's model. Jenkins and Alavi (1981) describes a multiple time series model building approach which utilizes the autocorrelation matrix, partial autocorrelation matrix and q-conditional autocorrelation matrix to identify the orders of a multivariate ARIMA process and then used it to model the interaction between the price and consumption time series for butter and margarine in the U.K. The state space approach to stochastic multivariate model building utilized in this paper differs from the above in that it utilizes an information criterion based on the minimization of entropy and cannonical analysis to identify and estimate the parameters of a generalized model which includes the multivariate ARMA model as a special case. This avoids the following problems cited by Jenkins (1981): (1) mixed ARIMA models are difficult to identify since the first q autocovariance matrices do not follow a fixed pattern; (2) the convergent properties of the procedure for obtaining initial estimates of moving average parameters are not satisfactory; and (3) a lack of knowledge regarding sampling properties of the determinants of the autocovariance, partial autocovariance and q-conditional partial autocovariance matrices.

The remainder of this paper is organized as follows: Section 2 briefly describes the nature of cotton-polyester interfiber competition. This information provides the logical basis for limiting the analysis to the post 1973 period. Section 3 describes the results obtained from a state space analysis of the cotton-polyester data. Section 4 discusses the implications of this model.

2. HISTORICAL BACKGROUND OF COTTON-POLYESTER INTERFIBER COMPETITION

Originally there were four major fibers: cotton, linen, silk, and wool. For the most part these fibers did not compete with one another to any major extent. Of the four, cotton was by far the predominant one.

For centuries cotton maintained this dominant position, accounting for roughly 75 percent of the world fiber consumption. In the 1920's the first man-made fibers (rayon and acetate) were introduced, followed by nylon in 1939. None of these fibers tended to present a serious challenge to cotton. However, in 1953 polyester was introduced in the U.S. by DuPont. During the next 20 years polyester developed into a major challenger to cotton's leadership and by 1973 had captured approximately 79 percent of markets which had previously belonged to cotton. Polyester was able to make these enormous inroads due primarily to the fact that polyester producers had a superior program for marketing their fiber.

In the mid-to-late 1960's a group of large cotton farmers in California became acutely aware of the threat that polyester posed for cotton. As a result they hired a prestigious management consulting firm to analyze the problem and to advise them on how to counter the polyester threat. This analysis revealed that cotton was far superior to polyester in terms of comfort and that consumers, almost without exception, rated comfort as one of the most important characteristics in items of apparel.

This analysis also revealed that cotton had a major weakness--namely the lack of a marketing program. Cotton farmers formerly had been mainly concerned with producing their product and there was no one or no organization responsible for marketing or advertising cotton.

The analysis of the consulting firm concluded that most of cotton's losses were not due to cotton being inferior to polyester as a textile fiber, but rather were due to the lack of a systematic program for marketing cotton. As a result, the firm proposed that cotton farmers create an organization with the sole purpose being to promote the use of cotton. Such an organization (Cotton Incorporated) was formed.

Two unique events occurred during the months of December 1973 and January 1974. One was the start of an advertising campaign launched by Cotton Incorporated which emphasized that "cotton is comfortable." The second event was the energy crisis with the appeal to the American people to turn thermostats up in the summer and down in the winter in order to conserve energy.

Seemingly the timing could not have been better for cotton. At the same time the President was telling the American people to shiver in the winter and sweat in the summer, Cotton Incorporated's advertising campaign was telling them that they could become more comfortable by wearing more cotton. The net effect of these two events was to give cotton a splendid opportunity to regain some of the market it had lost to polyester.

Prior to 1974 the price of cotton was influenced to a great extent by the Federal Government's price-support program for cotton. This program was created during the 1930's along with similar programs for other essential farm crops. The purpose of these programs was to prevent the prices of agricultural commodities from falling to levels which would cause producers of the essential crops to go bankrupt.

The major features of the cotton price support program included acreage controls and a non-recourse loan. The acreage control programs provided direct payments to farmers who plowed under a prescribed portion of their crop. The non-recourse loan program was administered by an agency of the Federal government, Commodity Credit Corporation (CCC). A farmer could borrow an amount equal to a specified percentage, which over the years ranged from 65 to 105 percent, of the anticipated value of the crop. If the price of cotton was lower than the loan value on June 30 of the next year and the loan had not been repaid, then the cotton was considered sold to the CCC at the loan price.

The cotton which the CCC acquired in this fashion became part of the CCC's cotton stockpile. In 1966 this stockpile contained a record high of over 6 billion pounds of cotton. During the late 60's and early 70's there was a movement to return to a free market agriculture. As a result of this movement the cotton stockpile was virtually eliminated by 1974.

To summarize, three events occurred in 1973 which marked the end of an era in cotton-polyester interfiber competition: (1) a program for marketing cotton was launched; (2) the energy crisis made comfort more important than it previously had been, and (3) dismanteling the commodity stockpile enabled the price of cotton to become more market responsive. For these reasons the scope of this study was limited to the period 1974 to 1980.

3. STATE SPACE ANALYSIS

The state space approach to building multivariate time series models differs from the approaches advocated by Box and Tiao (1978) and Jenkins and Alavi (1981) which restrict attention to only the input and output variables of a system. In contrast the state space approach introduces some intermediate variables, called state variables, which help to describe the internal state of a system.

The general form of the state space model is given by

$$v_{n+1} = Av_n + Bz_{n+1}$$

$$y_n = Cv_n$$

where

v_n = value of state vector at time n.

y_n = value of observation vector at time n.

z_n = value of white noise input vector at time n.

A = transition matrix.

B = input matrix.

C = observation matrix.

The state space model building approach consists of two major steps. The first is concerned with determining the most appropriate structure for the state space vector, while the second is concerned with determining the optimum values of the parameter estimates of the transition, input, and observation matrices. A program developed by Delong (1979) was utilized to perform the analysis necessary for developing a state space model for the monthly price and consumption time series for cotton and polyester.[1] The analysis to determine the structure of the state vector resulted in a 5 dimensional vector with the following components

Component	Symbol	Defining
1	p_{1t}	current value of price of cotton
2	p_{2t}	current value of price of polyester
3	c_{1t}	current value of cotton consumption
4	c_{2t}	current value of polyester consumption
5	$c_{1,t+1\|t}$	consumption made from origin t

The estimates of the parameters of the transition and input matrices and the standard errors of these estimates are shown in Tables 1 and 2. In these matrices the odd numbered rows contain the parameter estimates while numbers in parentheses in the even numbered rows contain the standard errors of estimates.

TABLE 1

TRANSITION MATRIX

	p_{1t}	p_{2t}	c_{1t}	c_{2t}	$c_{1,t+1\|t}$
p_{1t}	0.916092	-0.07779	-0.011699	0.0259659	0
	(0.048052)	(0.0810098)	(0.0177295)	(0.0168192)	(0)
p_{2t}	0.0069963	0.925767	0.000910	0.0085752	0
	(0.0229596)	(0.0387071)	(0.0084713)	(0.0080363)	(0)
c_{1t}	0	0	0	0	1
	(0)	(0)	(0)	(0)	(0)
c_{2t}	0.136131	-0.351758	-0.268788	0.889264	0.0548983
	(0.163231)	(0.29696)	(0.071329)	(0.0575019)	(0.109035)
$c_{1,t+1\|t}$	-0.04892	-0.279535	0.351313	-0.056357	0.321355
	(0.178259)	(0.338627)	(0.0996556)	(0.0634186)	(0.170096)

TABLE 2

INPUT MATRIX

	a_{p1}	a_{p2}	a_{c1}	a_{c2}
p_{1t}	1	0	0	0
	(0)	(0)	(0)	(0)
p_{2t}	0	1	0	0
	(0)	(0)	(0)	(0)
c_{1t}	0	0	1	0
	(0)	(0)	(0)	(0)
c_{2t}	0	0	0	1
	(0)	(0)	(0)	(0)
$c_{1,t+1\|t}$	0.672018	-2.37399	0.138065	0.244032
	(0.455912)	(0.94929)	(0.918807)	(0.136051)

4. IMPLICATIONS

In this section the state space model developed in the preceding section will be analyzed. In making this analysis all parameters with an absolute value of less than two standard errors will be considered to be zero.

After eliminating the nonsignificant terms the following equations are obtained for the price time series for cotton and polyester.

$$p_{1,t+1} - .92p_{1t} = a_{p1} \tag{1}$$

$$p_{2,t+1} - .93p_{2t} = a_{p2} \quad (2)$$

$$c_{1,t+1} - c_{1,t+1|t} = a_{p2} \quad (3)$$

$$c_{2,t+1} + .27c_{1t} - .89c_{2t} = a_{c2} \quad (4)$$

$$c_{1,t+1|t} - .35c_{1t} = -2.37a_{p2} \quad (5)$$

The first two equations indicate that both the cotton and polyester time series are very close to a random walk. This is consistent with the random walk hypothesis for commodity prices first advanced by Working (1934). However, Labys and Granger (1970) have noted that most subsequent studies have rejected this hypothesis.

The second, third and fifth equations are coupled together. The second and fifth equations indicate that the one step ahead forecast for cotton increases when the price of polyester increases and visa versa. If it can be assumed that consumption forecast results in actual consumption this suggests that price competition exists between cotton and polyester. The third equation indicates that the difference between the one step ahead forecast and the corresponding actual values for cotton consumption follow a randomalk which is to be expected for an adequate forecasting model.

The fourth equation indicates consumption of polyester is inversely related to the consumption of cotton and directly related to the consumption of polyester for the preceding time period.

5. SUMMARY

This paper has utilized state space analysis to develop a model of the interaction between price and consumption for cotton and polyester. The resulting model indicated that the prices of both fibers follow a random walk. This model also suggests that price competition exists between cotton and polyester.

FOOTNOTE

[1] The authors wish to thank the U.S. Department of Agriculture for providing this data.

REFERENCES

AKAIKE, H. (1976). Canonical correlation analysis of time series and the use of an information criterion. In *System Identification:Advances and Case Studies*, Eds. R. K. Mehra and D. G. Lainiotis, Academic Press, New York.

BOX, G.E.P. and TIAO, G.C. (1978). *Multiple Time Series Model Building and Forecasting*, University Associates, Princeton, New Jersey.

BOX, G.E.P. and TIAO, G.C. (1977). A canonical analysis of multiple time series, *Biometrika, 64*, 355-365.

JENKINS, G.M. and ALAVI, A.S. (1981). Some aspects of modelling and forecasting multivariate time series, *Journal of Time Series Analysis, 2*, 1-47.

LABYS, W.C. and GRANGER, C.W.J. (1970). *Speculation Hedging and Commodity Price Forecasts*, Heath Lexington Books, Lexington, Massachusetts.

QUENOUILLE, M.H. (1957). *Multiple Time Series*, Wiley, New York.

WORKING, H. (1934). A random difference series for use in the analysis of time series, *Journal of American Statistical Association, 29*, 11-24.

APPLIED TIME SERIES ANALYSIS
O.D. Anderson & M.R. Perryman (eds.)

A TIME SERIES ANALYSIS APPROACH FOR DESIGNING MILITARY COMPENSATION SYSTEMS

W. Robert Terry
Industrial Engineering, University of Toledo
and
Lee R. Bishop
Hill Air Force Base, U. S. Air Force

This paper illustrates how state space analysis can be used to develop a system for manipulating military pay to ensure that target manpower levels for each skill category are met. More specifically it describes how state space analysis can be used to develop a model for predicting what enlistment and reenlistment rates will be under alternative compensation programs.

1. INTRODUCTION

1.1. BACKGROUND

Maintaining a state of military readiness adequate to protect the vital interest of the U.S. involves a complex planning process. Plans are carefully formulated to produce the most cost-effective mix of manpower and hardware for maintaining the level of military readiness specified as a national objective.

The goal of the planning process is to formulate plans for the acquisition of manpower and hardware which will minimize the cost required to maintain the desired level of military readiness. Unfortunately, this goal is seldom if ever achieved. Hardware systems almost invariably cost more than budgeted, fail to meet performance specifications, and reach operational readiness much later than expected. Manpower acquisition goals almost without exception are not achieved. Total aggregate strength is frequently either too high or too low. When total aggregate strength targets are met the mix of skills can, and typically does, differ substantially from that desired. Invariably, there will be wasteful surpluses in certain fields and critical shortages in others.

The failure to achieve the goals specified in acquisition plans for hardware and manpower can cause the hardware-manpower mix to deviate substantially from that which would be optimal. Symptoms of such deviations are abundantly evident. Certain groups of personnel are not fully and effectively utilized due to shortages of key items of equipment. On the other hand certain items of expensive equipment can not be fully and effectively utilized due to shortages in key groups of personnel.

The ultimate solution to these problems will require that acquisition plans for both hardware and manpower be more closely coordinated. In order to do this it will be necessary to develop better means for predicting and controlling the hardware and the manpower acquisition processes.

1.2. PRECIS

The scope of this paper will be limited to specifying a better means for controlling the manpower acquisition process. Section 2 will critically analyze the present military compensation system. The primary purpose of this analysis will be to identify those areas in which the present system is most inadequate. These inadequacies indicate the need for a control system which will manipulate military pay to ensure that target levels for personnel in each skill category are met. Section 3 will argue that a key requirement for the development for such a system is the ability to accurately predict the enlistment and reenlistment rates. Section 4 will briefly describe the state space approach to developing models for predicting these quantities. Section 5 will present an example to illustrate the use of the state space approach to developing predictive models.

2. EVALUATION OF CURRENT MILITARY COMPENSATION SYSTEM

2.1. CRITERIA

To evaluate any system it is necessary to establish criteria. These criteria can be derived from the goals which the system is supposed to achieve. Since every system is part of some super system the goal of a system should be to help its supersystem achieve its goals.

The various branches of the armed forces constitute the supersystem for a military compensation system. The defense doctrine of the U.S. specifies that each of these branches should be capable of accomplishing a variety of missions. Strategic plans specify the resources which will be needed to maintain the readiness required for accomplishing these missions. Manpower plans, which are derived from strategic plans, specify quotas for the numbers of personnel in each of a number of skill categories which will be needed to maintain the required level of readiness.

It should be the responsibility of a military compensation system to ensure that the specified skill category quotas are met at all times. This suggests that one criterion for evaluating a military compensation system is its ability to meet these skill category quotas. Obviously, a less expensive compensation system should be preferred to a more expensive system if other relevant factors are equal. This suggests that cost should be a second criterion. The next two subsections evaluate the present military compensation system from the standpoint of its ability to meet skill category quotas and the cost associated with the system.

2.2. ABILITY TO MEET SKILL QUOTAS

The present military compensation system has historically performed very poorly in terms of its ability to meet skill quotas. This is evidenced by chronic shortages in some skill categories and chronic surpluses in others. These chronic shortages and surpluses suggested that the foundation on which the present system is based could be unsound and prompted an investigation to determine if this is the case. The results of this investigation are summarized below.

The present military compensation system is based on the premise that all persons with the same longevity in a rank should receive the same pay irrespective of their skill category. This practice of giving everyone with the same seniority in rank the same pay can create an imbalance of skills with chronic shortages in certain areas and chronic surpluses in others. These skill imbalances arise because giving everyone with the same seniority in rank the same pay fails to recognize a fundamental law of economics--the law of supply and demand. Failure to recognize the law of supply and demand can cause both shortages and surpluses to arise simultaneously. Military personnel with skills that are in short supply in the civilian sector could make more money as civilians. The lure of higher pay in the civilian sector no doubt causes many persons, who would not have done so otherwise, to leave the military, resulting in chronic shortages in these areas. On the other hand, military personnel with skills that are in surplus supply in the civilian sector would not be able to make as much money as civilians. The prospects of taking a cut in pay no doubt prevents many persons from leaving the military who might otherwise do so. This greater reluctance to leave the military coupled with the fact that skills that are in surplus supply in the civilian sector tend to be relatively easy to recruit, can create chronic surplus of certain skills in the military.

2.3. COST

The practice of paying everyone with the same longevity in grade the same thing, results in a poor allocation of resource. Persons with skills that are in surplus supply in the civilian sector can receive pay that is higher than competitive. This means that the armed services will pay more than is necessary to acquire these skills. On the other hand, persons with skills that are in short supply in the civilian sector will be able to do better as civilians. This can result in lower reenlistment rates for technically skilled personnel, which in turn will lower the level of readiness and cause increased training costs to be incurred.

3. DESIRED PERFORMANCE CHARACTERISTICS FOR A MILITARY COMPENSATION SYSTEM

3.1. INTRODUCTION

As noted in the previous section the goal for a military compensation system is to ensure that the target levels for each skill category are met in future time periods. In order to do this it will be necessary to develop a system for controlling the level of personnel in each skill category.

The need for control arises when enviornmental variables have the potential to impact the output of the controlled system. The ability to identify and to measure these environmental variables is a key factor in determining when feedback and feedforward control should be used. A pure feedforward control strategy, which consists of monitoring the environmental variables and manipulating the input variable to counteract the effects of the environmental variables, will be most appropriate when <u>all</u> of the environmental variables, which can influence system output, can be both identified and accurately measured. In contrast, a pure feedback control strategy, which consists of monitoring the deviations of the output from its target and then adjusting according to some function of the present and past values of these deviations, will be most appropriate when it is not possible to both identify and measure any of the environmental variables.

In practice it is often possible to both identify and measure some but not all of the environmental variables. In such situations a mixed strategy of feedback and feedforward control will be superior to either of the pure strategies discussed above. Such a feedback/feedforward control strategy will monitor both measurable enviromental variables and the deviations of the output variable from its target. The historical values of all of these variables will then be used to predict future values of deviation of an output variable from its target values. These predicted deviations will then be used to determine how the values of the input variables should be manipulated to counteract the effects of changes in the environmental variables.

3.2. MOST APPROPRIATE CONTROL STRATEGY FOR A MILITARY COMPENSATION SYSTEM

The preceding discussion indicates that the key consideration in selecting a control strategy is the ability of the control system designer to identify and to develop methods of measuring the environmental variables which could impact the output variable which are to be controlled. For the case of a military compensation system the primary variable which is to be controlled is the level of personnel in each skill category. In order to do this it will be necessary to control the enlistment and discharge rates for each skill category. To select the most appropriate strategy for controlling these rates, which for the sake of simplicity will be referred to as output variables, it will be necessary to consider the degree to which the environmental variables, which could influence these output variables, could be identified and measured.

There are certain measurable environmental variables, such as the percent unemployment and average pay rates in the civilian sector, which could influence both the discharge and enlistment rates. High unemployment will, all other things being equal, tend to encourage some civilians to join the military and to discourage some persons, presently in the military, from leaving, while low unemployment will tend to create opposite effects. Low civilian pay rates will, all other things being equal, tend to encourage some civilians to join the military and to discourage persons presently in the military from leaving, while high civilian pay rates will tend to create opposite effects. However, there are other unmeasurable environmental variables, such as the threat of war, which could influence both the enlistment rate and the reenlistment rate.

The above considerations suggest that a feedback/feedforward strategy should be the most appropriate choice for controlling the enlistment rate and reenlistment rate. In order to implement a feedback/feedforward control strategy it will be necessaery to develop a system for predicting future values of the deviation of the output variable from its target. The inputs to this prediction system will be the historical values of the measurable environmental variables and the deviations of the output variable from its target.

There are several criteria which an input variable should satisfy. The first is that there should be a reasonable basis for believing that a prospective input variable could be related to the output variable of interest. The purpose of this requirement is to minimize the risk of basing a model on a spurious correlation between a pair of input and output variables. Such relationships can result in models which fit historical data well, but perform poorly in predicting the future. The second is that there should be a reasonable basis for believing that the input-output relationship will be stable with respect to time. If such a relationship changes, then a model established on the basis of the prior relationship will become invalid. Such models could lead to unreliable and inaccurate predictions. The third is that the input variable should lead the output variable. This is necessary if the input variable is to be useful in predicting the output variable.

4. STATE SPACE ANALYSIS

4.1 BACKGROUND

The state space method, to be described in this section, was developed specifically to solve statistical model identification problems which arise in the design of controllers for multivariate stochastic processes. This method utilizes the concept of a state variable which can be defined conceptually as a vector which contains all of the information about the past and present behavior of a system which is useful in predicting its future behaviors.

The general form of the state space model is given by

$$v_{n+1} = Av_n + Bz_{n+1}$$

where v_n, y_n, and z_n respectively represent the values of the state vector, the output vector, and a white noise input vector at time n. A and B are respectively referred to as the transition and input matrices. Akaike (1974a) has shown that every state space model can be expressed as an ARMA model and, conversely, that every ARMA model can be expressed as a state space model.

The equivalence of the state space and ARMA representations suggests that one approach to developing a state space model would be to employ a multivariant generalization of the methods described in Box and Jenkins (1970), and to then transform the resulting ARMA model into its state space equivalent. However, it will be seen shortly that such an approach could create problems.

These problems arise from the fact that the Box and Jenkins type approach for identifying a multivariante ARMA model is based on the theoretical correlations properties of the ARMA model. For example, Jenkins and Alavi (1981) have suggested the following guidelines.

(1) Use a MA matrix of order q if the sample estimate of the correlation matrix is not significantly different from zero for all lags greater than q.

(2) Use an AR matrix of order p if the sample estimate of the partial correlation matrix is not significantly different from zero for all lags greater than p.

(3) Use an AR matrix of order p and a Matrix of order q if the sample estimate of the q - conditioned partial correlation matrix, i.e., partial correlation matrix for case where order of MA matrix is fixed at q, is not significantly different from zero for all lags greater than p.

Although the multivariate ARMA model building process is similar to the univariate process there are nevertheless, some important differences which complicate the model building process. The major problem is that unlike the univariate process, the multivariate process will not, in general, result in a unique multivariate ARMA model. This problem, which was recognized by Quenouille (1957, Chapter 3), and further discussed by Hannan (1976) who established necessry and sufficient conditions for a unique representation. Quenouille (1957, Chapter 2) also recognized that the covariance matrix will be singular when there are linear relationships among the variable and then recommended that the zero latent root of the covariance matrix be used to determine this relationship.

4.2. STATE SPACE MODEL BUILDING

The presence of the above problems apparently motivated Akaiki (1974) to develop an alternative approach for determining the most appropriate state space model for a given situation. The theoretical foundations on which this approach is based will first be briefly summarized below after which the approach itself will be briefly described.

4.2.1 THEORETICAL FOUNDATIONS

The theoretical basis for identifying the structure of the state space model was established by Akaike (1974). This paper suggested that a cannonical correlation analysis between the set of the present and past observations and the set of the present and future observations could be utilized to obtain a reasonable estimate of the structure for the state space vector. However, in order to implement this approach it is necessary to have a procedure of determining the maximum number of past and future observations to include in the cannonical correlation analysis. The theoretical basis for solving this problem is provided by Akaike (1974b).

This paper recommended that the quantity

$$\begin{aligned} AIC = &- 2 \log_e (\text{maximum likelihood}) \\ &+ 2 (\text{number of independently adjusted parameters}) \end{aligned}$$

be used as a criterion for evaluating the fit of a statistical model.

4.2.2 MODEL BUILDING PROCESS

The approach for developing a state space model, which is described in Akaike (1976), consists of the following steps. (1) A sequence of AR models of increasing order are fit and the order of the model with minimum AIC is used as the number of lags into the past to include in the cannonical correlation analysis. (2) Cannonical correlations of the past with an increasing number of steps into the future are calculated and the model with minimum AIC is used to determine the number of steps into the future to include in the state vector. (3) Preliminary estimates of the parameters of the transition, and input matrices are obtained from the cannonical correlation analysis. (4) The preliminary estimates of the matrix parameters are then used to obtain an infinite AR representation which in turn is used to obtain a sample estimate of the residual covariance matrix. (5) The sample estimate of residual covariance matrix is used to replace its theoretical counterpart in a log likelihood expression for the parameters of the transition, and input matrices. (6) The non-linear equations which result from maximizing this log likelihood expression are solved by the Newton Rapson method.

5. ILLUSTRATIVE EXAMPLE

An example will be used to illustrate the state space model building process. The data for this example were generated by simulating 300 observations with the following model:

$$\begin{bmatrix} 1 - .70B & 0 \\ 1.00B + .50B^2 & 1 - 1.00B + .25B^2 \end{bmatrix} \begin{bmatrix} x_t \\ y_t \end{bmatrix} = \begin{bmatrix} 1 + .40B & 0 \\ .95B & 1 + .50B \end{bmatrix} \begin{bmatrix} a_t \\ b_t \end{bmatrix}$$

where a_t and b_t are mutually uncorrelated standard normal random deviates.

Since both the AR and MA matrices are lower triangular, this model is a transfer function model.

A state space estimation procedure developed by Delong (1979) was used to develop the state space model. The cannonical correlation analysis to identify the structure of the state vector resulted in a 4 dimensional state vector with the following components:

Component	Symbol	Definition
1	x_t	Current value of x
2	y_t	Current value of y
3	$x_{t+1\|t}$	One step ahead forecast of x from origin t
4	$y_{t+1\|t}$	One step ahead forecast of y from origin t

The estimates of the parameters of the transition and input matrices and the standard errors of these estimates are shown in the matrices below. In these matrices the odd numbered rows contain the parameter estimates while the numbers in parentheses in the even numbered rows contain the standard errors of estimates.

TRANSITION MATRIX

	x_t	y_t	$x_{t+1\|t}$	$y_{t+1\|t}$
x_t	0 (0)	0 (0)	1 (0)	0 (0)
y_t	0 (0)	0 (0)	0 (0)	1 (0)
$x_{t+1\|t}$	-0.027329 (0.132086)	0 (0)	0.698485 (0.149775)	0 (0)
$y_{t+1\|t}$	-0.826904 (0.436267)	-0.158667 (0.101547)	-0.673202 (0.356557)	0.870523 (0.130682)

INPUT MATRIX

	a_t	b_t
x_t	1 (0)	0 (0)
y_t	0 (0)	1 (0)
$x_{t+1\|t}$	1.12509 (0.0597574)	0 (0)
$y_{t+1\|t}$	-0.726286 (0.110987)	1.1164 (0.059163)

The above matrices reveal that there is no feedback from y to x which is consistent with the prior expectation since a transfer function model was used to generate the data.

6. CONCLUSION

This paper has revealed that a controller should be part of a military compensation system. It has also revealed that a model for predicting future values of the deviation of manpower levels from their targets will be necessary for developing such a controller. It has also recommended that the state space approach be used for developing a predictive model since it is theoretically sensible from a statistical viewpoint and since modern control theory utilizes a state space representation.

REFERENCES

AKAIKE, H. (1974a). Markovian Representation of Stochastic Processes and its Application to the Analysis of Autoregressive Moving Average Processes, *Annals Institute of Statistical Mathematics,* 363-387.

AKAIKE, H. (1974b). A New Look at the Statistical Model Identification, *IEEE Transactions on Automatic Control,* AC-19, 716-722.

AKAIKE, H. (1976). Canonical Correlation Analysis of Time Series and the Use of an Information Criterion, in *Advances and Case Studies in System Identification.* Eds: R. K. Mehra and D. G. Lainiotis, Academic Press, New York.

BOX, G. E. P. and JENKINS, G. M. (1970). *Time Series Analysis: Forecasting and Control.* Holden-Day, San Francisco.

DELONG, D. (1979). The State Space Procedure, *Sasets,* SAS Institute, Cary, North Carolina.

HANNAN, E. J. (1976). The Identification and Parametrization of ARMAX and Statespace Forms, *Econometrica.* 44, 713-722.

JENKINS, G. M. and ALAVI, A. S. (1981). Some Aspects of Modelling and Forecasting Multivariate Time Series, *Journal of Time Series Analysis* 2, 1-47.

QUENOUILLE, M. H. (1957). *The Analysis of Multiple Time Series,* Charles Griffin, London.

APPLIED TIME SERIES ANALYSIS
O.D. Anderson & M.R. Perryman (eds.)

A TIME SERIES ANALYSIS APPROACH FOR DETERMINING THE IMPACT OF A TRAUMATIC ACCIDENT ON PRODUCTIVITY

W. Robert Terry
Industrial Engineering, University of Toledo
and
Lee R. Bishop
Hill Air Force Base, U. S. Air Force

This paper is concerned with the problem of determining the impact of a traumatic accident on productivity. It describes situations which could cause production dates to be autocorrelated and discusses how such autocorrelation could distort the level of significance for statistical tests which assume independent observations. It also illustrates how transfer function/intervention analysis could be used to estimate the impact of a traumatic accident on productivity in the presence of autocorrelated data.

1. INTRODUCTION

It has been said that "motherhood is a matter of fact, but that fatherhood is at best a matter of opinion." However, there is no doubt what fathered the sudden growth in the demand for safety engineers. It was the Occupational Safety and Health Act (OSHAct) which became effective on April 28, 1971. Before the OSHAct safety engineers were a "dime a dozen." In recent years they have been as "scarce as hen's teeth." However, as will be seen shortly, there are some dark clouds on the horizon.

The OSHAct was created during an era in which it was generally believed that avaricious organizations would run over the little man if government regulation did not prevent them from doing so. Today the mood of the country is drastically different: excessive government regulations are believed to be partially to blame for the persistently high rate of inflation and the inability of many domestic industries to compete effectively in international markets, and politicians are sounding the cry, "it is time to get government out of business."

This apparent readiness for less government regulations could lead to reforms in the OSHAct which could drastically change the way accident prevention programs are evaluated. At present the penalties of not having a safety program are so great that corporations can not afford not to have a safety program. Under its present form the OSHAct authorizes compliance officers to make "no notification inspections" and to issue citations for noncompliance of Occupational Safety and Health Administration (OSHA) standards which could result in stiff fines, a plant shutdown, and even jail sentences for the manager responsible for the violation. Under these circumstances there is little that can be done except to have a safety program. However, if, in the future, the OSHAct is revised so as to relax the standards or to reduce the penalties for non-compliance, then enterprise managers would be more free to evaluate a safety program on the basis of whether or not it would provide savings that would at least counterbalance any costs and financial losses to which the company might be subjected.

The prospects of having to justify their programs on an economic basis should be enough to cause most safety engineers to breakout in a "cold sweat." The problem is that many of the benefits of a safety program are difficult to quantify. The purpose of this paper is to illustrate how intervention analysis could be used to quantify such benefits. In particular it will focus on the problem of assessing the impact of a traumatic accident on productivity in future time periods. The second section will discuss the problems which could arise if classical statistical techniques, i.e. those which assume independent observations, were to be used to assess the impact of a traumatic accident on future productivity. The third section will explain how transfer function/intervention analysis might be used to overcome these problems. In the fourth section transfer function/intervention analysis will be used to assess the impact of a fatal accident on productivity. The results obtained from this analysis will be contrasted with those obtained from analyses based on classical statistics.

2. STATISTICAL PROBLEM IN ASSESSING LONG TERM IMPACT OF TRAUMATIC ACCIDENT ON PRODUCTION

2.1. INTRODUCTION

The task of assessing the long term impact of a traumatic accident on production involves two phases. The first is concerned with determining whether or not the accident had a significant impact on the level of the time series. The second is concerned with determining whether or not the accident had a significant impact on the rate at which the level of the time series is changing. Together, these two factors provide the information necessary for determining the amount of production lost as a result of the accident.

Culver (1979) has proposed using classical statistical techniques, i.e., those based on the assumption of independent observations, to assess the long term impact of a traumatic accident on production. In particular he proposed using the t-test of the difference between the means of two populations to determine whether or not the accident was associated with a significant loss in productivity. He also proposed fitting a regression model to the post accident data and then using this model to estimate the length of time that would be required to recover from the productivity loss which resulted from the accident.

The problem with the approach proposed by Culver is that both the t-test of the difference between two population means and regression analysis are based on the assumptions that the observations in the t-test and the error terms in regression analysis are independent. However, it will be seen shortly that not only are these assumptions unlikely to be satisfied in practice, but also that failure to satisfy these assumptions can have a dramatic effect on the level of significance associated with the t-test and regression analysis.

2.2. CAUSES OF DEPENDENT OBSERVATIONS

There are two factors which suggest that successive production observations might not be independent. First, there is the well known learning curve effect which suggests that the time to perform a particular task decreases as the number of repetitions of the task increase. This suggests that, all other things being equal, the rate of production should increase over time. This tendency for the rate of production to increase over time will tend to induce a positive autocorrelation into a set of production observations. A second consideration is that many production workers are afraid, and with good justification, that if they produce too much that management will increase the production standards. As a result of this fear workers in a given area will often adopt what they consider to be safe upper limits for production and bring enormous pressure to bear on those who violate these levels. Consequently, this means that if a person or a team produces more than the prescribed upper limit during the early part of a week, then they must produce less than the upper limit by an amount sufficient to bring the weekly production to within the upper limit. This type of behavior can induce negative autocorrelations into a set of production observations.

The nature of autocorrelation in a particular set of production data will depend on the relative strength of the above mentioned factors. If the learning effect is stronger than the safe upper limit effect, then the observations will have positive autocorrelations. If the reverse is true, then the observations will have negative autocorrelation. However, in actual practice it will not be possible to determine the relative strengths of these two factors. Consequently, it will not be possible to specify a priori whether the observations will have positive or negative autocorrelations and to determine the relative strength of such autocorrelations.

2.3. CONSEQUENCES OF FAILURE TO SATISFY ASSUMPTION OF INDEPENDENCE

The failure to satisfy the assumption of independence can have a dramatic impact on the level of significance and the power of various statistical tests. Cochran (1947) observed that a positive correlation between observations in analysis of variance will cause the actual variance of a treatment mean to be larger than would be the case for an independent series, while the estimated variance will be smaller than the variance for an independent series. If the autocorrelation is negative, then the actual variance will be smaller, while the estimated variance will be larger than they would be for an independent series. Box (1954) studied the effects of correlation between errors in the two way classification analysis of variance problems. Scheffe' (1959, p. 339) presented the following table to demonstrate that the effect of serial correlation, i.e., autocorrelation, on inferences about means can be serious.

TABLE 1

Effect of Serial Correlation (ρ) on True Probability of Nominal 95 Percent Confidence Interval for μ Not Covering True μ for Large n.

ρ	-0.4	-0.3	-0.2	-0.1	0	0.1	0.2	0.3	0.4
Probability	1.000	0.002	0.011	0.028	0.050	0.074	0.098	0.120	0.140

This table demonstrates that the true probability of nominal 95 percent confidence interval for μ not covering the true value of μ is equal to 0.050 for $\rho = 0$, i.e. independent observations. However, for $\rho > 0$ the probability of not covering is higher than it should be, while for $\rho < 0$ the probability of not covering is lower than it should be. This led Scheffe' (1959, p.338) to conclude that "the effect of serial correlation on inferences about means can be serious." Zellner and Tiao (1964) have concluded that violation of the assumption of independent error can have a dramatic effect on the inferences which can be legitimately drawn in a Bayesian regression analysis. Gastwirth and Rubin (1971) have concluded that relatively slight dependencies can have a strong influence on the level of significance of two nonparametric tests for the mean: the sign test and the Wilcoxon test. Box and Newbold (1971) have shown that the failure to correctly account for autocorrelated residuals in a regression analysis can lead to nonsensical results.

The works cited above all indicated that the failure to properly account for the autocorrelation inherent in a set of data will increase the risk that faulty conclusions will be drawn.

3. TRANSFER FUNCTION INTERVENTION ANALYSIS

3.1. INTRODUCTION

The preceding section has noted that there are two situations, i.e. learning curve effect and social pressures to restrain production, that could induce autocorrelation into a set of production data. This section also noted that use of statistical procedures, which does not directly account for the dependency in a given set of data, to assess the impact of unique, identifiable events, such as a traumatic accident, on production can dramatically increase the risk of making a wrong decision. The purpose of this section is to describe a technique, known as transfer/function intervention analysis, which provides a systematic procedure for accounting for such dependencies and for assessing the impact of isolated identifiable events such as traumatic accidents.

The general form of the transfer function intervention model (Box and Tiao, 1975) partitions the variation in an appropriate Box-Cox transformation (Box and Cox, 1964) of the original time series as the sum of three components: dynamic component, stochastic noise component, and white noise component. The transfer function intervention model building approach provides a systematic procedure (Box and Tiao, 1975) for determining models for the dynamic and stochastic noise components such that the residuals will be statistically indistinquishable from a white noise process.

3.2. DYNAMIC COMPONENT

The purpose of the dynamic component of the transfer function intervention model is to account for the manner in which the system, which is generating a time series, reacts to the occurrence of unique, identifiable events which will also be referred to as interventions. This reaction can begin immediately or it can be delayed for a variable number of periods. Once the reaction starts, the rate of reaction can be slow and gradual or it can be instantaneous and abrupt.

An intervention can create either a permament or a temporary change in the behavior of the time series. An intervention which creates a permanent change can be represented by a step function intervention variable

$$S_t^{(T)} = \begin{matrix} 0 & \text{for } t < T \\ 1 & \text{for } t \geq T \end{matrix}$$

where T represents the time at which the intervention occurred. On the other hand, an intervention which creates a temporary change can be represented by a pulse function intervention variable

$$P_t^{(T)} = \begin{matrix} 0 & \text{for } t \neq T \\ 1 & \text{for } t = T \end{matrix}$$

In the following discussion y_t will be used to denote the dynamic component of the discrete time series which describes the state of the system at time t. In order to adequately represent the impact of an intervention it will be necessary to utilize a model which permits the rate of change of y_t to be instantaneous and abrupt or slow and gradual or anything in between. It will also be necessary for the model to reflect the fact that the reaction can begin immediately or be delayed for a variable number of periods. As will be seen shortly, a first order linear difference equation satisfies the above requirements. In order to describe such a model it will be useful to employ the backshift operator, B, which is defined as $B^n y_{dt} = y_{t-n}$ for n = 1, 2, 3, ...

The first order linear difference equation for the response of y_t to a step function intervention is given by

$$y_t = \frac{\omega_s B^b S_t^{(T)}}{(1 - \delta_s B)} \qquad (1)$$

where $S_t^{(T)}$ is the step function intervention variable previously defined, b is the delay before the system starts reacting, and ω_s and δ_s are unknown parameters. The parameter δ_s determines the rate at which the system adjusts to the step function intervention. $\delta_s = 0$ implies an instantaneous rate of adjustment in which case the level of y_t abruptly changes by ω_s after the passage of b time periods. $\delta_s = 1$ implies that after a delay of b periods the level of the system y_t increases by a constant amount ω_s in each succeeding time period. In this case the level of the system will increase without bound. Both of the above cases are somewhat unrealistic. Most systems have a finite rate of adjustment and are bounded. This case is represented by $0<\delta_s<1$. In this case the quantity $\omega_s/(1-\delta_s B)$ represents the asymptotic change in the level of the system.

The first order linear difference equation for the response of y_t to a pulse function intervention is given by

$$y_t = \frac{\omega_p B^b {}_t^{(T)}}{(1 - \delta_p B)} \tag{2}$$

where $P_t^{(T)}$ and b have been previously defined and ω_p and δ_p are unknown parameters. The parameter δ_p determines the rate at which the system adjusts to the pulse function intervention. $\delta_p = 0$ implies an instantaneous rate of recovery, while $\delta_p = 1$ implies that there is no recovery. For $0<\delta_p<1$ the quantity $\omega_p/(1-\delta_p B)$ represents total displacement in the level of y_t created by the pulse intervention function.

In practice the reaction of a system to a unique, identifiable event could consist of both an initial shock component and a gradual adjustment component. The initial shock component results in an abrupt temporary effect which results when a unique, identifiable event occurs. The gradual adjustment component represents a permanent effect which results as the system successively adjusts to the new environment created by the occurrence of the unique identifiable event. In this case the dynamic component of the system's behavior might be described as

$$y_t = \frac{\omega_s B^b S_t^{(T)}}{(1 - \delta_s B)} + \frac{\omega_p B^b p_t^{(T)}}{(1 - \delta_p B)} \tag{3}$$

After the parameters of the above model have been estimated, it is highly unlikely that the residual errors will be independent, identically distributed (iid) random variables. If this occurs, then there will be a pattern of dependency in the residuals, which indicates that all of the useful information in the data has not been incorporated into the model. The purpose of the noise component, described in the next subsection is to account for the dependency pattern inherent in the residuals for the dynamic component of the transfer function intervention model.

3.3. STOCHASTIC NOISE COMPONENT

The stochastic noise component utilizes an ARIMA model (Box and Jenkins, 1970) to account for the dependency pattern which is inherent in the residuals for the dynamic component of the transfer function intervention model. The general form of such a model is

$$(1 - \sum_{i=1}^{q} \phi_i B^i)\, r_t = (1 - \sum_{i=1}^{p} \theta_i B^i)\, a_t \tag{4}$$

where r_t and a_t represent the residuals from the dynamic component model and a white noise time series respectively and the ϕ_i's and θ_i's are unknown parameters. The quantity $(1 - \sum_{i=1}^{p} \phi_i B^i)$, is referred to as an autoregressive operator and can be used to model a system which gradually adjusts to shocks. The quantity $(1 - \sum_{i=1}^{q} \theta_i B^i)$, is referred to as a moving average operator and can be used to model a system which abruptly adjusts, after possibly a finite delay, to shocks. The autoregressive and moving average operators together can provide the flexibility needed to account for a great deal of the pattern in the residuals.

5. IMPACT OF A FATAL ACCIDENT ON PRODUCTIVITY

The data reported by Culver (1979) were analyzed by the transfer function/intervention analysis approach described above. This analysis revealed that the occurrence of a fatal accident did not produce a significant change in either the level or the rate of change of the productivity time series. These results are in contrast to those obtained by Culver, who used the two sample t-test to conclude that there was a significant, at the 0.05 level, drop in production after the accident. The reason for the lack of agreement between the results of the t-test and the intervention analysis was found to lie in the fact that the autocorrelation in the productivity time series had distorted the level of significance of the t-test as indicated by Cochran (1947).

6. SUMMARY AND CONCLUSION

This paper was concerned with the problem of determining the impact of a traumatic accident on productivity. It described situations which could cause production data to be autocorrelated and discussed how such autocorrelation could distort the level of significance for statistical tests which are based on the assumption of independent observations. It also described an approach, transfer function/intervention analysis which could be used to estimate the impact of a traumatic accident on productivity in the presence of autocorrelated data.

REFERENCES

BOX, G. E. P. (1954). Some Theorems on Quadratic Forms Applied in the Study of Analysis of Variance Problems: II. Effects of Inequality of Variance and of Correlation Between Errors in the Two-Way Classification, *Ann, Math. Statist.* 25, 484.

BOX, G. E. P. and COX, D. R. (1964). An Analysis of Transformations, *J. Roy. Stat. Soc.* B, 26, 211-243.

BOX, G. E. P. and NEWBOLD, P. (1971). Some Comments on a Paper of Coen, Gomme, and Kendall, *J. Roy. Stat. Soc.* A, 134, 229-240.

COCHRAN, W. G. (1947). Some Consequences When the Assumptions for the Analysis of Variance Are Not Satisfied, *Biometrics*, 3, 22-38.

CULVER, D. (1979). Accident: Related Losses Make Cost Soar, *Industrial Engineering*, 11, 26-29.

GASTWIRTH, J. L. and RUBIN, H. (1971). Effect of Dependence on the Level of Some One-Sample Test, *J. Amer. Statist. Assoc.*, 66, 816-820.

SCHEFFE', H. (1959). *The Analysis of Vaxiance*, Wiley, New York.

ZELLNER, A. and TIAO, G. C. (1946). Bayesian Analysis of the Regression Model with Autocorrelated Errors, *J. Amex. Statist. Assoc.*, 59, 763.

APPLIED TIME SERIES ANALYSIS
O.D. Anderson & M.R. Perryman (eds.)

LEARNING AND INFLATION

William Robert Terry
Industrial Engineering, University of Toledo
and
Franz A. P. Frisch
Sea Systems Command, U. S. Navy
and
Shiv G. Kapoor
Mechanical Engineering, University of Illinois

This paper discusses the limitations of traditional approaches for modeling learning and inflation. It briefly describes an alternative model building approach, data dependent systems (DDS), which does not suffer from these limitations. It then illustrates the DDS approach by developing learning and inflation models for the durable goods section of the U. S. economy.

1. INTRODUCTION

A large scale project typically consumes large amounts of scarce resources and requires a considerable amount of calendar time to complete. Minimizing the total cost of such a project can be a key consideration. However, it will be seen shortly that the task of minimizing the total cost of such a project is complicated by effects of learning and inflation. Both of these effects can be influenced by the timetable or schedule for completing the project. Thus, specifying the schedule for completing the project is a key decision in planning a large scale project.

The effects of learning and inflation on the total cost of a project can be seen by considering what occurs when the schedule for completing the project is either compressed or elongated. Compressing the completion schedule will reduce the amount of time that inflation has to drive up the prices of resources. Compressing the schedule will also reduce the amount of time for learning to occur which can increase the amounts of resources required to accomplish the project. Elongating the completion schedule can have exactly the opposite effect; more time will be available for both learning and inflation to influence total cost.

The problem of specifying the completion schedule which will minimize total cost involves making a trade-off between the effects of learning and inflation. Unfortunately, the process of making this trade-off is complicated by the fact that the amounts of learning and inflation that occur in future time periods are stochastic variables. Thus, the first step toward determining the optimum completion schedule is to develop models of the stochastic generating processes for learning and inflation. The purpose of this paper will be to illustrate how time series analysis can be utilized to develop these models. The problem of utilizing these stochastic process models to determine the amount of calendar time for completing a project which will minimize expected total cost will be considered elsewhere.

The remainder of this paper is organized as follows. Sections 2 and 3 discuss limitations of traditional approaches for modeling learning and inflation. Section 4 briefly discusses an alternative modeling building appraoch, Data Dependent Systems (DDS), which does not suffer from these limitations. Section 5 presents an example in which the DDS approach is used to model actual learning and inflation data.

2. LEARNING

2.1. INTRODUCTION

It has often been demonstrated that the amounts of the various resources, e.g., manpower, equipment, energy, and material, required for the accomplishment of a task tends to decrease as the number of repetitions of the task increase. This phenomenon is referred to as the learning effect or learning for short. Cheney (1977, p. 19-20) notes that a wide variety of factors can be responsible for or contribute to the learning effect: "worker learning, designer learning, production planning, scheduling, sequencing of operations, synchronization of functions, ordering of materials in proper sizes and quantities, better use of materials to minimize waste, specialization, worker morale, rejection and rework reduction, increased lot sizes, and reduced quantity of engineering changes."

The rate at which learning occurs can vary widely among tasks. The rate of learning tends to be relatively high for manually paced tasks and relatively low for machine paced tasks. The rate of learning also tends to be relatively high for complex tasks and relatively low for simple tasks. Furthermore, the rate of learning is not constant for all resources.

2.2 MATHEMATICAL MODELS

The traditional form of the mathematical model for representing the learning effect is:

$$Y = AX^B \tag{1}$$

where Y represents a measure of the resource consumed, X represents the cumulative number of units produced, A is a parameter reflecting the amount of the resource required to produce the first unit, and B is a parameter reflecting the rate of learning. The quantity Y in equation (1) can be used to represent either the cumulative average amount consumed in producing the first X unit or the amount of resource consumed in producing unit X.

Equation (1) is referred to as a power-form learning model. This model implies that the amount of resource consumed, Y, decreases by a constant proportion as the cumulative production, X, increases proportionally. The power-form model can be transformed to a linear model by taking the logarithm's of both sides of equation (1):

$$\log Y = \log (AX^B) = \log A + B \log X \tag{2}$$

This transformation makes it possible to use ordinary least squares to estimate the values of A and B.

2.3. LIMITATIONS OF POWER FORM

The power form learning model lacks the flexibility needed to adequately model many real world situations. This lack of flexibility results from three factors. First, the power form model implies that learning can continue indefinitely and that the amount of resource required to accomplish a task approaches zero asymptotically as the number of repetitions increase without bound. Unfortunately, there are many real world situations in which this does not occur. Second, the use of regression analysis to estimate the values of the models parameters will produce biased estimates when residual errors are autocorrelated. Needless to say there is no underlying, immutable, natural law which says that residual errors will not be autocorrelated. Third, the power form model does not account for the fact that learning can be influenced by exogeneous factors such as weather, experience of work force, delivery lead time, production facility wearout, production run duration, duration of a break in production, and design changes. These factors are part of reality and ignoring them can lead to biased estimates.

The presence of the problems cited above suggests the need for an approach for developing learning models which is more flexible than the power form model/ linear regression analysis appraoch. Such an approach is the Data Dependent Systems approach which is described in Section 4.

3. INFLATION

3.1 INTRODUCTION

The prices which must be paid to obtain the resources needed for completing a large scale project can be expected to increase with the passage of time due to inflation. However, the prices for all resources in relatively short supply can be expected to inflate at a faster rate than those which are more abundant.

Obviously, reliable and accurate forecast of inflation could be most valuable information for decision makers responsible for the planning and execution of large scale projects. The traditional approach to forecasting inflation utilizes econometric models.

3.2. LIMITATIONS OF ECONOMETRIC MODELS

In theory a model for forecasting price indices could be obtained from either an econometric model or a time series model. However, a number of empirical tests have indicated that time series based forecasts were more accurate than those produced by econometric forecasts. Naylor, Seaks, and Wichern (1977) compared forecasts generated by the Box and Jenkins (1970) method with those of the Wharton model. This study, based on quarterly forecasts from 1963 to 1967, revealed that the Box-Jenkins method produced much more accurate forecasts. In a 1976 paper, Rausser and Oliveria used a Box-Jenkins model and an econometric model to predict wilderness area usage. They concluded:

> It is doubtful, in an application where it would be necessary to use forecasted values of the exogeneous variables, that the econometric predictions would be superior to the ARIMA (Box-Jenkins) values (p. 284).

Granger and Newbold (1975) cited two references, Bray (1971) and Cooper (1969), which indicated that time series model had provided more accurate forecasts than forecasts derived from econometric models. Granger and Newbold offered the following explanation as to why econometric models failed to produce better forecasts than those derived from time series models:

> If the statistician can consistently produce the better forecasts, one might well conclude that the economic theory is not very near to the truth or that the true theory has not been very well represented by the formulation actually used in the econometric model. (Granger and Newbold, 1975, p.2)

4. DATA DEPENDENT SYSTEMS

4.1. INTRODUCTION

The discussions of learning and inflation in sections 2 and 3 respectively have indicated the need for a model building approach which utilizes past data to determine an appropriate model. Two approaches which can utilize past data to determine an appropriate model are the Box and Jenkins (BJ) approach and the Data Dependent Systems (DDS) approach developed by Pandit (1979).

These approaches are similar in some respects. Both utilize a flexible family of models and a systematic procedure for selecting an appropriate member of the family. Although the BJ approach and the DDS approach are similar, there are nevertheless some important differences. The BJ approach is restricted to systems in which the variable of interest is observed continuously or at equally spaced intervals on an appropriate metric.

Data which is available for developing learning and inflation models is typically available at equally spaced time intervals. Historical data on resource consumption for similar tasks provide the basis for developing a learning model, while historical resource price data provide a basis for developing an inflation model. However, the mathematical analysis necessary for determining the completion schedule, which will minimize total cost, will be more tractable if both learning and inflation are treated as continuous processes. This factor favors use of the DDS approach since it can utilize data sampled at uniform intervals to develop a differential equation for describing the dynamic behavior of a continuous process. The next subsection discusses the theory which supports the DDS approach. This is followed by a brief outline of the DDS modeling procedure.

4.2. DDS THEOREMS

There are two theorems on which the DDS approach is based: the fundamental theorem of DDS and the uniform sampling theorem. The fundamental theorem of DDS provides a basis for formulating the modeling problem as an eigenvalue series expansion. It also provides the basis for the use of nonlinear least squares to successively fit stochastic differential equations of increasing orders until increasing the order further does not provide a statistically significant reduction in the residual sum of squares. In particular the fundamental theorem of DDS states that an arbitrary stationary stochastic system can be represented as closely as desired by a model of the form

$$\frac{d^nX(t)}{dt^n} + \alpha_{n-1}\frac{d^{n-1}X(t)}{dt^{n-1}} + \ldots + \alpha_1\frac{dX(t)}{dt} + \alpha_0X(t)$$

$$= b_{n-1}\frac{d^{n-1}Z(t)}{dt^{n-1}} + \ldots + b_1\frac{dZ(t)}{dt} + Z(t) \qquad (1)$$

where $Z(t)$ is a continuous white noise process with $E[Z(t)] = 0$ and $E[Z(t)Z(t-u)] = \delta(u)\,\sigma^2$, $X(t)$ is the system response to the white noise input, $\alpha_0, \alpha_1, \ldots, \alpha_{n-1}$ are autoregressive parameters (AR), $b_1, b_2, \ldots, b_{n-1}$ are moving average (MA) parameters, and $\delta(u)$ is the Dirac delta function. Pandit refers to the above model as an Autoregressive Moving average model of order n and n-1 and denotes it by AM(n, n-1).

The uniform sampling theorem states that when a continuous time stationary stochastic system is sampled at uniformly spaced intervals it can be represented by a stochastic difference equation of the ARMA form

$$X_t - \phi_1X_{t-1} - \phi_2X_{t-2} - \ldots - \phi_nX_{t-n}$$

$$= a_t - \theta_1a_{t-1} - \theta_2a_{t-2} - \ldots - \theta_{n-1}a_{t-n+1} \qquad (2)$$

where a_t is a discrete time white noise process with $E[a_t] = 0$ and $E[a_ta_{t-k}] = \delta_K\sigma_a^2$ where δ_K is the Kronecker delta which is zero for $K \neq 0$ and one for K=0. Pandit refers to the model in equation (2) as a Uniformly Sampled Autoregressive Moving-average model of order n,n-1, and uses USAM(n,n-1) to denote it. The uniform sampling theorem specifies the relationship between the parameters of the AM(n,n-1) model and the USAM(n,n-1) model. Details of the parametric relationships between these models are given in Pandit (1979).

Both the fundamental theorem and the uniform sampling theorems are based on the assumption that the time series to be analayzed is stationary. Fortunately, this assumption is not overly restrictive as Pandit (1977, p. 222) has observed that "nonstationary components, if present, can be removed by sinusodial, linear, or exponential trends."

4.3. DDS MODELLING STRATEGY

The DDS modeling process consists of fitting successively higher order USAM (n) models until an F test indicates that the increase in the order fails to produce a significant improvement in fit. The uniform sampling theorem is then used to determine the parameters of the corresponding AM(n,n-1) model. Further details of the DDS modeling process are available in Pandit (1979).

5. EXAMPLE APPLICATIONS

In this section the DDS approach will be utilized to develop stochastic differential equation models for the generating processes for learning and inflation in the durable goods producers sector of the U. S. economy. Data for this analysis were obtained from monthly data compiled by the U. S. Department of Labor, Bureau of Labor Statistics for the period January 71 to December 78.

5.1. DDS MODEL OF LEARNING

The discrete time series which was used in developing a learning model was calculated by

$$p_t = \frac{i_t}{e_t}$$

where i_t and e_t represent respectively the seasonally adjusted values of the production index and the number of employees on the payrolls for durable goods industries in the U.S.

An examination of a plot of the p_t time series revealed that it was nonstationary. This nonstationarity was accounted for by utilizing the approach described in Kapoor, Madhok, and Wu (1981). This approach utilized Wold's decomposition theorem (1938) to justify decomposing an observed time series y(t) into two components

$$y(t) = \mu(t,\psi) + x(t)$$

where $\mu(t,\psi)$ is a deterministic function for the mean with parameter vector ψ and x(t) is a zero mean stochastic series. The Kapoor, Mudhok and Wu appraoch utilizes a three stage model building process for determining the most appropriate models for $\mu(t,\psi)$ and x(t). In stage one an initial estimate of the parameter vector, ψ, is obtained by minimizing the sum of squares of residuals. Stage two employs the DDS approach to determine the order and initial parameters estimates of an USAM model of the residuals from stage one. Stage three uses Marquardt's algorithms to simultaneously estimate the parameters of the deterministic function $\mu(t,\psi)$ and the USAM model.

The application of the Kapoor, Madhok and Wu method to the learning time series for the durable goods industries resulted in a continuous time model with deterministic component

$$.9213 + .0025t.$$

and stochastic component given by

$$\frac{d^2x(t)}{dt} + 0.438\frac{dx(t)}{dt} + 0.062x(t) = z(t)$$

where z(t) is a continuous white noise series. All of the parameters of both the deterministic and stochastic components were significant at the 0.05 level.

5.2. DDS MODEL OF INFLATION

The wholesale price index for durable goods industries was used to develop a continuous time model for inflation in the durable goods industries. A plot of this discrete time series revealed that it was nonstationary. This nonstationarity was accounted for by utilizing the Kapoor, Madhok and Wu approach described in the previous section. This resulted in a continuous time model for inflation in the durable goods industries with deterministic component

$$113.25 + 0.9787t$$

and stochastic component given by

$$\frac{d^2x(t)}{dt^2} + 0.113\frac{dx(t)}{dt} + 0.0077x(t) = z(t)$$

where z(t) is a continuous white noise input. All of the parameters of both the determination and the stochastic components were significant at the 0.05 level.

6. CONCLUSION

Both the models for learning and for inflation developed for the durable goods industries required second order differential equations to describe the stochastic component. This indicates the presence of a strong dependence in the residuals which could bias the results of parameter estimates if ordinary least squares were used to estimate the parameters of models for learning and inflation.

REFERENCES

BRAY, J. (1974). Dynamic Equations for Economic Forecasting with GDP-Unemployment Relations and the Growth of GDP in the UK as an Example, *J. Roy. Stat. Soc.*, Vol. 134, 167-209.

CHENEY, W. F. (1977). Strategic Implications of the Experience Curve Effect for Avionics Acquisitions by the Department of Defense, Doctoral Dissertation, Purdue University.

COOPER, R. L. (1969). The Predictive Performance of Quarterly Econometric Models, Doctoral Dissertation, University of California.

GRANGER C. W. J., and NEWBOLD, P. (1975). Econometric Forecasting: The Atheists Viewpoint, in *Modeling the Economy*, Ed: G. A. Renton, Heinemann Education Books, London.

KAPOOR, S. G., MADHOK, P. and WU, S. M. (1981). Modeling and Forecasting Sales Data by Time Series Analysis, *Journal of Marketing Research*, XVIII, 94-100.

NAYLOR, T. H., SEAKS, T. G., and WICHERN, D. W. (1977). Box-Jenkins Methods: An Alternative to Econometric Models, *Int. Stat. Rev.*, Vol. 40, 123-137.

PANDIT, S. M. (1977). Stochastic Linearization by Data Dependent Systems, Trans. of ASME, *J. Systems Dynamics and Control*, 221-226.

PANDIT, S. M. (1980). Data Dependent Systems and Exponential Smoothing, in *Analysing Time Series*, (Proceedings of International Conference held at Nottingham University, March 1979). Ed: O. D. Anderson, North Holland, Amsterdam, 217-238.

RAUSSER, G. C., and OLIVERIA, R. A. (1975). An Economic Analysis of Wilderness Area Use, *JASA*, Vol. 71, 276-285.

TERRY, W. R. and BISHOP, L. R. (1981). A Time Series Analysis Approach for Designing Military Compensation Systems," in *Applied Time Series Analysis*, (Proceedings of International Conference held at Houston, Texas, August 1981). Eds: O. D. Anderson and M. R. Perryman, North-Holland, Amsterdam), to appear.

WOLD, H. O. A. (1954). *A Study in the Analysis of Stationary Time Series*. 2nd ed., Almqvist & Wiksell, Stockholm.

APPLIED TIME SERIES ANALYSIS
O.D. Anderson & M.R. Perryman (eds.)

A TIME SERIES ANALYSIS APPROACH FOR EVALUATING THE IMPACT OF SAFETY PROGRAMS

W. Robert Terry
Industrial Engineering, University of Toledo
and
Dennis L. Price
Industrial Engineering Operations Research, Virginia Tech.

Safety programs are implemented in an operating environment where the ability to conduct controlled experiments is limited. This inability to conduct controlled experiments complicates the task of evaluating the impact of such a program. This paper describes how transfer function/ intervention analysis can be used to cope with the threats to valadity which result from the inability to conduct controlled experiments.

1. INTRODUCTION

On December 29, 1970 Congress passed Public Law 91-596 which has become known as the Occupational Safety and Health Act (OSHA) of 1970. The purpose of the law was to assure safe and healthful working conditions for working men and women; by authorizing enforcement of the standards developed under the Act; by assisting and encouraging the states in their efforts to assure safe and healthful working conditions; by providing for research, information, education, and training in the field of occupational safety.

Since the law's passage, a rather rigorous set of standards has been enforced throughout the United States by OSHA compliance officers with authority to impose stiff fines for noncompliance and to shut down operations for repeated violators.

The threat of such penalties has made it relatively easy to justify large expenditures for safety programs. As a result the past ten years have seen an enormous proliferation of safety programs throughout industry. However, in the near future the survival of many of these programs could be in jeopardy as a result of a dramatic change in the sentiments of the general public towards government regulation of business.

In the era when OSHA was passed, there was widespread support for the thesis that government regulations were needed to protect individuals from the unconscionable acts of large organizations. Today the pendulum has swung back in the other direction. Now there appears to be wide-spread support for the thesis that excessive governmental regulations are responsible in part for inflation and for the inability of U.S. businesses to compete effectively in the international arena.

These anti-regulatory sentiments suggest that the various OSHA programs will be scrutinized more closely than ever before. Those programs which cannot present convincing evidence that they have been successful are likely to be eliminated. Unfortunately it will not be possible to conduct controlled experiments to obtain evidence that a safety program has been successful. Thus, it will be crucial, if truly effective safety programs are to be kept, to provide convincing evidence of success at little or no cost.

The purpose of this paper is to present an approach which utilizes transfer function/intervention analysis to evaluate natural experiments, i.e. experiments in which the investigator lacks the ability to randomly assign subjects to experimental and control groups and to control the experimental environment to the extent necessary to rule out the existence of rival causal mechanisms. Section 2 presents a taxonomy of rival causal mechanisms which can not be ruled out in a natural experiment. Section 3 describes some of the problems which can arise in the analysis of natural experiments and how transfer function/ invention analysis can be used to cope with these problems.

2. RIVAL CAUSAL HYPOTHESES - THREATS TO VALIDITY

2.1. INTRODUCTION

In a natural experiment the investigator can intervene to measure, but can not intervene to control experimental variables or to randomly assign subjects to treatment groups. The inability to exercise experimental control or to randomly select subjects can make it difficult to rule out certain rival causes which could also account for an observed effect. The potential for these rival causes to account for an observed effect constitutes a threat to the validity of natural experiments. It will be seen in this subsection, a key factor in substantiating a causal relationship in such situations is the ability to prove that none of the other factors, i.e. rival causal mechanisms, which could have caused the effect were present.

2.2 SUBSTANTIATING CAUSAL RELATIONSHIPS IN COMPLEX SITUATIONS

In a simple situation it will be possible to identify the complete set of all possible causal factors which are assumed to be binary in nature, and to understand in detail the process which transforms a particular causal factor into a specified effect. However, most real world situations will be more complex. There can be an incomprehensible number of causal factors which can be present at a number of different levels. The transformation process, which converts a particular cause into a specified effect, can be inperceivably complex. The various causal factors can operate in either a noncompensatory mode, which does not permit tradeoffs between or among various levels of the causal factors, or a compensatory mode in which such tradeoffs are permissible.

The above mentioned differences between simple and complex situations make it necessary to employ different approaches to substantiating the existence of causal relationships in these situations. In a simple situation deductive logic can be used to establish the necessary and sufficient conditions for the existence of a cause-effect relationship. In contrast, an inferential approach must be used to obtain evidence for supporting the existence of a causal relationship in a complex situation.

The types of situations which will typically be encountered in evaluating the impact of a safety program will typically be far too complex to be handled by deductive logic. Thus, the investigator responsible for evaluating such a program will have to rely primarily on evidence which can be obtained through an inferential process.

There are three conditions which must be met in order to infer the existence of a causal relationship. The first is referred to as "associative variation." Two types of associative variation are: (1) association between two variables which reflects the extent to which the presence of a specified level of a variable is associated with the presence of a particular level of the other variable, and (2) association between the changes of two variables which reflects the extent to which a change in the level of one variable is associated with a change in the level of the other. Note, however, that the presence of associative variation is not sufficient for the existence of a causal relationship since since the variation in both variables could be caused by some extraneous variable. The second condition which must be satisfied is that changes in the hypothesized causal factor must not occur after changes in the hypothesized effect has occurred. However, the fact that the hypothesized cause precedes the effect and the presence of associative variation are not jointly sufficient to establish the existence of a causal relationship, since it is possible that changes in the cause prior to changes in the effect were purely coincidental. The third condition for establishing a causal relationship is that the possibility that the presence of other possible causal factors, i.e. rival causal mechanisms, which could have produced the observed effect must be ruled out.

3. STRATEGIES FOR COPING WITH THREATS TO VALIDITY OF A NATURAL EXPERIMENT

3.1. INTRODUCTION

The possible presence of one or more rival causal mechanisms, which could have produced the observed effect, constitutes a threat to the validity of evidence obtained from a natural experiment. Such threats to the validity of a natural experiment result from the inability of the investigator to control experimental conditions and to randomly assign subjects to treatment groups. The inability to randomly assign subjects to treatment groups prevents the use of a random assignment process to ensure that the compositions of the various treatment groups are probabilistically equivalent. The lack of probabilistically equivalent treatment groups opens the possibility that differences between and among treatment groups were caused by the differences in the composition of the groups rather than the differences in the treatments which they received. On the other hand, the inability to control experimental conditions precludes isolating subjects from the influence of other causal variables. This opens the possibility that differences between and among treatment groups were caused by differences in the levels of the other causal variables to which the various treatment groups were exposed. The purpose of this section will be to present a basic strategy for dealing with the above mentioned threats to validity. For the purposes of this discussion the application of a treatment, i.e., change in the level of a causal variable, will be referred to an intervention.

3.2. BEFORE AFTER REPEATED MEASURES DESIGN WITHOUT CONTROL GROUP

The before after repeated measures design without a control group involves observing the behavior of one experimental group for a number of time periods both before and after the intervention. The logical basis for a causal inference is that, in the absence of other causal factors, any abrupt change in the behavior of the time series of observations that coincides with the intervention is the result of that intervention.

The observations before the intervention provide a basis for establishing the behavioral patterns of the time series in the absence of the intervention. In the absence of other causal factors, the observations after the intervention provides a basis for establishing the behavioral pattern of the time series in the presence of the intervention. The difference between these two behavioral patterns represents the effect of the intervention.

The absence of an equivalent control group can complicate the problem of determining if other causal factors were present. Nevertheless evidence regarding the presence or absence of other causal factors can still be obtained. One way to do this is to keep a log of all possible events which could affect the behavior of the hypothesized effect variable. This log should note changes in the way records are kept and in the composition of the experimental group as well as changes in the levels of other causal variables.

The ability to generalize the results of a single group-single intervention study will be dependent on the extent to which the characteristics of the group are similar to those of other populations. The estimated effect of the intervention can also be biased by the omission of relevant variables.

3.3. BEFORE AFTER REPEATED MEASURES DESIGN WITH CONTROL GROUP

In the before after repeated measures design with a control group an active intervention is introduced into the experimental group while a placebo intervention is introduced into the control group. The behavior of each group is observed for a number of time periods both before and after each of the interventions. Since the control group and the experimental group are equivalent, it is reasonable to expect that a change in the level of one of the other causal factors will influence both the control groups and the experimental groups in roughly the same fashion. This will make it possible to estimate the impact of the various interventions by comparing the behavior of each experimental group with that of the control group.

4. STATISTICAL ANALYSIS OF NATURAL EXPERIMENTS

4.1. INTRODUCTION

Section 2 has discussed some of the major ways in which the inability to conduct randomized controlled experiments, i.e. natural experiments, can threaten the validity of the evidence obtained. Section 3 has discussed the basic strategies for controlling these threats. All of these strategies involve making a number of equally spaced observations before and after the introduction of an intervention. The combined set of before and after observations can be regarded as interrupted or mongrel time series. The purpose of this section is to describe an appropriate method for analyzing such data. The second subsection examines the nature of the processes which could generate the data needed for the evaluation of the impact of a safety program and concludes that the resulting observations are likely to be dependent and that such observations could violate the assumption of independence needed for valid application of classical statistical methods such as the t-test for the difference in means, ANOVA, ANCOVA, etc. The third subsection illustrates how transfer function/intervention analysis, which explicitly recognizes the dependency inherent in a time series, can be used to analyze the data obtained from natural experiments which could be used to evaluate the impact of a safety program.

4.2. DEPENDENT DATA PROBLEM

In a natural experiment a well defined group, the composition of which is stable during the time interval of interest, is observed at a number of equally spaced time intervals. As noted in section 2, there may be a number of rival causal mechanisms which can influence the behavior of the group in addition to the possible influence of the hypothesized causal factor. The presence of one or more of the rival causal factors has the potential for inducing dependencies into the behavior of a group.

This potential for inducing dependencies results from the fact that a change in the level of rival causal factors constitutes an exogeneous shock which is not accounted for by the system which defines the relationship between the hypothesized cause and the hypothesized effect. In a statistical model of this hypothesized causal system the effects caused by changes in the level of rival causal factors will be indistinguishable from the model's error term.

There are three classical ways in which a system can respond to one of these exogeneous shocks. The first occurs when the strength of the shock is strong enough to permanently alter the structure of the system. In this case the shock will, in the absence of other shocks, cause a permanent change in the systems behavior. The second occurs when the system makes a series of successively smaller partial adjustments to the exogeneous shock. In this case the system will, in the absence of other shocks, more or less gradually return to the pattern of behavior which prevailed prior to the shock. The third occurs when the system makes a more or less total and complete adjustment to the exogeneous shock. Typically a nontrivial amount of time will be required to formulate and implement plans for making this one shot adjustment. In this case the system will more or less abruptly return to its preshock pattern of behavior once the adjustment is implemented.

The failure to correctly account for the dependencies in a set of data could cause the levels of significance of a statistical test, used to test various hypotheses about a models parameters, to be distorted. This could cause the investigator to draw faulty conclusions about the hypothesized causal relationship. This risk can be reduced by utilizing a model which correctly accounts for the dependencies that are inherent in the data. Unfortunately, the knowledge, needed to specify a priori a model which accounts for the dependencies in a set of data, will not be available, especially in situations in which it is necessary to use natural experiments. This creates the need for a systematic approach to model building which is capable of identifying not only an appropriate model to account for whatever dependencies are present, but also an appropriate model for the relationship between the hypothesized cause and hypothesized effect. Fortunately, such an approach has been developed by Box and Tiao (1965 and 1975). The basic model for this approach, which is typically referred to as either intervention or impact analysis, will be briefly described in the following subsection.

4.3. TRANSFER FUNCTION INTERVENTION ANALYSIS

The transfer function intervention analysis model building process, developed by Box and Tiao (1965 and 1975), recognizes that there can be three sources of variations which can influence the behavior of a time series. The first source represents that portion of the variability caused by changes in the levels of one or more causal variables. These causal variables can be either continuous or discrete. The second source represents that portion of variability which results from the way in which the system which generated the time series data reacts to an exogeneous shock. In particular the system's reaction can induce dependencies into the data. The nature of these dependencies will depend on the degree to which the three classical reaction patterns, i.e. permanent change, gradual adjustment, and abrupt adjustment, are present in the system's reaction. The third source represents that portion of the variability which can not be accounted for by either changes in the levels by hypothesized causal variables or the dependencies which are induced by the reaction of the system to exogeneous shocks.

The model building process consists of four major types of activities. The first involves tentatively identifying an autoregressive integrated moving average (ARIMA) model to account for the dependencies which exist in the data. The second consists of tentatively identifying a transfer function/intervention (TFI) model to account for the impact produced by changes in the levels of one or more causal variables. This model describes the future values of the hypothesized causal variables in terms of a linear combination of the present and past values of the hypothesized causal variables. The third involves simultaneously estimating the parameters of the ARIMA and TFI models tentatively identified in the first and second steps above. The fourth involves conducting diagnostic tests to evaluate the adequacy of the estimated model and to determine, if necessary, how an inadequate model should be modified to rectify its deficiencies.

The TFI and ARIMA components of the combined TFI/ARIMA model are generalized models which possess the flexibility needed to adequately model a wide variety of causal relationships.

5. SUMMARY

This paper has described how natural experiments could be used to obtain evidence that a safety program has been successful. In particular it has discussed how rival causal hypothesis could threaten the validity of conclusions drawn from natural experiments, strategies for coping with these threats, and how transfer function/intervention analysis could be used to analyze data obtained from natural experiments.

REFERENCES

(1) BOX, G. E. P. and TIAO, G. C. (1965). A Change in Level of Nonstationary Time Series, *Biometrika*, 181-192.

(2) BOX. G. E. P. and TIAO, G. C. (1975). Intervention Analysis with Applications to Economic and Environmental Problems, *Journal of American Statistical Association*, 70-92.

APPLIED TIME SERIES ANALYSIS
O.D. Anderson & M.R. Perryman (eds.)

AN APPROXIMATION APPROACH TO SEASONAL ADJUSTMENT OF TIME SERIES

Cheng Jun Tian

Shanxi University, China
and Columbia University, U.S.A.

Seasonal adjustment is discussed from the viewpoint of numerical approximation. Not only are some probabilistic methods explained and extended correspondingly, but the successive average method is also proposed as a new iterative algorithm. Some relative properties of this method are studied in depth; in particular a convergence theorem is proved. The successive average method is simpler and easier to apply than some other methods; several numerical examples are given.

1. INTRODUCTION

Seasonal adjustment, one of the most important topics in the field of time series analysis, has been studied in depth for a long time. In most of the literature, seasonal adjustment is taken to consist of the estimation and removal of a seasonal component $s(t)$ from an observable series $x(t)$, expressable (possibly after transformation) as $x(t) = s(t) + n(t)$, $n(t)$ being the nonseasonal component with whose interpretation or analysis $s(t)$ is presumably interfering. Up to now several methods have been proposed, for example, the widely-used periodogram method (Grenander and Rosenblatt, 1957), the iterated moving average method (Leong, 1962; Kukkonen, 1968), changing seasonal pattern (Hannan, 1964), spectrum analysis of seasonal adjustment (Godfrey and Karreman, 1967), autoregressive spectral or maximum entropy spectral analysis (Newman, 1977; Cao and Luo, 1979), the variance analysis periodic extrapolated method (Wuham Central Climate Bureau, et al, 1974; Tian, 1981), the census-11 method (Shiskin, Young and Musgrave, 1967), an approach to modelling seasonally stationary time series (Parzen and Pagano, 1979), and so on.

Although these methods are all different, their practical utility, in some sense, is limited to the same restriction. The main cause is that the reasoning of these methods is founded on certain probabilistic assumptions and these assumptions are verified with difficulty by a group of observations with finite length. Prazen (1978) considers "it is still an unsolved problem how to carry out in practice for an arbitrary observed time series a decomposition into trend, seasonal or cyclic, and residual component." Pierce (1978) points out seasonal adjustment models are never more than approximation, so that theoretically incompatible models can and do produce results uncomfortably close to each other and uncomfortably far from the "truth". Akaike (1974) gives the following frank opinion that the practical utility of hypothesis testing procedure as a method of statistical model building or identification must be considered quite limited. To develop useful procedures of identification, a more direct approach to the control of the error or loss caused by the use of the identified model is necessary.

From this viewpoint, we will discuss the seasonal adjustment problem by means of numerical approximation. Not only are some probabilistic methods explained, but a new iterative algorithm, the successive average method, is also proposed. Therefore we can study some properties of the successive average method, especially, the convergence theorem. Since the conclusion of the convergence theorem is proved, it becomes necessary to reconsider the problem of evaluating the estimate of seasonal component by residual. In this paper, some numerical examples are given which demonstrate the practical utility of the successive average method.

2. APPROXIMATE EXPLANATION OF SEASONAL ADJUSTMENT 2.1 FORMULATION

We regard the observed series $x(1), x(2), \ldots, x(m)$ as a point $\underset{\sim}{x}$ in m-dimensional Euclidian space R^m with the norm

$$\|\underset{\sim}{x}\| = [\sum_{t=1}^{m} x^2(t)]^{\frac{1}{2}} .$$

The series $\underset{\sim}{f} = (f(1), f(2), \ldots, f(m))^T$ is a series with implicit period p if $f(i)$ is the value of a periodic function $f(t)$ at $t = i$, $i=1,2,\ldots,m$. In particular, f is a periodic series if the implicit period, p, is an integer. We consider the seasonal adjustment of time series as approximation from some subset A of R^m to a point $\underset{\sim}{x}$ in the sense of norm. Incidentally, for reasonableness of seasonal adjustment we should require $0<p\le m$, even $0<p\le\frac{m}{2}$ in most applications.

Given $\underset{\sim}{f}_i$ with period p_i, $i=1,\ldots,r$, define

$$(2.1) \qquad E_1 = \{\sum_{i=1}^{r} a_i \underset{\sim}{f}_i : \ a_i \text{ is an arbitrary real number}, \ i=1,\ldots,r\}.$$

Notice that $E_1 \subset R^m$. The approximation problem from E_1 to $\underset{\sim}{x}$ may be solved by the well-known least square method. The solution is

$$(2.2) \qquad \hat{\underset{\sim}{x}} = \underset{\sim}{B}\,\underset{\sim}{c} = \sum_{i=1}^{r} c(i)\,\underset{\sim}{f}_i , \qquad \underset{\sim}{c} = (\underset{\sim}{B}^T\underset{\sim}{B})^{-1}\underset{\sim}{B}^T\underset{\sim}{x} ,$$

where

$$\underset{\sim}{c} = \begin{bmatrix} c(1) \\ \vdots \\ c(r) \end{bmatrix}, \qquad \underset{\sim}{x} = \begin{bmatrix} x(1) \\ \vdots \\ x(m) \end{bmatrix}, \qquad \underset{\sim}{f}_i = \begin{bmatrix} f_i(1) \\ \vdots \\ f_i(m) \end{bmatrix} \quad \text{and}$$

$$\underset{\sim}{B} = \begin{bmatrix} f_1(1), & f_2(1), & \ldots, & f_r(1) \\ \ldots & & & \\ f_1(m), & f_2(m), & \ldots, & f_r(m) \end{bmatrix} .$$

The square error is

(2.3) $$\|\underset{\sim}{x}-\hat{\underset{\sim}{x}}\|^2 = \|\underset{\sim}{x} - \underset{\sim}{B}(\underset{\sim}{B}^T\underset{\sim}{B})^{-1}\underset{\sim}{B}^T\underset{\sim}{x}\|^2 .$$

In particular, if

$$\underset{\sim}{f}_i = \begin{bmatrix} x(1-i) \\ \cdot \\ \cdot \\ \cdot \\ x(m-i) \end{bmatrix}$$

then $\underset{\sim}{c}$ is a coefficient-vector of autoregressive equation.

2.2 OPTIMAL SOLUTION UNDER THE ORTHOGONAL CONDITION

If the orthogonal condition is added, i.e. the inner product satisfies

$$(\underset{\sim}{f}_i, \underset{\sim}{f}_j) = \underset{\sim}{f}_i^T \underset{\sim}{f}_j = \sum_{t=1}^{m} f_i(t)\, f_j(t) = 0 \quad \text{for} \quad i \neq j ,$$

then

$$\underset{\sim}{B}^T\underset{\sim}{B} = ((\underset{\sim}{f}_i, \underset{\sim}{f}_j)) = \begin{bmatrix} \|\underset{\sim}{f}_1\|^2 & & & 0 \\ & \|\underset{\sim}{f}_2\|^2 & & \\ & & \ddots & \\ 0 & & & \|\underset{\sim}{f}_r\|^2 \end{bmatrix} .$$

From the previous equation and taking cognizance of (2.2) we obtain

(2.4) $$c(i) = \frac{1}{\|f_i\|^2} \sum_{t=1}^{m} f_i(t)\, x(t) ,$$

so that the square error in (2.3) can be expressed as

(2.5) $$\|\underset{\sim}{x}-\hat{\underset{\sim}{x}}\|^2 = \|\underset{\sim}{x}\|^2 - \sum_{i=1}^{r} c^2(i)\|\underset{\sim}{f}_i\|^2 .$$

From (2.4) we have

(2.6) $$\|\underset{\sim}{x}-c(i)\, \underset{\sim}{f}_i\|^2 = \|\underset{\sim}{x}\|^2 - \|c(i)\underset{\sim}{f}_i\|^2 = \|\underset{\sim}{x}\|^2 - c_i^2\|\underset{\sim}{f}_i\|^2 .$$

From (2.5) and (2.6) we have

(2.7) $$\|\underset{\sim}{x}\|^2 - \|\underset{\sim}{x}-\hat{\underset{\sim}{x}}\|^2 = \sum_{i=1}^{r} c^2(i)\, \|\underset{\sim}{f}_i\|^2 = \sum_{i=1}^{r} [\|\underset{\sim}{x}\|^2 - \|\underset{\sim}{x}-c(i)\underset{\sim}{f}_i\|^2].$$

The last expression indicates that the contribution of $\underset{\sim}{x}$ to approximate error is equal to the sum of contributions of $c(i)\underset{\sim}{f}_i$ to error, i=1,...,r. Therefore, the error contributions of

$c_i \underset{\sim}{f}_i$, $i=1,\ldots,r$ have additivity for any arbitrary $\underset{\sim}{x} \in R^m$ and orthogonal system $\underset{\sim}{f}_i$, $i=1,\ldots,r$.

In a special case of $m=2k+1$ and

$$f_i(t) = \cos \lambda_i t \quad , \quad i=0,1,\ldots,k \ ,$$

$$f_{k+i}(t) = \sin \lambda_i t \ , \quad i=1,\ldots,k \quad .$$

where

$$\lambda_i = \frac{2\pi i}{m} \ ,$$

we have

$$\|\underset{\sim}{f}_0\|^2 = m, \quad \|\underset{\sim}{f}_i\|^2 = \frac{m}{2} \ , \ i\neq 0,$$

and

$$\hat{x}(t) = \frac{a_0}{2} + \sum_{i=1}^{k} (a_i \cos \lambda_i t + b_i \sin \lambda_i t) \tag{2.8}$$

where

$$a_i = \frac{2}{m} \sum_{t=1}^{m} x(t) \cos \lambda_i t, \quad i=0,1,..,k \ , \tag{2.9}$$

$$b_i = \frac{2}{m} \sum_{t=1}^{m} x(t) \sin \lambda_i t, \quad i=1,\ldots,k \ .$$

From (2.6) we know the error contribution of $\frac{a_0}{2} \underset{\sim}{f}_0$ is

$$\|\underset{\sim}{x}\|^2 - \|\underset{\sim}{x} - \frac{a_0}{2} \underset{\sim}{f}_0\|^2 = (\frac{a_0}{2})^2 \|\underset{\sim}{f}_0\|^2 = m \, \overline{x}^2 \ .$$

According to (2.11) we know the error contribution of the term with period $\frac{m}{i}$, $a_i \cos \lambda_i t + b_i \sin \lambda_i t$, is

$$\|\underset{\sim}{x}\|^2 - \|\underset{\sim}{x} - (a_i \underset{\sim}{f}_i + b_i \underset{\sim}{f}_{k+i})\|^2 = a_i^2 \|\underset{\sim}{f}_i\|^2 + b_i^2 \|\underset{\sim}{f}_{k+i}\|^2$$

$$= \frac{m}{2}(a_i^2 + b_i^2).$$

Notice that the function

$$I(i) = a_i^2 + b_i^2 \tag{2.10}$$

is generally called the periodogram.

Combining the last two expressions, we can therefore have the intuitive explanation of the periodogram. In applications, when

$$\|\underset{\sim}{x} - \hat{\underset{\sim}{x}}\|^2 = \|\underset{\sim}{x}\|^2 - m\overline{x}^2 - \frac{m}{2} \sum_{i=1}^{k} I(i)$$

is no larger than a preassigned error limit, we may use the harmonic polynomial as an approximate. If we want to reduce as many harmonic terms as possible, we may rearange the harmonic terms

$$a_i \cos \lambda_i t + b_i \sin \lambda_i t, \quad i=1,\ldots,k \ ,$$

according to the magnitude of $I(i)$ and choose enough first terms so that the sum of the corresponding values of the periodogram attains enough error contribution. Actually, the procedure is just the same as in the periodogram method. Therefore, we may apply the periodogram method to a more general case in the way of numerical approximation.

In the case of $m=2k$, we have to change the following for (2.8),

$$\hat{x}(t) = \frac{a_0}{2} + \sum_{i=1}^{k-1} (a_i \cos \lambda_i t + b_i \sin \lambda_i t) + a_k(-1)^t ,$$

where

$$a_k = \frac{1}{m} \sum_{t=1}^{m} x(t)\ (-1)^t ,$$

$a_0, \ldots, a_{k-1}$, $b_1, \ldots, b_k$ are the same in (2.9), and give the similar explanation. Here, we do not attempt to show the details (see Anderson, 1971, Chap. 4).

2.3. OPTIMAL SOLUTION UNDER THE ADDITIONAL CONDITION

We say that the approximate errors of $\underset{\sim}{u}_i \in R^m$, $i=0,1,\ldots,n$ to $\underset{\sim}{x}$ satisfy the additional condition if

$$\|\underset{\sim}{x}_i\|^2 = \|\underset{\sim}{x}_i - \underset{\sim}{u}_i\|^2 + \|\underset{\sim}{u}_i\|^2, \quad i=0,1,\ldots,n , \tag{2.11}$$

where

$$\underset{\sim}{x}_{i+1} = \underset{\sim}{x}_i - \underset{\sim}{u}_i \tag{2.12}$$

$$\underset{\sim}{x}_0 = \underset{\sim}{x} ,$$

holds.

From (2.11) we obtain

$$\|\underset{\sim}{x}\|^2 - \|\underset{\sim}{x} - \sum_{i=1}^{p} \underset{\sim}{u}_i\|^2 = \|\underset{\sim}{x}_0\|^2 - \|\underset{\sim}{x}_{p+1}\|^2 = \sum_{i=0}^{p} (\|\underset{\sim}{x}_i\|^2 - \|\underset{\sim}{x}_{i+1}\|^2) \tag{2.13}$$

$$= \sum_{i=0}^{p} \|\underset{\sim}{u}_i\|^2 , \quad \text{for every } p \le n.$$

Denote

$$E_2 = \{\sum_{i=0}^{n} \underset{\sim}{u}_i : \text{ the approximate errors of } \underset{\sim}{u}_i,\ i=0,1,\ldots,n ,$$
$$\text{to } \underset{\sim}{x} \text{ satisfy the additional condition}\}.$$

Let us now discuss the approximation problem of E_2 to $\underset{\sim}{x}$. In fact, we may equivalently regard the approximation as the optimal control of a dynamic system (2.12) with the objective function $J_{n+1} = \|\underset{\sim}{x}\|^2 - \|\underset{\sim}{x}_{n+1}\|^2$. Since (2.13) holds, we may obtain the control series $\underset{\sim}{u}_0^*, \underset{\sim}{u}_1^*, \ldots, \underset{\sim}{u}_n^*$ such that J_{n+1} attains the maximum by using dynamic programming. According to the optimal principle of dynamic programming, a multi-stages optimization problem may be regarded as many one-stage optimization problems and the maximum of the objective function must satisfy

(2.15) $$J^*_{n+1} = \max_{\underset{\sim}{u}_0} \{\max_{\underset{\sim}{u}_1,\ldots,\underset{\sim}{u}_n} J_{n+1}\} = \max_{\underset{\sim}{u}_0} \{\max_{\underset{\sim}{u}_1}\{\ldots\{\max_{\underset{\sim}{u}_n} J_{n+1}\}\}\}.$$

We thus obtain an algorithm of the optimal control series $\underset{\sim}{u}^*_0, \underset{\sim}{u}^*_1, \ldots, \underset{\sim}{u}^*_n$.

Turning back to the approximation problem we discussed, (2.15) also an algorithm of the optimization problem. Clearly, it is rather difficult to perform this algorithm in a general case. However, when the form of $\underset{\sim}{u}_i$, i=1,2,...,r is further restricted, the algorithm (2.15) becomes a performable one. In particular, for an orthogonal system $\{\underset{\sim}{f}_i\}$, i=1,2,...,r, we denote

$$D_1 = \{c(1)\underset{\sim}{f}_1, c(2)\underset{\sim}{f}_2, \ldots, c_{(r)}\underset{\sim}{f}_r\}$$

where, c(i), i=1,2,...,r, are defined in (2.4).

Denote

$$E_3 = \{\sum_{i=0}^{n} \underset{\sim}{u}_i : \underset{\sim}{u}_i \in D_1\}, \quad n<r .$$

Clearly,

$$E_3 \subset E_2$$

According to the magnitude of norm we rearrange $c_i\underset{\sim}{f}_i$, i=1,2,...,r, as follows

$$\underset{\sim}{h}_1, \underset{\sim}{h}_2, \ldots, \underset{\sim}{h}_r .$$

From (2.15) we have

$$\underset{\sim}{u}^*_i = \underset{\sim}{h}_{i+1}, \quad i=0,1,2,\ldots,n.$$

The optimal approximate of E_3 to $\underset{\sim}{x}$ therefore is

$$\hat{\underset{\sim}{x}} = \sum_{i=1}^{n+1} \underset{\sim}{h}_i ,$$

which is the generalized form of the periodogram method in the case of a general orthogonal system.

3. SUCCESSIVE AVERAGE METHOD

The average method has been well adopted to periodic analysis for quite a while. Chinese scientists developed this kind of method and proposed the variance analysis periodic extrapolated method, but the statistical testing of the hypothesis used by this method to the successive average method from the viewpoint of numerical approximation.

3.1 ALGORITHM OF SUCCESSIVE AVERAGE METHOD

Denote

$$\overline{x} = \frac{1}{m}\sum_{t=1}^{m} x(t),$$

$$S(\underset{\sim}{x}) = \sum_{t=1}^{m} [x(t)-\overline{x}]^2 ,$$

(3.1) $$\overline{x}^{(p)}(i) = \frac{1}{m_i}[x(i) + x(i+p)+\ldots+x(i + \overline{m_i-1}\, p)],$$

$$1 < p \le \frac{m}{2}, \quad 1\le i\le p ,$$

where m_i is the largest integer such that $i + (m_i - 1)p \leq m$,

$$(3.2) \qquad S^{(p)}(\underset{\sim}{x}) = \sum_{i=1}^{p} m_i [x^{(p)}(i) - \bar{x}]^2 ,$$

$$(3.3) \qquad F^{(p)}(\underset{\sim}{x}) = \frac{S^{(p)}(\underset{\sim}{x})/_{p-1}}{[S(\underset{\sim}{x}) - S^{(p)}(\underset{\sim}{x})]/_{m-p}} , \quad 2 \leq p \leq k,$$

where $k = [\frac{m}{2}]$ is the largest integer which is no larger than $\frac{m}{2}$,

$$(3.4) \qquad M(\underset{\sim}{x}) = \max_{2 \leq p \leq k} S^{(p)}(\underset{\sim}{x}) , \quad \text{and}$$

$$(3.5) \qquad N(\underset{\sim}{x}) = \max_{k-2 \leq p \leq k} S^{(p)}(\underset{\sim}{x}) .$$

The successive average method is an iterative method as follows. Start from all these initial values

$$\hat{x}_0(t) = u_0(t) = \bar{x} ,$$

$$x_1(t) = x(t) - u_0(t), \quad t=1,2,\ldots,m .$$

At the i-th stage $i=1,2,\ldots$

(i) Compute $S^{(p)}(\underset{\sim}{x}_i)$, $p=2,3,\ldots,k$,

if $M(\underset{\sim}{x}_i) = o$ then end, or else

(ii) find p_i such that

$$S^{(p_i)}(\underset{\sim}{x}_i) > 0,$$

(iii) define

$$u_i(t) = \bar{x}_i^{(p_i)}(j) , \quad t=j \bmod p_i ,$$

(iv) $\hat{x}_i(t) = \hat{x}_{i-1}(t) + u_i(t)$,

$$x_{i+1}(t) = x_i(t) - u_i(t), \quad t=1,2,\ldots,m.$$

The computation result up to the n-th stage of the successive average method is as follows

$$\hat{\underset{\sim}{x}}_n = \sum_{i=0}^{n} \underset{\sim}{u}_i$$

which is an approximate of $\underset{\sim}{x}$. Notice that $\hat{\underset{\sim}{x}}_n$ is a periodic superposed series.

Its square error is

$$Q_n = \|\underset{\sim}{x} - \sum_{i=1}^{n} \underset{\sim}{u}_i\|^2 = S(\underset{\sim}{x}-\hat{\underset{\sim}{x}}_n) .$$

Naturally, we would combine the terms with the same period at every stage and end the computation when the magnitude of Q_n at some stage attains a preassigned bound.

The successive average method has several computation schemes, because (ii) in the algorithm may have several forms. For example,

(I) find p_i such that $\dfrac{S^{(p_i)}(\underset{\sim}{x}_i)}{p_i} = \max\limits_p \dfrac{S^{(p)}(\underset{\sim}{x}_i)}{p}$,

(II) find p_i such that $F^{(p_i)}(\underset{\sim}{x}_i) = \max\limits_p F^{(p)}(\underset{\sim}{x}_i)$,

(III) find p_i such that $S^{(p_i)}(\underset{\sim}{x}_i) = M(\underset{\sim}{x}_i)$,

(IV) find p_i such that $S^{(p_i)}(\underset{\sim}{x}_i) = N(\underset{\sim}{x}_i)$.

Notice that scheme (II) is the generalized form of the variance analysis periodic extrapolated method in the sense of numerical approximation. Moreover, while using the successive average method, if periods $q_1,\ldots,q_r$ are given, then we should restrict p_i equal to one of q_j, $j=1,\ldots,r$. In the case of $r=1$, we obviously have

$$u(t) = x^{(p)}(j),\ t=j \bmod p, \text{ and}$$

$$S^{(p)}(\underset{\sim}{x}) = S(\underset{\sim}{u}) = \sum_{i=1}^{p} m_i[x^{(p)}(i) - \overline{x}]^2 .$$

The preceding formulas happen to correspond to the formulas of estimates $m_n(j)$ and σ_n^2 obtained by Parzen and Pagano (1979) under the assumption of seasonal stationality.
Especially,
the seasonal means is $m_n(j) = \overline{x}^{(p)}(j)$, and
the average seasonal variance is $\sigma_n^2 = \frac{1}{m}[S(\underset{\sim}{x}) - S^{(p)}(\underset{\sim}{x})]$.

3.2 PROPERTIES OF THE SUCCESSIVE AVERAGE METHOD

From (3.6) we have

$$(3.9) \qquad \sum_{t=1}^{m} u_i(t) = \sum_{j=1}^{p_i} m_i \cdot \frac{1}{m_i} \sum_{t=j \bmod p_i} x(t) = \sum_{t=1}^{m} x_i(t) ,$$

Taking cognizance of $\overline{x}_i = 0$, $\overline{u}_i = 0$, $i=1,2,\ldots$, it follows instantly that

$$S^{(p_i)}(\underset{\sim}{x}_i) = S(\underset{\sim}{u}_i) , \text{ and}$$

$$\|\underset{\sim}{x}_{i+1}\|^2 = \sum_{t=1}^{m} [x_{i+1}(t)]^2$$

$$= \sum_{t=1}^{m} [x_i(t) - u_i(t)]^2$$

$$= S(\underset{\sim}{x}_i) + S^{(p_i)}(\underset{\sim}{x}_i) - 2 \sum_{j=1}^{p_i} m_i [\overline{x}_i^{(p_i)}(j)]^2$$

$$= \|\underset{\sim}{x}_i\|^2 - \|\underset{\sim}{u}_i\|^2 \ .$$

We thus obtain the following property

Property 1. $\underset{\sim}{u}_i$, $i=1,2,\ldots,$ generated by the successive average method, satisfy the additional condition, i.e.

(3.10) $$\|\underset{\sim}{x}_{i+1}\|^2 = \|\underset{\sim}{x}_i\|^2 - \|\underset{\sim}{u}_i\|^2 \ .$$

This implies that

$$E_4 = \{\sum_{i=0}^{n} \underset{\sim}{u}_i : \ \underset{\sim}{u}_i, \ i=0,1,\ldots,n \text{ are generated by the successive average method}\} \subset E_2 \ .$$

The determination of the optimal approximate $\hat{\underset{\sim}{x}} = \sum_{i=0}^{n} \underset{\sim}{u}_i^*$ of E_4 to $\underset{\sim}{x}$ becomes the problem of how to determine periods $p_1^*, p_2^*, \ldots, p_n^*$. Obviously, the last period should satisfy

$$S^{(p_n^*)}(\underset{\sim}{x}_n) = M(\underset{\sim}{x}_n) \ .$$

Generally, it is difficult to determine other periods from (2.15), unless we directly compare the computation results of $k\ (k-1)^{n-2}$ variant choices of $p_1, \ldots, p_{n-1}$ by using the exhaustion. Naturally, this direct comparision is not available. Moreover, the approximate error cannot be guaranteed because of the restriction on n. If we remove the restriction on n, then we will find that such an interesting property as the successive average method has good asymptotic behavior when iterative time increases.

From (3.7) and (3.10) we have

(3.11) $$\|\underset{\sim}{x}_{n+1}\|^2 = \|\underset{\sim}{x} - \hat{\underset{\sim}{x}}_n\|^2 = \|\underset{\sim}{x}\|^2 - \sum_{i=0}^{n} \|\underset{\sim}{u}_i\|^2 \ .$$

It follows instantly that

Property 2.

$$\lim_{n\to\infty} \|\underset{\sim}{u}_n\| = 0, \text{ and}$$

$$\lim_{n\to\infty} \|\underset{\sim}{x}_n\| \downarrow c \geq 0 \ .$$

This indicates that the successive average is a successive approximate method.

3.3 CONVERGENCE THEOREM

Lemma. For arbitrary $\underset{\sim}{x}\in R^m$, when $m=2k$, $k=5$ or $m=2k+1$, $k\geq 7$, the following conditions are equivalent

(i) $\underset{\sim}{x} = 0$,

(ii) $M(\underset{\sim}{x}) = 0$, and

(iii) $N(\underset{\sim}{x}) = 0$.

Proof. Trivially, $\underset{\sim}{x}=0 \Rightarrow M(\underset{\sim}{x})=0 \Rightarrow N(\underset{\sim}{x})=0$.

It suffices to show $N(\underset{\sim}{x})=0 \Rightarrow \underset{\sim}{x}=0$. We now prove the conclusion of the lemma in the two cases of $m=2k$, $k\geq 5$ and $m=2k+1$, $k\geq 7$ respectively.

(1) In the case of $m=2k$, $k\geq 5$

From $\sum_{i=1}^{k} [x(i) + x(i+k)]^2 = S^{(k)}(\underset{\sim}{x}) = 0$, we have

$$x(i) = -x(i+k), \quad i=1,2,\ldots,k \tag{3.12}$$

Similarly, from $S^{(k-1)}(\underset{\sim}{x}) = 0$, we have

$$\begin{aligned} x(1) + x(k) + x(2k-1) &= 0 \\ x(2) + x(k+1) + x(2k) &= 0 \\ x(3) + x(k+2) &= 0 \\ \cdots\cdots \\ x(k-1) + x(2k-2) &= 0 \end{aligned} \tag{3.13}$$

and combining (3.12) with (3.13) we obtain

$$x(2) = x(3) = \ldots = x(k-1), \text{ say } a , \tag{3.14}$$

$$\text{and } x(k+2) = x(k+3) = \ldots = x(2k-1) = -a . \tag{3.15}$$

Similarly, from $S^{(k-2)}(\underset{\sim}{x}) = 0$ we have

$$\begin{aligned} x(1) + x(k-1) + x(2k-3) &= 0 , \\ x(2) + x(k) + x(2k-2) &= 0 , \\ x(3) + x(k+1) + x(2k-1) &= 0 , \\ \cdots \end{aligned} \tag{3.16}$$

When $k>5$, i.e. $2k\leq 3(k-2)$, $S^{(k-2)}(\underset{\sim}{x}) = 0$ can be expressed by the system of equations with the form of (3.16). In the case of $k=5$, the system of equations becomes

$$(3.17) \quad \begin{aligned} x(1) + x(k-1) + x(2k-3) + x(2k) &= 0 \\ x(2) + x(k) + x(2k-2) &= 0 \\ x(3) + x(k+1) + x(2k-1) &= 0 \end{aligned}$$

Combining (3.16-2)*, (3.17-2) and (3.12), $x(k) = 0$ follows. From (3.12), we have $x(2k) = 0$. It implies we may use (3.16) for (3.17). Combining (3.14), (3.15) and (3.16-3) we have

$$x(k+1) = 0.$$

From (3.14), (3.15) and (3.16-1) we have

$$x(1) = 0,$$

and via (3.13-1),

$$x(2k-1) = 0,$$

and hence, $a = 0$.

Therefore, $x(i) = 0$, $i=1,2,\ldots,2k$, i.e. the conclusion of the lemma holds.

(2) In the case of $m=2k+1$, $k\geq 7$

In similar fashion to (1), $S^{(k-i)}(\underset{\sim}{x}) = 0$, $i=0,1,2$ can be expressed by the following systems of equations.

$$(3.18) \quad \begin{aligned} x(1) + x(k+1) + x(2k+1) &= 0, \\ x(2) + \ldots + x(k+2) &= 0, \\ \cdots & \\ x(k) + x(2k) &= 0, \end{aligned}$$

$$(3.19) \quad \begin{aligned} x(1) + x(k) + x(2k-1) &= 0, \\ x(2) + x(k+1) + x(2k) &= 0, \\ x(3) + x(k+2) + x(2k+1) &= 0, \\ \cdots & \\ x(k-1) + x(2k-2) &= 0, \quad \text{and} \end{aligned}$$

$$(3.20) \quad \begin{aligned} x(1) + x(k-1) + x(2k-3) &= 0 \\ x(2) + x(k) + x(2k-2) &= 0 \\ x(3) + x(k+1) + x(2k-1) &= 0 \\ x(4) + x(k+2) + x(2k) &= 0 \\ \cdots & \end{aligned}$$

*(3.16-2) offers the second equation in system (3.16).

From (3.18), $x(i) = -x(i+k)$, $i=2,3,\ldots,k$, and from (3.19), $x(i) = -x(i+k-1)$, $i=4,5,\ldots,k-1$.

Hence,

(3.21) $x(3) = x(4) =\ldots= x(k-1)$, say b, and

(3.22) $x(k+3) = x(k+4) =\ldots= x(2k-1) = -b$.

From (3.21) and (3.20-3),

(3.23) $x(k+1) = 0$,

and from (3.21) and (3.20-1),

(3.24) $x(1) = 0$.

From (3.24) and (3.18-1), $x(2k+1) = 0$, and from (3.22) and (3.19-1), $x(k) = -x(2k-1) = b$. From the last equation and (3.20.2), $x(2)=0$ and from (2.18-2) and (2.19-2), $x(k+2) = 0$, $x(2k) = 0$. Hence, $b=0$. Consequently, $x(i) = 0$, $i=1,2,\ldots,k+1$. Q.E.D.

<u>Theorem</u>. When $m=2k$, $k\geq 5$ or $m=2k+1$, $k\geq 7$, for an arbitrary scheme of the successive average method, if

$$\lim_{n\to\infty} N(\underset{\sim}{x}_n) = 0$$

then

$$\lim \hat{\underset{\sim}{x}}_n = \underset{\sim}{x} .$$

Proof. According to property 2, $\lim_{n\to\infty} \|\underset{\sim}{x}_n\|$ exists, say c. Since

$$\underset{\sim}{x}_{n+1} = \underset{\sim}{x} - \hat{\underset{\sim}{x}}_n = \underset{\sim}{x} - \sum_{i=0}^{n} \underset{\sim}{u}_i ,$$

it suffices to prove $c=0$.

By the way of contradiction, suppose $c>0$. Let $\delta>0$ such that $c-\delta>0$.
We may find an integer N such that

(3.25) $\left| \|\underset{\sim}{x}_n\| - c \right| < \delta, \ \forall \ n>N$.

From the definition of δ, for an arbitrary $\underset{\sim}{y}$

$$\underset{\sim}{y} \in \{\underset{\sim}{z} : \left| \|\underset{\sim}{z}\| - c \right| \leq \delta\} = A$$

we have $\underset{\sim}{y}\neq 0$.
According to the Lemma we thus assert

$$N(\underset{\sim}{y})\neq 0 .$$

Hence we may choose a positive number $\epsilon(\underset{\sim}{y})$ such that

$$0 < \epsilon(\underset{\sim}{y}) < N(\underset{\sim}{y}) .$$

Since $N(\underset{\sim}{z})$ is a continuous functional of $\underset{\sim}{z}$, for an arbitrary point $\underset{\sim}{y}$, we have a positive number, $\lambda(\underset{\sim}{y})$, such that

(3.26) $N(\underset{\sim}{z}) > N(\underset{\sim}{y}) - \epsilon(\underset{\sim}{y}) > 0$

if $\|\underset{\sim}{z}-\underset{\sim}{y}\| > \lambda(\underset{\sim}{y})$.

Notice that all of neighborhoods $\{\underset{\sim}{z} : \|\underset{\sim}{z}-\underset{\sim}{y}\| < \lambda(\underset{\sim}{y})\}$, $\underset{\sim}{y}\in A$ construct a covering of A. Since A is a compact set in R^m, so we can find a finite number of neighborhoods,

$$\{z: \|z-y\| < \lambda(y_j)\}, \quad j=1,2,\ldots,r,$$

which cover A, i.e.

$$\bigcup_{j=1}^{r} \{z: \|z-y_j\| < \lambda(y_j)\} \supset A.$$

Denote

$$d = \min_j \{N(y_j) - \epsilon(y_j)\} > 0.$$

For an arbitrary $y \in A$, there exists i such that

$$y \in \{z: \|z-y_i\| < \lambda(y_i)\}.$$

Taking cognizance of (3.26),

$$(3.27) \qquad N(y) > N(y_i) - \epsilon(y_i) \geq d > 0.$$

From (3.25) we have, when $n>N$,

$$x_n \in A.$$

From (3.27) we have

$$N(x_n) > d,$$

and hence

$$\lim_{n\to\infty} N(x_n) \geq d > 0, \quad \text{Q.E.D.}$$

Remark: The requirement of m in lemma and theorem cannot be improved. Counter examples in all possible cases are given as follows:

$$m = 4, \quad x^T = (1, 1, -1, -1),$$
$$m = 5, \quad x^T = (1, 1, 0, -1, -1),$$
$$m = 6, \quad x^T = (1, 0, -1, -1, 0, 1),$$
$$m = 7, \quad x^T = (1, 1, 1, 0, -1, -1, -1),$$
$$m = 8, \quad x^T = (1, 0, 0, -1, -1, 0, 0, 1),$$
$$m = 9, \quad x^T = (1, 1, 2, 1, 0, -1, -2, -1, -1),$$
$$m = 11, \quad x^T = (1, 2, 3, 3, 2, 0, -2, -3, -3, -2, -1),$$
$$m = 13, \quad x^T = (1, 1, 2, 2, 2, 1, 0, -1, -2, -2, -2, -1, -1),$$

$M(x)$ at each of preceding points is equal to zero.

Denote

$$E = \{\sum_{i=1}^{n} f_i: \text{ the period of } f_i \text{ is } p_i,\ 2 \leq p_i \leq k,\ i=1,2,\ldots,n\}.$$

The periods of u_i, generated by the successive average method, satisfy $2 \leq p_i \leq k$, hence

$$E_4 \subset E.$$

The theorem indicates that an arbitrary point in R^m is a limiting point of E, i.e. the closure of E is therefore the whole space R^m. We will furthermore show the theorem holds for all previous schemes.

Corollary. When m satisfies the condition of the lemma, for schemes (I), (II), (III) and (IV),

$$\lim_{n\to\infty} \hat{\underset{\sim}{x}}_n = \underset{\sim}{x}$$

holds.

Proof. According to the theorem it suffices to prove

$$\lim_{n\to\infty} N(\underset{\sim}{x}_n) = 0$$

for each scheme.

(i) For scheme (IV),

$$S(\underset{\sim}{u}_n) = S^{(p_n)}(\underset{\sim}{x}_n) = N(\underset{\sim}{x}_n).$$

From property 2,

$$S(\underset{\sim}{u}_n) \to 0$$

i.e. $N(\underset{\sim}{x}_n) \to 0$.

(ii) For scheme (III),

$$S(\underset{\sim}{u}_n) = S^{(p_n)}(\underset{\sim}{x}_n) = M(\underset{\sim}{x}_n) .$$

In a similar fashion of (i),

$$M(\underset{\sim}{x}_n) \to 0 .$$

Taking into account $N(\underset{\sim}{x}_n) \le M(\underset{\sim}{x}_n)$,

$$N(\underset{\sim}{x}_n) \to 0 .$$

(iii) For scheme (II), we have

$$S(\underset{\sim}{u}_n) = S^{(p_n)}(\underset{\sim}{x}_n)$$

and $S^{(p_n)}(\underset{\sim}{x}_n) \to 0$.

If $S^{(q_n)}(\underset{\sim}{x}_n) = M(\underset{\sim}{x}_n)$,

then

$$\frac{S^{(p_n)}(\underset{\sim}{x}_n)/p_n-1}{[S(\underset{\sim}{x}_n) - S^{(p_n)}(\underset{\sim}{x}_n)]/m-p_n} = F^{(p_n)}(\underset{\sim}{x}_n) = \max_p F^{(p)}(\underset{\sim}{x}_n)$$

$$\ge \frac{M(\underset{\sim}{x}_n)/q_n-1}{[S(\underset{\sim}{x}_n)-M(\underset{\sim}{x}_n)]/m-q_n} .$$

Removing the terms, we finally obtain

$$S^{(p_n)}(\underset{\sim}{x}_n) \ge \frac{(p_n-1)(m-q_n)}{(q_n-1)(m-p_n)} \cdot \frac{S(\underset{\sim}{x}_n) - S^{(p_n)}(\underset{\sim}{x}_n)]M(\underset{\sim}{x}_n)}{S(\underset{\sim}{x}_n) - M(\underset{\sim}{x}_n)} .$$

Since

$$2 \le p_n,\ q_n \le \frac{m}{2} , \qquad S^{(p_n)}(\underset{\sim}{x}_n) \le M(\underset{\sim}{x}_n) ,$$

$$S^{(p_n)}(\underset{\sim}{x}_n) > \frac{\frac{m}{2}}{\frac{m}{2}\cdot m} \cdot \frac{[S(\underset{\sim}{x}_n) - M(\underset{\sim}{x}_n)]M(\underset{\sim}{x}_n)}{S(\underset{\sim}{x}_n) - M(\underset{\sim}{x}_n)}$$

$$= \frac{1}{m} M(\underset{\sim}{x}_n) \geq \frac{1}{m} N(\underset{\sim}{x}_n) \quad .$$

Hence

$$\lim N(\underset{\sim}{x}_n) = 0 \; .$$

(iv) For scheme (I)

$$\frac{S^{(p)}(\underset{\sim}{x}_n)}{p} \leq \frac{S^{(p_n)}(\underset{\sim}{x}_n)}{p_n} \;, \quad p=2,3,\ldots k \;,$$

$$S^{(p)}(\underset{\sim}{x}_n) \leq p\cdot \frac{S^{(p_n)}(\underset{\sim}{x}_n)}{p_n} \leq \frac{m}{4} \cdot S^{(p_n)}(\underset{\sim}{x}_n), \quad p=2,3,\ldots,k$$

$$N(\underset{\sim}{x}_n) \leq \frac{m}{4} S^{(p_n)}(\underset{\sim}{x}_n) \;,$$

Hence $N(\underset{\sim}{x}_n) \to 0$. Q.E.D.

These theoretical results show that the successive average method is a good approximate algorithm. Meanwhile, we should notice that the convergence theorem does not impose any extra restrictions on $x(1), x(2),\ldots,x(m)$, except the requirement of m which is often satisfied in practice. Precisely, any group of observations of a time series with a longer length, m, may be approximated by a periodic superposed series with the periods not beyond $\frac{m}{2}$, which may be generated by the successive average method to achieve an arbitrarly preassigned bound of the residual. Naturally, it is necessary to reconsider the problem of how to evaluate an estimate of the seasonal component of time series by residual, but in general, it is not sufficient to evaluate an estimate of the seasonal component by residual. We must furthermore analyze the numerical results of the successive average method and all preceding methods, and also distinguish real periods from psuedo periods by comparison or other inspection. As a suggestion, we may simultaneously use the successive average method for several partial groups of a group of observations and compare these numerical results. When several partial groups have the same period, we may regard the time series as having real periods.

3.4 APPLICATIONS OF THE SUCCESSIVE AVERAGE METHOD

In general, scheme (I), or (II), is a performable algorithm for the seasonal adjustment. We only use the scheme (I), in the following examples.

In order to measure the effect of approximation, we introduce

$$\rho_n = \frac{S(\hat{\underset{\sim}{x}}_n)}{S(\underset{\sim}{x}_n)}$$

which is the ratio of variances. Generally, it is convenient to end the operation of the successive average method by measuring the value of ρ .

Example 1. An artificial series

$$s(t) = f_1(t) + f_2(t) + f_3(t),$$

where

$$f_1(t+4) = f_1(t),\ f_1(1) = f_1(2) = -1,\ f_1(3) = f_1(4) = 1,$$

$$f_2(t+5) = f_2(t),\ f_2(1) = -1,\ f_2(2) = -0.5,\ f_2(3) = 0,$$

$$f_2(4) = 0.5,\ f_2(5) = 1,$$

$$f_3(t+10) = f(t),\ f_3(1) = 1,\ f_3(2) = -1,\ f_3(3)=f_3(4)=\ldots=f_3(10) = 0.$$

We now apply the successive average method to series $x(t)$, $t=1,2,\ldots,20$ and have numerical results as following.

$p_1 = 4,\quad \rho_1 = 0.578,\quad \underset{\sim}{u}_1^T = (-1.2, -0.8, 0.8, 1.2),$

$p_2 = 5,\quad \rho_2 = 0.967,\quad \underset{\sim}{u}_2^T = (-1.5, 0, 0, 0.5, 1),$

$p_3 = 10,\quad \rho_3 = 1.000,\quad \underset{\sim}{u}_3^T = (-.03, 0.3, 0.2, -0.2, 0.2, 0.3, -0.3, -0.2, -0.2).$

Incidentally, the same periodic superposed series can be expressed in various forms that cause some difficulties for the nonparametric method for seasonal adjustment.

Example 2. Sunspot numbers, yearly (1749-1924).
We have applied the successive average method to several partial series of sunspot numbers series, and list the results of three partial series as follows:

TABLE

Numerical Results of Sunspot Numbers

	m	p_1	ρ_1	p_2	ρ_2	p_3	ρ_3
1749-1924	179	11	0.165	10	0.293	12	0.399
1749-1910	165	11	0.179	10	0.282	12	0.404
1770-1864	105	10	0.186	12	0.367	11	0.454

We thus decide that 11, 10 and 12 are periods of sunspot numbers series and 11 is the main period. This conclusion agrees with the classical results.

We now give three different seasonal adjusted AR-models.

$m=179$, $\bar{x}=44.637$, $S(x)=0.21104\times10^6$

(i) AR model without seasonal adjustment

$y(t) = x(t)-\bar{x}$,

AR models are fitted to $y(t)$, $t=1,\ldots,m$ and the order of AR model is determined by Akaike's criteria.

The minimum of the final predicted error $\emptyset$ is attained at order 8.

$\emptyset_8 = 237.2$ and $Q = 0.3636\times10^5$

(ii) Seasonal adjusted AR model with a seasonal term

$p_1 = 11$, $y(t) = x(t) - \bar{x} - u_1(t)$.

The minimum of the final predicted error $\emptyset$ is $\emptyset_8 = 214.5$, and $Q = 0.33534\times10^5$.

(iii) Seasonal adjusted AR model with three seasonal terms

$p_1 = 11$, $p_2 = 10$, $p_3 = 12$,

$y(t) = x(t) - \bar{x} - u_1(t) - u_2(t) - u_3(t)$,

The minimum of the final predicted error $\emptyset$ is $\emptyset_8 = 183.7$ and $Q = 0.293\times10^5$.

Consequently, either the final predicted error $\emptyset$ or the sum of residual squares shows that previous seasonal adjusted models are better than the original model without seasonal adjustment.

<u>Example 3</u>. For the convenience of the readers who wish to check the results by themselves, the seasonal adjusted AR models are fitted to data A-C given in the book by Box and Jenkins, see Tian and Lü (1981). We do not plan to show details here.

I am grateful to Professors H. Robbins, H. Levene, Y.S. Chow and T.L. Lai for their help and encouragement. I am glad to record my thanks to Mr. Zhi-jun Lü.

Research supported in part by the National Science Foundation (U.S.A.) under grant NSF MCS 78-09179.

REFERENCES:

AKAIKE, H. (1974). A new look at the Statistical model identification, <u>IEEE Trans. on Autom. Contr.</u>, Vol AC-19, 716-723.

ANDERSON, T.W. (1971). <u>The Statistical Analysis of Time Series</u>, John Wiley & Sons, INC. New York.

BOX, G.E.P. and JENKINS, G. (1970, 1976). Time Series Analysis, forecasting and control, San Francisco, Holden-Day.

CAO, H.S. and LUO, C.L. (1979). The maximum entropy spectral analysis of climate historic series, Kiexue Tongbao, 8, 351-355.

CLEVELAND, W.P. and TIAO, G.C. (1976). Decomposition of seasonal time series: a model for the Census X-11 program, Journal of the American Statistical Association, 71, 581-587.

GODFREY, M.D. and H.F. KARREMAN, (1967). A spectrum analysis of seasonal adjustment, in Essays in Mathematical Economics in Honor of Osker Morgenstern (ed. M. Shubik), Princeton, 1967, 367-421.

GRENANDER, U. and ROSENBLATT, M. (1957). Statistical Analysis of Stationary Time Series, John Wiley & Sons, New York.

HANNAN, E.J. (1964). The estimation of a changing seasonal pattern, Journal of the American Statistical Association, 59, 1063-1077.

KUKKONEN, P. (1968). Analysis of seasonal and other short-term variations with applications to finish economic time series, Bank of Finland Institute for Economic Research, Helsinki.

LEONG, Y.S. (1962). Use of an iterated moving average in measuring seasonal variations, Journal of the American Statistical Association, 57, 149-171.

NERLOVE, Y.S. (1962). Spectral analysis of seasonal adjustment procedures, Econometrica, 32, 241-286.

NEWMAN, W.I. (1977). Extension to the maximum entropy method. IEEE Trans. Inform. Theory, I.T. 23:1, 87-91.

PARZEN, E. (1974). Some recent advances in time series modelling. IEEE Trans. on Autom. Contr. vol. AC-19, 723-730.

PARZEN, E. (1978). Time series modelling, spectral analysis, and forecasting, Direction in Time Series, ed by Brillinger, D.R. and Tiao, G.C., Iowa State University, 80-111.

PARZEN, E. and PAGANO, M. (1979). An approach to modelling seasonally stationary time series. Journal of Econometrics, 9, 137-153.

PIERCE, D.A. (1978). Some recent development in seasonal adjustment, Direction in Time Series, ed by Brillinger, D.R. and Tiao, G.C., Iowa State University, 123-146.

SHISKIN, J., YOUNG, A.H. and MUSGRAVE, J.C. (1967). The X-11 variant of the census method-11 seasonal adjustment program. Technical paper No. 15, U.S. Bureau of the Census.

TIAN, C.J. (1981). Convergence of variance analysis periodic method, Kiexue Tongbao, 3 (In Chinese).

TIAN, C.J. and LÜ, Z. (1981). Seasonal adjustment of time series with the successive average method, Technical Report, Columbia University.

WU, S.M. and PANDIT, S.M. (1979). Time Series and System Analysis, Modelling and Application, Wisconsin University.

WUHAM CENTRAL CLIMATE BUREAU, et al (1974). Applications of the variance analysis periodic method to the weather forecasting, Acta Mathematica, 17:3 (In Chinese).

APPLIED TIME SERIES ANALYSIS
O.D. Anderson & M.R. Perryman (eds.)

A METHOD FOR DETRENDING CORRELATED RANDOM DATA

George Trevino[1]

Senior Research Associate[2]
Atmospheric Sciences Laboratory
White Sands, New Mexico, 88002
U.S.A.

A simple nonprejudiced method for approximating the linear trend in correlated random data is developed and discussed. The applicability of the method to a given set of random data does not require detailed knowledge of the functional form of the correlation function of the data, and is therefore different from the well-known method of "generalized least squares". It is shown that whenever there is "significant" correlation in a given sample of random data the use of the method to determine the linear trend provides a better estimate than does simple least squares regression to both the slope and the intercept of the trend. Illustrative examples, using atmospheric data, are presented to demonstrate the feasibility as well as the versatility of the method.

INTRODUCTION

In many fields of applied science the statistical analyst is often faced with the task of obtaining, among other statistics, the linear trend[3] in a given sample of random data. Accordingly, the analyst may then take one of two approaches — on one hand, if the data is known (or can be assumed) to be uncorrelated, i.e. the correlation function is a "delta" function, the task at hand is quite straightforward since the method of least squares regression is then directly applicable to the determination of the linear trend; on the other hand, if the data is assumed to be correlated, the task correspondingly becomes more complex, and the analyst is then forced to make some assumption about the functional form of the correlation function so that the method of generalized least squares may then be applied. In some fields of applied science however, particularly in the field of atmospheric science, a given set of random data is rarely uncorrelated and moreso, if the data is taken from an ensemble of data which may be safely assumed to be statistically nonhomogeneous as well as nonstationary (such as atmospheric data), the shape of the correlation function unfortunately varies from point to point in space and the analyst can therefore only "guess" as to what its appropriate functional form may be.

In the following sections a simple nonprejudiced method for obtaining the linear trend in a given set of correlated random data, data taken over a finite interval of time (0,T), is formulated. Illustrative examples are included to demonstrate the feasibility of the method (when applied to random data which clearly exhibits some trend) as well as the versatility of the method (when applied to data which exhibits no trend at all).

FORMULATION OF THE METHOD

Consider a set of random data which is defined by the random varying function $x(t)$ over the finite but sufficiently large domain $D:\{0,T\}$, data whose mean value,

$\mu(t) = \langle x(t) \rangle$, is known to be linear, or can be reasonably approximated as linear, in the indicated domain, i.e. data such that $\mu(t) \simeq \mu_1 t + \mu_0$, $0 \leq t \leq T$, the parameters μ_1 and μ_0 being constant. It is desired to approximate both $\mu(t)$ and the correlation function, $Q_x(t_1,t_2) = \langle x(t_1)\, x(t_2) \rangle$, of the data by performing some sort of time-average on the single function x(t) and subsequently determine the conditions under which the mean-square error between μ and Q_x, and the respectively obtained time-average approximations to μ and Q_x, is approximately zero. Note that in the case of μ the conventional time-average,

$$\underline{\mu}(T) = \frac{1}{T} \int_0^T x(t)dt, \tag{1}$$

is meaningless since not only does the mean value of the random variable, $\underline{\mu}(T)$, in general not equal μ but also $\underline{\mu}(T)$ as defined in Equ. (1) depends only on T while the mean value of interest depends on t; obviously, in the case of Q_x, a correspondingly similar time-average will also be meaningless. However, by utilizing the fact that the defining parameters of μ are constants, it is then possible to accurately approximate these parameters, in the mean-square sense, by some relatively simple integrations of x(t), provided that a single <u>weak</u> condition is satisfied by the covariance function, $C_{\dot{x}}(t_1,t_2) = \langle [\dot{x}(t_1) - \langle \dot{x}(t_1) \rangle][\dot{x}(t_2) - \langle \dot{x}(t_2) \rangle] \rangle$, of the derivative, $\dot{x} = dx/dt$, of the given random data. Having the approximation to μ, the approximation to Q_x will immediately follow.

As an approximation to the parameter μ_1 the random variable

$$\underline{\mu}_1(T') = \frac{1}{T'} \int_0^{T'} (\Delta t)^{-1}[x(t + \Delta t) - x(t)]dt \tag{2}$$

is defined; in this definition $T' = T - \Delta t$ and the magnitude of Δt is to be determined shortly. Note that the expected value of $\underline{\mu}_1(T')$ is equal to μ_1 which indicates that $\underline{\mu}_1(T')$ is an <u>unbiased</u> estimate of μ_1 and consequently that the mean-square error between $\underline{\mu}_1(T')$ and μ_1 can be expressed as

$$E_1 = \langle \underline{\mu}_1^2 \rangle - \langle \underline{\mu}_1 \rangle^2; \tag{3}$$

in terms of the covariance function, $C_x(t_1,t_2) = \langle [x(t_1) - \langle x(t_1) \rangle][x(t_2) - \langle x(t_2) \rangle] \rangle$, of the given random data E_1 can be equivalently expressed as

$$E_1 = \frac{1}{T'^2} \iint_0^{T'} (\Delta t)^{-2}[C_x(t_1 + \Delta t, t_2 + \Delta t) - C_x(t_1 + \Delta t, t_2) - C_x(t_1, t_2 + \Delta t) + C_x(t_1,t_2)]dt_1 dt_2. \tag{4}$$

In order to determine the magnitude of E_1 and also to ascertain the conditions under which $E_1 \approx 0$, the covariance functions appearing in the integrand of Equ. (4) are expanded in a Taylor's series about $\Delta t = 0$, a procedure which yields for E_1 the infinite series

$$E_1 = \frac{1}{T'^2} \iint_0^{T'} \frac{\partial^2 C_x}{\partial t_1 \partial t_2} dt_1 dt_2 + \frac{1}{2T'} \left(\frac{\Delta t}{T'}\right) \iint_0^{T'} \left(\frac{\partial^3 C_x}{\partial t_1^2 \partial t_2} + \frac{\partial^3 C_x}{\partial t_1 \partial t_2^2}\right) dt_1 dt_2$$

$$+ \frac{1}{6} \left(\frac{\Delta t}{T'}\right)^2 \iint_0^{T'} \left(\frac{\partial^4 C_x}{\partial t_1^3 \partial t_2} + \frac{3}{2} \frac{\partial^4 C_x}{\partial t_1^2 \partial t_2^2} + \frac{\partial^4 C_x}{\partial t_1 \partial t_2^3}\right) dt_1 dt_2 + \ldots \tag{5}$$

This expression clearly illustrates that the magnitude of E_1 depends among other things upon the value of the ratio $(\Delta t/T')$; if this ratio is "small", i.e. if $\Delta t \ll T'$ it then follows that

$$E_1 \simeq \left(\frac{1}{T'}\right)^2 \iint_0^{T'} C_{\dot{x}}(t_1, t_2) dt_1 dt_2, \tag{6}$$

where

$$C_{\dot{x}}(t_1, t_2) = \frac{\partial^2 C_x}{\partial t_1 \partial t_2} . \tag{7}$$ [4]

Assuming the given random data to be covariance stationary, i.e. that $C_x(t_1, t_2) = C_x(t_2 - t_1) = C_x(\tau)$, requires that

$$C_{\dot{x}}(t_1, t_2) = - \frac{d^2 C_x(\tau)}{d\tau^2} = C_{\dot{x}}(\tau) \tag{8}$$

and the expression for E_1 becomes, after transformation of coordinates,

$$E_1 \simeq \frac{2}{T'} \int_0^{T'} (1 - \tau/T') C_{\dot{x}}(\tau) d\tau. \tag{9}$$ [5]

The necessary (weak) condition for E_1 to approximate zero is then simply that $C_x(\tau)$ itself approach zero rapidly enough over the domain $D':\{0,T'\}$, a condition which is likely to be satisfied for most physical phenomena, provided that T' is sufficiently large. Specifically E_1 reduces to

$$E_1 \simeq \frac{2}{T'^2} C_x(0), \tag{10}$$

which clearly approximates zero for large values of T , and it therefore follows that $\underline{\mu}_1(T')$ is also a consistent estimate of μ_1.

As an approximation to the parameter μ_0 the function $y(t) = x(t) - \underline{\mu}_1 t$ and the random varialbe

$$\underline{\mu}_0(T') = \frac{1}{T'} \int_0^{T'} y(t)dt \tag{11}$$

are both defined. From the definition of $\underline{\mu}_0(T')$ the expected value of $\underline{\mu}_0(T')$ is necessarily equal to μ_0, indicating that $\underline{\mu}_0(T')$ is an unbiased estimate of μ_0, and the mean-square error between $\underline{\mu}_0(T')$ and μ_0 is accordingly

$$E_0 = \langle \underline{\mu}_0^2 \rangle - \langle \underline{\mu}_0 \rangle^2, \tag{12}$$

an expression which ultimately reduces to

$$E_0 \simeq \frac{2}{T'} \int_0^{T'} (1 - \tau/T')C_x(\tau)d\tau + (T'/2)^2 E_1. \tag{13}$$

Note here that while the first term on the right hand side of Equ. (13) reduces to zero for sufficiently large T′ the second term on the right does not; indeed, Equ. (10) gives $(T'/2)^2 E_1 = C_x(0)/2$, hence

$$E_0 \simeq 2C_x(0)\left(\frac{L_t}{T'}\right) + C_x(0)/2, \tag{14}$$

where L_t is the integral time scale of the data and is defined as

$$C_x(0)L_t = \int_0^{T'} C_x(\tau)d\tau. \tag{15}$$

The result that E_0 does not approximate zero, even for excessively large values of T′, is altogether not too disturbing in view of the fact that the slope of $\mu(t)$ has been accurately determined. The situation at hand is simply that $\underline{\mu}_0(T')$ is an unbiased but inconsistent estimate of μ_0 and consequently that $\mu(t)$ can only be determined to within an additive constant, i.e. that

$$\mu(t) \simeq \underline{\mu}_1 t + \underline{\mu}_0 \pm \sqrt{E_0}.$$

COMPARISON WITH LEAST-SQUARES REGRESSION

For continuous data the least-squares regression scheme begins with the form

$$E = \int_0^T [x(t) - \mu(t)]^2 dt \tag{16}$$

where E represents the sum of the squared deviations from the mean value. It can be shown that the conditions for E to be minimum, in the case where $\mu(t)$ is assumed to be linear, are

$$\int_0^T t\mu(t)dt = \int_0^T t \;\; x(t)dt \tag{17}$$

and

$$\int_0^T \mu(t)dt = \int_0^T x(t)dt, \tag{18}$$

and the resulting (unbiased) estimates for μ_1 and μ_0 are respectively

$$\underline{\tilde{\mu}}_1 = \frac{6}{T^2} \int_0^T (\frac{2t}{T} - 1) \;\; x(t)dt \tag{19}$$

and

$$\underline{\tilde{\mu}}_0 = \frac{4}{T} \int_0^T (1 - \frac{3t}{2T}) \;\; x(t)dt. \tag{20}$$

These estimates produce the corresponding mean-square errors

$$\varepsilon_1 = \frac{24}{T^2} C_x(0)(\frac{L_t}{T}) = 12(\frac{L_t}{T})E_1 \tag{21}$$

and

$$\varepsilon_0 = 8C_x(0)(\frac{L_t}{T}) = \{\frac{16(L_t/T)}{4(L_t/T) + 1}\}E_0. \tag{22}$$

Note that ε_1 and ε_0 (the mean-square errors obtained with the method of least-squares regression) are respectively less than E_1 and E_0 (the mean-square errors obtained with the proposed method) whenever there is relatively no correlation in the data over the total domain D, viz. whenever $(L_t/T) < 1/12$; however, when $(L_t/12) > 1/12$, the proposed method gives better estimates to both the slope and the intercept. What this implies then is that if in the actual acquisition of (correlated) random data the time interval (0,T) over which the data is taken is "large enough," when compared to the integral scale of the data, then the method of least squares will provide sufficiently accurate estimates to the linear trend in the data, regardless of the shape of the correlation function, while if the time interval is not "large enough" then the method proposed herein will provide the best estimates to the linear trend. The reason for this is that the method of linear regression fits a straight line to random data itself and it turns out that this straight line fit is the best estimate to the mean value of the data only when the data is uncorrelated while the method developed herein fits a straight line to the mean value of the data and not to the data itself.

APPROXIMATION OF THE CORRELATION FUNCTION

The correlation function, as defined in Section 2, can be expressed in terms of the covariance function as

$$Q_x(t_1,t_2) = \mu(t_1)\mu(t_2) + C_x(\tau) \tag{23}$$

where, recall, it has been assumed that the data is covariance stationary. In this form it is clear that the time variations in the intensity, $Q_x(t,t) = \mu^2(t) + C_x(0)$, of the given data are due solely to time variations in the mean value of the data. The error in approximating Q_x is therefore composed of the error in approximating μ plus whatever error occurs in approximating C_x. The error in approximating μ has already been discussed; it can be shown (Papoulis (1965), Bendat and Piersol (1971)) that the error in approximating C_x is given by

$$\varepsilon = \frac{1}{T'^2} \iint_0^{T'} [R_\chi(\xi) - C_x^2(\tau)] d\xi_1 d\xi_2 \tag{24}$$

where $\xi = \xi_2 - \xi_1$, $R_\chi(\xi) = \langle\chi(t)\chi(t+\xi)\rangle$, $\chi(t) = x'(t)x'(t+\tau)$, $x'(t) = x(t) - \langle x(t)\rangle$, and $C_x(\tau) = \langle\chi(t)\rangle$. Note that for sufficiently large values of T', $\varepsilon \to 0$ since for large ξ, $R_\chi(\xi) \simeq \langle\chi(t)\rangle\langle\chi(t+\tau)\rangle = C_x^2(\tau)$.

ILLUSTRATIVE EXAMPLES AND DISCUSSION

The method derived herein was applied to the determination of the linear trend in some given samples of atmospheric velocity data. The data used was obtained from the Global Atmospheric Research Program - Atlantic Tropical Experiment (GATE) of the National Center for Atmospheric Research. Specifically, the data represents measurements of the atmospheric velocity components u and v (measured positive eastward and northward, respectively) obtained by flying an instrumented airplane through the atmosphere, in the general area of Puerto Rico during the early afternoon hours of 15 December 1972 (Pennell and LeMone (1974)). Only data obtained during the time intervals that the aircraft was flying along a straight path, i.e. the aircraft was neither "veering" nor turning although the aircraft could be climbing, or descending through the atmosphere, was used. The data was visually inspected to determine if there were any isolated excessive peaks (which were assumed to represent bad data points) and if so, these values were deleted.

The presence of any trend(s) in the data was examined by performing a preliminary analysis following the procedure suggested in Bendat and Piersol (loc. cit.), i.e. the simple time average

$$\bar{x}(t,\Delta T) = \frac{1}{\Delta T} \int_{t}^{t+\Delta T} x(\xi)d\xi, \tag{25}$$

where x can be either u or v, was evaluated for a constant but short (i.e. $\Delta T < T$) value of ΔT and a necessarily varying value of t. The results of this procedure, together with the results of the application of the proposed method, are discussed in the following.

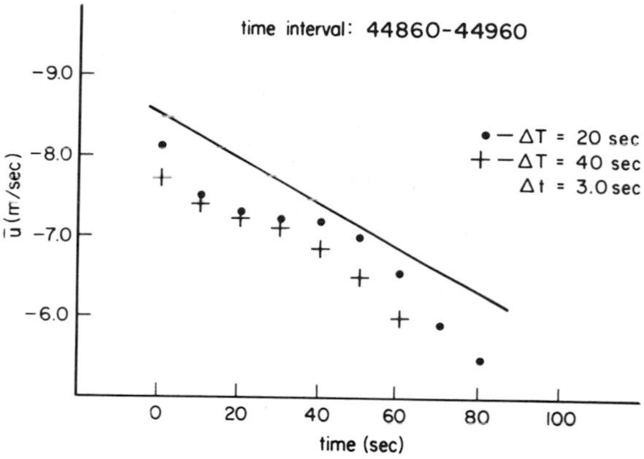

Figure 1
Linear Trend in Atmospheric Velocity Data

A particular result of the preliminary analyses for which a linear trend is suggested is depicted in Figure 1. The respective computed linear approximations to the mean value is $\mu(t) \simeq 0.026t - 8.35$; this mean value is the mean value of the original data and <u>not</u> the "mean value" of the points depicted in the figure.

A particular result of the preliminary analyses which suggested no discernible trend, i.e. the data appeared to be "stationary," is depicted in Figure 2. The proposed method was applied to this data and the resulting mean value was $\mu(t) \simeq -2.8$ = constant. The resulting normalized covariance functions were determined in both cases and these are illustrated in Figures 3 and 4.

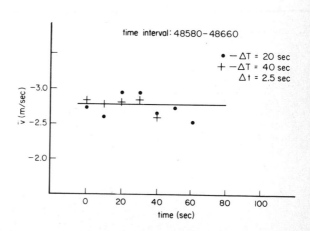

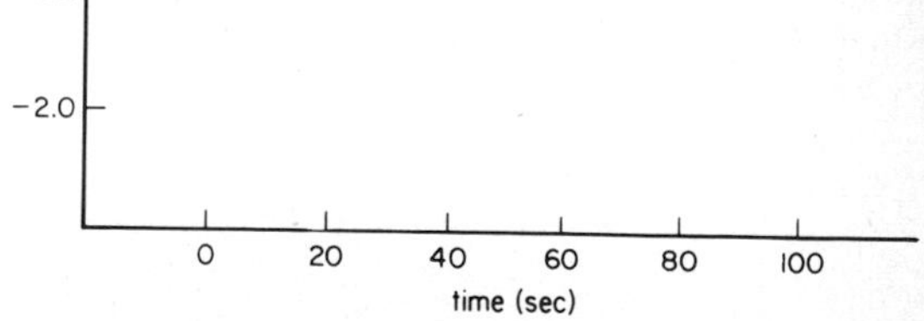

Figure 2
No Trend in Atmospheric Velocity Data

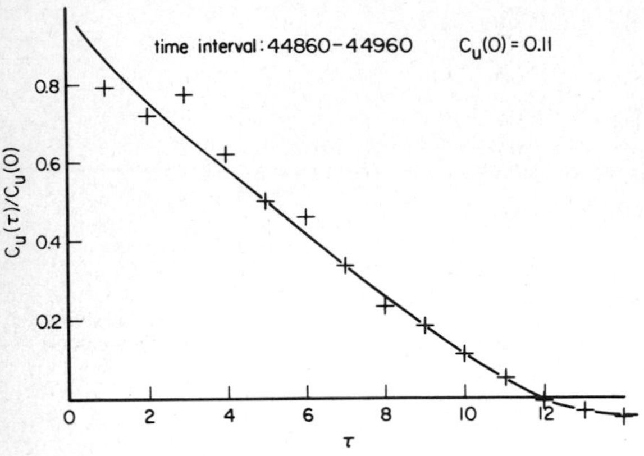

Figure 3
Covariance Function for Data with Linear Trend

In the specific case where a linear approximation to μ was used it appears that the slope approximation is extremely accurate (compare the respective value of $C_x(0)$ and the length of time interval, T, covered); the approximation to the intercept is also equally accurate. To sustain this latter assertion note that E_0, from Equ. (14), can be rewritten as

$$E_0 \simeq \frac{C_x(0)}{2} \{1 + 4 (\frac{L_t}{T})\}. \quad (26)$$

Assuming a "conservative" value of $(L_t/T) \simeq 1/8$ produces $E_0 \simeq 0.75\ C_x(0)$ whence a comparison of the square root of the determined value of E_0 from Figure 3 with the corresponding value of $\underline{\mu}_0$ affirms the above. It can also be shown that the computed mean value of the data which appears to be stationary is also extremely accurate.

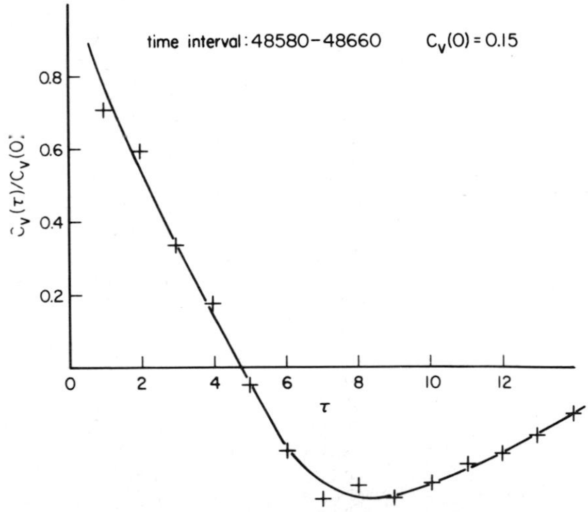

Figure 4
Covariance Function for Data with No Trend

FOOTNOTES

[1]Associate Professor of Physics and Mathematics, Del Mar College, Corpus Christi, Texas (on leave of absence).

[2]Senior Research Associateship sponsored by National Research Council.

[3]By "linear trend" is meant any frequency component whose wavelength is significantly longer than the length of the data record itself.

[4]The restriction that the ratio $(\Delta t/T')$ be "small" may be somewhat relaxed since it is not expected that the higher-order derivatives of C_x will contribute much to the value of E_1.

[5]The assumption of covariance stationary data automatically equates to zero all terms in Equ. (5) which involve odd-order derivatives of C_x; hence, the first term neglected in obtaining Equ. (9) is the product of $(\Delta t/T')^2$ and the integral of the fourth derivatives of C_x.

REFERENCES

[1] Papoulis, A., Probability, Random Variables, and Stochastic Processes (McGraw-Hill, New York, 1965).

[2] Bendat, J. S. and Piersol, A. G., Random Data: Analysis and Measurement Procedures (Wiley-Interscience, New York, 1971).

[3] Pennell, W. T. and LeMone, M. A., An Experimental Study of Turbulence Structure in the Fair-Weather Trade Wind Boundary Layer, Jrnl. Atmos. Sci. 37 (1974) 1121-1128.

APPLIED TIME SERIES ANALYSIS
O.D. Anderson & M.R. Perryman (eds.)

FORECASTING FROM THE FINAL EQUATIONS OF AN ECONOMETRIC MODEL AND THE AUTOREGRESSIVE PROCESSES

Yien-I Tu

Department of Economics
University of Arkansas
Fayetteville, Arkansas
U.S.A.

The final-form equations derived from estimated linear structural-form equations are used for forecasting the endogenous variables beyond sample-period. The exogenous variables and the residuals calculated from the structural form are estimated by autoregressive processes and incorporated when deriving the final-form. Forecasts prepared by several procedures are presented and compared with those prepared from the reduced-form. Furthermore, the forecasts are evaluated by several criteria.

INTRODUCTION

A model of three equations has been estimated and is used for post-sample period forecasting. Specifically, the derived final equations are used in preparing the forecasts of the three endogenous variables. All the endogenous variables have the same autoregressive processes. The exogenous variables and the estimated residuals are also analyzed by the appropriate autoregressive processes. The projections of the exogenous variables and the residuals beyond the sample-period from these processes are used as the inputs in the final equations for forecasting.

THE FINAL EQUATIONS

A generalized dynamic linear econometric in the structural-form may be defined as:

$$A(0)Y_t+A(L)Y_{t-1}+B(L)X_t=C(L)U_t \tag{1}$$

In equation (1), A(0) is a non-singular nxn matrix of parameters associated with a nxl column vector of the current endogenous variables Y_t. A(L) is a nxn matrix of polynomials in the lagged operator L with appropriate degree. B(L) is a nxm matrix of the polynomials, X_t is a mxl vector of the exogenous variables. C(L) is a nxn diagonal matrix of the polynomials. U_t is a nxl vector of random disturbances with the usual assumptions such as white noises. All the polynomials of L in equation (1) are assumed to be of finite degree.

For the purpose of forecasting, usually the reduced-form equation derived from equation (1), such as:

$$Y_t=-A(0)^{-1}[A(L)Y_{t-1}+B(L)X_t-C(L)U_t] \tag{2}$$

is used. $A(0)^{-1}$ is the inverse of A(0).

The final equations (or transfer functions as called by Zellner and Palm[2]) derived from equation (1), after appropriately redefining A(L), are contained in the following expression:

$$|A(0)+A(L)|Y_t=-Adj[A(0)+A(L)][B(L)X_t-C(L)U_t] \tag{3}$$

In equation (3), $|A(0)+A(L)|$ is the determinant of the sum of A(0) and A(L) after redefining A(L). Adj[A(0)+A(L)] is the adjoint matrix. Equation (3) may also be used for forecasting. The left-hand side represents a unique autoregressive process for all the endogenous variables included in the model. The right-hand side represents a mixture of the autoregressive processes of the exogenous variables and the random disturbances. For forecasting the current endogenous variables, the inputs consist of the lagged endogenous and exogenous variables and the random disturbances; the projected current exogenous variables and random disturbances. The random disturbances may be dropped, if forecasting the mean values of the endogenous variables is of primary interest.

THE ESTIMATED EQUATIONS

A linear dynamic model describing the interrelationship between the tax revenues, the total expenditures and the tax rate of Arkansas has been estimated by 3SLS and Box-Jenkins method for the period of 1958-74. The estimated equations are shown as the following:

$$\begin{aligned}
&X_t-48.1613R_t=.0592Y_t=.0309LN_t+351.1481=(1.8839L-.2095L^2-1.2327L^3+.5583L^4)\hat{U}_t\\
&E_t-2.1863X_{t-1}-.0509LF_t+.3433LM_t+21.9634=(1.0313L+.4049L^3+.0963L^3-.5325L^4)\hat{V}_t \qquad (4)\\
&R_t-.0060X_{t-1}+.0021E_{t-1}-5.9927=(1.1096L+.4578L^2-.2444L^3-.3230L^4)\hat{W}_t
\end{aligned}$$

Equation (4) corresponds to equation (1). In this equation, X_t, R_t and E_t are the endogenous variables; Y_t, N_t, F_t and M_t are the exogenous variables; $\hat{U}_t$, $\hat{V}_t$ and $\hat{W}_t$ are the estimated random disturbances. The forecasts from the reduced-form equations derived from equation (4) have been reported elsewhere[1]. However, the forecasts by the final equations derived from equation (4) will be discussed.

The final equations derived from equation (4) are shown as the following:

$$\begin{aligned}
(-1+.2890L-.2211L^2)X_t=\;&0.9582T_t-.0309LN_t+.0051L^2F_t-.0347L^2M_t+60.3114\\
&+(-1.8839L-.2095L^2-1.2327L^3+.5583L^4)U_t+(-49.6687L\\
&-19.5005L^2-4.6379L^3+25.6459L^4)\hat{V}_t+(.1122L^2+.0463L^3\\
&-.0247L^4=.0327L^5)\hat{W}_t\\
(-1+.2890L-.2211L^2)E_t=\;&-.1294LY-.0676L^2N_t-.0147L^2F_t-.0509LF_t+.3433LM_t\\
&-.0992L^2M_t+152.3292+(-4.1188L^2+.4589L^3+2.6951L^4\\
&-1.2206L^5)\hat{U}_t+(-108.5908L^2-42.6340L^3-10.1399L^4 \qquad (5)\\
&+56.9696L^5)\hat{V}_t+(-1.1096L-.1371L^2+.3767L^3+.2524L^4\\
&-.0933L^5)\hat{W}_t\\
(-1+.2890L-.2211L^2)R_t=\;&.0003L^2Y_t-.0004LY_t-.0002L^2N_t+.0001L^3N_t+.0001L^2F_t\\
&-.0007L^2M_t-5.5472+(-.0113L^2+.01L^3+.0064L^4-.0090L^5\\
&+.0026L^6)U_t+(-1.0313L-.4049L^2-.0963L^3+.5325L^4)\hat{V}_t\\
&+(.0023L^2+.0010L^3-.0005L^4-.0007L^5)\hat{W}_t
\end{aligned}$$

Equation (5) corresponds to equation (3). The left-hand side is a unique second-order autoregressive process for each of the endogenous variables. The right-hand side is a mixture of the autoregressive processes of the exogenous variables and the estimated disturbances. The maximum lag of the mixed processes is six periods. Equation (5) is used for generating the forecasts to be discussed below.

THE POST-SAMPLE PERIOD FORECASTS

The post-sample period forecasts are experimented for 1975-78. Both one-period and multi-period forecasts have been performed. The autoregressive processes of the exogenous variables and the residuals used in preparing the forecasts and a more detailed discussion concerning the one-period and multi-period forecasts can be found in my previous paper [1].

The data and the forecasts of the endogenous variables from the final equations together with those generated from the reduced-form equations are shown in Table 1 and Table 2. It has been found that the forecasts by the method of multi-period forecasting show a pattern of explosive fluctuations, if the residual terms ($\hat{U}_t$, $\hat{V}_t$ and $\hat{W}_t$) are included in equation (5). It therefore does not seem worthwhile to add these terms for forecasting the endogenous variables, when the final equation is used. However, this was not so when the reduced-form equation was used for forecasting [1].

TABLE 1

THE OBSERVED ENDOGENOUS VARIABLES

Year	X_t	E_t	R_t
1975	681·3	1,325·8	7·2
1976	750·9	1,419·7	7·2
1977	843·3	1,561·8	7·2
1978	962·7	1,799·8	7·4

TABLE 2

COMPARISONS OF FORECASTS

Final Equations

	One-Period			Multi-Period		
Year	$\hat{X}_t$	$\hat{E}_t$	$\hat{R}_t$	$\hat{X}_t$	$\hat{E}_t$	$\hat{R}_t$
1975	651·51	1184·40	7·66	651·51	1184·40	7·66
1976	738·81	1324·46	7·52	729·73	1333·30	7·84
1977	795·75	1393·82	7·75	796·70	1477·12	7·93
1978	868·45	1583·37	8·19	860·48	1618·08	8·03

Reduced-Form Equations

Year	$\hat{X}_t$	$\hat{E}_t$	$\hat{R}_t$	$\hat{X}_t$	$\hat{E}_t$	$\hat{R}_t$
1975	653·80	1141·80	6·70	653·80	1141·80	6·70
1976	667·10	1383·10	7·24	721·43	1254·49	7·35
1977	749·90	1533·10	7·52	789·96	1392·33	7·49
1978	853·80	1727·60	7·77	854·84	1497·15	7·56

THE EVALUATION OF FORECATS

The forecasts are evaluated by Root Mean Square Error (RMSE), Mean Absolute Percentage Error (MAPE) and Theil's "U" Statistics. These statistics and the

Relative Efficiency (RE) of RMSE and MAPE between one-period and multi-period forecasts are illustrated in Table 3. As Table 3 indicates that one-period forecasts for X_t and R_t out perform their multi-period forecasts; while the multi-period forecasts for E_t out perform their one-period forecasts. The pattern of these results is just opposite to that of the forecasts derived from the reduced-form equations [1].

TABLE 3

EVALUATION OF FORECASTS

	RSME			MAPE			U	
Variables	One-Period	Multi-Period	RE	One-Period	Multi-Period	RE	One-Period	Multi-Period
X	55·18	59·05	·93	·054	·058	·93	·50	·45
E	167·70	137·86	1·22	·105	·085	1·24	1·24	·29
R	·56	·62	·90	·073	·085	·86	·86	6·28
Average	-	-	-	·077	·076	-	-	-

Furthermore, the Relative Error of Forecasting (REF), a ratio of RMSE to observed value [1], for the endogenous variables have also been prepared. They are shown in Table 4. The average REF derived from the final equations for all the variables is about 8.47%, which is slightly larger than that (8.1%) derived from the reduced-form equations. This error may also be compared with the averages of MAPE as shown in Table 3. The overall average of MAPE is 7.7%.

TABLE 4

RELATIVE ERROR OF FORECASTING (REF)

Variables	One-Period	Multi-Period	Average
X	·0693	·0742	·0718
E	·1107	·0910	·1009
R	·0767	·0859	·0813
Average	·0856	·0837	·0847

The "U" statistics for R(tax rate) are greater than 1 and large. This suggests that the forecasts from the final equation is worse than the naive forecasting. This result is consistent with the observations, since the tax rate change varies little during the period 1975-78 as shown in Table 1.

CONCLUSIONS

This experiment has demonstrated the methodology of using the final equations, the autoregressive processes of the exogenous variables and the residuals for forecasting the endogenous variables beyond the sample-period. It has been found that the residuals in the final equations tend to be explosive with fluctuations, when the method of multi-period forecasting is employed. Therefore, the projected residuals have not been added to the forecasts of the endogenous variables.

By the criteria of RMSE, MAPE and Theil's "U" statistics, the overall accuracy of the forecasts from the final equations is about the same magnitude whether the one-period or multi-period method is used. When the forecasts from the final equations compared with those derived from the reduced-form equations, the overall accuracy also is about the same. However, the accuracy of the forecasts may be different from variable and from period to period.

REFERENCES

[1] TU, YIEN-I (1981), Econometric Modeling and Time Series Analysis: An Empirical Example. In Time Series Analysis, eds: O.D. Anderson and M.R. Perryman, North-Holland, Amsterdam & New York, 589-598.

[2] ZELLNER, A. and F. PALM (1974), Time Series Analysis and Simultaneous Equation Econometric Models, Journal of Econometrics 2, 17-54.

APPLIED TIME SERIES ANALYSIS
O.D. Anderson & M.R. Perryman (eds.)

EDGEWORTH SERIES EXPANSIONS FOR THE DISTRIBUTION OF THE SERIAL CORRELATION COEFFICIENT

John S. White

Mechanical Engineering Department
University of Minnesota
Minneapolis, MN 55455

Edgeworth series expansions to order $0(1/T^2)$ are given for the density and distribution functions of the sample serial correlation coefficient.

$$\hat{\alpha} = \sum_1^T x_t x_{t-1} / \sum_1^T x_{t-1}^2$$

where x_t $(t = \ldots, -1, 0, 1, \ldots)$ is a stationary time series defined by the first-order auto-regression equation

$$x_t = \alpha\, x_{t-1} + u_t$$

where α is an unknown parameter, $|\alpha| < 1$, and the u_t are NID$(0, \sigma^2)$. This extends earlier results of White (1961) and Phillips (1978). A generalization of the Daniels (1956) estimator

$$\alpha^* = \sum_1^T x_t x_{t-1} / (\tfrac{1}{2}x_0^2 + \sum_2^T x_{t-1}^2 + \tfrac{1}{2}x_T^2)$$

is also considered.

INTRODUCTION

Let x_t $(t = \ldots, -1, 0, 1, \ldots)$ be a stationary first order auto-regressive Gaussian time series. That is, the process x_t is generated by the stochastic difference equation.

$$x_t = \alpha\, x_{t-1} + u_t \qquad t = \ldots, -1, 0, 1, \ldots, \tag{1.1}$$

where α is an unknown parameter $(|\alpha| < 1)$ and the u_t are NID $(0, \sigma^2)$. We note that u_t and x_s are independent for $s < t$.

Since the process is stationary we have

$$E(x_t) = 0,\ \mathrm{Var}(x_t) = \sigma_t^2 = \sigma^2/(1-\alpha^2),\ \mathrm{Cov}(x_t x_{t-n}) = \alpha^n \sigma_t^2 \tag{1.2}$$

For a finite sample, say

$$X' = (x_0, x_1, x_2, \ldots, x_T)$$

the density function of X is (e.g., White, 1961)

$$f(X) = (1 - \alpha^2)^{\frac{1}{2}} \quad K \exp(-X'RX/2\sigma^2)$$

where

$$K = 1/(\sigma\sqrt{2\pi})^{T+1}$$

$$X'RX = (1-\alpha^2)x_0^2 + \sum_1^T (x_t - \alpha x_{t-1})^2$$

$$= x_0^2 + (1+\alpha^2) \sum_1^{T-1} x_t^2 + x_T^2 - 2\alpha \sum_1^T x_t x_{t-1}$$

Various estimators for α have been proposed. White (1961) has noted that the maximum likelihood estimator for α is a unique root of a cubic equation, but further analytic results seem to be out of the question. The least squares estimator

$$\hat{\alpha} = \sum_1^T x_t x_{t-1} / \sum_1^T x_{t-1}^2$$

$$= \frac{x_0 x_1 + x_1 x_2 + \ldots + x_{T-1} x_T}{x_0^2 + x_1^2 + \ldots + x_{T-1}^2}$$

and Daniels (1956) modification

$$\alpha^* = \frac{x_0 x_1 + x_1 x_2 + \ldots + x_{T-1} x_T}{\frac{1}{2} x_0^2 + x_1^2 + \ldots + x_{T-1}^2 + \frac{1}{2} x_T^2}$$

are both asymptotically equivalent to the maximum likelihood estimator and, as such, are candidates for further investigation.

MOMENT GENERATING FUNCTIÓN

Following Leipnik (1958), we consider the quadratic forms

$$P_1 = x_0^2, \; P_2 = x_1^2 + \ldots + x_{T-1}^2, \; P_3 = x_T^2 \tag{2.1}$$

$$Q = x_0 x_1 + \ldots + x_{T-1} x_T.$$

Thus

$$X'RX = P_1 + (1+\alpha^2) \; P_2 + P_3 - 2\alpha Q$$

$$\hat{\alpha} = Q/(P_1 + P_2)$$

The joint moment generating function of P_1, P_2, P_3 and Q is then

$$M(z_1, z_2, z_3, z_4) = E(\exp(P_1z_1 + P_2z_2 + P_3z_3 + Qz_4)) \quad (2.2)$$

$$= \int e^{P_1z_1+P_2z_2+P_3z_3+Qz_4} f(X)\,dX$$

$$= (1 - \alpha^2)^{\frac{1}{2}} K \int e^{-X'SX/2\sigma^2}\,dX$$

$$= (1 - \alpha^2)^{\frac{1}{2}} \quad |S|^{-\frac{1}{2}}$$

where $|S|$ is the determinant of S and

$$X'SX = X'RX - 2\sigma^2(P_1z_1 + P_2z_2 + P_3z_3 + Qz_4)$$

$$= (1-2\sigma^2z_1)x_0^2 + (1+\alpha^2-2\sigma^2z_2)\sum_1^{T-1} x_t^2 + (1-2\sigma^2z_3)x_T^2$$

$$-2(\alpha+\sigma^2z_4)\sum_1^T x_t x_{t-1}.$$

Hence S is a (T+1) by (T+1) matrix of the form

$$S = \begin{bmatrix} a & b & 0 & 0 & . & . & . \\ b & c & b & 0 & . & . & . \\ 0 & b & c & b & . & . & . \\ . & . & . & . & b & c & b \\ . & . & . & . & 0 & b & d \end{bmatrix} \qquad \begin{aligned} a &= 1 - 2\sigma^2z_1 \\ b &= -(\alpha + \sigma^2z_4) \\ c &= 1 + \alpha^2 - 2\sigma^2z_2 \\ d &= 1 - 2\sigma^2z_3 \end{aligned} \quad (2.3)$$

It may be shown, e.g., by induction, that

$$|S| = \frac{(a-s)(d-s)r^T}{r-s} + \frac{(a-r)(d-r)s^T}{s-r} \quad (2.4)$$

where r and s are roots of the quadratic equation

$$y^2 - cy + b^2 = 0,$$

$$r, s = \tfrac{1}{2}(c \pm (c^2 - 4b^2)^{\frac{1}{2}}).$$

Following White (1958), we start our investigation of $\hat{\alpha}$ and α^* by considering

$$\hat{\alpha}-\alpha = \frac{Q-\alpha(P_1 + P_2)}{P_1 + P_2} = \hat{C}/\hat{V}$$

and obtaining the joint moment generating function of $(\hat{C},\hat{V})$. Since we are dealing primarily with ratios, we may w.l.o.g. set $\sigma^2 = 1$.

$$\begin{aligned} E(\exp(\hat{C}t+\hat{V}u)) &= E(\exp(Qt+P_1(u-\alpha t) + P_2(u-\alpha t)) \\ &= M(u-\alpha t,\ u-\alpha t,\ 0,\ t) \\ &= \sqrt{1-\alpha^2}\ |S|^{-\frac{1}{2}} = \hat{m}(t,\ u) \end{aligned} \tag{2.5}$$

where the elements of S are

$$a = 1-2u + 2\alpha t,\ b = -(\alpha+t),\ c = 1+\alpha^2 -2u + 2\alpha t,\ d = 1 \tag{2.6}$$

ASYMPTOTIC GENERATING FUNCTIONS

It is well known (Mann and Wald, 1942) that $\hat{\alpha}$ has an asymptotic normal distribution with mean and variance (White(1961))

$$E(\hat{\alpha}) = \alpha - 2\alpha/T + 0(1/T^2)$$

$$\text{Var}(\hat{\alpha}) = (1 - \alpha^2)/T + (10\alpha^2 - 2)/T^2 + 0(1/T^3).$$

This suggests that the distribution of the reduced statistic

$$\sqrt{\frac{T}{1-\alpha^2}}\ (\hat{\alpha} - \alpha) = \sqrt{\frac{T}{1-\alpha^2}}\ \frac{(Q - \alpha(P_1+P_2))}{P_1+P_2}$$

be investigated.

Setting

$$g^2 - g^2(T) = T/(1-\alpha^2)$$

we have

$$\sqrt{\frac{T}{1-\alpha^2}}\ (\hat{\alpha} - \alpha) = \frac{(Q - \alpha(P_1+P_2))/g}{(P_1+P_2)/g^2} = C/V$$

The joint moment generating function of C and V is

$$\begin{aligned} m(t,u) &= E(\exp(Ct + Vu)) \\ &= E(\exp(\hat{C}t/g + \hat{V}u/g^2)) \\ &= \hat{m}(t/g,\ u/g^2) = (1-\alpha^2)^{\frac{1}{2}}|S|^{-\frac{1}{2}} \end{aligned} \tag{3.1}$$

$$= \left[\frac{(a-s)(d-s)}{(1-\alpha^2)(r-s)} r^T + \frac{(a-r)(d-r)}{(1-\alpha^2)(s-r)} s^T \right]^{-\frac{1}{2}}$$

where, from (2.5) and (2.6)

$$a = 1 - 2u/g^2 + 2\alpha\ t/g, \quad b = -(\alpha + t/g) \tag{3.2}$$

$$c = 1 + \alpha^2 - 2u/g^2 + 2\alpha t/g, \quad d = 1$$

$$r,s = \tfrac{1}{2}(c \pm (c^2 - 4b^2)^{\frac{1}{2}})$$

It turns out that a convenient variable for the asymptotic expansions is

$$q = (T(1-\alpha^2))^{\frac{1}{2}} = g(T)(1-\alpha^2)$$

and, since $1-\alpha^2$ is a factor in a number of intermediate results, we set $\beta = 1-\alpha^2$ in many preliminary steps. For example:

$$q = \sqrt{T\beta}, \qquad T = q^2/\beta$$

From (3.2) we have

$$c^2 - 4b^2 = \beta^2(1-4\alpha t/q - 4(\beta u^2+(1+\alpha^2)t)/q^2+0(1/q^3)).$$

A formal expansion using the binomial theorem gives

$$(c^2 - 4b^2)^{\frac{1}{2}} = \beta(1-4\alpha t/q - 2(t^2 + (1+\alpha^2)u)/q^2+0(1/q^3)) = r-s \tag{3.3}$$

From (3.2) we have

$$r = 1-\beta(t^2+2u)/q^2 + 0(1/q^3) \tag{3.4}$$

$$s = \alpha^2 + 2\alpha\beta/q + \beta(t^2+2\alpha^2 u)/q^2 + 0(1/q^3)$$

We note immediately that, since $|\alpha| < 1$,

$$s^T = \alpha^{2T}(1+0(1/q))^T = o(1)$$

and hence we may approximate $m(t,u)$ as

$$m(t,u) \sim \left[\frac{(a-s)(d-s)}{\beta(r-s)} r^T \right]^{-\frac{1}{2}} \tag{3.5}$$

From (3.2) and (3.4) we have

$$a-s = \beta(1-(t^2+2u)/q^2 + 0(1/q^3)) \tag{3.6}$$

$$d-s = \beta(1-2\alpha t/q - 2(t^2+ \alpha^2 u)/q^2 + 0(1/q^3))$$

$$1/(r-s) = (1/\beta)(1+2\alpha t/q + (2t^2 + 4\alpha^2 t^2+2u)/q^2 + 0(1/q^3))$$

and hence

$$\frac{(a-s)(b-s)}{\beta(r-s)} = 1 + 0\ (1/q^3)$$

From the formal Taylor series expansion

$$\log(1+z) = z - z^2/2 + z^3/3 \pm \ldots$$

We have

$$\log r = \beta(-(t^2+2u)/q^2 + 0(1/q^3)) \tag{3.7}$$

$$\log r^T = T \log r = -(t^2+2u) + 0(1/q)$$

The cumulant generating function is then

$$K(t,u) = \log m(t,u) \tag{3.8}$$

$$\sim -\tfrac{1}{2} \log \frac{(a-s)(b-s)}{\beta(r-s)} r^T$$

$$\sim -\tfrac{1}{2} (\log(1+0(1/q)) - (t^2+2u)+0(1/q))$$

$$\sim t^2/2 + u + 0(1/q).$$

From the limiting form of $K(t,u)$, we see that, to terms of order $0(1/q)$, C and V are independent, C is N (0,1) and Prob (V=1) =1. Hence the well known result

$$C/V = \sqrt{\frac{T}{1-\alpha^2}}\ (\hat{\alpha} - \alpha) \to N(0,1).$$

The preceding analysis was given to indicate the method. From (3.7) it is obvious that to obtain $\log r^T$ and hence $K(t,u)$ to order $0(1/q^k)$ requires that the earlier computations be taken to terms of order $0(1/q^{k+2})$. Table I gives the expansion of $K(t,u)$ (3.8) to terms of order $0(1/q^4) = 0(1/T^2)$.

MOMENTS AND CUMULANTS

The moments and cumulants of $g(\hat{\alpha}-\alpha) = C/V$ may be obtained from the joint moment generating function

$$m(t,u) = E(\exp(tC + uV)).$$

We then have, at least formally

$$\left.\frac{\partial^k m(t,u)}{\partial t^k}\right|_{t=0} = E(C^k e^{uV}) \tag{4.1}$$

$$\int_{-\infty}^{o} u^{k-1} m(t,u)\,du = E(e^{tC} \int_{-\infty}^{o} u^{k-1} e^{uV} du) \tag{4.2}$$

$$= (-1)^k \ (k-1)!\ E(e^{tC}/V^k)$$

Combining (4.1) and (4.2) and setting $D = \partial/\partial t$ we have (4.3)

$$E((C/V)^k) = (-1)^k \int_{-\infty}^{o} \frac{u^{k-1}}{(k-1)!} D^k m(o,u)\,du$$

To actually evaluate (4.3) we note that from Table I a formal expansion of $m(t,u) = \exp(K(t,u))$ may be obtained as

$$m(t,u) = e^u \Sigma a_{ij} \frac{t^i}{i!} \frac{u^j}{j!} \tag{4.4}$$

where $a_{ij} = 0(1/q^j)$.

Hence, to terms of order $0(1/q^n)$,

$$D^k m(o,u) = e^u \sum_{j=0}^{n} a_{kj}\ u^j/j!$$

and

$$E((C/V)^k) = \sum_{j=0}^{n} a_{kj} (-1)^j \binom{j+k-1}{j} \tag{4.5}$$

It turns out that the n + 2 cumulant of C/V is of order $0(1/q^n)$. Since we will be dealing with approximations to order $0(1/T^2) = 0(1/q^4)$, the first six moments and cumulants of C/V will be required. These are listed in Table II.

HERMITE POLYNOMIALS AND EDGEWORTH SERIES

The polynomial set $H_n(x)$, $n=0,1,2,\ldots$, defined by the generating function

$$g(x,t) = e^{xt-t^2/2} = \sum_{n=0}^{\infty} H_n(x)\ t^n/n! \tag{5.1}$$

is one version of the Hermite polynomial set (e.g., Kendall and Stuart (1969), pp. 155-158.)

From the form of (5.1) it is obvious that $H_n(x)$ is indeed a polynomial of degree n. The first few Hermite polynomials are

$$H_0(x) = 1,\ H_1(x) = x,\ H_2(x) = x^2-1,\ \ldots \tag{5.2}$$

We shall only exploit a few of the many known properties of the Hermite polynomials. First, the Hermite polynomials are orthogonal with respect to the reduced normal distribution. That is, let X be a random variable with density and distribution functions

$$f(x) = e^{-x^2/2}/\sqrt{2\pi},\ F(x) = \int_{-\infty}^{x} f(t)\,dt \tag{5.3}$$

then

$$E(H_j(X)H_k(X)) = \int H_j(x)\ H_k(x)f(x)\,dx = 0 \quad i \neq j \tag{5.4}$$

$$E(H_j^2(X)) = j!$$

Two additional properties of the Hermite polynomials are obtained by differentiating the generating function. Differentiating with respect to x gives

$$dg(x,t)/dx = tg(x,t) = \Sigma\ H_n'(x)\ t^n/n!$$

or, identifying coefficients of powers of t,

$$nH_{n-1}(x) = H_n'(x). \tag{5.5}$$

Differentiating with respect to t gives

$$(-t+x)g(x,t) = \Sigma\ H_n(x)t^{n-1}/(n-1)!$$

which yields the recursion relation

$$H_{n+1}(x) = xH_n(x) - nH_{n-1}(x) \tag{5.6}$$

Combining (5.5) and (5.6) we have

$$d(H_n(x)f(x))/dx = (nH_{n-1}(x) - xH_n(x))\ f(x)$$

$$= -H_{n+1}(x)\ f(x)$$

or, by integration,

$$\int_{-\infty}^{x} H_{n+1}(y)f(y)dy = -H_n(x)f(x) \qquad (5.7)$$

Since the set $H_n(X)$, n=0, 1, 2,... spans the space of all polynomials, any polynomial

$$P_n(x) = \sum_{j=0}^{n} b_j x^j$$

may also be expressed as

$$P_n(x) = \sum_{j=0}^{n} a_j H_j(x). \qquad (5.8)$$

The coefficients a_j may easily be computed using the orthogonality relation (5.4)

$$E(H_k(X)P_n(X)) = \sum_{j=0}^{n} a_j\ E(H_j(X)H_k(X)) = a_j\ j!$$

$$a_j = E(H_k(X)\ P_n(X))/j! \qquad (5.9)$$

$$E(e^{-t^2/2+Xt}\ P_n(X)) = \sum_{k=0}^{\infty} \sum_{j=0}^{n} a_j\ E(H_j(X)H_k(X))\ t^k/k!$$

$$= \sum_{j=0}^{n} a_j t^j \qquad (5.10)$$

We note that the coefficient a_j depend explicitly on n, the degree of $P_n(x)$. In particular let P(x) have a (formal) power series expansion

$$P(x) = \sum_{j=0}^{\infty} b_j x^j$$

and let

$$P_n(x) = \sum_{j=0}^{n} b_j x^j = \sum_{j=0}^{n} a_j(n)\ H_j(x)$$

An expansion of P(x) in Hermite polynomials is formally given by

$$P(x) = \sum_{j=0}^{\infty} b_j x^j = \sum_{j=0}^{\infty} a_j(\infty) H_j(x) \tag{5.11}$$

Where, by analogy with (5.9)

$$a_j(\infty) = E(H_j(X)P(X))/j!$$

$$\sum_{j=0}^{\infty} a_j(\infty)t^j = E(e^{-t^2/2+Xt}P(X)). \tag{5.12}$$

It should be noted that while the expansion (5.8) always exists, (5.11) requires the existence of

$$E(e^{Xt}P(X)) = \int e^{xt}P(x)f(x)dx,$$

at least for t in a neighborhood of t=0.

The Edgeworth series is a special case of Hermite polynomial expansions. Let Y be a random variable with distribution function

$$\text{Prob } [Y < x] = G(x)$$

and density function

$$g(x) = dG(x)/dx.$$

The Gram-Charlier Type A series for g(x) is the formal expansion

$$P(x) = g(x)/f(x) = \sum_{j=0}^{\infty} a_j H_j(x),$$

$$g(x) = \sum_{j-0}^{\infty} a_j H_j(x) \ f(x), \tag{5.13}$$

where f(x) is the normal density function (5.3).
The corresponding distribution function is, by (5.7),

$$\begin{aligned} G(x) &= \int_{-\infty}^{x} g(y)dy \\ &= \int^{x} f(y)dy + \sum_{1}^{\infty} a_j \int^{x} H_j(y)f(y)dy \\ &= F(x) - \sum_{j=1}^{\infty} a_j H_{j-1}(x)f(x). \end{aligned} \tag{5.14}$$

If the moment generating function

$$M_Y(t) = E(e^{Yt}) = \int e^{xt}g(x)dx$$

exists, then by (5.12) we have

$$\Sigma a_j t^j = E(e^{-t^2/2+Xt}\ P(X)) \tag{5.15}$$

$$= \int e^{-t^2/2+xt}\ (g(x)/f(x))f(x)dx$$

$$= e^{-t^2/2}M_Y(t)$$

$$a_j = E(H_j(X)P(X))/j! \tag{5.16}$$

$$= \int H_j(x)g(x)dx/j!$$

$$= E(H_j(Y))/j!$$

In practice, one is frequently interested in approximating $g(x)$ by a truncated expansion

$$g_n(x) = \sum_{j=0}^{n} a_j H_j(x)\ f(x) \tag{5.17}$$

If the first n moments

$$E(Y^j) = \int x^j g(x)dx \qquad j \leq n$$

exist, then a_j may be obtained from (5.16). We shall consider (5.17) as a formal approximation to $g(x)$ without regard to convergence.

DISTRIBUTION AND DENSITY FUNCTIONS

The Edgeworth series for the density function of

$$Y = g(T)\ (\hat{\alpha}-\alpha) = \sqrt{\frac{T}{1-\alpha^2}}\left(\frac{\Sigma x_t x_{t-1}}{\Sigma x_{t-1}{}^2} - \alpha\right)$$

may be obtained directly from the cumulants as listed in Table II. Let the cumulant generating function of Y be

$$K(t) = \Sigma\ k_j t^j/j!$$

Then, from (5.15)

$$E(e^{-t^2/2+Yt}) = e^{K(t)\ -\ t^2/2} = \sum_{j=0}^{\infty} a_j t^j$$

The Type A series for the density function of Y is by (5.13).

$$\phi(x) = \Sigma\ a_j H_j(x)\ f(x). \tag{6.1}$$

It turns out that if K(t) is truncated to $0(1/q^n)$ then

$$a_j = 0 \qquad j > 3n$$

and hence only the first 3n Hermite polynomials appear. The expansion of $\phi(x)$ to order $0(1/q^4)$ then contains powers of x up to x^{12}. This expansion is given in Table III as

$$\phi(x) = f(x) + f(x)\ \Sigma A_j/q^j \tag{6.2}$$

where A_j is a polynomial of degree 3j in x.
The corresponding distribution function is

$$\Phi(x) = F(x) - \Sigma\ a_j H_{j-1}(x)\ f(x) \tag{6.3}$$

The distribution function is also given in Table III in the form

$$\Phi(x) = F(x) + f(x)\ \Sigma B_j/q^j \tag{6.4}$$

OTHER ESTIMATORS

Following Leipnik (1958), we consider a general class of estimators

$$\bar{\alpha} = \Sigma\ x_t x_{t-1}/\bar{V} \tag{7.1}$$

$$\bar{V} = p\ x_o + \sum_1^{T-1} x_t^2 + q\ x_t^2,\ p + q = 1 \tag{7.2}$$

$$= pP_1 + P_2 + qP_3$$

We note that for p = 1

$$\bar{\alpha} = \hat{\alpha}$$

the least squares estimator (1.5), and for $p = q = ½$

$$\bar{\alpha} = \alpha^*,$$

the Daniels estimator (1.6).

Since

$$E(x_t^2) = \sigma^2/(1-\alpha^2)$$

we have, for all p, q,

$$E(\bar{V}) = T\ \sigma^2/(1-\alpha^2).$$

It will be convenient to define δ by

$$p = (1+\delta)/2$$

and, for reasons which will become apparent later,

$$\gamma = 1-\delta^2$$

We then write

$$\bar{\alpha} = \bar{\alpha}(\gamma)$$

and have

$$\hat{\alpha} = \bar{\alpha}(0), \quad \alpha^* = \bar{\alpha}(1) \tag{7.3}$$

Let $g(T)(\bar{\alpha}(\gamma) - \alpha) = \bar{C}/\bar{V}(\gamma)$

The joint moment generating function of $\bar{C}$ and $\bar{V}(\gamma)$

$$E(\exp(t\bar{C}/g + u\bar{V}(\gamma)/g^2) = \bar{m}(t,u)$$

is of the same form as (3.1).
That is

$$\bar{m}(t,u) = (1-\alpha^2)^{\frac{1}{2}} |S|^{-\frac{1}{2}}$$

where, as before,

$$S = \frac{(a-s)(d-s)r^T}{r-s} + \frac{(a-r)(d-r)s^T}{s-r}$$

with

$$a = 1-(1+\delta)(u/g^2-\alpha t/g)$$

$$b = -(\alpha+t/g)$$

$$c = 1+\alpha^2-2(u/g^2-\alpha t/g)$$

$$d = 1-(1-\delta)(u/g^2-\alpha t/g)$$

and r and s are the roots of

$$y^2 - cy + b^2 = 0$$

We note that b and c do not depend on δ and, hence, r and s are as in (3.4). Carrying through the computations as before, we find that the cumulant generating function

$$\bar{K}(t,u) = \log \bar{m}(t,u)$$

differs from the cumulant generating function associated with $\hat{\alpha}$ by a multiple of δ^2. Indeed, setting $\gamma = 1-\delta^2$, we have

$$K(t,u) = K(t,u) - \gamma\left(\frac{\alpha^2 t^2}{2q^2} + \frac{\alpha^3 t^3 + \alpha^3 tu - \alpha tu}{q^3} + \ldots\right)$$

Similar results then hold for the moments and cumulants of $\bar{\alpha}(\gamma)$. For example:

$$E(\bar{\alpha}(\gamma)) = E(\hat{\alpha}) + \gamma(\alpha^3 - \alpha^2)/q^3 + \ldots$$

$$\mathrm{Var}(\bar{\alpha}(\gamma)) = \mathrm{Var}(\hat{\alpha}) - \gamma(\alpha^2/q^2) + \ldots$$

and we note that asymptotically the variance is a minimum for $\gamma = 1$, that is for the Daniels estimator (1.6).

Table III gives adjustments to the cumulants in terms of $\gamma = 1-\delta^2$.

Table I

Joint Cumulant Generating Function

$E(\exp(tC+uV) = \exp(K(t,u))$, $K(t,u) = \Sigma K_j/q^j$, $q = (T(1-\alpha^2))^{½}$.

j	K_j
0	$u + t^2/2$
1	$\alpha t^3 + \alpha tu$
2	$(7\alpha^2 + 3)t^4/4 + (4\alpha^2 + 2)t^2u + (\alpha^2 + 1)u^2$
3	$(3\alpha^3+4\alpha)t^5 + (8\alpha^3 + 12\alpha)t^3u - \alpha t^3 + (4\alpha^3+8\alpha)tu^2 - 2\alpha tu$
4	$(31\alpha^4+85\alpha^2+10)t^6/6 + (16\alpha^4+48\alpha^2+6)t^4u - (8\alpha^2+1)t^4/2$
	$(12\alpha^4+42\alpha^2+6)t^2u^2 - (9\alpha^2+1)t^2u + (4\alpha^4+16\alpha^2+4)u^3/3 - 2\alpha^2u^2$

Table II

Moments and Cumulants of $(T/(1-\alpha^2)^{\frac{1}{2}}(\hat{\alpha}-\alpha)$

$\hat{\alpha} \quad \Sigma x_t x_{t-1}/\ \Sigma x_{t-1}^2,\ q = (T(1-\alpha^2))^{\frac{1}{2}}$

$$\mu_1 = k_1 = -2\alpha/q + \alpha(-4\alpha^2+6)/q^3$$

$$\mu_2 = 1 + (14\alpha^2-2)/q^2 + (76\alpha^4-104\alpha^2+8)q^4$$

$$k_2 = \sigma^2 = 1+(10\alpha^2-2)/q^2 + (60\alpha^4-80\alpha^2+8)/q^4$$

$$\mu_3 = -12\alpha/q + (-192\alpha^3+84\alpha)/q^3$$

$$k_3 = -6\alpha/q + (-112\alpha^3+54\alpha)/q^3$$

$$\mu_4 = 3 + (198\alpha^2-18)/q^2 + (3924\alpha^4-2688\alpha^2+120)/q^4$$

$$k_4 = (66\alpha^2-6)/q^2 + (1920\alpha^4-1320\alpha^2+60)/q^4$$

$$\mu_5 = -90\alpha/q + (-4440\alpha^3+1170\alpha)/q^3$$

$$k_5 = (-1080\alpha^3 + 240\alpha)/q^3$$

$$\mu_6 = 15 + (2700\alpha^2-180)/q^2 + (127500\alpha^4-58440\alpha^2+1860)/q^4$$

$$k_6 = (23640\alpha^4-8760\alpha^2+240)/q^4$$

Table III

Adjustments to Table II

$$k_1(\gamma) = k_1 + (\alpha^3-\alpha)\ \gamma/q^3$$

$$k_2(\gamma) = k_2 - \alpha^2\gamma/q^2 - (17\alpha^4-16\alpha^2+3)\ \gamma/q^4$$

$$k_3(\gamma) = k_3 + (12\alpha^3-6\alpha)\ \gamma/q^3$$

$$k_4(\gamma) = k_4 - 6((34-\gamma)\alpha^4-24\alpha^2+2)\gamma/q^4$$

$$k_5(\gamma) = k_5$$

$$k_6(\gamma) = k_6$$

Table IV

Edgeworth Series

$$G(x) = \text{Prob}((T/(1 - \alpha^2))^{\frac{1}{2}}(\hat{\alpha}-\alpha) < x)$$

$$F(x) \sim N(0,1),\ q = (T(1 - \alpha^2))^{\frac{1}{2}}$$

$$g(x) = f(x) + f(x)\ \Sigma\ A_j/q^j$$

$$G(x) = F(x) + f(x)\ \Sigma\ B_j/q^j$$

j	A_j	B_j
1	$-\alpha x^3 + \alpha x$	$\alpha x^2 + \alpha$
2	$\alpha^2 x^6/2$	$-\alpha^2 x^5/2$
	$-(11\alpha^2+1)x^4/4$	$(\alpha^2+1)x^3/4$
	$+(2\alpha^2+1)x^2/2$	$-(\alpha^2-1)x/4$
	$-(\alpha^2-1)/4$	
3	$-\alpha^3 x^9/6$	$\alpha^3 x^8/6$
	$+(9\alpha^3+\alpha)x^7/4$	$-(11\alpha^3+3\alpha)x^6/12$
	$-(23\alpha^3+7\alpha)x^5/4$	$(\alpha^3+\alpha)x^4/4$
	$+(5\alpha^3+9\alpha)x^3/4$	$-(2\alpha^3+\alpha)x^2/8$
	$-(\alpha^3+3\alpha)x/4$	$-(\alpha^3+7\alpha)/4$
4	$+\alpha^4 x^{12}/24$	$-\alpha^4 x^{11}/24$
	$-(25\alpha^4+3\alpha^2)x^{10}/24$	$+(14\alpha^4+3\alpha^2)x^9/24$
	$+(217\alpha^4+62\alpha^2+1)x^8/32$	$-(49\alpha^4+26\alpha^2+7)x^7/32$
	$-(265\alpha^4+172\alpha^2+7)x^6/24$	$+(31\alpha^4+142\alpha^2+7)x^5/96$
	$+(27\alpha^4+134\alpha^2+13)x^4/16$	$-(7\alpha^4+94\alpha^2+43)x^3/96$
	$-(2\alpha^4+23\alpha^2+7)x^2/8$	$+(\alpha^4-2\alpha^2-15)x/32$
	$+(\alpha^4-2\alpha^2-15)/32$	

REFERENCES

DANIELS, H. E. (1956) The approximate distribution of serial correlation coefficients. *Biometrika*, 43, 169-185.

KENDALL, M. G., & STUART, A. (1969) The Advanced Theory of Statistics, Vol. I. Hafner Publishing Company, New York.

LEIPNIK, R. B. (1958) Moment generating functions of quadratic forms in serially correlated normal varialbes. *Biometrika*, 45, 198-210.

PHILLIPS, P. C. B. (1977) Approximations to some finite sample distributions associated with a first-order stochastic difference equation. *Econometrika*, 45, 463-485.

PHILLIPS, P. C. B. (1978) Edgeworth and saddlepoint approximations in the first-order noncircular autoregression. *Biometrika*, 65, 91-98.

WHITE, J. S. (1958) The limiting distribution of the serial correlation coefficient in the explosive case. *Ann. Math. Statistics*, 29, 1188-1197.

WHITE, J. S. (1961) Asymptotic expansions for the mean and variance of the serial correlation coefficient. *Biometrika*, 48, 85-94.

APPLIED TIME SERIES ANALYSIS
O.D. Anderson & M.R. Perryman (eds.)

OPTIMUM DESIGN AND ANALYSIS OF STATIONARY BIVARIATE TIME-SERIES EXPERIMENTS

J.S. Williams

Department of Statistics
Colorado State University
Fort Collins, Colorado
U.S.A.

Sufficient conditions are presented for continued-covariate and changeover designs for one treatment factor to be most informative solutions of the optimum-design problem for bivariate time series with a design linear-model trend and stationary Normal residuals. The general solutions for maximum-likelihood estimators of the parameters of the design linear model and residual series are presented and then are reduced to simple forms. The complement the results of Wu et al (1972) for regular balanced crossover and switchback and for contined-covariate and changeover designs.

INTRODUCTION

We continue our report of results of an investigation of optimum and restricted-optimum designs and of analyses for bivariate time-series experiments. These are extensions of earlier findings by Wu et al (1972) for series with bivariate white-noise residuals and by Williams and Iyer (1981a) for series with stationary bivariate residuals.

The definition of compromise designs, an algorithm to calculate a design that is at least as informative as a compromise design, and sufficient conditions for regular balanced crossover and switchback assignments to be both compromise and most informative designs were presented by Williams and Iyer (1981a). The results for crossover and switchback designs, which apply to experiments in which there is no pretreatment phase of the bivariate series, were predicted by Wu et al (1972) who conjectured that the numbers of different possible assignments of a treatment within a period for a most informative design will be nearly the same for experiments with or without pretest records and with either bivariate white-noise residuals or stationary bivariate residuals and autocorrelations that vanish rapidly with increasing lag order. The outcome of a numerical study by Williams and Iyer (1981b) also are in agreement with the conjecture for experiments with short to intermediate-length pretest records. In this paper, we derive conditions for which the conjecture is true when there is a sufficiently long pretest record. In these cases, a continued-covariate or changeover design is most informative for experiments with either bivariate white-noise residuals or stationary bivariate residuals with autocorrelations that satisfy the conditions.

The analysis of data from designed bivariate time-series experiments is given here for the case when the residual series is stationary bivariate Normal. The theoretical development of the analysis is presented in two steps. The first contains a description of maximum-likelihood estimators of parameters that measure treatment effects. These are given as functions of a series record and estimators of variance and covariance parameters of the corresponding residual series. The second stage contains an expression for a likelihood function evaluated at the coordinates of the maximum-likelihood estimators of treatment-effect parameters and, therefore, written as a function that can be maximized numerically for estimates of the variance and covariance parameters of the residual series.

characteristics of the expressions for the estimators are given for regular balanced crossover and switchback designs and for continued-covariate and changeover designs.

THE STATISTICAL MODEL

Two experimental units followed for K=m+N time periods receive applications of two levels of a principal treatment factor in a manner specified by a vector $\tilde{a}^t = (\tilde{a}_1^t : \tilde{a}_2^t)$ where a_{ij} is 0 or 1 if respectively, unit 1 in time period j is subject to the first (control) or second level. The first m elements of both partitions are zero in correspondence with a pretreatment (historical) phase during which both units are subject to the control level of the principal treatment factor. When means of untreated units are equal, then there may be a secondary, noninteracting treatment factor with two levels, one assigned to each unit.

Let $\tilde{Y}_i (K\times 1)$ denote the response vector of unit i. The differential effects of treatments on response are assumed to be additive, and the residual series is assumed to be stationary. Hence for each variate

$$\mathcal{E}(\tilde{Y}_i) = \tilde{j}_K \mu_i + \tilde{a}_i \tau, \quad i = 1, 2,$$

where μ_i is the expected response of the ith unit when the control level of the principal factor is applied.

Let $\tilde{Y}^t = (\tilde{Y}_1^t : \tilde{Y}_2^t)$. Then the dispersion, or variance-covariance, matrix for the finite-length record of the bivariate series is

$$\tilde{V} = \left[\begin{array}{c|c} \tilde{V}_{11} & \tilde{V}_{12} \\ \hline \tilde{V}_{21} & \tilde{V}_{22} \end{array}\right], \quad \tilde{V}_{21} = \tilde{V}_{12}^t.$$

The $\tilde{V}_{hi}$ are Toeplitz matrices. The series record is called symmetric if $\tilde{V}_{12}$ is symmetric and asymmetric otherwise.

It is convenient in the study of estimators and tests for $\mu_1 - \mu_2$ and τ, and for the approximation of optimum-design solutions for inference on τ, to work with Fourier coefficients of $\tilde{Y}$. Therefore let

$$\tilde{\Phi} = ((\phi_{k\ell})) \text{ where } \phi_{k\ell} = K^{-\frac{1}{2}} \exp[i2\pi(k-1)(\ell-1)/K], \; k, \ell=1, \ldots, K,$$

$$\bar{\tilde{\Phi}} = ((\bar{\phi}_{k\ell})) \text{ where } \bar{\phi}_{k\ell} = K^{-\frac{1}{2}} \exp[-i2\pi(k-1)(\ell-1)/K], \; k, \ell=1, \ldots, K,$$

and

$$\tilde{Z} = (\bar{\tilde{\Phi}}^t \otimes \tilde{I}_2)\tilde{Y}.$$

Then $K^{-\frac{1}{2}}\tilde{Z}_i = K^{-\frac{1}{2}}\bar{\tilde{\Phi}}^t \tilde{Y}_i$, i = 1, 2, are Fourier coefficients of the ith variate. There also exist Fourier coefficients of the design which correspond to those of the series. The first of these (ℓ=1) are

$$a_1 = K^{-1} \sum_{k=1}^{K} a_{1k} \text{ and } a_2 = K^{-1} \sum_{k=1}^{K} a_{2k}$$

where a_1 and a_2 are the respective fractions of time periods in which the non-control level is applied to units 1 and 2. The remaining coefficients will be denoted by

$$K^{-\frac{1}{2}}\, \underset{\sim}{b} = K^{-\frac{1}{2}}(\underset{\sim}{\Phi}_2^t \otimes \underset{\sim}{I}_2)\underset{\sim}{a}$$

where $\underset{\sim}{\Phi}_2$ is the K × (K-1) matrix formed of the final K-1 columns of $\underset{\sim}{\Phi}$. The dispersion matrix of scaled Fourier coefficients of the series record is then

$$(\overline{\underset{\sim}{\Phi}}^t \otimes \underset{\sim}{I}_2)\underset{\sim}{V}(\underset{\sim}{\Phi} \otimes \underset{\sim}{I}_2) = [\overline{\underset{\sim}{\Phi}}^t \underset{\sim}{V}_{hi}\underset{\sim}{\Phi}] = \underset{\sim}{\Lambda} \text{ (say)}.$$

When the elements in $\underset{\sim}{V}_{hi}$ are absolutely summable, the $\underset{\sim}{\Lambda}_{hi}$ have expansions of the form

$$\underset{\sim}{\Lambda}_{hi} = \text{diag}(\underset{\sim}{\lambda}_{hi}) + O(K^{-1})$$

where the $\underset{\sim}{\lambda}_{hi}$ are vectors of characteristic roots of the $\underset{\sim}{V}_{hi}$.

We do not use the distribution of $\underset{\sim}{Z}$ in our analysis. Instead we use the density function of $\underset{\sim}{Y}$ expressed as a function of $\underset{\sim}{Z}$, a_1, a_2, $\underset{\sim}{b}$ and the unknown parameters μ_1, μ_2, τ, and $\underset{\sim}{\Lambda}$. This distinction should be kept in mind whenever the notation

$$\underset{\sim}{Z} \sim \text{CMVN}\left\{\left[\begin{array}{c} \underset{\sim}{u}_1 K^{\frac{1}{2}}(\mu_1 + a_1\tau) + \underset{\sim}{b}_1\tau \\ \hline \underset{\sim}{u}_1 K^{\frac{1}{2}}(\mu_2 + a_2\tau) + \underset{\sim}{b}_2\tau \end{array}\right], \underset{\sim}{\Lambda}\right\}$$

is used. In this, $\underset{\sim}{u}_1$ is a K×1 vector with 1 in the first position and zeros elsewhere. The logarithm of a likelihood function of μ_1, μ_2, τ, and $\underset{\sim}{V}$ based on the Normal density of $\underset{\sim}{Y}$ and expressed in terms of $\underset{\sim}{Z}$, a_1, a_2, and $\underset{\sim}{b}$ is

$$\ell nL = -\frac{1}{2}\ell n|\underset{\sim}{\Lambda}| - \frac{1}{2}\,\overline{\underset{\sim}{d}}^t \underset{\sim}{\Lambda}^{-1}\underset{\sim}{d}$$

where

$$\underset{\sim}{d} = \left\{\underset{\sim}{Z} - \left[\begin{array}{c} \underset{\sim}{u}_1 K^{\frac{1}{2}}(\mu_1 + a_1\tau) + \underset{\sim}{b}_1\tau \\ \hline \underset{\sim}{u}_1 K^{\frac{1}{2}}(\mu_2 + a_2\tau) + \underset{\sim}{b}_2\tau \end{array}\right]\right\}.$$

OPTIMUM DESIGNS

Efficient inference for τ when $\underset{\sim}{Y}$ is Normal is the criterion used to define an optimum design for one of these bivariate time-series experiments. An analysis will be efficient if the information about τ in $\underset{\sim}{Y}$ after adjustment for μ_1 and μ_2, viz.

$$\text{AdjI}(\tau) = \overline{\underset{\sim}{b}}^t \underset{\sim}{\Lambda}_{22}^{-1}\underset{\sim}{b},\quad \underset{\sim}{\Lambda}_{22} = (\overline{\underset{\sim}{\Phi}}_2^t \otimes \underset{\sim}{I}_2)\underset{\sim}{V}(\underset{\sim}{\Phi}_2 \otimes \underset{\sim}{I}_2),$$

is a maximum with respect to choice of $\underset{\sim}{a}$. A design for which AdjI(τ) is a maximum is called most informative (Wu et al, 1972). The adjusted information can be expanded in the form

$$
\mathrm{AdjI}(\tau) = \begin{cases} 2\sum_{j=1}^{\frac{1}{2}(K-1)} \bar{c}_j^t \Lambda_{22j}^{-1} c_j + O(K^{-1}), & \text{when K is odd,} \\ 2\sum_{j=1}^{\frac{1}{2}(K-2)} \bar{c}_j^t \Lambda_{22j}^{-1} c_j + \bar{c}_{K/2}^t \Lambda_{22K/2}^{-1} c_{K/2} + O(K^{-1}), & \text{when K is even,} \end{cases}
$$

where for each j

$$
c_j = (u_j^t \otimes I_2) b \text{ and } \Lambda_{22j}^{-1} = (u_j^t \otimes I_2)[\mathrm{diag}(\lambda_{22hi})]^{-1}(u_j \otimes I_2).
$$

A design is called a compromise solution if it maximizes one of the leading terms in the expansion and in the set of all such designs also maximizes $\mathrm{AdjI}(\tau)$ (Williams and Iyer, 1981a,b). In general, a compromise design is restricted optimum rather than optimum, but there are important cases where it is most informative.

Four designs are particularly important in the application of optimum-design theory to bivariate time-series experiments. These are:

1. The regular balanced crossover designs (m=0), in which a design kernel

	Period 1	Period 2			
Unit 1	1	0		0	1
			or		
Unit 2	0	1		1	0

is repeated N/2 times.

2. The regular balanced switchback designs (m=0), in which a design kernel

1	0		0	1
		or		
1	0		0	1

is repeated N/2 times.

3. The continued-covariate designs (m>0), in which the treatment assignment, which is applied after the historical phase, is

	Period m+1	...	Period K				
Unit 1	1	...	1		0	...	0
				or			
Unit 2	0	...	0		1	...	1

4. The changeover design (m>0), in which the treatment assignment, which is applied after the historical phase, is

1	...	1
1	...	1

Wu et al (1972) proved that when the residual series is bivariate white noise, then the regular balanced crossover design is most informative if m=0 and the response covariance is nonnegative and that a continued-covariate design is most informative if m is sufficiently large and the response covariance between variates is no less than the larger of the response variances divided by two. It is easy to evaluate their general results and show that the regular balanced switchback and changeover designs are most informative respectively for m=0 and for m sufficiently large with the respective complementary bounds on the response covariance. Williams and Iyer (1981a) presented sufficient conditions on $\underset{\sim}{V}$ for the regular balanced crossover and switchback designs to be both compromise and most informative solutions. They also showed that these conditions hold for symmetric series for which

$$\underset{\sim}{V} = (\underset{\sim}{C} \otimes \underset{\sim}{\Sigma}) \text{ or } (\underset{\sim}{I}_N \otimes \underset{\sim}{\Sigma}) + (\underset{\sim}{C} \otimes \underset{\sim}{\nu}\underset{\sim}{\nu}^t)$$

where $\underset{\sim}{C}$ is the N×N correlation matrix of the record of a circular stationary series, $N^{-\frac{1}{2}}(1, -1, \ldots, 1, -1)$ is the characteristic vector corresponding to the smallest root of $\underset{\sim}{C}$, $\underset{\sim}{\Sigma}$ is a 2×2 positive definite dispersion matrix, and $\underset{\sim}{\nu}$ is a 2×1 vector of real coefficients.

Our first set of new results are sufficient conditions for continued-covariate and changeover designs to be most informative. These are quite restricted conditions, but they can be checked easily to determine for a particular series whether one of the two designs is optimum.

The following elementary lemmas form the basis of the sufficient conditions. These are stated in terms of a 2N×2N positive definite matrix $\underset{\sim}{M}$ and a set A of all possible 2N×1 design vectors for a bivariate time-series experiment. The matrix and vectors are partitioned into corresponding N dimensional components indexed by 1 and 2.

Lemma 3.1 Let $\underset{\sim}{m}$ be a characteristic vector associated with a maximum characteristic root of $\underset{\sim}{M}$. If $\underset{\sim}{j}_{2N}^t\underset{\sim}{m} > 0$ and $\underset{\sim}{u}_i^t\underset{\sim}{m} \in [0, \underset{\sim}{j}_{2N}^t\underset{\sim}{m}/(2N-1)]$, i = 1, ..., 2N, then $\underset{\sim}{j}_{2N}^t\underset{\sim}{M}\underset{\sim}{j}_{2N} = \max_A \underset{\sim}{a}^t\underset{\sim}{M}\underset{\sim}{a}$.

Proof: Notice that not every element of $\underset{\sim}{m}$ can be zero because $\underset{\sim}{m}^t\underset{\sim}{m} = 1$. Thus, the conditions of the lemma can be satisfied only if $\underset{\sim}{j}_{2N}^t\underset{\sim}{m} \neq 0$, and there is no loss of generality in taking $\underset{\sim}{j}_{2N}^t\underset{\sim}{m} > 0$.

Let S be a sphere of radius $(2N-1)^{\frac{1}{2}}$ centered at the origin of Euclidean N space. Then the surface of S includes the points with coordinates $\underset{\sim}{j}_{2N} - \underset{\sim}{u}_i$, i=1, ...,2N. All elements of A except $\underset{\sim}{j}_{2N}$ are associated with points on or within S. Let C be a convex set generated by the points with coordinates $\underset{\sim}{j}_{2N}-\underset{\sim}{u}_i$, i=1, ..., 2N. Then the vertices of C lie on the surface of S, and all other elements of C are interior points of S. Observe as well that C is the set of all points with coordinates in [0,1] and that sum to 2N-1.

The elements of $(2N-1)^{\frac{1}{2}}\underset{\sim}{m}$ are the coordinates of the intersection of a minor axis of the ellipsoids $\underset{\sim}{x}^t\underset{\sim}{M}\underset{\sim}{x} = Q$, $0 < Q < \infty$, with S. If $\underset{\sim}{u}_i^t\underset{\sim}{m} \in [0, \underset{\sim}{j}_{2N}^t\underset{\sim}{m}/(2N-1)]$, i=1, ..., 2N, then the minor axis intersects C at the point with coordinates $[(2N-1)/\underset{\sim}{j}_{2N}^t\underset{\sim}{m}]\underset{\sim}{m}$. From this and the fact that points of C are also points of S, it follows that

$$\frac{(2N-1)^2\underset{\sim}{m}^t\underset{\sim}{m}}{(\underset{\sim}{j}_{2N}^t\underset{\sim}{m})^2} \leq (2N-1)\underset{\sim}{m}^t\underset{\sim}{m},$$

which in turn with $\underset{\sim}{m}^t\underset{\sim}{m} = 1$ implies that $(\underset{\sim}{j}_{2N}^t\underset{\sim}{m})^2 \geq 2N-1$.

The orthogonal projection of the point with coordinates $\underset{\sim}{j}_{2N}$ onto the minor axis has coordinates

$$\frac{\underset{\sim}{j}^t_{2N}\underset{\sim}{m}}{\underset{\sim}{m}^t\underset{\sim}{m}}\underset{\sim}{m} = (\underset{\sim}{j}^t_{2N}\underset{\sim}{m})\underset{\sim}{m}$$

and squared length $(\underset{\sim}{j}^t_{2N}\underset{\sim}{m})^2$. The projection is a point which lies on the surface of or outside of S because $(\underset{\sim}{j}^t_{2N}\underset{\sim}{m})^2 \geq 2N-1$. From this, it follows that the tangent (2N-1)-flat to the ellipsoid $\underset{\sim}{x}^t\underset{\sim}{M}\underset{\sim}{x} = (\underset{\sim}{j}^t_{2N}\underset{\sim}{m})^2\underset{\sim}{m}^t\underset{\sim}{M}\underset{\sim}{m}$ at the projection, which contains the point associated with $\underset{\sim}{j}_{2N}$, either touches S $((\underset{\sim}{j}^t_{2N}\underset{\sim}{m})^2 = 2N-1)$ or lies outside of S $((j^t_{2N}m)^2 > 2N-1)$. Therefore $\underset{\sim}{j}^t_{2N}\underset{\sim}{M}\underset{\sim}{j}_{2N}$ is at least equal to $(\underset{\sim}{j}^t_{2N}\underset{\sim}{m})^2\underset{\sim}{m}^t\underset{\sim}{M}\underset{\sim}{m}$, the value of the ellipsoid at the projection. This in turn is no smaller than $(2N-1)\underset{\sim}{m}^t\underset{\sim}{M}\underset{\sim}{m}$, the value of the ellipsoid at the intersection of the minor axis with S. This in turn is no smaller than the value of the ellipsoid at any other point in S because S is inscribed in $\underset{\sim}{x}^t\underset{\sim}{M}\underset{\sim}{x} = (2N-1)\underset{\sim}{m}^t\underset{\sim}{M}\underset{\sim}{m}$. This completes the proof.

Lemma 3.2 The conditions

1. $\underset{\sim}{j}^t_N\underset{\sim}{M}_{11}\underset{\sim}{j}_N \geq \underset{\sim}{a}^t_1\underset{\sim}{M}_{11}\underset{\sim}{a}_1, \ \underset{\sim}{a}^t_2\underset{\sim}{M}_{22}\underset{\sim}{a}_2 \ \forall \ \underset{\sim}{a} \in A$;

2. $-2\underset{\sim}{u}^t_k\underset{\sim}{M}_{12}\underset{\sim}{a}_2 \geq \underset{\sim}{a}^t_2\underset{\sim}{M}_{22}\underset{\sim}{a}_2 \ \forall \ \underset{\sim}{a} \in A, \ k = 1, \ldots, N$,

are sufficient for $\underset{\sim}{j}^t_N\underset{\sim}{M}_{11}\underset{\sim}{j}_N = \max_A \underset{\sim}{a}^t\underset{\sim}{M}\underset{\sim}{a}$.

Proof: First if $\underset{\sim}{a}_1 = \underset{\sim}{0}$, then $\underset{\sim}{a}^t\underset{\sim}{M}\underset{\sim}{a} = \underset{\sim}{a}^t_2\underset{\sim}{M}_{22}\underset{\sim}{a}_2$. Thus if condition 1. obtains, it follows that

$$\underset{\sim}{j}^t_N\underset{\sim}{M}_{11}\underset{\sim}{j}_N \geq \underset{\sim}{a}^t\underset{\sim}{M}\underset{\sim}{a} \ \forall \ \underset{\sim}{a} \in A \ni \underset{\sim}{a}_1 = \underset{\sim}{0}.$$

Next if $\underset{\sim}{a}_1 \neq \underset{\sim}{0}$ and condition 2. holds, then

$$-2\underset{\sim}{a}^t_1\underset{\sim}{M}_{12}\underset{\sim}{a}_2 \geq \underset{\sim}{a}^t_2\underset{\sim}{M}\underset{\sim}{a}_2 \ \forall \ \underset{\sim}{a} \in A \ni \underset{\sim}{a}_1 \neq \underset{\sim}{0},$$

because each nonzero $\underset{\sim}{a}_1$ is a sum of $\underset{\sim}{u}_k$ vectors. From this inequality, it follows that

$$\underset{\sim}{a}^t_1\underset{\sim}{M}_{11}\underset{\sim}{a}_1 \geq \underset{\sim}{a}^t_1\underset{\sim}{M}_{11}\underset{\sim}{a}_1 + \underset{\sim}{a}^t_2\underset{\sim}{M}_{22}\underset{\sim}{a}_2 + 2\underset{\sim}{a}^t_1\underset{\sim}{M}_{12}\underset{\sim}{a}_2 = \underset{\sim}{a}^t\underset{\sim}{M}\underset{\sim}{a} \ \forall \ \underset{\sim}{a} \in A \ni \underset{\sim}{a}_1 \neq \underset{\sim}{0}.$$

Therefore with these inequalities and condition 1., we have

$$\underset{\sim}{j}^t_N\underset{\sim}{M}_{11}\underset{\sim}{j}_N \geq \underset{\sim}{a}^t\underset{\sim}{M}\underset{\sim}{a} \ \forall \ \underset{\sim}{a} \in A \ni \underset{\sim}{a}_1 \neq \underset{\sim}{0},$$

which complements the inequalities for $\underset{\sim}{a}_1 = \underset{\sim}{0}$ and completes the proof.

The sufficient conditions for changeover and continued-covariate designs to be most informative are set out in the following two corollaries to these lemmas. Each contains the condition m > 0, which is required for estimable τ when $\mu_1 \neq \mu_2$. Each is stated in terms of a matrix product $(\underset{\sim}{\Phi}_{2,N} \otimes \underset{\sim}{I}_2)\Lambda^{-1}_{22}(\overline{\underset{\sim}{\Phi}}^t_{2,N} \otimes \underset{\sim}{I}_2)$ in which $\underset{\sim}{\Phi}_{2,N}$ is the N × (K-1) matrix formed of the final N rows of $\underset{\sim}{\Phi}_2$.

Corollary 3.1: A changeover design is most informative if m > 0 and if a maximum characteristic vector associated with a maximum characteristic root of $(\Phi_{2,N} \otimes I_2)\Lambda_{22}^{-1}(\bar{\Phi}^t_{2,N} \otimes I_2)$ has nonnegative elements, none of which exceeds the sum of the elements of the vector divided by 2N-1.

Proof: Observe that

$$b^t\Lambda_{22}^{-1}b = a^t(\Phi_2 \otimes I_2)\Lambda_{22}^{-1}(\bar{\Phi}_2^t \otimes I_2)a = a_0^t(\Phi_{2,N} \otimes I_2)\Lambda_{22}^{-1}(\bar{\Phi}_2^t \otimes I_2)a_0$$

where a_0 is an element of A. With this identification, it is clear that the corollary is a special case of Lemma 3.1. Therefore

$$j_{2N}^t(\Phi_{2,N} \otimes I_2)\Lambda_{22}^{-1}(\bar{\Phi}^t_{2,N} \otimes I_2)j_{2N} = \max_A a_0^t(\Phi_{2,N} \otimes I_2)\Lambda_{22}^{-1}(\bar{\Phi}^t_{2,N} \otimes I_2)a_0.$$

This completes the proof.

Corollary 3.2: A continued-covariate design in which unit 1 is treated is most informative if m > 0, if a maximum characteristic vector associated with a maximum characteristic root of $[(\Phi_{2,N} \otimes I_2)\Lambda_{22}^{-1}(\bar{\Phi}^t_{2,N} \otimes I_2)]_{11}$ has nonnegative elements, none of which exceeds the sum of the elements of the vector divided by N-1, and if

$$N^{-1}j_N^t[(\Phi_{2,N} \otimes I_2)\Lambda_{22}^{-1}(\bar{\Phi}^t_{2,N} \otimes I_2)]_{11}j_N, \; - 2u_k^t[(\Phi_{2,N} \otimes I_2)\Lambda_{22}^{-1}(\bar{\Phi}^t_{2,N} \otimes I_2)]_{12}u_\ell \geq \gamma_2^2$$

for all k, ℓ = 1, ..., N. In this γ_2^2 is a maximum characteristic root of

$$[(\Phi_{2,N} \otimes I_2)\Lambda_{22}^{-1}(\bar{\Phi}^t_{2,N} \otimes I_2)]_{22}.$$

Proof: First Lemma 3.1 can be applied here because of the conditions on the maximum characteristic vector. With the result of the lemma, we obtain

$$j_N^t[(\Phi_{2,N} \otimes I_2)\Lambda_{22}^{-1}(\bar{\Phi}^t_{2,N} \otimes I_2)]_{11}j_N \geq a_{01}^t[(\Phi_{2,N} \otimes I_2)\Lambda_{22}^{-1}(\bar{\Phi}^t_{2,N} \otimes I_2)]_{11}a_{01} \; \forall a_0 \in A$$

where $a_0^t = (a_{01}^t \vdots a_{02}^t)$. Next from the first inequality condition in the corollary, we obtain

$$j_N^t[(\Phi_{2,N} \otimes I_2)\Lambda_{22}^{-1}(\bar{\Phi}^t_{2,N} \otimes I_2)]_{11}j_N \geq N\gamma_2^2 \geq a_{02}^t[(\Phi_{2,N} \otimes I_2)\Lambda_{22}^{-1}(\bar{\Phi}^t_{2,N} \otimes I_2)]_{22}a_{02} \; \forall a_0 \in A.$$

These two sets of inequalities form a special case of condition 1. of Lemma 3.2.

From the second inequality condition of the corollary, we obtain $\forall a_0 \in A$, the inequality

$$-2u_k^t[(\Phi_{2,N} \otimes I_2)\Lambda_{22}^{-1}(\bar{\Phi}^t_{2,N} \otimes I_2)]_{12}a_{02} \geq n_{\cdot 1}\gamma_2^2 \geq a_{02}^t[(\Phi_{2,N} \otimes I_2)\Lambda_{22}^{-1}(\bar{\Phi}^t_{2,N} \otimes I_2)]_{22}a_{02}$$

where $a_{02}^t j_N = a_{02}^t a_{02} = n_{\cdot 1}$. This is a result of each a_{02} being a sum of u_ℓ vectors. These inequalities form a special case of condition 2. of Lemma 3.2.

The result of applying Lemma 3.2 here is that

$$j_N^t[(\Phi_{2,N} \otimes I_2)\Lambda_{22}^{-1}(\bar{\Phi}^t_{2,N} \otimes I_2)]_{11}j_N = \max_A a_0^t(\Phi_{2,N} \otimes I_2)\Lambda_{22}^{-1}(\bar{\Phi}^t_{2,N} \otimes I_2)a_0,$$

which is the equality required to complete the proof.

An expression for $(\Phi_2 \otimes I_2)\Lambda_{22}^{-1}(\bar{\Phi}_2^t \otimes I_2)$, which involves only real variables, is is given in the next section.

It is not as easy for the changeover and continued-covariate as it was for the balanced crossover and switchback to find particular structures of $\underset{\sim}{V}$ for which the available sufficient conditions for the designs to be most informative are met. However there are some features required for $\underset{\sim}{V}$ that can be deduced from a search for conditions for $\underset{\sim}{j}_{2N}$ or for $\underset{\sim}{j}_N$ to be proportional to a characteristic vector associated with a maximum characteristic root of $(\underset{\sim}{\Phi}_{2,N} \otimes \underset{\sim}{I}_2)\underset{\sim}{\Lambda}_{22}^{-1}(\underset{\sim}{\bar{\Phi}}_{2,N}^{t} \otimes \underset{\sim}{I}_2)$ or of one of the diagonal partitions of this matrix. First, the matrix $(\underset{\sim}{\Phi}_2 \otimes \underset{\sim}{I}_2)\underset{\sim}{\Lambda}_{22}^{-1}(\underset{\sim}{\bar{\Phi}}_2^{t} \otimes \underset{\sim}{I}_2)$ is a conditional inverse of the dispersion matrix of $\underset{\sim}{Y} - (\underset{\sim}{j}_K \otimes \hat{\underset{\sim}{\mu}})$, the series record after linear adjustment for the unit means μ_1 and μ_2. The effects of the adjustment on the dispersion structure of the series in any fixed interval, such as the treatment phase of the record, diminishes as the length of the series record increases whenever $\hat{\underset{\sim}{\mu}}$ is mean-squared consistent for $\underset{\sim}{\mu}$. Thus the effects can be minimal when a treatment phase is preceded by a long historical phase. Next the matrix $(\underset{\sim}{\Phi}_{2,N} \otimes \underset{\sim}{I}_2)\underset{\sim}{\Lambda}_{22}^{-1}(\underset{\sim}{\Phi}_{2,N} \otimes \underset{\sim}{I}_2)$ is the inverse of the dispersion matrix for the treatment-phase partition of $\underset{\sim}{Y} - (\underset{\sim}{j}_K \otimes \hat{\underset{\sim}{\mu}})$, after a second linear adjustment for the correlative effects of the adjusted residuals in the historical-phase partition. A diagonal partition $[(\underset{\sim}{\Phi}_{2,N} \otimes \underset{\sim}{I}_2)\underset{\sim}{\Lambda}_{22}^{-1}(\underset{\sim}{\bar{\Phi}}_{2,N}^{t} \otimes \underset{\sim}{I}_2)]_{kk}$, $k = 1$ or 2, is the inverse of the dispersion structure of one variate series, twice adjusted, after a third adjustment for the correlative effects of the twice adjusted residuals of the second variate in the treatment phase. If the relative contribution of very low frequency terms to one of these adjusted series is minimal, then a scaled characteristic vector of a minimum characteristic root of the dispersion matrix will be close to $\underset{\sim}{j}_{2N}$ or to $\underset{\sim}{j}_N$. Such a vector is associated with a maximum characteristic root of an inverse, a condition required in both Corollary 3.1 and 3.2.

MAXIMUM-LIKELIHOOD ESTIMATORS AND LIKELIHOOD-RATIO TESTS

The principal objectives of an analysis of data from a bivariate time-series experiment are to estimate τ and sometimes $\delta = \mu_1 - \mu_2$ and to find tests for hypotheses concerning one or both of these parameters. To achieve any of these goals, it is also necessary to estimate $\mu_1 + \mu_2$ and the parameters that determine the elements of $\underset{\sim}{V}$.

It is convenient for the purpose of deriving estimators to choose a slightly different parameterization of the problem and partitions of $\underset{\sim}{Z}$ than were used for presentation of the model and designs. Therefore, let

$$\underset{\sim}{\xi}^t = (\mu_1 + a_1\tau,\ \mu_2 + a_2\tau)$$

and

$$\underset{\sim}{Z}_1 = K^{-\frac{1}{2}}(\underset{\sim}{j}_K^t \otimes \underset{\sim}{I}_2)\underset{\sim}{Y} \text{ and } \underset{\sim}{Z}_2 = (\underset{\sim}{\bar{\Phi}}_2^t \otimes \underset{\sim}{I}_2)\underset{\sim}{Y}.$$

Then $\underset{\sim}{d}$ can be partitioned as follows

$$\underset{\sim}{d}_1 = \underset{\sim}{Z}_1 - K^{\frac{1}{2}}\underset{\sim}{\xi} \text{ and } \underset{\sim}{d}_2 = \underset{\sim}{Z}_2 - \underset{\sim}{b}\tau,$$

and rows and columns of $\underset{\sim}{\Lambda}$ can be reorganized and partitioned into

$$\begin{pmatrix} \Lambda_{11} & \Lambda_{12} \\ \Lambda_{21} & \Lambda_{22} \end{pmatrix}$$

in correspondence with the partitions of d. Notice here that Λ_{22} is the same matrix that was defined in the section on optimum designs. Finally let $\Lambda_{11} - \Lambda_{12}\Lambda_{22}^{-1}\Lambda_{21}$ be denoted by $\Lambda_{11.2}$, and notice that $|\Lambda| = |\Lambda_{11.2}||\Lambda_{22}|$.

The logarithmic likelihood function, reexpressed in the newly established notation, is

$$-\frac{1}{2}\ln|\Lambda_{11.2}| - \frac{1}{2}\ln|\Lambda_{22}| - \frac{1}{2}(\bar{d}_1^t - \bar{d}_2^t\Lambda_{22}^{-1}\Lambda_{21})\Lambda_{11.2}^{-1}(d_1 - \Lambda_{12}\Lambda_{22}^{-1}d_2) - \frac{1}{2}\bar{d}_2^t\Lambda_{22}^{-1}d_2.$$

The structure of Λ determines much of how well δ and τ can be estimated. It is necessary to specify more than just Toeplitz form for the V_{hi}, h, i = 1, 2, in order to investigate the effects of structure. We will consider here dispersion matrices which for all K have elements that are differentiable functions of q real parameters $\theta_1, \ldots, \theta_q$. When it is convenient, this functional dependence will be indicated by writing $\Lambda(\theta)$ in place of Λ. A further restriction will be that there exist estimators $\tilde{\theta}_1, \ldots, \tilde{\theta}_q$ that converge in probability to $\theta_1, \ldots, \theta_q$ at a rate specified by a function $\phi(K)$, i.e. the sequences

$$\phi(K)|\tilde{\theta}_k - \theta_k|, \; k = 1, \ldots, q,$$

should be bounded with probability one. We will also consider other rate functions $\psi(K)$ which satisfy the following inequalities:

$$\sup_K \max_{1 \le k \le q} \frac{\max|\mathrm{ch}R_{(k)}^{-1}(\theta_0)|}{\psi_1(K)} < \infty$$

and

$$\sup_K \max_{1 \le k, \ell \le q} \frac{\max|\mathrm{ch}R_{(k,\ell)}^{-1}(\theta_0)|}{\psi_2(K)} < \infty.$$

In these "ch" denotes a characteristic root of the matrix that follows it, $R^{-1}(\theta_0) = \Lambda^{\frac{1}{2}}(\theta_0)\Lambda^{-1}(\theta)\Lambda^{\frac{1}{2}}(\theta_0)$ where $\Lambda^{\frac{1}{2}}$ is a symmetric square-root factor of Λ, and $R_{(k)}^{-1}(\theta_0)$ and $R_{(k,\ell)}^{-1}(\theta_0)$ are respective first- and second-order matrix partial derivatives with respect to θ_k and θ_ℓ evaluated at $\theta = \theta_0$. This structure includes autogressive, moving average form for V_{hi}, h, i = 1, 2, as well as many other stationary (and nonstationary) forms for the partitions of V.

The condition that $\theta_1, \ldots, \theta_q$ be estimable is a nontrivial restriction on applications of the analysis. For example, if the spectral distribution functions of the residual series are discontinuous at zero, then the variance of $\hat{\delta}$ is a nonidentifiable parameter of the distribution of Y. To illustrate this point with a familar example, consider m=0, a balanced crossover design, and $V = C \otimes \Sigma$ where $C_N = [(1-\rho)I_N + \rho j_N j_N^t]$, $\rho > 0$. In this, V is the dispersion structure of linear combinations of two uncorrelated stationary series, each of which is the sum of a random variable with variance $\rho\sigma_{ii}$ and an uncorrelated white-noise series with variance $(1-\rho)\sigma_{ii}$. Therefore there is a jump at the origin of height ρ in each spectral distribution function. The variance of any unbiased estimator of δ is $(\sigma_{11}+\sigma_{22}-2\sigma_{12})\{[(1-\rho)/n]+\rho\}$ where n is a known number. In this, $(\sigma_{11}+\sigma_{22}-2\sigma_{12})\times(1-\rho)$ is estimable, but $(\sigma_{11}+\sigma_{22}-2\sigma_{12})\rho$ is not. Therefore δ can be estimated, but there is no test for hypotheses concerning δ. This is just a variation of the nonidentifiability problem associated with the testing of a main-plot effect in a split-plot experiment when there are only two main plots, and therefore, there is no main-plot residual sum of squares. Problems like this one can be avoided if the spectral distribution functions of the variate series are continuous and belong to families indexed by parameters of low dimension.

Existence of maximum-likelihood estimators of $\theta_1, \ldots, \theta_q$ will be assumed for all of the remaining discussion in this paper.

Efficient Estimators and Tests for Treatment Effects

The first of our second set of new results is the following theorem.

Theorem 4.1: Maximum-likelihood estimators of δ and τ are

$$\hat{\delta} = K^{-\frac{1}{2}}(1,-1)(\underset{\sim}{Z}_1 - \hat{\underset{\sim}{\Lambda}}_{12}\hat{\underset{\sim}{\Lambda}}_{22}^{-1}\underset{\sim}{Z}_2) - K^{-\frac{1}{2}}[K^{\frac{1}{2}}(a_1-a_2) - (1,-1)\hat{\underset{\sim}{\Lambda}}_{12}\hat{\underset{\sim}{\Lambda}}_{22}^{-1}\underset{\sim}{\bar{b}}]\hat{\tau}$$

and

$$\hat{\tau} = \frac{\underset{\sim}{\bar{b}}^t\hat{\underset{\sim}{\Lambda}}_{22}^{-1}\underset{\sim}{Z}_2}{\underset{\sim}{\bar{b}}^t\hat{\underset{\sim}{\Lambda}}_{22}^{-1}\underset{\sim}{b}}$$

where $\hat{\underset{\sim}{\Lambda}}_{12}$, $\hat{\underset{\sim}{\Lambda}}_{22}$, and $\hat{\underset{\sim}{\Lambda}}_{11.2}$ are evaluated at the $\underset{\sim}{\theta}$ coordinates of the minimum of

$$\ell n|\underset{\sim}{\Lambda}_{11.2}| + \ell n|\underset{\sim}{\Lambda}_{22}| + \underset{\sim}{\bar{Z}}_2^t(\underset{\sim}{\Lambda}_{22}^{-1} - \frac{1}{\underset{\sim}{\bar{b}}^t\underset{\sim}{\Lambda}_{22}^{-1}\underset{\sim}{b}}\underset{\sim}{\Lambda}_{22}^{-1}\underset{\sim}{b}\underset{\sim}{\bar{b}}^t\underset{\sim}{\Lambda}_{22}^{-1})\underset{\sim}{Z}_2.$$

If $\phi(K)|\hat{\theta}_k - \theta_k|$ is bounded with probability one for each θ coordinate of the minimum, and if

1. $$\lim_{K\to\infty}\frac{\psi_1(K)}{\phi(K)} = 0 \text{ and } \sup_K K^{\frac{1}{2}} \max_{1\le k\le q} \frac{\max|ch[R_{(k)}^{-1}(\hat{\underset{\sim}{\theta}}) - R_{(k)}^{-1}(\underset{\sim}{\theta})]|}{\psi_1(K)} < \infty$$

or

2. $$\lim_{K\to\infty}\frac{\psi_2(K)}{\phi(K)} = 0 \text{ and } \sup_K \frac{K^{\frac{1}{2}}}{\phi(K)} \max_{1\le k,\ell\le q} \frac{\max|ch[R_{(k,\ell)}^{-1}(\hat{\underset{\sim}{\theta}}) - R_{(k,\ell)}^{-1}(\underset{\sim}{\theta})]|}{\psi_2(K)} < \infty,$$

then critical regions for tests of fixed size for H_1: $\delta=\delta_0$, H_2: $\delta\ge\delta_0$, H_3: $\delta\le\delta_0$ and H_4: $\tau=\tau_0$, H_5: $\tau\ge\tau_0$, H_6: $\tau\le\tau_0$ that are asymptotically similar and uniformly most powerful unbiased are

$$\left\{ \frac{(1,-1)\hat{\Lambda}_{11.2}\binom{1}{-1}}{K} + \frac{[K^{\frac{1}{2}}(a_1-a_2)-(1,-1)\hat{\Lambda}_{12}\hat{\Lambda}_{22}^{-1}b)]^2}{K\, \bar{b}^t\hat{\Lambda}_{22}^{-1}b} \right\}^{-\frac{1}{2}}(\hat{\delta}-\delta_0) \begin{cases} <-r_1 \text{ or } >r_1 & \text{for } H_1, \\ <-r_2 & \text{for } H_2, \\ >r_3 & \text{for } H_3, \end{cases}$$

and

$$(\bar{b}^t\hat{\Lambda}_{22}^{-1}b)^{\frac{1}{2}}(\hat{\tau}-\tau_0) \begin{cases} <-r_4 \text{ or } >r_4 & \text{for } H_4, \\ <-r_5 & \text{for } H_5, \\ >r_6 & \text{for } H_6. \end{cases}$$

In these, the r constants are quantiles of the standard Normal distribution and are determined by the sizes of the tests.

Remark: The information function of δ adjusted for τ is the square of the term preceding $(\hat{\delta}-\delta_0)$ in the inequality definitions of critical regions. The choice of a design affects only the second term in the sum. In general, the most informative design for τ is not most informative for δ, but the amount of information on δ is irrelevant or less important than the amount of information on τ.

Remark: The theorem is a straight-forward generalization of the properties of maximum-likelihood procedures derived by Wu et al (1972) for bivariate white-noise residuals. In that case, $\phi(K) = K^{\frac{1}{2}}$, and the $\psi(K)$ are constants.

Remark: Numerical minimization for the determination of maximum-likelihood estimates of θ parameters is the most difficult aspect of the application of results of this theorem. Therefore, it is useful to note that first-order approximations of $\hat{\delta}$ and $\hat{\tau}$ and the test statistics can be obtained by replacing $\hat{\Lambda}$ everywhere by $\Lambda(\tilde{\theta})$ where the $\tilde{\theta}_i$ are any consistent estimators of the θ_i. The resulting statistics are asymptotically equivalent to the functions of the maximum-likelihood estimators if $\phi(K)$ satisfies either of condition 1 or 2 with $\tilde{\theta}$.

Proof: First consider the logarithmic likelihood function over the subset of the parameter space where $\tau = \hat{\tau}$ and $\theta = \hat{\theta}$. Only the first quadratic term varies with ξ in this subset. It follows immediately then that this term vanishes and the logarithmic likelihood function achieves a maximum if and only if

$$d_1 - \hat{\Lambda}_{12}\hat{\Lambda}_{22}^{-1}\hat{d}_2 = 0.$$

This is zero if and only if

$$\hat{\xi} = K^{-\frac{1}{2}}(Z_1 - \hat{\Lambda}_{12}\hat{\Lambda}_{22}^{-1}\hat{d}_2) = K^{-\frac{1}{2}}[Z_1 - \hat{\Lambda}_{12}\hat{\Lambda}_{22}^{-1}(Z_2 - b\hat{\tau})].$$

The stated form of $\hat{\delta}$ follows immediately from this because δ is a function of ξ and τ.

The same form of argument applies in the derivation of $\hat{\tau}$. Only the second quadratic term in the logarithmic likelihood function varies with τ in the set where $\xi = \hat{\xi}$ and $\theta = \hat{\theta}$ (the first quadratic vanishes). Therefore, the logarithmic likelihood function achieves a maximum when

$$\bar{d}_2^t\hat{\Lambda}_{22}^{-1}d_2 = (\bar{Z}_2^t - \tau\bar{b}^t)\hat{\Lambda}_{22}^{-1}(Z_2 - b\tau)$$

is minimized, that is when τ has the stated form of $\hat{\tau}$.

It follows from the derivations of the estimators $\hat{\xi}$ and $\hat{\tau}$ that the form of the logarithmic likelihood function at its maximum with respect to variation in ξ, τ, and θ is

$$-\tfrac{1}{2}\ln|\hat{\Lambda}_{11.2}| - \tfrac{1}{2}\ln|\hat{\Lambda}_{22}| - \tfrac{1}{2}(\bar{Z}_2^t - \tau\bar{b}^t)\hat{\Lambda}_{22}^{-1}(Z_2 - b\hat{\tau}).$$

When substitution is made for $\hat{\tau}$, the quadratic can be reduced to

$$\bar{Z}_2^t(\hat{\Lambda}_{22}^{-1} - \frac{1}{\bar{b}^t\hat{\Lambda}_{22}^{-1}b}\hat{\Lambda}_{22}^{-1}b\bar{b}^t\hat{\Lambda}_{22}^{-1})Z_2.$$

Therefore the θ coordinates at the points of the maximum value of the logarithmic likelihood function are the coordinates at which

$$\ln|\Lambda_{11.2}| + \ln|\Lambda_{22}| + \bar{Z}_2^t(\Lambda_{22}^{-1} - \frac{1}{\bar{b}^t\Lambda_{22}^{-1}b}\Lambda_{22}^{-1}b\bar{b}^t\Lambda_{22}^{-1})Z_2$$

is a minimum.

Tests that are uniformly most powerful unbiased and similar are easy to derive when θ is known. The critical regions are those given in the theorem after $\hat{\Lambda}$ has been replaced by Λ everywhere, including in $\hat{\delta}$ and $\hat{\tau}$. Williams (1975) has shown that the conditions on $\phi(K)$, $\psi(K)$, and the matrix derivatives of $R^{-1}(\theta)$ are sufficient for the difference between the test statistics set out in the theorem and those obtained by replacing $\hat{\Lambda}$ everywhere by Λ to be $O[\psi(K)/\phi(K)]$ in probability. His conditions were developed for a distribution-free description of the problem. With the normality constraint, they can be strengthened from "in probability" to "with probability one". This completes the proof.

Calculation of Statistics

The functions of complex exponentials and Fourier coefficients set forth in Theorem 4.1 can be easily rewritten in terms of Y, a, and V. These are the second of our new results for inference. The key identification that is needed for this and application of Corollaries 3.1 and 3.2 is

$$(\Phi_2 \otimes I_2)\Lambda_{22}^{-1}(\bar{\Phi}_2^t \otimes I_2) = V^{-1} - V^{-1}(j_K \otimes I_2)[(j_K^t \otimes I_2)V^{-1}(j_K \otimes I_2)]^{-1}(j_K^t \otimes I_2)V^{-1}.$$

Then it follows that

$$\ln|V| + Y^tV^{-1}Y - Y^tV^{-1}(j_K \otimes I_2)[(j_K^t \otimes I_2)V^{-1}(j_K \otimes I_2)]^{-1}(j_K^t \otimes I_2)V^{-1}Y$$

$$- \frac{\{Y^tV^{-1}a - Y^tV^{-1}(j_K \otimes I_2)[(j_K^t \otimes I_2)V^{-1}(j_K \otimes I_2)]^{-1}(j_K^t \otimes I_2)V^{-1}a\}^2}{a^tV^{-1}a - a^tV^{-1}(j_K \otimes I_2)[(j_K^t \otimes I_2)V^{-1}(j_K \otimes I_2)]^{-1}(j_K^t \otimes I_2)V^{-1}a}$$

has a global minimum at $\theta = \hat{\theta}$. The treatment-effect estimators reduce to

$$\hat{\tau} = \frac{a^t\hat{V}^{-1}Y - a^t\hat{V}^{-1}(j_K \otimes I_2)[(j_K^t \otimes I_2)\hat{V}^{-1}(j_K \otimes I_2)]^{-1}(j_K^t \otimes I_2)\hat{V}^{-1}Y}{a^t\hat{V}^{-1}a - a^t\hat{V}^{-1}(j_K \otimes I_2)[(j_K^t \otimes I_2)\hat{V}^{-1}(j_K \otimes I_2)]^{-1}(j_K^t \otimes I_2)\hat{V}^{-1}a}$$

and

$$\hat{\delta} = (1, -1)[(j_K^t \otimes I_2)\hat{V}^{-1}(j_K \otimes I_2)]^{-1}(j_K^t \otimes I_2)\hat{V}^{-1}(Y - a\hat{\tau}).$$

For the test statistics involving $\hat{\tau}$ and $\hat{\delta}$, the squares of the respective multipliers of $(\hat{\tau}-\tau_0)$ and $(\hat{\delta}-\delta_0)$, are estimated information numbers, and reduce to

$$a^t\hat{V}^{-1}a - a^t\hat{V}^{-1}(j_K \otimes I_2)[(j_K^t \otimes I_2)\hat{V}^{-1}(j_K \otimes I_2)]^{-1}(j_K^t \otimes I_2)\hat{V}^{-1}a$$

and the reciprocal of

$$(1,-1)[(j_K^t \otimes I_2)\hat{V}^{-1}(j_K \otimes I_2)]^{-1}\binom{1}{-1} + \frac{\{(1,-1)[(j_K^t \otimes I_2)\hat{V}^{-1}(j_K \otimes I_2)]^{-1}(j_K^t \otimes I_2)\hat{V}^{-1}a\}^2}{a^t\hat{V}^{-1}a - a^t\hat{V}^{-1}(j_K \otimes I_2)[(j_K^t \otimes I_2)\hat{V}^{-1}(j_K \otimes I_2)]^{-1}(j_K^t \otimes I_2)\hat{V}^{-1}a}.$$

Relationship of Efficient Analysis to Optimum Design

The relationship that we have already noted between continued-covariate and changeover designs extends to balanced crossover and switchback designs as well when N is even. In addition, this feature of the designs results in a common linear adjustment for effects of a covariate series in each analysis.

Let

$$T = \begin{pmatrix} 1 & 1 \\ 1 & -1 \end{pmatrix},$$

and recall that for one continued-covariate design

$$a^t = (a_1^t : 0^t),\ a_1^t = (0^t : j_N^t),\ \text{and}\ \bar{b}^t = (\bar{b}_1^t : 0^t)$$

where $\bar{b}_1^t = a_1^t(\Phi_2 \otimes I_2)$ and at least one, and as many as N, elements of b_1 are nonzero. Then we have the following results:

1. Balanced Crossover, N even,

$$a^t(I_N \otimes T) = (N^{\frac{1}{2}}u_1^t\bar{\Phi}^t : N^{\frac{1}{2}}u_{N/2+1}\bar{\Phi}^t),\ \bar{b}^t(I_N \otimes T) = (0^t : N^{\frac{1}{2}}u^t_{N/2+1}),$$

2. Balanced Switchback, N even,

$$a^t(I_N \otimes T) = (N^{\frac{1}{2}}(u_1^t + u^t_{N/2+1})\bar{\Phi}^t : 0^t),\ b^t(I_N \otimes T) = (N^{\frac{1}{2}}u^t_{N/2+1} : 0^t),$$

3. Continued Covariate,

$$a^t = (a_1^t : 0^t),\ \bar{b}^t = (\bar{b}_1^t : 0^t),$$

4. Changeover,

$$a^t(I_N \otimes T) = (2a_1^t : 0^t),\ \bar{b}^t(I_N \otimes T_2) = (\bar{b}_1^t : 0^t).$$

The relationship is the common structure of $\bar{b}^t$ or $\bar{b}^t(I_N \otimes T)$ where there is one zero and one nonzero partition, either for the original series or its replacement by the pair composed of the sum and difference of the original series.

The effect of a zero partition on the analysis can be seen by considering a single case. It is easiest to use the continued covariate to avoid the notational complications associated with a transformed record of sums and differences. The vector $\underset{\sim}{d}_2$ and matrix $\underset{\sim}{\Lambda}_{22}$ in the expression for a likelihood function can be partitioned in the following manner:

$$\underset{\sim}{d}_2 = \begin{pmatrix} \underset{\sim}{d}_{21} \\ \hline \underset{\sim}{d}_{22} \end{pmatrix} = \begin{pmatrix} \underset{\sim}{Z}_{21} - \underset{\sim}{b}_1 \tau \\ \hline \underset{\sim}{Z}_{22} \end{pmatrix} \text{ and } \underset{\sim}{\Lambda}_{22} = [(\underset{\sim}{\Lambda}_{22})_{hi}].$$

These can then be used to expand $\bar{\underset{\sim}{d}}{}_2^t \underset{\sim}{\Lambda}_{22}^{-1} \underset{\sim}{d}_2$ into a sum of two terms,

$$[\bar{\underset{\sim}{d}}{}_{21}^t - \bar{\underset{\sim}{d}}{}_{22}^t (\underset{\sim}{\Lambda}_{22})_{22}^{-1} (\underset{\sim}{\Lambda}_{22})_{21}] \underset{\sim}{\Lambda}_{22}^{11} [\underset{\sim}{d}_{21} - (\underset{\sim}{\Lambda}_{22})_{12} (\underset{\sim}{\Lambda}_{22})_{22}^{-1} \underset{\sim}{d}_{22}] + \bar{\underset{\sim}{d}}{}_{22}^t (\underset{\sim}{\Lambda}_{22})_{22}^{-1} \underset{\sim}{d}_{22},$$

in which it is evident that $\underset{\sim}{d}_{22}$ acts as a covariate for $\underset{\sim}{d}_{21}$ and that $\bar{\underset{\sim}{b}}{}^t \underset{\sim}{\Lambda}_{22}^{11} \underset{\sim}{b}$ is the adjusted information function evaluated at $\underset{\sim}{a}^t = (\underset{\sim}{a}^t : \underset{\sim}{0}^t)$. Since $\underset{\sim}{d}_{21}$ and $\underset{\sim}{d}_{22}$ are linear functions of $\underset{\sim}{Y}_1$ and $\underset{\sim}{Y}_2$ respectively, it follows that the record of the second series serves as a covariate in the analysis for τ of the record of the first series when these continued-covariate designs are used.

The same result holds for the series records of sums and differences when a changeover design is used. Similar results hold for the series records of sums and differences when N is even and balanced crossover and switchback designs are used, although there is an additional adjustment for these which should not be missed. The nonzero partition of $\bar{\underset{\sim}{b}}{}^t (\underset{\sim}{I}_N \otimes \underset{\sim}{T})$ has zero everywhere except in the N/2 + 1 st position. Therefore responses from all frequencies from the 2nd ($\ell - 1 = 1$) through the N/2 nd ($\ell - 1 = N/2$) of the corresponding series also serve as covariates.

REFERENCES:

WILLIAMS, J.S. (1975). Lower bounds on convergence rates of weighted least squares to best linear unbiased estimators. In A Survey of Statistical Design and Linear Models. Ed: J.N. Srivastava, North-Holland, Amsterdam.

WILLIAMS, J.S. and IYER, H.K. (1981a). Optimum design of stationary bivariate time-series experiments. In: Time Series Analysis. Eds: O.D. Anderson and M.R. Perryman, North-Holland, Amsterdam.

WILLIAMS, J.S. and IYER, H.K. (1981b). Optimum and restricted-optimum designs for stationary multivariate time-series experiments. Technical Report No. 45, Department of Statistics, Colorado State University.

WU, S.C., WILLIAMS, J.S. and MIELKE, P.W. (1972). Some designs and analyses for temporally independent experiments involving correlated bivariate responses. Biometrics 28, 1043-1061.

APPLIED TIME SERIES ANALYSIS
O.D. Anderson & M.R. Perryman (eds.)

BASE 2-3 FAST FOURIER TRANSFORMS

E. A. Yfantis

Department of Mathematical Sciences
University of Nevada, Las Vegas
Las Vegas, Nevada 89154
U.S.A.

L. E. Borgman

Statistics Department
University of Wyoming
Laramie, Wyoming
U.S.A.

Several computer programs based on the Fast Fourier Transform (FFT) Algorithm, developed by Cooley and Tukey, are available. The Cooley and Tukey algorithm for the FFT requires the number of data N be a power of 2. Here this algorithm is extended to the case when the number N of data is a power of 3 i.e. $N=3^k$, as well as when the number N of measurements is of the form $N=2^m \cdot 3^n$, where m, n are positive integers. Subroutines that implemented the Cooley and Tukey, Fast Fourier Transform to the base 2 (FFT2) algorithm, as well as subroutines implementing the two new Fast Fourier Transforms, namely the Fast Fourier Transform to the base 3 (FFT3) and the Fast Fourier Transform to the base 2,3 (FFT23), are available from the authors. Finally, an application to ocean engineering is presented.

INTRODUCTION

The Fast Fourier Transform (FFT) has many applications in engineering, economics, geosciences, and statistics. It is useful especially in the analysis of time series or space series data. Several researchers have contributed to the development of mathematical theories and algorithms related to the FFT (Cooley and Tukey (1965)), (Cooley et al (1967)), (Cochran et al (1967)), (Sande (1968)), (Singleton (1969)), (Fisher (1970)), (Kameko and Lieu (1970)), (Ahmed, Rao and Shultz (1971)), (Hudspeth and Borgman (1979)), (Yfantis and Borgman (1981)).

Available FFT programs are based on the Cooley and Tukey algorithm which require that the number of data N to be analyzed is a power of 2, or N is of the form $N=2^k$ (for example, if N=2048, then k=11, i.e. $N=2^{11}$). If the sequence of numbers to be analyzed is not a power of 2, then one must either discard some measurements, so that the remaining number of measurements is a power of 2 or one could pad the sequence with zeros so that the total number of measurements is a power of 2. For example, if 3072 data are available for analysis, one could discard 1024 measurements so that $N=2048=2^{11}$ or pad the sequence with 1024 zeros (i.e. $x_{3072}=x_{3073}=\dots x_{4095}=0$) so that $N=4096=2^{12}$. Discarding measurements amounts to loss of information, which implies less stable estimates. Also, frequencies determined after adding false data are not true frequencies. So, in the second situation, if N is the number of actual measurements available and T is the period $Df=\frac{1}{T}$ and after padding the sequence with zeros the number of measurements is $N1 = 2^k$, then the new increment in frequency denoted by D'f would be $D'f = \frac{N}{N1} \cdot Df$.

Although the new FFT subroutines require more coding than the FFT2, their speed of execution and accuracy is comparable to that of the FFT2.

THE FFT3-TRANSFORM

The FFT Transform has worked out for $N=3^k$, k is any natural number. The work is displayed with an example. Let $N=27=3^3$. Let $W_0, W_1, \ldots, W_{26}$ be the finite sequence to be transformed and

$$A_m = \Delta t \sum_{n=0}^{N-1} W_n e^{-i2\pi mn/N}, \qquad m=0,1,\ldots,26,\ N=27 \tag{1}$$

be the finite Fourier transform of W_n, $n=0,1,\ldots,26$, then the number of complex multiplications required in the above formula is N^2.

If

$$T_m = \sum_{n=0}^{N-1} W_n e^{-i2\pi mn/N} = \frac{A_m}{\Delta t}, \qquad m=0,1,\ldots,26 \tag{2}$$

then the development of the FFT3 algorithm goes as follows. Both n and m can be expressed in trinary form.

$$n = n_2\cdot 3^2 + n_1\cdot 3^1 + n_0\cdot 3^0 = \text{trinary number } (n_2,n_1,n_0), \tag{3}$$

$$m = m_2\cdot 3^2 + m_1\cdot 3^1 + m_0\cdot 3^0 = \text{trinary number } (m_2,m_1,m_0), \tag{4}$$

T_m and W_n can just as well be written as $T(m_2,m_1,m_0)$ and $W(n_2,n_1,n_0)$ where the triple indices in the parenthesis refer to the corresponding trinary digits. Thus eq. (2) can be rewritten as

$$T(m_2,m_1,m_0) = \sum_{n_0=0}^{2} \sum_{n_1=0}^{2} \sum_{n_2=0}^{2} W(n_2,n_1,n_0)\, e^{-\frac{i2\pi mn}{N}}, \tag{5}$$

where the n and m in the exponential are given by eqs. (3) and (4). In the following development, the relation

$$e^{-i2\pi k} = 1, \tag{6}$$

for k an integer will be used repeatedly.

Now eq. (5) can be written

$$T(m_2,m_1,m_0) = \sum_{n_0=0}^{2} \{ \sum_{n_1=0}^{2} [\sum_{n_2=0}^{2} W(n_2,n_1,n_0)\exp(-i\tfrac{2\pi}{27}(m_2\cdot 3^2 + m_1\cdot 3 + m_0)n_2\cdot 3^2)]$$

$$\cdot \exp(-i\tfrac{2\pi}{27}(m_2\cdot 3^2 + m_1\cdot 3 + m_0)n_1\cdot 3)\}\cdot\exp(-i\tfrac{2\pi}{27}(m_2\cdot 3^2 + m_1\cdot 3 + m_0)n_0). \tag{7}$$

The bracketed quantity in eq. (7) can be simplified to

$$[\ ] = \sum_{n_2=0}^{2} W(n_2,n_1,n_0)\ \exp(-\frac{i2\pi m_0 n_2}{3}), \tag{8}$$

since the rest of the parts of the exponential are all unified by eq. (6). The bracketed quantity is a function of m_0, n_1 and n_0 but not n_2 since that index is summed over. Let the bracketed quantity be denoted by $T^{(1)}(m_0,n_1,n_2)$. Then

$$T(m_2,m_1,m_0) = \sum_{n_0=0}^{2} [\sum_{n_1=0}^{2} T^{(1)}(m_0,n_1,n_0)\exp(-i\tfrac{2\pi}{27}(m_2\cdot 3^2 + m_1\cdot 3 + m_0)n_1\cdot 3)]$$

$$\cdot \exp(-i\tfrac{2\pi}{27}(m_2\cdot 3^2 + m_1\cdot 3 + m_0)n_0), \tag{9}$$

The bracketed quantity in the above equation reduces

$$[\ ] = \sum_{n_1=0}^{2} T^{(1)}(m_0,n_1,n_0)\exp(-i\frac{2\pi}{9}(3m_1+m_0)n_1), \tag{10}$$

this is a function of m_0, m_1 and n_0 but not n_1. Let the bracketed quantity above be denoted by $T^{(2)}(m_0,m_1,n_0)$. Hence

$$T(m_2,m_1,m_0) = \sum_{n_0=0}^{2} T^{(2)}(m_0,m_1,n_0)\exp(-i\frac{2\pi}{27}(m_2\cdot 3^2+m_1\cdot 3+m_0)n_0). \tag{11}$$

The steps in the calculation of the finite Fourier Transform have been reduced to

$$W(n_2,n_1,n_0) \longrightarrow T^{(1)}(m_0,n_1,n_0) \longrightarrow T^{(2)}(m_0,m_1,n_0) \longrightarrow$$

$$T^{(3)}(m_0,m_1,m_2) \longrightarrow T(m_2,m_1,m_0). \tag{12}$$

Note that although n starts out on the left of eq. (12) in normal increasing order, m ends on the right side in scrabbled order, in the sense that m_0 takes the place of n_2 in the argument of $T^{(1)}$, then m_1 takes the place of n_1 in $T^{(2)}$, and finally, m_2 takes the place of n_0 in $T^{(3)}$. If n is given by $n = n_2\cdot 3^2 + n_1\cdot 3 + n_0$, then the m that carries through on the same line is given by $m = m_2\cdot 3^2 + m_1\cdot 3 + m_0 = n_0\cdot 3^2 + n_1\cdot 3 + n_2$. This operation is called digit reversal.

In our example $N=27=3^3$, and there are $\log_3 N=3$ columns, namely, $T^{(1)},T^{(2)},T^{(3)}$ for computing T_m, $m=0,1,2,\ldots,N-1$. Hence the number of complex multiplications in computing T_m is

$$\text{FFT number of operations} = 2\cdot N\cdot \log_3 N = 2\cdot 27\cdot 3 = 162\,, \tag{13}$$

$$\text{Direct transform number of operations} = N^2 = 27^2 = 729\,. \tag{14}$$

The extension of the procedure to larger N for base 3 is straightforward. Suppose $N=3^k$. There are N initial complex numbers, W, which form the left sequence, or column that we start with. There are k intermediate functions, or columns, $T^{(1)},T^{(2)},\ldots,T^{(k)}$ involved in computing T_m. Each column requires 2N multiplications. Hence

$$\text{FFT number of operations} = 2N\log_3 N - 2Nk,$$

$$\text{Direct transform number of operations} = N^2,$$

$$\text{Fractional savings in operations} = \frac{2N\log_3 N}{N^2} = \frac{2k}{N}\,. \tag{15}$$

For selected choices of N, the savings are given in Table (I).

N	k	FFT Fractional Saving
9	2	4/9
27	3	2/9
81	4	8/81
243	5	10/243
729	6	4/243
59049	10	20/59049

Table (I) Fractional Savings Using the FFT3 ($N=3^k$), relative to Direct Transform.

The procedure used in the previous example is generalized for the case when $N=3^k$, k=1,2,... . The $T^{(1)}$ sequence can be considered to consist of one block (NB=1) of length 27 (LB=3^3). The multipliers used in the block are (1,1,1) for the first third of the block, $(1, e^{-i\frac{2\pi}{3}}, e^{-i\frac{4\pi}{3}})$ for the second third of the block, and $(1, e^{-i\frac{4\pi}{3}}, e^{-i\frac{2\pi}{3}})$ for the third third of the block. Let $T_i^{(0)} = W_i$ for $0 \le i \le N-1$. The first third of the block is obtained by the formula

$$T_i^{(1)} = T_i^{(0)} + m_1 T^{(0)}_{i+\frac{LB}{3}} + m_1^2 T^{(0)}_{i+\frac{2LB}{3}}, \tag{16}$$

the second third of the block is obtained by the formula

$$T^{(1)}_{i+\frac{LB}{3}} = T_i^{(0)} + e^{-i\frac{2\pi}{3}} m_1 T^{(0)}_{i+\frac{LB}{3}} + e^{-i\frac{4\pi}{3}} m_1^2 T^{(0)}_{i+\frac{2LB}{3}}, \tag{17}$$

while the third third of the block is obtained by the formula

$$T^{(1)}_{i+\frac{2LB}{3}} = T_i^{(0)} + e^{-i\frac{4\pi}{3}} m_1 T^{(0)}_{i+\frac{LB}{3}} + e^{-i\frac{2\pi}{3}} m_1^2 T^{(0)}_{i+\frac{2LB}{3}}, \tag{18}$$

where $m_1=1$, and i ranges over the integers for the first third of the block. The same multipliers m are used in each third of a block. The only changes are in the second third of the block and third third of the block, i.e. in the second third of the block m_1 is multiplied by $e^{-i\frac{2\pi}{3}}$ and the m_1^2 by $e^{-i\frac{4\pi}{3}}$, while in the third third of the block, the m_1 is multiplied by $e^{-i\frac{4\pi}{3}}$, and the m_1^2 by $e^{-i\frac{2\pi}{3}}$. The $T^{(2)}$, sequence or column consists of three blocks, (NB=3), each of length 9 (LB=3^2). The multipliers used for the three blocks are $m_1=1$, $m_2=e^{-i\frac{2\pi}{9}}$, $m_3=e^{-i\frac{4\pi}{9}}$, respectively. As before, the values are obtained by the formulas

$$T_i^{(2)} = T_i^{(1)} + mT^{(1)}_{i+\frac{LB}{3}} + m^2 T^{(1)}_{i+\frac{2LB}{3}}, \text{ first third of block} \tag{19}$$

$$T^{(2)}_{i+\frac{LB}{3}} = T_i^{(1)} + e^{-i\frac{2\pi}{3}} \cdot mT^{(1)}_{i+\frac{LB}{3}} + e^{-i\frac{4\pi}{3}} \cdot m^2 T^{(1)}_{i+\frac{2LB}{3}}, \text{ second third of block} \tag{20}$$

$$T^{(2)}_{i+\frac{2LB}{3}} = T_i^{(1)} + e^{-i\frac{4\pi}{3}} \cdot mT^{(1)}_{i+\frac{LB}{3}} + e^{-i\frac{2\pi}{3}} \cdot m^2 T^{(1)}_{i+\frac{2LB}{3}}, \text{ third third of block} \tag{21}$$

where m is the multiplier of the block, i.e. $m=m_1$ for the first block, $m=m_2$ for the second block, $m=m_3$ for the third block, and i ranges over the integers for the first third of the block. The same pattern continues for the $T^{(3)}$ column where NB=9 and LB=3. The multipliers for each of the nine blocks are

$$m_1=\exp(0)=1,\ m_2=\exp(-i\tfrac{2\pi}{9}),\ m_3=\exp(-i\tfrac{4\pi}{9}),\ m_4=\exp(-i\tfrac{2\pi}{27}),\ m_5=\exp(-i\tfrac{8\pi}{27}),$$

$$m_6=\exp(-i\tfrac{14\pi}{27}),\ m_7=\exp(-i\tfrac{6\pi}{27}),\ m_8=\exp(-i\tfrac{10\pi}{27}),\text{ and } m_9=\exp(-i\tfrac{16\pi}{27}). \tag{22}$$

The nine multipliers given above represent a basic list. When NB=1, only the first of the list was used. When NB=3, the first, second and third multipliers entered into the computations. Finally, NB=9 required all nine. The nine multipliers have another interesting structure. All are of the form $e^{-i\frac{2\pi k}{3N}}$. Furthermore, as shown in Table (II), k is a number obtained by reversing the trinary digits for the integers 0,1,2,3,4,5,6,7 and 8.

Digit	Trinary Number	Reversed Trinary Number	k	Corresponding Multipliers
0	000	000	0	e^0
1	001	100	9	$e^{-i2\pi/9}$
2	002	200	18	$e^{-i4\pi/9}$
3	010	010	3	$e^{-i2\pi/27}$
4	011	110	12	$e^{-i8\pi/27}$
5	012	210	21	$e^{-i14\pi/27}$
6	020	020	6	$e^{-i4\pi/27}$
7	021	120	15	$e^{-i10\pi/27}$
8	022	220	24	$e^{-i16\pi/27}$

Table (II) Relation of Multipliers to the First Nine integers for N=27.

If $N=3^k$, NC denotes the column number, and the number of blocks in each column is given by $NB=3^{NC-1}$. The length of each block is

$$LB = \frac{N}{NB} = \frac{3^k}{3^{NC-1}} = 3^{k-NC+1}, \tag{23}$$

Now, let IB designate the sequence number for the blocks within the column (i.e. IB=1 indicates the first block, IB=2 indicates the second block, etc.)

The formulas corresponding to eqs. (19), (20) and (21) would be

$$T_i^{(NC)} = T_i^{(NC-1)} + m_{IB} T_{i+\frac{LB}{3}}^{(NC-1)} + (m_{IB})^2 T_{i+\frac{2LB}{3}}^{(NC-1)}, \quad \text{first third of block IB} \tag{24}$$

$$T_{i+\frac{LB}{3}}^{(NC)} = T_i^{(NC-1)} + e^{-i\frac{2\pi}{3}} m_{IB} T_{i+\frac{LB}{3}}^{(NC-1)} + e^{-i\frac{4\pi}{3}} (m_{IB})^2 T_{i+\frac{2LB}{3}}^{(NC-1)}, \quad \text{second third of block IB} \tag{25}$$

$$T_{i+\frac{2LB}{3}}^{(NC)} = T_i^{(NC-1)} + e^{-i\frac{4\pi}{3}} \cdot m_{IB} T_{i+\frac{LB}{3}}^{(NC-1)} + e^{-i\frac{2\pi}{3}} (m_{IB})^2 T_{i+\frac{2LB}{3}}^{(NC-1)}, \quad \text{third third of block IB} \tag{26}$$

where i ranges over the first third of the block.

The execution of a Fast Fourier transform reduces pivotally to the creation of a list of reversed bit numbers. Such a list provides both the multipliers for each block and the information for unscrambling the final column of transformed values into normal order. One way to generate the reversed digit list is the following: suppose one has the reversed bit list for $N=3^{k-1}$, then the list for $N=3^k$ will be three times as long. The first third of the new list can be obtained if one multiplies the elements of the old list by three, the second third of the new list can be obtained by adding one to each element of the first third of the new list, while the third third of the new list can be obtained by adding two to each element of the first third of the new list. This procedure is illustrated in Table (III). This method is based on the changes in the reversed digit trinary numbers when going from one k value to the next. The procedure has the advantages of speed and simplicity but the disadvantage of requiring computer storage of the lists as they build up. Some storage can be saved by only computing the first third of the last column needed. This is sufficient because the first third provides all the multipliers needed. As we pointed out, for the final bit reversal step the second third can be computed as needed by adding one to the first third of the list, while the third third can be obtained by adding two to the first third of the list.

k	
0	0
1	0,1,2
2	0,3,6,1,4,7,2,5,8
3	0,9,18,3,12,21,6,15,24,1,10,19,4,13,22,7,16,25,2,11,20,5,14,23,8,17,26

Table (III) Example of the Generation of a Reversed Bit, Recursively from the List for the Next Lower k Value.

The subroutine FFT3 is given by Yfantis and Borgman (1981) for doing the Fourier transform in cases when $N=3^k$. The trinary digit reversal sequence is generated initially by the procedure shown in Table (III). The sines and cosines are developed and stored at the same time. The FFT3 algorithm uses the values from these lists when they are needed in the computations. If the subroutine is called a second time to transform a new sequence of numbers of the same length, the lists from the previous call are available for the new computations.

TWO WAY FFT

Let $N=N_1 \cdot N_2$ and let $X_0,X_1,\ldots,X_{N-1}$ be the N measurements. Let

$$A(m) = \sum_{n=0}^{N-1} X_n e^{-i\frac{2\pi mn}{N}}, \tag{29}$$

Ahmed, Rao (1975).

$$\begin{aligned} A(m_1,m_2) &= A(m_1N_2+m_2) = \sum_{n_1=0}^{N_1-1}\sum_{n_2=0}^{N_2-1} X(n_2N_1+n_1)\exp(-i2\pi(N_1n_2+n_1)\cdot(N_2m_1+m_2)/N_1N_2) \\ &= \sum_{n_1=0}^{N_1-1}\sum_{n_2=0}^{N_2-1} X(n_2N_1+n_1)\exp(-i2\pi n_2m_1)\cdot\exp(-i\frac{2\pi n_2m_2}{N_2})\cdot\exp(-i\frac{2\pi m_1n_1}{N_1}) \\ &\cdot\exp(-i\frac{2\pi n_1m_2}{N}) = \sum_{n_1=0}^{N_1-1}[(\sum_{n_2=0}^{N_2-1} X(n_2N_1+n_1)\exp(-i\frac{2\pi n_2m_2}{N_2}))\cdot\exp(-i\frac{2\pi n_1m_2}{N_2})] \\ &\cdot\exp(-i\frac{2\pi n_1m_1}{N_1}). \end{aligned} \tag{30}$$

The bracketed part is a function of m_2 and n_1 but not n_2 because we sum over n_2. Let

$$[\;] = T^{(1)}(m_2,n_1), \tag{31}$$

and

$$A(m_1,m_2) = \sum_{n_1=0}^{N_1-1} T^{(1)}(m_2,n_1)\exp(-i\frac{2\pi n_1m_1}{N_1}) = T^{(2)}(m_2,m_1), \tag{32}$$

or

$$A(m_1N_2+m_2) = T^{(2)}(m_2N_1+m_1), \tag{33}$$

where $0 \le n_1 \le N_1-1$, $0 \le n_2 \le N_2-1$, $0 \le m_1 \le N_1-1$, $0 \le m_2 \le N_2-1$ which means that when we find the T's in the two way FFT, we must reverse their order by using the above relation.

Based on the theory presented in this section, a subroutine which implements the two way FFT is given by Yfantis and Borgman (1981).

EXAMPLE OF ANALYSIS OF WAVE DATA

3456 wave data, showing the water level elevation of the sea caused by Hurricane Carla on September 9, 1961 at 2100 hr., were analyzed. The $\Delta t=0.20$, $N=3456$, $\Delta f=\frac{1}{T}=\frac{1}{N\cdot\Delta t}=0.001447$, the Nyquist frequency for this choice of Δt is $\frac{1}{2\cdot\Delta t}=\frac{1}{0.4}=2.5\ \text{sec}^{-1}$ or $f_{NY}=2.5\ \text{sec}^{-1}$ which corresponds to a period of 0.4 sec. If η_n, $n=0,1,2,\ldots,3455$, is the given sequence of water level elevations, then using the FFT23 we found

$$T_m = \sum_{n=0}^{3455} \eta_n e^{-i\frac{2\pi mn}{N}}, \tag{34}$$

the Fourier transform is $A_m=\Delta t\cdot T_m$ and the spectral lines were computed from

$$\hat{P}(f_m) = \frac{|A_m|^2}{N\cdot\Delta t} = \frac{A_m \cdot \bar{A}_m}{691.2}, \tag{35}$$

where $|A_m|$ denotes the complex modulus of A_m. The spectral density was then estimated by a moving average of the spectral lines

$$\hat{\hat{P}}(f_m) = \sum_j W_j\ \hat{P}(f_{m-j})/\sum_j W_j, \tag{36}$$

where

$$W_j = \exp(-j^2/18). \tag{37}$$

These weights are thus Gaussian Smoothers with a standard deviation of $J=3\cdot\Delta f=0.004341\ \text{sec}^{-1}$. (Borgman (1973)). The spectral density estimate averaged from the lines is given in Figure (A).

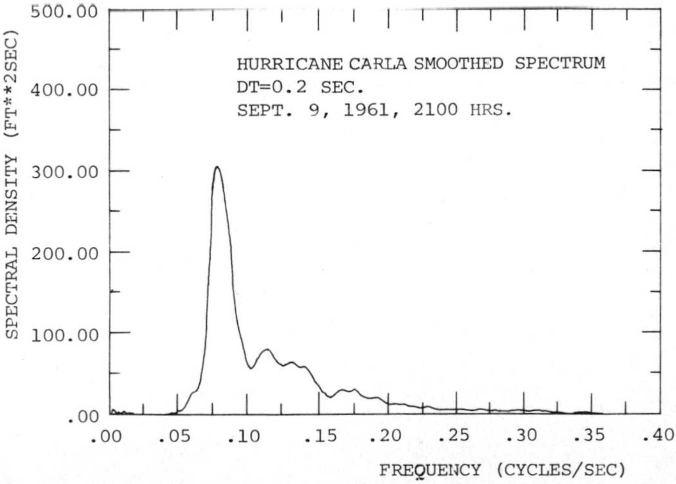

Figure (A).

Before finishing this subsection it should be helpful to summarize how a user of the FFT could benefit from it. If the number of data available to the user is a power of 2, the user only needs to use the subroutine FFT2 along with his main program; if it is a power of 3, he only needs to use the FFT3 along with his main program; if it is of the form $N=2^k \cdot 3^\ell$, where k,ℓ are natural numbers he needs to use the FFT2, FFT3 and FFT23 along with his main program.

The above theory can be extended to three or four way FFT's where the number of data is of the form $N=2^k \cdot 3^\ell \cdot 5^m \cdot 7^n$, where the exponents are non-negative integers.

REFERENCES

AHMED, N., RAO, K.R. and SHULTZ, R.B. (1971). A generalized discrete transform. Proc. IEEE, V.59, p.1360-1362.

BORGMAN, L.E. (1973). Statistical properties of Fast Fourier transform coefficients computed from real valued covariance-stationary periodic random sequencies. Res. paper No.23, College of Commerce and Industry, Univ. Wyoming, p.2-20.

COCHRAN, W.T. et al (1967). What is the Fast Fourier transform? Proc. IEEE, V.55, p.1664-1674.

COOLEY, T.W. and TUKEY, T.W. (1965). An algorithm for the machine calculation of complex Fourier series. Mathematics of Computation, V.19, p.297-306.

COOLEY, T.W. et al (1967). Historical notes of the Fast Fourier transform. Proc. IEEE, V.55, p.1675-1677.

FISHER, T.R. (1970). FORTRAN program for Fast Fourier transform. Naval Res. Lab., Report No. 7041, Washington, D.C., p.2-20.

HUDSPETH, R.T. and BORGMAN, L.E. (1979). Efficient FFT simulation of digital time sequences. Jour. Engineering Mechanics Division, V.105, No. EM2, p.223-235.

KAMEKO, T. and LIU, B. (1970). Accumulation of round off error in Fast Fourier transform. Jour. ACM, V.17, p.637-654.

SAND, G. (1968). Arbitrary radix one dimensional Fast Fourier transform subroutines. Univ. Chicago, Illinois, p.1-20.

SINGLETON, R.C. (1969). An algorithm for computing the mixed radix Fast Fourier transform. IEEE transactions audio and electronics. V.17, p.99-103.

YFANTIS, E.A. and BORGMAN, L.E. (1981). Fast Fourier transforms 2-3-5. Computers and Geosciences, V. 7, p. 99-108.

APPLIED TIME SERIES ANALYSIS
O.D. Anderson & M.R. Perryman (eds.)

TIME SERIES ANALYSIS AND FORECASTING (TSA&F): A Personal View

O.D. ANDERSON
9 Ingham Grove, Lenton Gardens, Nottingham NG7 2LQ, England

We discuss the nature of TSA&F and question the exagerated influence of Mathematical Statistics on its development. We give the aims of the TSA&F Special Interest Group and trace the growth of TSA&F Activities. Finally we seek an answer as to where the TSA&F Movement is heading.

1. The Nature of TSA&F as a Discipline

A Time Series is some sequence of observations proceeding through time (or space) where the actual order in which the values occur has importance, in that they may be related temporally (or spatially).

The theory and practice of Time Series is very different from that for the "remainder" of Statistics:

(1) Everywhere else, observations are assumed (or, by appropriate Design, arranged) to be independent. Not so with Time Series.

(2) Everywhere else, asymptotic theory tends generally to be meaningful in practice. Not so with Time Series.

The lack of independence is inevitable, due to the nature of all but trivial Time Series; and it is, in fact, this very temporal (or spatial) dependence which is of interest and importance.

In Statistics, asymptotic theory is relevant (and useful) when one is dealing with fairly large samples. With Time Series, "fairly large" very frequently means "far larger than is generally supposed". For instance, elsewhere, a sample size of n=100 may be more than sufficiently large; but, with Time Series, we could need a length, of observed realisation, n>1000.

Sometimes one has such large n (in the physical sciences, say), more often not. In economics, for example, an n of 40 or 50 is, typically, wishful thinking.

Time Series Analysis has very wide ranging applications - from oil exploration, through estimating unemployment and forecasting sales, to forestalling cardiac arrest. Thus the study of Time Series is much more than just that of a sub-branch of Statistics. Time Series, as a subject, is a wide collection of techniques and experience, spanning virtually all the quantitative sciences and reaching into all their corners where measurements are made; and, as such, has more the nature and status now of an emerging discipline in its own right. Rather like Operations Research a few decades ago.

A major object of many Time Series analyses is Forecasting; and, for this reason, we link the two in TSA&F. Of course, Time Series Analysis may be carried out for other purposes; and it is not the only route to obtaining Forecasts. Indeed, Forecasting itself might rightfully be considered as a discipline, whose methods include those of Time Series. (See section 7, below.)

2. Is TSA&F being Strangled at Birth?

What is important to the vast majority of people working with TSA&F, be they Accountants or Zoologists, is analysing the actual situations which they need to investigate.

Theory is only of prime concern to the relatively few academic statisticians, who unfortunately seem to think that Time Series is merely another branch of mathematical statistics. They then tend to judge the work of the majority by how far it measures up to their own discipline's particular (and here, I believe, inappropriate) standards of theoretical rigour, rather than by the criterion of its practical effectiveness - which is the really important aspect.

This apparent attitude of the academic statisticians is indeed "unfortunate" since the views of that minority, which contains many highly respected individuals, are disproportionately influential. Indeed, their prestige provides the statisticians with a near monopoly of expression through the established respectable (statistical) channels, and the effect (I believe) is an overbearing stiffling of the emergent discipline.

If the academically orientated statisticians continue to have the only serious say in Time Series, then of course it will remain a sub-branch of Statistics. Science is based on Consensus of Opinion. Whilst only (overspecialised?) statisticians are regarded as being sufficiently competent to be accorded a voice in where the subject goes and what it is, then of course the statisticians will hold the Consensus; but, for the freshly developing discipline, it is more of a strangle-hold, which at very least is stunting growth.

How does the statistical establishment exert such a restrictive (if not downright damaging) influence? By arrogance, in two ways:

(a) By perpetuating the myth that formal statistical expertise is the factor of overwhelming importance when faced with data.

(b) By an implicit assumption that the value of what people have to say (and indeed their intrinsic ability) is very highly correlated with their facility to express themselves clearly in English. (Foreigners, as a species, are patently stupid - they frequently cannot even speak our language.)

Let us consider two intelligent specialists, S (a mathematical Statistician) and, say, B (a Biologist), who have, independently, been hard at work on some project involving biological data. They both write up papers - SP and BP, respectively, say - and submit these (naughtily) to the same two Journals simultaneously - a leading academic statistical one, sj, and a top applied biology publication, bj. I suggest that the following reports from the referees are not too unrealistic in their essence.

SP submitted to sj:	An excellent and interesting applied paper, with important biological conclusions. Accept.
BP " " sj:	Devoid of any methodological innovation or interest, a trivial paper without merit. Reject.
SP " " bj:	Pretentious theoretical treatment lacking in biological insight or understanding. Reject.
BP " " bj:	A very useful case study providing substantial contributions to biological knowledge. Accept.

The thing to especially note are the unwarranted value judgements (which are not restricted to the Statistical journal's reports). The referees appear to extrapolate how well the submissions meet their own specialist demands to areas perhaps outside their competence; and, although they may avoid doing this when they favour the paper, my experience (as an editor) suggests that they rarely refrain from it when their specialist instincts have been adversely aroused. Certainly, referees tend to regard only the values of their own discipline as holding any validity.

Indeed, when an expert in field X shows shortcomings in area Y, it seems to be more-or-less normal for a referee from field Y to assume the X-man an idiot, and fail to realise that X may still have something useful to say, if he (the Y's-guy) would only listen.

At present, respectable time series papers are given their respectability almost solely from their ability to satisfy the criteria required by statistical journals. (It is rather like only allowing the Arts to be experienced through the eyes. For most people, reading the score is a poor substitute for going to the concert.) It is thus not very surprising that virtually all the so-called top authorities on Time Series are, apart from anything else, strong theoretical statisticians. But few of them provide much hard evidence of having any real practical experience. Academic case-studies abound to bolster theoretical niceties, but these are rarely tempered with reality. It tends to be theory in search of examples (for granting it credibility), rather than the solution of actual problems given the practical restraints imposed by little time and restricted budgets.

These "top authorities" favour publishing their material in the prestigious statistical journals, and so set the standards for everyone else. Other outlets, at best, are regarded as second-rate. And indeed they must be so, given this distorted emphasis on statistical prowess, because they are full of "BP's" which, judged by statisticians, are nowhere nearly up to "sj" standards.

Of course, Statisticians accept "Biologists" as being competent in "Biology" (although they may well regard this as a lower, less-exacting calling). But what they cannot stomach is the Biologist being a shade bit more, when that extra is Statistics. A little knowledge is a dangerous thing, they say, and imply that Statistics should be left to its High Priesthood. (Of course, taken to a logical conclusion, this implies that no-one should ever start investigating anything.)

I would suggest that this attitude of belittling is most prevalent between closely connected fields, for instance Statistics and Operations Research. Statisticians tend to regard the leading OR journals as second rate (because statistical content is frequent, but does not usually reach the esoteric level of the best (?) statistical journals); whereas the genuine OR practitioners find academic Statisticians divorced from reality and their journals irrelevant to the handling of the "dirty" complex problems, for which they need to produce working solutions quickly. (Certainly I regard many of the time series papers published in the "leading" statistical literature as worthless; although, going back to the original working papers of some case-studies, one does sometimes find valuable aspects - presumably vetoed by the referees for not being amenable to slick facile academic resolution.)

3. Other Routes to Salvation: Not just Mathematical Statistics

To analyse data satisfactorily, one must have a feel for the particular situation under study, which can only be obtained from a thorough appreciation of the background giving rise to it. That is, one does not just look at data, but studies all relevant aspects of the system from which the data were derived. There are many ways of approaching problems, and the more open-minded and multi-skilled the analyst is, the better.

However, the three factors of prime importance are general ability, experience, and an outlay of thought about the current problem. Methodology, in practice, is most valuable as a means of focusing the analyst's attention on important issues and in concentrating his mind to thinking along certain productive lines. The end result of a methodological training should be the absorption of certain modes of thought almost into the analyst's unconscious.

This then gives him (or her) a point of departure from which to start tackling real problems. He has some logical frame to which he can relate his growing experience, and so build to ever more effective understanding of how data actually behaves. But that is not the only way and we give some analogies:

A bridge engineer first learns as a student how to formally calculate stresses and strains, and eventually obtains an intuitive feel as to what sort of values to expect. He may well, with time, even forget how to do the calculations, but this does not matter in practice. However, someone who builds up similar expertise from experience alone, without a formal training, can be just as effective - although he may have to overcome a number of artificial barriers before his skill is recognised.

Until quite recently, acupuncture was not taken seriously in the West. What value could it possibly have, when not even the Chinese professed an understanding as to why it worked? Without a well-developed theory underpinning its practice, respectable medical science could not consider becoming involved. The point that acupuncture appeared, in a number of areas, much more effective than western remedies was completely ignored. Or, rather, western science refused to assess such claims scientifically. The preoccupation with the lack of theory completely obliterated any chance for the evidence from acupuncture practice to make an impact. As no immediate reason for why it worked was evident, the question of did it work was hardly asked - not being based on theory, it was assumed to be bogus.

Now the Chinese themselves were more interested in how it worked, rather than why, and through the ages had developed a very extensive practical knowledge of acupuncture. Cultural prejudice is surely a poor way of reaching conclusions, and there is currently little doubt, even in Western Medicine, that acupuncture is a valid and valuable means of treatment. However, this is perhaps only so because it was allowed to develop unhampered by western science, and indeed far in advance of it. Imagine what would have happened if some mad untrained "quack" had first tried curing addictions in, say, 1981 by sticking needles into people's ears.

Good Time Series (or Statistical) practice does not necessarily require a massive methodological foundation. There is no substitute for clarity and incisiveness of thought, and "statisticians" do not hold a monopoly there. Darwin's synthesis of natural selection from the vast amounts of data he had collected represents surely one of the monumentally great statistical achievements. And yet he would need to dress his analysis up in much more fancy clothes, if he wished a modern statistical journal to consider it at all seriously nowadays.

In the UK, there are government statisticians who have been trained as Economists (E) and others with backgrounds in Mathematical Statistics (MS). They approach data with different knowledge and skills, but there is no evidence that the MS outperform the E, in fact perhaps the reverse. The MS tends to retain the idea that dealing with real data is rather beneath his true calling - a suitable activity for his less gifted colleagues, perhaps, but not for him. He will dream of returning to some ivory tower, where his merit and superiority can be established.

The most prevalent example of this sort of thing is that of the pure Mathematician, who assumes that the Engineer is at the very lowest end of the scale for which he forms the pinnacle. This must be so, students entering Engineering have far poorer grades in Maths than he had. Ergo, they are less intelligent. It does not occur to him that he might perform rather poorly in the perhaps no-less demanding tests of engineering aptitude.

Luckily the work of Box and Jenkins has given pragmatic Time Series Analysis a welcome boost. The MS regard Box-Jenkins as unsound in principle - one allows the individual analyst to use his judgement and experience, rather than leaving everything to objective hypothesis testing. However, the stature of George Box and Gwilym Jenkins as theoreticians has lent considerable respectability to their "B-J" approach. The MS may mutter ominously, but B-J papers find their way into the sacred "sj" pages.

The B-J approach has proved itself to be a remarkably useful tool in practice. Much of the fine detail of its theory is questionable, but the establishment has not (as it may have wished) been able to squash it on such technicalities. The result has been a very marked improvement in forecasting practice, following an astonishingly swift and widespread adoption of the Box-Jenkins ideas - which, in essence, are just those of any good analyst. (Look at the data and its background, infer a plausible model from a class one can handle, estimate the necessary parameters and then assess whether or not the fitted model is satisfactory for one's purposes; and, if not, modify the previous inference and reestimate and reverify the fresh model now proposed.) It is surely no coincidence that George Box continually advocates an iterative conversation between theory and practice, whilst Gwilym Jenkins recommends falling in love with one's data.

Again, the crystal ball approach to Forecasting should not be completely discounted. The good Gypsy really does use a sound methodology. She sums up the person before her with a high degree of skill, and projects a very persuasive future for them. That is, she works with the data available at time now, makes a few sensible deductions as to the future and then wraps her forecast up in a way likely to appeal to her client. Most forecasters know they also have to work with the past and present, but they are often less successful in selling their conclusions (predictions) to those who should be heeding them.

4. Aims of the TSA&F Special Interest Group

The main object is to help spread ideas and circulate information in this subject area, and to encourage practitioners to work together by affording them good opportunities for making contact with each other.

Apart from this, the group tries to foster active participation from a much wider class of TSA&F people, than just those representing the English-speaking schools of Statistics. We wish to bring in members of every nationality and working in all disciplines, where it is (or should be) important to analyse Time Series.

5. The Early Development of TSA&F

This started with a predominantly national 1976 Conference on Forecasting, convened by the present author and held at Cambridge University, England.

At the time, only the F of TSA&F was thought to hold sufficient appeal as the theme for a major meeting, and the main effort was an attempt to build links between the then feuding camps of statistical and econometric forecasters, and also to approach the problems facing managers and other users of forecasts. The intention was to emphasise the practical issues involved.

This flavour is reflected in the very popular Proceedings volume, Anderson 1979, which includes a major paper by G.M. Jenkins - later published independently by him as a monograph. The other contributors included: R.J. Ball, G.A. Barnard, E.M.L. Beale, Sir Paul Chambers, C.W.J. Granger, P.J. Harrison, Sir Maurice Kendall, P. Newbold, M.B. Priestley and P. Whittle.

The success of this venture allowed the planning of a more ambitious event in 1978, again at Cambridge. Here the emphasis was on the TSA (to complement the 1976 Meeting) and a truly international conference resulted. This time the authors included H. Akaike, M.S. Bartlett, G.E.P. Box, E.B. Dagum, C.W.J. Granger, R.E. Kalman, Sir Maurice Kendall, M.B. Priestley, J. Shiskin and J.W. Tukey, and many of the papers were subsequently published.

With interest well-established, TSA&F was officially launched at the beginning of 1979.

6. The Current State of TSA&F

The first project was the creation of a quarterly Newsletter, TSA&F News, which started publication in January 1979. Its present circulation is 2800 copies going to 69 countries.

Later, in June 1980, a monthly bulletin, the TSA&F Flyer, with an even larger print-run, was introduced to supplement the Newsletter and allow the speedy spread of more pressing news. It also now acts as a general publicity broadsheet.

The publication flag ship, a new Journal of Time Series Analysis appeared in October 1980, with an editorial board consisting of H. Akaike, T.W. Anderson, D.R. Brillinger, C.W.J. Granger, E.J. Hannan, P. Newbold, E. Parzen, M.B. Priestley, E.A. Robinson, M. Rosenblatt, G.C. Tiao, G. Tunnicliffe Wilson, A.M. Walker and A.M. Yaglom - although, given the views expressed by the author in previous sections, it is clear that the academic emphasis of this board is not what was originally intended.

Other publications have included four further proceedings volumes, which feature papers by H. Akaike, K. Astrom, E.B. Dagum, M. Deistler, F. Eicker, E.J. Godolphin, E.J. Hannan, L.D. Haugh, P.A.W. Lewis, L. Ljung, C.L. Mallows, R.D. Martin, E. Parzen, D.A. Pierce, V.S. Pugachev, P.M. Robinson, P. Shaman, G.C. Tiao, H. Tong, G. Wahba and A.M. Yaglom. The corresponding conferences have been held in Nottingham (UK), 1979, on Guernsey (Channel Islands), 1979, and in Houston (Texas), 1980, all of which were International Time Series Meetings (ITSM), and also a Public Utilities Forecasting Conference (PUFC) in Nottingham, 1980.

The next Proceedings to appear will be that for the 4th ITSM, held at Valencia (Spain), June 1981. This is expected to include ten plenary papers by T.W. Anderson, R.J. Bhansali, W.S. Cleveland, M. Deistler, R.E. Kalman, P. Newbold, M.B. Priestley, P. Shaman, E.A. Robinson and H. Tong, as well as contributions by many other well-known or up and coming authors.

Further ITSM are the present one held at Houston (Texas), August 1981 and then three more that are planned. These are (1) a specialised meeting on Time Series Methods in the Hydrosciences to take place in Burlington, Ontario, October '81 and then general ITSM to be held (2) in Dublin (Ireland), 15-19 March, 1982, and (3) at Cincinnati (Ohio), satelliting the ASA Annual Meetings there, August '82. An International Forecasting Conference (IFC) has also been projected for (4) Valencia (Spain), 24-28 May 1982. Paper submissions for Conferences (2), (3) and (4) and offers to serve on their Organising Committees (mainly as Session Chairs) will be welcomed.

Other concrete services include a dozen TSA&F Instructional Courses, which have been run in conjunction with Universities in Argentina, Finland, Spain and the UK; and some thirty instructional and research seminars, which have been given in Argentina, Austria, Belgium, Brazil, Canada, Czechoslovakia, Finland, Ireland, Netherlands, Norway, Poland, Spain, UK, USA, Venezuela and West Germany. Future courses and seminars are scheduled for Campinas (Brazil), Dar es Salaam (Tanzania), Merida (Venezuela), Montreal (Canada), Philadelphia (USA), Rio de Janeiro and Sao Paulo (Brazil).

Amongst the TSA&F Special Interest Group, an inner core of especially motivated people provides a world-wide Network for information collection and distribution, spanning much of the globe and a great many parent subject disciplines. In January 1981, the Group and Network formed the basis for a non-profit making professional TSA&F Society. See the Newsletter, TSA&F News, and Information Bulletin, TSA&F Flyer, again for details.

7. Some Offshoots of TSA&F

The F of TSA&F is primarily Time Series Forecasting and, as such, of course only covers a small subarea of Forecasting in the wide sense. As Forecasting is a major aim of much Time Series Analysis, it is not surprising that interest in a more general range of Forecasting activities should be swift to develop.

Indeed this has now happened. At the small ITSM Conference held on Guernsey, British Isles, October 1979, talk about a Meeting in Canada began. In May 1981, the result of this was an International Forecasting Symposium at Quebec, chaired by Robert Carbone of Laval University, who had been to Geurnsey.

Amongst the Technical Programme Committee were other people from the Guernsey Meeting: Allan Andersen (Australia), Oliver Anderson (UK), Jan de Gooijer (Netherlands) and Spyros Makridakis (France), with many others who had attended other ITSM's or subscribe to TSA&F, such as: J. Scott Armstrong (USA), Estela Dagum (Canada), Jean-Pierre Indjehagopian (France), Raman Mehra (USA), Guy Melard (Belgium), Douglas Montgomery (USA), David Pack (USA) and Otto Tomasek (Canada).

Speakers at this Symposium also included a large number from Guernsey and other ITSM's and a substantial contingent of TSA&F people.

The event (which is now viewed as the first of a series to be organised) was used to launch a Journal of Forecasting, much more in the spirit of the original proposals for the Journal of Time Series Analysis - the emphasis being on usefulness in practice rather than esoteric academic quality - and, following the TSA&F Society, an International Institute of Forecasters has been formed.

It is reasonable to argue that, to a considerable extent these three new Forecasting activities derive directly from TSA&F and ITSM.

Even closer to our own projects, as described in this article, are the objectives of the Interaction Committee on Time Series inaugurated at the 1980 American Statistical Association Meetings under the chairmanship of Emanuel Parzen - although, presumably, these will follow a more academic course.

8. The Future of TSA&F

It is hoped that an Institute for the study of TSA&F will be endowed over the next few years. Whereas the Society is mainly concerned with the people in TSA&F, the Institute would concentrate on the subject, devoting itself to teaching, training, writing and research. At present there is a strong demand from many isolated workers around the world for somewhere to come and study and exchange ideas, and the Institute could perhaps supply a setting for this.

The sequence of ITSM will continue, and will soon include an annual Society Conference. As already mentioned, the 1982 events will be held in Dublin and Cincinnati.

Also the F of TSA&F will continue to be emphasised from time to time. The next occasion is the 1982 Valencia IFC, again listed above.

We have given our own personal views about the TSA&F Movement, but do not wish to steer its course alone. We invite all who are interested to have their say and so influence to where the boat is headed.

BIBLIOGRAPHY

ANDERSON, O.D. (1979). Ed: *Forecasting*. North-Holland: Amsterdam.

ANDERSON, O.D. (1980a). Ed: *Time Series*. North-Holland: Amsterdam.

ANDERSON, O.D. (1980b). Ed: *Analysing Time Series*. North-Holland: Amsterdam.

ANDERSON, O.D. (1980c). Ed: *Forecasting Public Utilities*. North-Holland: Amsterdam.

ANDERSON, O.D. (1982). Ed: *Time Series Analysis: Theory and Practice 1*. North-Holland: Amsterdam.

ANDERSON, O.D. & PERRYMAN, M.R. (1981). Eds: *Time Series Analysis*. North-Holland: Amsterdam.

ANDERSON, O.D. & PERRYMAN, M.R. (1982). Eds: *Applied Time Series Analysis*. North-Holland: Amsterdam. In preparation.

A.H. EL-SHAARAWI & ANDERSON, O.D. (1982). Eds: *Time Series Methods in the Hydrosciences*. In preparation.

TSA&F NEWS (Quarterly Newsletter). Ed: O.D. Anderson. ISSN 0143-0505.

TSA&F FLYER (Monthly Information Bulletin). Ed: O.D. Anderson. ISSN 0260-9053.